T0192951

ASCE Manuals and Reports of Engineering Practice No. 77
WEF Manual of Practice FD-20

Design and Construction of Urban Stormwater Management Systems

Prepared by
The Urban Water Resources Research
Council of the American Society of Civil
Engineers and the Water Environment
Federation (formerly the Water Pollution
Control Federation)

Published by the
American Society of Civil Engineers
345 East 47th Street
New York, New York 10017-2398
and the
Water Environment Federation
601 Wythe Street
Alexandria, VA 22314-1994

ABSTRACT

Design and Construction of Urban Stormwater Management Systems (Manual of Practice No. 77) updates relevant portions of the ASCE/WPCF Manual of Practice No. 37, *Design and Construction of Sanitary and Storm Sewers*. This update is necessary due to the many changes taking place in the field such as the use of microcomputers and the need to control the quality of runoff as well as the quantity. In order to broaden the base of experience reflected in the Manual, each chapter was prepared by one or more authors with experience and expertise in the particular subject area. Thus, the Manual aids the practicing engineer by presenting a brief summary of currently accepted procedures relating to the following areas: 1) Financial services; 2) regulations; 3) surveys and investigations; 4) design concepts and master planning; 5) hydrology and water quality; 6) storm draining hydraulics; and 7) computer modeling.

Library of Congress Cataloging-in-Publication Data

Design and construction of urban stormwater management systems/prepared by the Urban Water Resources Research Council of the American Society of Civil Engineers and the Water Enviroment Federation (formerly the Water Pollution Control Federation).

    p.   cm.—(ASCE manuals and reports engineering practice: no. 77) (WEF manual of practice: FD-20)
    Includes bibliographical references and index.

    ISBN: 978-1-57278-251-8

    1. Urban runoff—Management. 2. Storm sewers—Design and construction. I. American Society of Civil Engineers. Urban Water Resources Research Council. II. Water Environment Federation. III. Series: ASCE manuals and reports of engineering practice: no. 77. IV. Series: WEF manual of practice: no, FD-20.
TD657.D47   1992
628'.21—dc20   92-36519
                                  CIP

WATER ENVIRONMENT FEDERATION

The Water Environment Federation is a nonprofit, educational organization composed of member and affiliated associations throughout the world. Since 1928, WEF has represented water quality specialists, including civil, design, and environmental engineers, biologists, bacteriologists, local and national government officials, treatment plant operators, laboratory technicians, chemists, industrial technologists, students, academics, and equipment manufacturers and distributors.

For information on membership, publications, and conferences contact

Water Environment Federation
601 Wythe Street
Alexandria, VA 22314-1994
(703) 684-2400

AMERICAN SOCIETY OF CIVIL ENGINEERS

The American Society of Civil Engineers (ASCE) offers civil engineering professionals many opportunities for technical advancement, networking, and leadership and technical skill training. Also available to members are major savings on educational seminars, conferences, conventions, and publications. First class, low-cost insurance programs are among the most competitive available.

Members may participate on a national level, networking with colleagues on forums to advance the profession. This participation affords the opportunity to develop leadership skills and to expand personal contacts. On the local level, chapters (called Sections and Branches) act as advocates in the public interest on local issues and present seminars and programs relevant to the needs of the local community.

MANUALS OF PRACTICE
FOR WATER POLLUTION CONTROL

(*As developed by the Water Environment Federation*)

The WEF Technical Practice Committee (formerly the Committee on Sewage and Industrial Wastes Practice of the Federation of Sewage and Industrial Wastes Associations) was created by the Federation Board of Control on October 11, 1941. The primary function of the committee is to originate and produce, through appropriate subcommittees, special publications dealing with technical aspects of the broad interests of the Federation. These manuals are intended to provide background information through a review of technical practices and detailed procedures that research and experience have shown to be functional and practical.

Water Environment Federation
Technical Practice Committee Control Group

F.D. Munsey, *Chair*
L.J. Glueckstein, *Vice-Chair*
A.J. Callier
C. Lowery
R.W. Okey
T. Popowchak

Authorized for Publication by the Board of Control
Water Environment Federation
Quincalee Brown, *Executive Director*

MANUALS AND REPORTS ON ENGINEERING PRACTICE

(As developed by the ASCE Technical Procedures Committee, July 1930, and revised March 1935, February 1962, April 1982)

A manual or report in this series consists of an orderly presentation of facts on a particular subject, supplemented by an analysis of limitations and applications of these facts. It contains information useful to the average engineer in his everyday work, rather than the findings that may be useful only occasionally or rarely. It is not in any sense a "standard," however; nor is it so elementary or so conclusive as to provide a "rule of thumb" for nonengineers.

Furthermore, material in this series, in distinction from a paper (which expresses only one person's observations or opinions), is the work of a committee or group selected to assemble and express information on a specific topic. As often as practicable the committee is under the direction of one or more of the Technical Divisions and Councils, and the product evolved has been subjected to review by the Executive Committee of that Division or Council. As a step in the process of this review, proposed manuscripts are often brought before the members of the Technical Divisions and Councils for comment, which may serve as the basis for improvement. When published, each work shows the names of the committees by which it was compiled and indicates clearly the several processes through which it has passed in review, in order that its merit may be definitely understood.

In February 1962 (and revised in April, 1982) the Board of Direction voted to establish:

A series entitled 'Manuals and Reports on Engineering Practice,' to include the Manuals published and authorized to date, future Manuals of Professional Practice, and Reports on Engineering Practice. All such Manual or Report material of the Society would have been refereed in a manner approved by the Board Committee on Publications and would be bound, with applicable discussion, in books similar to past Manuals. Numbering would be consecutive and would be a continuation of present Manual numbers. In some cases of reports of joint committees, bypassing of Journal publications may be authorized.

Authorized for Publication by the Publications Committee in the name of the Board of Direction Edward O. Pfrang, Executive Director

AVAILABLE* MANUALS AND REPORTS OF ENGINEERING PRACTICE

Number

10	Technical Procedures for City Surveys
13	Filtering Materials for Sewage Treatment Plants
14	Accommodation of Utility Plant Within the Rights-of-Way of Urban Streets and Highways
31	Design of Cylindrical Concrete Shell Roofs
33	Cost Control and Accounting for Civil Engineers
34	Definitions of Surveying and Associated Terms
35	A List of Translations of Foreign Literature on Hydraulics
36	Wastewater Treatment Plant Design
37	Design and Construction of Sanitary and Storm Sewers
40	Ground Water Management
41	Plastic Design in Steel—A Guide and Commentary
42	Design of Structures to Resist Nuclear Weapons Effects
45	Consulting Engineering—A Guide for the Engagement of Engineering Services
46	Report on Pipeline Location
47	Selected Abstracts on Structural Applications of Plastics
49	Urban Planning Guide
50	Report on Small Craft Harbors
51	Survey of Current Structural Research
52	Guide for the Design of Steel Transmission Towers
53	Criteria for Maintenance of Multilane Highways
54	Sedimentation Engineering
55	Guide to Employment Conditions for Civil Engineers
57	Management, Operation and Maintenance of Irrigation and Drainage Systems
58	Structural Analysis and Design of Nuclear Plant Facilities
59	Computer Pricing Practices
60	Gravity Sanitary Sewer Design and Construction
62	Existing Sewer Evaluation and Rehabilitation
63	Structural Plastics Design Manual
64	Manual on Engineering Surveying
65	Construction Cost Control
66	Structural Plastics Selection Manual
67	Wind Tunnel Model Studies of Buildings and Structures
68	Aeration—A Wastewater Treatment Process
69	Sulfide in Wastewater Collection and Treatment Systems
70	Evapotranspiration and Irrigation Water Requirements
71	Agricultural Salinity Assessment and Management
72	Design of Steel Transmission Structures
73	Quality in the Constructed Project—a Guide for Owners, Designers, and Constructors
74	Guidelines for Electrical Transmission Line Structural Loading
75	Right-of-Way Surveying
76	Design of Municipal Wastewater Treatment Plants
77	Design and Construction of Urban Stormwater Management Systems
78	Structural Fire Protection

---

*Numbers 1, 2, 3, 4, 5, 6, 7, 8, 9, 11, 12, 15, 16, 17, 18, 19, 20, 21, 22, 23, 24, 25, 26, 27, 28, 29, 30, 32, 38, 39, 43, 44, 48, 56, and 61 are out of print.

HISTORY OF THE MANUAL

The Urban Water Resources Research Council (UWRRC) of the American Society of Civil Engineers (ASCE) has long been a leader in the transfer of urban drainage technology to the engineering community. A major part of their efforts has been the publication of more than a dozen books related to the general subject of urban storm drainage, most of which are the proceedings of a series of Engineering Foundation Conferences. Many of these books are used as standard references in the field.

The last Manual of Practice related to urban storm drainage was Manual of Practice 37, "Design and Construction of Sanitary and Storm Sewers," published jointly in 1969 by ASCE and the Water Environment Federation (formerly the Water Pollution Control Federation). In 1982, it was supplanted, for sanitary sewers, by Manual of Practice Number 60, "Gravity Sanitary Sewer Design and Construction," again jointly published by ASCE and WEF.

In an attempt to fill the void created by the publication of Manual of Practice 60 (which did not address storm drainage), the UWRRC established a Task Committee, chaired by Mr. Richard Lanyon, to begin drafting a manual of practice on urban storm drainage. The chairmanship of the Task Committee was subsequently assumed by Mr. Jonathan E. Jones, who has seen the Manual through to its completion.

To broaden the base of experience to be reflected in the manual, each chapter was prepared by one or more authors with experience and expertise in the particular subject area. The chapters were also extensively reviewed by the authors and other experts in the field, as well as other interested parties, including members of the regulatory and public works communities, and specialists in such diverse disciplines as law, planning, landscape architecture, meteorology, and ecology.

A draft of the complete manual was prepared and distributed to an ASCE Senior Review Committee and the Technical Practice Committee of the Water Environment Federation. Many valuable suggestions for improvement were received. After revision based on these reviews, portions of the Manual were distributed to both committees for a second review.

The final draft of the Manual was approved for publication by WEF and ASCE in 1991.

CHAIRMAN OF THE TASK COMMITTEE

Jonathan E. Jones

PRINCIPAL AUTHORS AND REVIEWERS OF THE MANUAL OF PRACTICE

(* indicates principal author)

Michael Bealey*
Bent Christensen*
John France*
John M. Hamilton*
Paul A. Hindman*
Mark R. Hunter*
D. Earl Jones, Jr.*
Eric Livingston*
David Lloyd*
Robert Marmon
Herbert G. Poertner*
Jan E. Roshalt
Ronald L. Rossmiller*
William P. Ruzzo*
Erez Sela*
Michael B. Sonnen
L. Scott Tucker*
Stuart G. Walesh*
Kevin Weiss
David E. Westfall*
Donald Woodward

John L. Blanchard*
Victor F. Coletti*
Mark W. Glidden*
Joseph Hill*
Wayne C. Huber*
Edward Jankiewicz*
David F. Kibler*
Harold Leedom
Jiri Marsalek
Bruce M. McEnroe*
Clifford Randall
Larry A. Roesner*
Richard M. Rudolph*
James E. Scholl*
W. Joseph Shoemaker*
Jon H. Sorenson*
Ben Urbonas*
John J. Warwick
David C. Wells
William Whipple, Jr.*
Kenneth R. Wright*

James Wulliman

ADDITIONAL AUTHORS AND REVIEWERS

David Anderson
David Balmforth
Jay Beaumont
Dan Brock
Patricia Bubar
Nancy C. Bramlett
Lamont W. Curtis
Thomas N. Debo
Jacques W. Delleur
Jonathan French
James P. Heaney
Donald Hey
Robert H. Hoffmaster
Kenneth Kienow
Alan Leak
Byron Lord
William Macaitis
Nathan D. Maier
David Maunder
Theodore Mikalsen
William Mitzelfeld
Kenneth E. Moss

Dennis E. Arbogast
Susan Banks
Carey Brand
Susan K. Bank
Chenchayya T. Bathala
Christopher Burke
Paul A. DeBarry
Robert Deeds
Patricia K. Flood
Gary R. Haynes
Richard J. Heggen
Steve Hickox
Carl Johnson
John R. Lavigne, Jr.
Raymond K. Linsley
Gordon Lutes
John R. Mackie
H. Rooney Malcom
Richard H. McCuen
Gary R. Minton
Herbert Moore
G.R. Olieger

ADDITIONAL AUTHORS AND REVIEWERS (continued)

MEMBERS OF THE ASCE SENIOR REVIEW COMMITTEE

WEF TASK FORCE ON STORMWATER MANAGEMENT

TECHNICAL EDITOR

FOREWORD

This Manual of Practice for the Design and Construction of Urban Stormwater Management Systems updates relevant portions of ASCE/WEF Manual of Practice No. 37, "Design and Construction of Sanitary and Storm Sewers," published in 1969. This update was undertaken by the Urban Water Resources Research Council of ASCE for several reasons:

    (a) ASCE and WEF published, in 1982, a revised Manual of Practice (No. 60) on "Gravity Sanitary Sewer Design and Construction," which did not address storm drainage.

    (b) There have been a number of changes in the field of urban storm drainage, including (1) the virtually universal use of the microcomputer for data organization and analysis, and for the analysis and design of urban storm drainage systems, and (2) the increasing importance of designing urban storm drainage systems for the control of runoff quality as well as quantity.

This Manual is intended to aid the practicing engineer by presenting a brief summary of currently accepted procedures. It is not intended to substitute for engineering experience and judgment, nor is it a replacement for more detailed standard texts and references in the field.

The Manual recognizes that the practice of urban storm drainage is dynamic and rapidly changing, with new techniques, materials, and equipment continuously being introduced, and emphasizes that practitioners in the field must constantly be aware of new developments and modify their practice accordingly. The UWRRC invites comments and recommendations for improvement for possible inclusion in future editions.

The authors recognize that many women professionals are involved in all aspects of the planning, design, and construction of urban stormwater management systems. The use of the masculine gender pronoun throughout the manual has been for the sake of simplicity and brevity, and no other inferences should be drawn.

ACKNOWLEDGEMENTS

The Urban Water Resources Research Council would like first to acknowledge the efforts of Mr. Richard Lanyon of the Metropolitan Sanitary District of Greater Chicago, and Mr. Jonathan E. Jones of Wright Water Engineers, Inc. of Denver, Colorado, without whose leadership, energy and enthusiasm this Manual would never have been completed.

Most of the real work was done by the individual chapter authors, and their work is gratefully acknowledged. Mr. Harry C. Torno worked tirelessly as Technical Editor of the entire book.

Special appreciation is extended to the staff of Wright Water Engineers, Inc. in Denver, Colorado who have worked on virtually every aspect of manuscript preparation.

Finally, the Council would like to thank the following organizations, which generously donated funds to assist with the preparation of the text and figures in the Manual of Practice:
- The Engineering Foundation
- U.S. Department of Transportation/Federal Highway Administration
- U.S. Geological Survey
- U.S. Environmental Protection Agency
- U.S. Soil Conservation Service
- Camp Dresser & McKee Inc.
- Michael Baker Engineering
- CH2M-Hill
- American Concrete Pipe Association

TABLE OF CONTENTS

TABLE OF CONTENTS

TABLE OF CONTENTS

TABLE OF CONTENTS

TABLE OF CONTENTS

Chapter 1

EVOLUTION OF URBAN STORMWATER MANAGEMENT

I. INTRODUCTION

Urban stormwater management is the conceptualization, planning, design, construction, and maintenance of stormwater control facilities in urban/urbanizing drainage basins, and includes all related political, social, and economic considerations. While this definition does not necessarily involve new construction, it includes such facilities as open channels, curbs and gutters, storm sewers, detention/retention ponds and associated structures, water quality enhancement measures, special structures (energy dissipators, transitions, inlets, etc.) and others. While the distinctions between stormwater management and floodplain planning and management are frequently unclear, they may be defined as:

(a) Stormwater (runoff) management is the planned set of public policies and activities undertaken to regulate runoff under various specified conditions within various portions of the urban drainage system (McPherson 1970). It may establish criteria for control of peak flows or volumes, for runoff detention and retention, or for control of pollution, and may specify criteria for the relative elevations among various elements of the drainage system. Stormwater management is primarily concerned with limiting future flood damages and environmental impacts due to development, whereas flood control aims at reducing the extent of flooding that occurs under current conditions (Walesh 1987).

(b) Floodplain management is the regulation of the nature and location of construction on (or other occupancy of) lands subject to inundation, so that foreseeable (probable) flooding damages will have an average annual value (risk) smaller than some preselected amount (Federal Emergency Management Agency 1983).

Federal and state policies and guidance are abundant on floodplain management and flood control, but little Federal and state conceptual design policy guidance exists for local drainage and stormwater man-

1

agement. It is essential for the drainage designer to recognize that local policies, principles, and criteria will be the primary forces governing design (even when such criteria are at odds with suggested procedures in this Manual of Practice).

II. NEED FOR THE MANUAL

The American Society of Civil Engineers' (ASCE) most recent Manual of Practice (MOP) pertaining directly to stormwater management is MOP No. 37, "Design and Construction of Sanitary and Storm Sewers" (ASCE 1969), which was published jointly with the Water Environment Federation (formerly The Water Pollution Control Federation) (as WPCF MOP No. 9). In 1982, ASCE and WEF updated MOP No. 37 by publishing "Gravity Sanitary Sewer Design and Construction," ASCE MOP No. 60 (ASCE 1982).

Much of MOP No. 37 is as relevant now as when it was written, and anyone designing storm sewers should be aware of its content, some of which has not been duplicated or updated in this Manual. There was a perceived need to update MOP No. 37, however, because MOP No. 60 does not adequately address urban storm drainage, and because there have been a number of changes in storm drainage practice over the past two decades, including the following:

(a) The proliferation of stormwater retention and detention ponds. A technique that was not widely used thirty years ago is now a common feature of small and large developments and drainage master plans (Malcolm 1982).

(b) Increasing emphasis on the preservation and improvement of natural drainageways. Stream master planning, with an emphasis on "designing with nature," is widely practised (Rickert and Spieker 1971; ASCE 1978; Sheaffer et al. 1982).

(c) Widespread use of the terms "minor" and "major" (see Glossary) in drainage system planning and design to describe, respectively, the carefully planned, designed and constructed urban storm drainage system; and the overflow path(s) followed by the rarer large flows that exceed the capacity of the minor system.

(d) The use of personal computers and their accompanying hydrologic, hydraulic, and data management software. For a relatively modest expenditure, any engineer can now have a system powerful enough to complete calculations in minutes that would have taken weeks twenty years ago. This has also caused problems, particularly where there has not been adequate computer program (model) calibration and verification (Huber 1986), and where the engineer has blindly relied on computer-generated results without the review normally applied in more conventional methods of design.

(e) The increasing importance of evaluation and mitigation of adverse impacts of urban runoff on receiving water quality. Quality enhancement measures are becoming routine components of stormwater control.

(f) Increased structural complexity of facilities commonly used in drainage systems. Structural and geotechnical engineers are now routinely retained to provide specialized consulting during design.

(g) Changes in construction materials and methods. For example, soil cement, roller compacted concrete, structural plastics and geotextiles, which were rarely used in drainage projects when MOP No. 37 was published, are now common materials in stormwater management projects. Construction sequencing and management have advanced considerably with the advent of personal computers.

Additional reasons for updating the Manual include:

(a) Professional liability concerns increasingly are influencing design practice. Every designer faces the threat of litigation if his design is perceived as less than the state-of-the-art or—practice. This has resulted in increased attention to the preparation of comprehensive, understandable, and practical plans and specifications (Wells 1984).
(b) Useful sources of information for the designer have multiplied, and now include federal, state, and local governments, universities, and commercial sources (for, for instance, rainfall, streamflow, and land use data). Computer-aided drafting and design (CADD) is rapidly changing the design process.
(c) Regulatory involvement has increased, along with the need to protect the environment, encourage public participation in decision making, and involve specialists from many disciplines in the design process (Spirn 1984).

Denver, Colorado—Attractive, multi-purpose wet retention pond with an adjoining golf course and trail. The water quality is good, and the design emphasizes safety. Pollutants from the golf course wash into the pond and are immobilized in sediments on the pond bottom.

(d) Engineers are increasingly recognizing the need to adopt risk-based decision making (McCuen 1985), rather than relying on fixed, arbitrary design standards (such as a "design storm"). The evaluation of upstream and downstream risks before complying with a general design standard is not only prudent but may be necessary to limit liability and protect the public welfare (Jones 1988).

Despite changes in the practice of urban stormwater management over the past two decades, and the use of many innovative techniques, the design of urban drainage systems remains a procedure bound by tradition. Common sense has frequently been overridden by adherence to arbitrary standards. A major goal of this Manual is to encourage creativity and the application of broadly-based, practical thinking. This is particularly important in view of the ever-rising expectations that urban storm drainage practitioners are facing, such as those related to the U. S. Environmental Protection Agency's stormwater permit program.

III. ORGANIZATION OF THE MANUAL

The chapters are arranged to permit the reader to follow sequentially through the planning and design process. Chapter 2 addresses the very important legal, regulatory, and financial factors that drive the drainage decision-making process (City of Stillwater 1979). Required surveys and information sources are delineated in Chapter 3. Chapter 4 deals with design concepts (including conceptualization of the major and minor drainage systems) and master planning.

With a conceptual drainage plan in mind, the engineer next estimates flows and makes a preliminary evaluation of likely water quality issues. These subjects are covered in Chapter 5. Routing through conveyances, ponds, spill-ways, culverts, and other hydraulic structures is necessary, and Chapter 6 outlines the relevant hydraulic principles.

Chapter 7 provides brief descriptions of modeling techniques and of the most commonly used models, along with discussion of how to properly interpret computer-generated results, with emphasis on the need for calibration and verification.

Chapter 8 summarizes the conveyance design procedure. Chapter 9 addresses special structures and appurtenances such as channel protection, manholes and junctions, check dams, energy dissipators, and other such structures, emphasizing their proper design and maintenance. Chapter 10 deals with the special problems of combined storm and sanitary sewer systems. A key component of nearly every new drainage system is stormwater detention or retention facilities, including measures to improve runoff quality, and they are covered in Chapter 11. Chapter 12 provides general guidelines for integrating runoff quality controls into drainage planning and design.

Chapter 13 discusses the wide variety of materials available for constructing storm sewers and open channels, as well as the important subject of maintenance. Chapter 14 reviews structural principles and design criteria. Chapter 15 addresses the preparation of construction contract documents (plans and specifications). Factors that influence construction such as phasing, safety, site preparation, de-watering, environmental regulations, and quality control are discussed in Chapter 16.

The Manual concludes with an appendix containing two case studies that illustrate some of the techniques described.

IV. THE INTENDED AUDIENCE

This Manual is intended primarily for engineers charged with the responsibility of planning and designing stormwater management facilities. It also should be useful to other engineers, as well as architects, attorneys, planners, and landscape architects, environmental scientists, concerned citizens, and others.

This Manual is not a drainage criteria manual and does not provide extensive detail. It is intended to provide the designer with an introduction to procedures for designing urban stormwater management systems. When additional detail is either helpful or required, the text provides appropriate references. Readers are reminded that this manual provides a summary of the state-of-practice, rather than the state-of-the-art, of urban stormwater management and design in the United States. Emphasis has been placed on practical, easy to use methods of identifying problems and solutions, rather than on theoretical or experimental procedures which have not been well established.

V. THEMES AND PRINCIPLES THAT DRIVE THE DESIGN EFFORT

Certain themes in the manual transcend individual chapters, and may result in some redundancy. These include:

(a) Technically sound drainage projects that are supported by the community evolve through the combined efforts of a large and diverse group of specialists and interest groups, and the engineer should draw on those resources in the planning and design process (Jones and Rossmiller 1990).

(b) Drainage systems are enhanced by facilities that are multi-purpose in nature. Drainageways, ponds, and other facilities can be attractive, provide recreational opportunities, remove pollutants from runoff, provide wildlife habitat, and fulfil other functions, while meeting their primary drainage and flood control objectives (Spirn 1984; Poertner 1980; ASCE 1985).

(c) Common sense on the part of the designer is essential. Rigid adherence to particular standards may be undesirable and unwise. Where arbitrary standards are imposed by statute or ordinance, it may be impractical to seek relief, but the engineer should notify the regulatory authorities

Reston, Virginia—Wet pond surrounded by single family and multi-family housing. Note boat dock, fountain, and well-lighted pedestrian access.

if such standards may result in adverse conditions or effects. The engineer must consider the broad picture and evaluate risks on a case-by-case basis.

(d) Despite the widespread availability and adoption of computer models, there is no substitute for experience and judgment. Model calculations must be checked for correctness, and the designer must constantly assess the reasonableness of his results to assure that facilities will function properly (Linsley et al. 1982; Debo and Small 1989; Huber 1990).

(e) Attention to project details is vitally important. An entire chapter is devoted to special structures and appurtenances, and discussions of maintenance, public safety, usability, and long-term performance are provided throughout.

(f) The ASCE Manual of professional practice entitled "Quality in the Constructed Project—A Guideline for Owners, Designers, and Constructors," defines quality as meeting the owner's or user's requirements. Specifically, this publication notes that quality in design and construction may be better understood as:

(1) Meeting the requirements of the owner as to functional adequacy, completion on time and within budget, life cycle costs and operation and maintenance.

(2) Meeting the requirements of the design professional as to the provision of a well-defined scope of work, an adequate budget, and timely decisions by both the owner and the design professional.

(3) Meeting the requirements of the constructor as to provision of contract plans, specifications, and other documents prepared in

sufficient detail to permit the constructor to prepare priced proposals or competitive bids; timely decisions by the owner and design professional on authorization and processing of change orders; fair and timely interpretation of contract requirements from field design and inspection staff; and a contract for performance of work on a reasonable schedule that permits a reasonable profit.

(4) Meeting the requirements of regulatory agencies (the public) regarding safety and health, protection of public property including utilities, environmental considerations, and conformance with applicable laws, regulations, codes, and policies.

(g) Whether or not computer models are used, designers must constantly assess the reasonableness of such factors as:

(1) Frequency of inundation of properties.
(2) Design depths and velocities.
(3) Computed peak discharges.
(4) Stormwater detention volume requirements.
(5) Unit costs of storm sewers and open channels.
(6) Estimated pollutant removal percentages.
(7) Project benefits compared to project costs.
(8) Maintainability.

(h) Many laws and regulations still promote piecemeal, crisis-solving approaches aimed at managing resources within political constraints, or that solve a specific problem without adequate attention to related aspects. Authorities regulating stormwater and other environmental impacts frequently overlap, and their policies are not always consistent. Therefore local stormwater problems often are addressed without evaluating the potential for adverse impacts on upstream or downstream areas. An example would be construction of a relief sewer in an upstream area without evaluating its downstream impacts. Another example would be the uncoordinated development of stormwater management facilities by local land developers, each of whom is responsible for maintaining peak runoff and pollution loads from the site at predevelopment levels. The problems with the piecemeal approach are well recognized (Livingston et al. 1988), and include:

(1) It results in only partial resolution of major flooding problems.
(2) It creates new downstream flooding problems.
(3) Improperly located detention basins may actually increase peak flows rather than reduce them.
(4) Costs of uncoordinated remedial solutions are likely to be much greater than the overall costs incurred had an adequate program been implemented in the first place.

This Manual encourages comprehensive watershed planning, also referred to as "master drainage planning." This approach identifies the most appropriate control measures and optimum locations for the control of watershed-wide runoff and pollution impacts (Schueler 1987). The designer must recognize that individual site development plans or retrofits of drainage facilities in urban areas may be only a part of a larger system and that he must integrate his design into that larger system, particularly with respect to creating upstream or downstream hazards (Urbonas and Glidden 1982; Roesner 1990). Comprehensive watershed planning may indicate that a single stormwater detention facility to control runoff from several land development projects is preferable to on-site detention, and that nonstructural measures such as parkland acquisition, infiltration measures such as porous pavement or swales, and flood-proofing to supplement structural control measures may be preferable to more traditional approaches (APWA 1981). Comprehensive watershed planning offers such advantages as (Urban Drainage and Flood Control District 1984):

(a) Reductions in capital and O&M costs.
(b) A basis for setting priorities for resource allocation.
(c) Reduction in downstream flooding and erosion, particularly in multi-jurisdictional watersheds.
(d) Capability to consider nonstructural measures.
(e) Increased opportunities for recreational and other multi-purpose runoff controls.
(f) Contributions to local land use planning.
(g) Enhanced reuse of stormwater.
(h) Maximum justifiable land occupancies.

It should also be noted that watershed planning can have profound land-use implications. Consider the following examples:

(a) In a newly developing area, the decision-makers who are conducting the master plan want to maximize undeveloped land in the vicinity of major drainageways. One way to do this would be to establish or improve a comparatively small channel that has capacity for, say, the 5-year runoff event. Runoff from larger events would leave the channel and pass downstream as overland flow. Development might then be prohibited within the overland flow zone (floodplain).
(b) The regulation of wetlands under Section 404 of the Clean Water Act can be a very effective land management tool (see Chapter 2).

A major disadvantage of watershed planning is that local governments must perform advanced planning studies to locate and develop preliminary designs for regional stormwater management facilities, and that local governments must finance, design and construct regional stormwater facilities before the majority of future urban development occurs, with reimbursement by developers over build out periods which may range from five to much more than twenty years.

VI. GLOSSARY

The following definitions apply throughout the Manual (although they may not be defined in precisely this manner in other texts).
Storm drainage systems or urban drainage systems—The physical facilities that collect, store, convey, and treat runoff in urban areas. These

facilities normally include detention and retention facilities, streets, storm sewers, inlets, open channels, and special structures such as inlets, manholes, and energy dissipators.

Urban area—Land associated with, or part of, a defined city or town. This Manual generally applies to urban or urbanizing, rather than rural, areas.

Floodplain planning/floodplain management—Technical and non-technical studies, policies, management strategies, statutes and ordinances that collectively manage flood plains along rivers, streams, major drainageways, outfalls, or other conveyances. The federal government normally plays a major role in floodplain planning and management, whereas in urban stormwater management and design, local governments dominate the decision-making process.

Major drainageway—A readily recognizable natural or improved channel that conveys runoff that exceeds the capacity of the minor drainage system, including emergency overflow facilities.

Outfall facility—Any channel, storm sewer, or other conveyance receiving water into which a storm drain or storm drainage system discharges.

Major system—The portion of the total drainage system that collects, stores, and conveys runoff that exceeds the capacity of the minor system. The major system is usually less controlled than the minor system, and will function regardless of whether or not it has been deliberately designed and/or protected against encroachment, including when the minor system is blocked or otherwise inoperable. It may be collinear with, or separate from, the minor system. It should be noted that there are those who object to the use of the terms "major" and "minor" to describe portions of the drainage system, perhaps because these terms imply that the minor system is less important. Other terms (primary system, convenience or basic system, overflow system, major/primary drainage ways, subordinate system, etc.), have been suggested. Major/minor are used in this Manual because they seem to be the most widely used terms.

Minor or primary system—The portion of the total drainage system that collects, stores and conveys frequently-occurring runoff, and provides relief from nuisance and inconvenience. This system has traditionally been carefully planned and constructed, and normally represents the major portion of the urban drainage infrastructure investment. The degree of inconvenience the public is willing to accept, balanced against the price it is willing to pay, typically establishes the discharge capacity or design recurrence frequency of a minor system. Minor systems include roof gutters and on-site drainage swales, curbed or side-swaled streets, stormwater inlets, underground storm sewers, open channels and street culverts. The minor system is considered to end at the point where there are no adverse backwater effects from downstream conditions during discharges smaller than the minor system design flow (note that, in general, this Manual makes no recommendations regarding the return frequencies that may be used in designing either the major or minor systems, and acknowledges that there has been no satisfactory rationale (in an engineering sense) advanced for the selection of any return frequency for any component of the urban drainage system.

Storm sewer (or storm drain)—Usually, buried pipe that conveys storm drainage. It may include open channel elements and culverts, particularly when drainage areas are large.

Special structures—Those components of urban drainage systems that can be thought of as "features" or "appurtenances" such as manholes, inlets, energy dissipators, transitions, channel slope protection, detention ponds and dams, and outlet works.

Stormwater detention—The temporary storage of stormwater runoff in ponds, parking lots, depressed grassy areas, rooftops, buried underground tanks, etc., for future release. Used to delay and attenuate flow.

Stormwater retention—Storage designed to eliminate subsequent surface discharge. Wet ponds are the most common type of retention storage (though wet ponds may also be used for detention storage).

Master drainage plan—The plan that an engineer/designer formulates to manage urban stormwater runoff for a particular project or drainage area. It typically addresses such subjects as characterization of site development, grading plan, peak rates of runoff, and volumes for various return frequencies, locations, criteria and sizes of detention ponds and conveyances, measures to enhance runoff quality, salient regulations and how the plan addresses them, and consistency with secondary objectives such as public recreation, aesthetics, protection of public safety, and groundwater recharge. It is usually submitted to regulatory officials for their review.

"Standard-based" design—Design of urban stormwater management facilities based on some specified set of regulatory standards. An example is the stipulation in local drainage policies that culverts for a given subdivision all be designed to pass the 10-year flood before road overtopping.

"Risk-based" design—Design of urban stormwater management facilities not only on the basis of local standards, but also on the basis of the risk (cost) of the flow exceeding a selected design. Virtually all stormwater management projects have some component of risk which is inherent in selection of a design return frequency. Risk may also account for special upstream or downstream hazards that would be posed by adherence to some recommended standard. For example, the designer of culverts in a subdivision might choose to upsize particular culverts from a 10-year to a 50-year basis to protect properties, or to make other provisions to secure emergency discharge capacity.

Multiple-purpose facility—An urban stormwater facility that fulfils multiple functions such as enhancement of runoff quality, erosion con-

Breckenridge, Colorado—Wet pond immediately adjacent to a hotel. The impoundment created by the low dam in the foreground serves as a magnet for tourists.

trol, wildlife habitat, or public recreation, in addition to its primary goal of conveying or controlling runoff.

Conveyance structure—A pipe, open channel, or other facility that transports runoff from one location to another.

Drainage criteria—Specific guidance provided to the engineer/ designer to carry out drainage policies. An example might be the specification of local design hydrology ("design storm").

Finally, the reader should remember that while this is a manual of practice, it should not be viewed as guidance that will apply forever (or even until the next update), but rather as the best current information available. The technology and practice of urban storm drainage is changing so rapidly that the engineer will have to adjust his thinking, and his plans and designs, as criteria, techniques and materials change.

VII. REFERENCES

"Urban stormwater management." (1981). *APWA Special Report No. 49.*

American Society of Civil Engineers (1969). "Design and construction of sanitary and storm sewers." *ASCE Manual of Practice No. 37, WPCF MOP. No. 9,* New York, NY.

American Society of Civil Engineers (1969). "An analysis of national and basic information needs in urban hydrology." ASCE Urban Water Resources Research Program, ASCE, New York, NY.

American Society of Civil Engineers (1978). "Water problems of urbanizing areas." *Proceedings of an Engineering Foundation Conference,* New York, NY.

American Society of Civil Engineers (1982). "Gravity sanitary sewer design and construction." *ASCE Manual of Practice No. 60, WPCF MOP No. RD-5,* New York, NY.

"Stormwater detention outlet control structures." (1985). A report of the task committee on the design of outlet control structures committee on hydraulic structures of the hydraulics division of the American Society of Civil Engineers, ASCE, New York, NY.

American Society of Civil Engineers (1990). *Quality in the constructed project: A guide for owners, designers and contractors.* ASCE, New York, NY.

City of Stillwater. (1979). *Drainage criteria manual.* Stillwater, Oklahoma.

Debo, T.N. and Small, G.N. (1989). "Hydrologic calibration: The forgotten aspect of drainage design." *Public Works,* 120 (1).

FEMA (1983). "Questions and answers on the national flood insurance program." *FIA-2.* Federal Emergency Management Agency. Washington, D.C.

Huber, W.C. (1986). "Modelling urban runoff quality: State-of-the-art." *Urban runoff quality—impact and quality enhancement technology, Proceedings of an Engineering Foundation Conference,* ASCE, New York, NY.

Huber, W.C. (1990–1991) "Overview of contemporary stormwater quality models." Course Notes from: Urban Stormwater Quality Management, American Society of Civil Engineers Program for Continuing Education, New York, NY.

McCrory, J.A., James, L.D., and Jones, D.E. Jr. (1976). "Dealing with variable flood hazard." *Journal of the Water Resources Planning and Management Division,* 102 (WR2).

Jones, D.E. (1967). "Urban hydrology—A redirection." *Civil Engineering,* 37(8).

Jones, J.E. (1986). "Urban runoff impacts on receiving waters." *Urban runoff quality—impact and quality enhancement technology: Proceedings of an Engineering Foundation Conference,* ASCE, New York, NY.

Jones, J.E. (1988). "An overview of the case of William P. Rooney et. al. versus Union Pacific Railroad." *Proceedings of the Wyoming American Society of Civil Engineers Annual Convention,* ASCE, New York, NY.

Jones, J.E. (1990). "Multipurpose stormwater detention ponds." *Public Works*, 121 (13).

Jones, J.E. and Rossmiller, R.L. 1989–1990. "Understanding and applying stormwater management techniques." Course Notes from: Department of Engineering Professional Development, University of Wisconsin, Madison WI.

Linsley, R., Kohler, M. and Paulhus, J. (1982). *Hydrology for engineers*, 3rd edition, McGraw-Hill, New York, NY.

Linsley, R.L. (1986) "Flood estimates—How good are they?" *Water Resources Research*, 22 (9), 1595–1645.

Livingston, E. et al. (1988) *The Florida development manual—A guide to sound land and water management*. State of Florida, Department of Environmental Regulation, Tallahassee, FL.

McCuen, R. (1985). *Statistical methods for engineers*. Prentice-Hall, Inc., Englewood Cliffs, NJ.

McPherson, M.B. et al. (1970) *Prospects for metropolitan water management*. ASCE, New York, NY.

Malcolm, R.H. (1982) "Some detention design ideas." *Proceedings of an Engineering Foundation Conference on stormwater detention facilities: Planning, design and maintenance*. W. DeGroot, ed., ASCE, New York, NY.

Poertner, H.G. (1980) *Stormwater management in the United States—A study of institutional problems, solutions and impacts*. Office of Water Research and Technology, Washington, D. C.

Rickert, D.A. and Spieker, A.M. (1971) "Real Estate Lakes—Water in the Urban Environment." *U.S. Geological Survey Circular (601-G)*, Washington, D.C.

Roesner, L. (1990–1991). Urban stormwater quality management." Short Course Notes from: American Society of Civil Engineers, Continuing Education Department, New York, NY.

Schueler, T.R. (1987). *Controlling urban runoff: A practical manual for planning and designing urban BMPs*. Department of Environmental Programs, Metropolitan Washington Council of Governments, Washington, D.C.

Sheaffer, J.R. et al. (1982). *Urban storm drainage management*. Marcel Dekker, Inc., New York, NY.

Spirn, A.W. (1984). *The granite garden—Urban nature and human design*. Basic Books, Inc., New York, NY.

Urban Drainage and Flood Control District (1984). *Urban storm drainage criteria manual*. Denver Regional Council of Governments, Denver, CO.

Urbonas, B. and Glidden, M. (1982) "Development of simplified detention sizing relationships." *Proceedings of an engineering foundation conference on stormwater detention facilities: Planning design and maintenance*. W. DeGroot, Ed., ASCE, New York, NY.

Urban Land Institute, American Society of Civil Engineers, and the National Association of Home Builders (1975). *Residential stormwater management: Objectives, principles and design considerations.* ASCE, New York, NY.

Walesh, S.G. (1987) "Course Notes on Stormwater Detention Facility Design." American Society of Civil Engineers Continuing Education Program ASCE, New York, NY.

Water Pollution Control Federation (1989). "Combined sewer overflow pollution abatement." *WPCF Manual of Practice FD-17,* Alexandria, VA.

Wells, Esq., D.G. (1984). *Colorado construction contracts.* Professional Education Systems, Inc.

Chapter 2

FINANCIAL, LEGAL AND REGULATORY CONCERNS

I. INTRODUCTION

The storm drainage component of the urban infrastructure has historically received less attention than other elements such as water supply, streets, airports, and wastewater collection and treatment. Out of sight, out of mind. As urbanization has intensified in the last 40 years and extended into previously rural areas, the consequences of poor stormwater management have become obvious. Urbanization causes significant increases in runoff peaks and volumes. If these increases are not anticipated and if adequate facilities are not provided, relatively small storms can flood streets, interrupt traffic, and cause property damage. Large storms can flood creeks and rivers, reclaim floodplains, cause significant property damage, and threaten lives. Construction of urban stormwater systems often has lagged behind other urban infrastructure systems.

The importance of good stormwater management has been acknowledged by planners and engineers, and more importantly by those responsible for urban development. Important factors have included the mortgage insurance programs of the U. S. Department of Housing and Urban Development that stimulated the adoption of drainage design standards for subdivision development, creation of the Federal Flood Insurance Program that has provided impetus for most local governments to adopt and enforce floodplain regulations, and court decisions that have placed responsibility for drainage problems on those who could be identified as causing the problem. Many local governments have adopted storm drainage criteria, and some have begun charging fees for providing a drainage service. Most recently, the adoption by Congress of the Water Quality Act of 1987 will require the nation's larger urban areas to prepare and implement stormwater quality management plans.

It has become essential for the engineer to understand the financial, legal, and regulatory requirements and design constraints. Since these issues can dominate the drainage system process, they must be addressed in concert with the technical aspects of the assignment.

The responsibility for management of the urban storm drainage system is typically not clearly defined, although some generalizations can be made. One problem has been that poor drainage control is obvious only during a time of heavy rainfall, whereas other urban systems such as water supply, transportation, and sewage disposal are more obvious when inadequate or missing. A subdivision cannot be built without streets, for example, but the lack of good drainage may not be so obvious, until storm runoff problems occur.

The responsibility for urban drainage rests primarily at the local level, and drainage facilities should be an inseparable part of urban infrastructure planning and design. The private sector generally is called on to provide storm drainage as a part of new developments (although operation and maintenance remain local government responsibilities). Federal and state agencies have tended to have greater influence and involvement with larger drainage systems, and increasingly are setting minimum requirements (or superseding local criteria), particularly in relation to floodplain management and water quality. Local governments, however, usually end up with the financial responsibility and liability for the entire system.

The importance of urban drainage system maintenance cannot be over-emphasized. The funding of maintenance is a problem, however, and in many communities deteriorated public facilities threaten the provision of basic community services, including flood protection (Choate and Walter 1983). A local tax base usually is necessary to support maintenance on a continuing basis to keep facilities in operating condition. Storm drainage facilities on private property are generally the responsibility of the owner. "Responsibility" involves financial, legal, and regulatory concerns, and the remainder of the chapter will address these three areas.

II. FINANCIAL

The most important financial considerations are who pays and what is the source of funds. Economic evaluation of master plan alternatives and preparation of construction cost estimates are covered in Chapter 8. Basic funding sources include (1) financing from outside the local area such as federal and state assistance; (2) financing from within the local area, either directly from local governments, from improvement districts, or from private interests such as developers. Financing can be difficult to obtain, and many well-conceived design and construction programs are never implemented because of the lack of funding. Operation and maintenance costs must be considered, and there must be a commitment to maintenance by local government or some other entity that can be held responsible and that has the necessary financial capability.

The review and development of financing methods for a project or community generally involves engineers as a part of the team. They are involved in preparing preliminary plans and cost estimates, defining drainage basins and hydrologic responses, identifying floodplains, assessing water quality impacts, and defining frequency and extent of flooding problems. This section reviews available financial options, which should not be considered in isolation, but viewed as a possible package of several options. The feasibility of different financing approaches varies widely from one area to another, and the nuances of each situation must be carefully considered, particularly in relation to relevant statutory authorities.

A. State and Federal Sources

Funding support for design and construction of urban storm drainage systems is limited at the state or federal level. Flood control, which deals with major drainageway or river flooding, receives some support from federal programs of the Corps of Engineers (COE) and the Soil Conservation Service (SCS). Little other funding is available, except from sources such as revenue sharing or block grants, for addressing storm drainage problems.

The following describes funding opportunities that existed as of 1990. Regulations implementing federal and state programs are constantly changing, and federal and state sources of funding change as programs are modified, eliminated, or added.

1. U.S. Army Corps of Engineers (COE)

The COE has been in the flood control business essentially since 1936 (Public Works Historical Society 1988), and has several programs that can aid in addressing urban flood control and erosion problems (U. S. Army Corps of Engineers 1990). The COE is authorized to provide technical assistance to local communities and states to support their efforts to control flooding and reduce erosion, and also is authorized to construct small flood control projects without specific Congressional approval. Large flood projects require the authorization of Congress.

There are several technical assistance programs. The Shore and Streambank Protection program (Section 14, Flood Control Act of 1946, as amended) can help design projects to prevent or repair damages that occur from shoreline and stream bank erosion. Floodplain management services (Section 206, Flood Control Act of 1960, as amended) are available to help local communities identify flood hazards and plan for wise use of floodplain lands. The Channel Renovation Program (Section 942, Water Resources Development Act of 1986, as amended) can provide designs, plans, specifications, and other technical assistance for renovation of navigable streams and their tributaries.

Construction programs include the Small Flood Control Projects programs (Section 205, Flood Control Act of 1948, as amended) which enables the COE to construct small flood control projects. This program focuses on solving local flooding problems in urban areas. The Emergency Streambank and Shoreline Protection Program (Section 14, Flood Control Act of 1946, as amended) allows the COE to construct emergency streambank and shoreline protection to prevent erosion or flooding from damaging highways, bridge approaches, hospitals, churches, schools, and other non-profit or public facilities, such as wastewater treatment facilities. The COE can clear stream channels to increase flow capacity, decrease flooding, and reduce damage from debris carried by flood flows through the Channel Clearing for Flood Control Program (Section 208. Flood Control Act of 1954, as amended). Finally, the Congress can authorize and fund individual COE projects, but this is a lengthy process, with the average project taking over 20 years to get under way.

The Water Resources Act of 1986 provides for COE and local sponsor cooperation, and defines minimum financial requirements for local sponsors. Local sponsors are afforded more active involvement in defining and planning the projects, but a greater local financial commitment is required.

2. Soil Conservation Service (SCS)

The SCS is authorized to provide technical and financial help to local organizations for floodplain management assistance, watershed protection, emergency flood relief, and flood warning assistance under Public Law 83-566, commonly called the Small Watershed Program (Soil Conservation Service 1984). While this program is agriculturally oriented, in some cases it is possible to address urban flooding and erosion control problems.

Projects are initiated by local sponsors, which can include municipalities, counties, and flood control or other special districts, who submit an application to the state agency designated to approve watershed applications. The applications are reviewed by the State and the SCS. If the project has a high priority the SCS Chief can authorize planning assistance, and the SCS and sponsors can proceed with project planning. Depending on the amount of federal assistance involved, construction approval of a project is made by the SCS state conservationist, by the SCS Chief, or by Congress. The SCS pays the full cost of construction for flood protection, and shares the cost for other purposes. Major obligations of local sponsors are to acquire lands, easements, and rights-of-way, and to share in certain construction costs.

3. Environmental Protection Agency (EPA)

There has been little EPA funding for urban storm drainage in recent years, except in limited amounts for implementation of projects and programs established under the nonpoint source pollution control program (Section 319 of the Federal Water Quality Act (FWQA) of 1987). The FWQA authorized federal funds to capitalize state revolving fund (SRF) programs and to provide a transition from the former Federal Construction Grant Program to a new approach for assisting communities in construction of publicly owned treatment works (POTWs). From the SRF, states can provide loans and other types of financial assistance (but not grants) to local communities and inter-municipal and interstate agencies, for the construction of POTWs and implementation of nonpoint source programs and projects. Some urban stormwater quality control projects may qualify for SRF assistance.

4. Community Development Block Grant Program

The Community Development Block Grant (CDBG) Program is a grant program to assist local communities with specific projects, funded through the Department of Housing and Urban Development (HUD). There are two types of CDBG programs. One provides annual CDBGs on a formula basis to entitled communities to carry out a wide range of community development activities directed toward neighborhood revitalization, economic development, and improved community facilities and services (U. S. Department of Housing and Urban development 1989). Metropolitan cities and urban counties (generally those with populations in excess of 50,000) are entitled to receive annual grants. Entitlement communities develop their own programs and funding priorities.

The other CDBG program is a non-entitlement program for states and small cities. Grants are awarded competitively based on established selection criteria. Each state has the option to administer the block grant funds provided for its non-entitlement areas (most states exercise this option). Typically grants are awarded competitively based on established selection criteria.

Drainage system projects are eligible for CDBG funds if they meet basic program requirements. Local matching funds are not required, however the competition for these funds is intense, and they will probably become more difficult to obtain in the future.

5. Federal Emergency Management Agency (FEMA)

FEMA is the designated leader of inter-agency task forces that are formed to provide assistance following emergencies or major disasters. This assistance can include low interest loans to the private sector and partial grants to local governments for repair or replacement of damaged public facilities. (Federal Emergency Management Agency 1989)

A major FEMA activity is the administration of the National Flood Insurance Program (NFIP), which is also discussed under the Regulatory section of this Chapter. FEMA also has limited funds available for the purchase of flood-damaged property under the provisions of Section 1362 of the Flood Disaster Protection Act. If a structure has been damaged repeatedly by flooding it may qualify for acquisition under this program.

6. *National Weather Service (NWS)*

The NWS has authority to assist local interests to develop and implement local advance flood warning programs.

7. *State Funds*

It may be possible in some states to obtain state funding support for drainage and flood control projects, although states are generally poor sources. Assistance may take the form of planning advice, floodplain delineation, flash flood and contingency planning assistance, revolving loan funds, or construction of facilities.

B. Local Sources

Traditionally, urban stormwater drainage has been primarily the responsibility of local government. Although there is increasing regulatory pressure by state and federal agencies to resolve urban drainage issues, the burden of financing urban drainage programs will continue to fall primarily on local governments and private developers. Demands and requirements for more involved and complex urban drainage programs are increasing, which is causing the costs associated with providing and maintaining urban drainage systems to rise.

Financing needs can be categorized into two basic areas: (1) existing development, or problems needing remedial treatment, and (2) new development, or situations offering the opportunity to prevent future problems. Financing needs include both annual operating costs (including planning, administration, regulatory enforcement, maintenance, and operations) and capital improvements. Some practices for financing urban storm drainage projects and programs are listed in Table 2.1. In most cases the sponsors will use a combination of these funding sources. Some of the approaches lend themselves to financing annual operating costs, and some to capital expenditures.

TABLE 2.1.   Methods of Financing Urban Storm Drainage Programs

Development Status (1)	Source of Funds (2)	Primarily O&M or Capital (3)
Existing developments	General tax fund	O&M or Capital
	Service charge or fees	O&M
	Special assessment	Capital
	Bonds	Capital
	Private funds	Capital
	Stormwater Utilities	O&M or Capital
New developments	Developer fees	Capital
	Developer provided facilities	Capital
	Dedications	Capital
	Floodplain regulations	Capital

1. General Tax Revenues

All local governments support their operations with basic tax revenues, including property taxes, sales taxes, fees, licenses, etc. There is always competition for these funds, and it is difficult to raise large amounts of capital for drainage improvement and maintenance. When flooding problems exist on major drainageways in developed areas, structural improvements, rights-of-way, engineering, and construction are particularly costly. General fund revenues usually are best suited for financing operation and maintenance, and less suited for funding large capital projects. There are exceptions. For example, a portion of a sales tax can be dedicated to capital projects including drainage, or funds can be annually escrowed into a sinking fund to pay for an improvement. General funds are in most cases the only source of money for planning, plan review, inspection, mapping, and similar activities.

2. Dedicated Ad Valorem Taxes

Most flood control districts and a few city and county governments have the authority to levy taxes on improved property specifically for storm drainage and flood control. The state of Illinois recently passed legislation allowing a county-wide tax for stormwater management in several urbanized areas. While ad valorem taxes are a significant source of revenue, the impetus for many jurisdictions to treat stormwater as a "utility" reflects the desire to allocate program costs on criteria more directly related to a property's contribution of runoff than to assessed valuation. There is no clear correlation between property value and contribution of runoff to the drainage system, whereas a strong correlation exists between the amount of impervious surface and impacts on the stormwater system. Increased equity is therefore a principal element in the decision to move toward stormwater service charges, and away from ad valorem taxes.

3. Service Charges or Fees

Many local governments are turning to service charges or fees for financing urban drainage programs because of competition for limited general fund tax dollars. The fee should be related to service provided, the most common basis being area of impervious surface. The principle is that each property owner pays a fee for the service of handling the drainage originating from that property. This is fair because uphill properties generate the same runoff as similarly developed downhill properties. Funding does not fall entirely on those who experience flooding problems, but is distributed equitably to all those who contribute to the problem. Such fees are particularly useful for industrial facilities, which discharge large amounts of water to the urban stormwater system.

Service charges have been adopted by large and small cities, as well as counties. In the Denver metropolitan area four cities have a drainage service charge. They are Denver (509,800), Boulder (80,400), Aurora (223,200) and Littleton (31,100). Drainage fees involve billing each individual property in the city. In some cases, such as Boulder and Littleton, the drainage fee supports a utility.

Some other cities in the United States with service charges include Tulsa, Oklahoma; Bellevue, Washington; Tampa, Florida; Cincinnati, Ohio; and Louisville, Kentucky. Generally, a uniform method of determining fees is applied throughout the service area, although fees will differ for various classes of property, usually based on the area of a property and its percentage of impermeable area.

There are numerous combinations and variations of stormwater service charge rate structures available (and that have been upheld in the courts). The three approaches most often used are:

(a) Amount of Impervious Surface. Under this approach, rates are set in direct proportion to the measured, estimated, or assumed extent of impervious area for each parcel of land.
(b) Density of Development. Rates are determined by a runoff coefficient which is deemed appropriate for the type of land and the nature of the improvements to each parcel.
(c) Flat Fee. This mechanism uses a constant or uniform fee for each property within pre-existing classes, or can be applied on a community-wide basis. Flat fees are used mainly because of their administrative simplicity.

Charges for stormwater management reflect a rationale that those who contribute to stormwater problems should bear the costs of mitigative services. This approach is regarded by most administrators and the courts as an appropriate technique for financing stormwater programs. Public acceptance is critical, however, and public awareness and problem recognition are the most important elements in the successful implementation of service charges.

Generally, operation and maintenance costs are paid from drainage fees. Drainage fees can also be used to finance capital projects through revenue bonds or on a pay-as-you-go basis. Capital needs vary from basin to basin and it would be most equitable to have the portion of the drainage fee for capital projects reflect the specific needs of the basin. Some disadvantages of drainage fees can be the need to obtain large amounts of detailed data about the location and size of properties and their impervious areas, the need for a system to bill each individual property, and the resistance of many property owners, particularly those who live uphill and do not have a drainage problem (although they contribute to drainage problems and benefit from improved urban drainage).

The most important advantage of a fee is that a dedicated and dependable source of funding can be made available, enabling the local government to think and plan for the long term. The drainage system can be viewed as a utility whose components can be considered together in their entirety. Fees also provide dependable funding for operation and maintenance, a critical part of an effective urban drainage program. Service charges are coming into more general use as urban drainage system needs become more pronounced and as more pressure is placed on general revenue sources. Since a drainage fee is not considered a tax, it can be charged to all landowners, public and private, from whose property storm waters flow. This includes traditionally nontaxable institutions such as churches, colleges, government facilities, and even state and local highway departments in some cases.

4. Special Assessments

A special assessment is a mandatory charge on selected properties for an identified improvement which benefits the property owners and which is undertaken in the public interest. Special assessment projects may be undertaken by general purpose governments or special purpose districts. The authority for local governments to levy special assessments is derived through enabling state legislation. The requirements and procedural details for their establishment vary from state to state, and from city to city.

Special assessments can generally be initiated in one of the following ways:

(a) By local government legislative body (council, commissioners, aldermen, etc.) action with consent of property owners, usually expressed in the form of a petition.
(b) By local government action, which may be stopped by opposing petition or remonstrance.
(c) By local government action without property owner consent and without possibility of remonstrance.

The benefit that accrues to each property by virtue of the project provides the foundation for levying a special assessment. The amounts assessed must be proportional to and not more than the benefits received, and must not exceed the cost of the project. Problems with special assessments include determining who is specially benefited, and determining the amount of special benefit to be received by each property owner. Benefits may include handling the discharge of surface waters from uphill properties through lower properties in a quantity greater than would naturally flow because the uphill owner made some of their property impermeable; adaptability of property to a more profitable use; alleviation of health and sanitary hazards; reduction in property maintenance costs; increase in convenience or decrease in inconvenience; and recreational improvements. As with service charges, special assessments may be made on properties that are normally non-taxable.

5. Bonds

Bonds are not an additional source of revenue per se, because they create an equivalent liability that must be met from pledged future revenues. Bonds provide a mechanism for local governments to borrow capital needed to finance public works improvements. They generally must be authorized by a vote of the electorate.

There are two basic types of bonds, general obligation (GO) bonds and revenue bonds. General obligation bonds are backed by the full faith and credit of the unit of government issuing the bonds. They may in fact, be repaid from fees, but if the fee revenue is inadequate, general tax revenues would have to be used. Revenue bonds are repaid entirely from revenues such as services charges or fees, and are not guaranteed by the issuing entity. The advantage of GO bonds is that their interest rates (and the amount required as "reserve") are lower.

6. Stormwater Utilities

Stormwater utilities are becoming increasingly popular as a means of financing and managing stormwater systems. The concept is the same as for other common utilities, such as water and telephone, which provide a service to their customers. A stormwater utility constructs capital improvements and operates and maintains the facilities in its system. Their operations are financed by regularly billing their customers for these services. Fees may be based on various criteria, including those enumerated in preceding sections. Administrative appeals and a formal variance procedure are typically accommodated. Utilities are particularly attractive because they provide a single unit with its own staff, regulations, budget and sources of revenue which can focus on solutions to drainage problems.

7. Development Fees and Developer-Provided Facilities

Developers have typically been required to provide on-site drainage facilities such as curbs and gutters, inlets, storm sewers, and detention ponds, however developer-provided major system facilities are less common.

A method of financing flood management projects impacting an entire basin is to spread the cost of required facilities over the entire basin. The rationale is that a development should finance those regional improvements that are necessitated by the cumulative development. A method of accomplishing this is with a development fee—a unit fee based upon acreage involved for all developments in the basin. The amount of the fee would depend on the cost of facilities, including right-of-way required in the basin, and the fee would vary from basin to basin. A basin development fee should be charged only for facilities required because of development, and not to finance improvements required to solve previously existing problems.

A master drainage plan defining needed facilities for each basin must be developed to determine the amount of the basin fee. One of the difficulties with drainage development fees is that an improvement defined in the master drainage plan may be needed before adequate funds are available in the basin fund to finance the project. This may require governments to provide or loan the basin fund enough monies to build the necessary facilities. Depending on future development in the basin, local governments may or may not recover their money.

8. Fee in Lieu of On-Site Detention/Retention

In-lieu-of fees are derived from system development charges specific to stormwater management. These fees can be either a regulatory requirement or a development option that affords the opportunity to construct on-site detention/retention facilities in accordance with established design criteria (i.e. local or county), or to pay a fee into a fund dedicated to the construction of a regional detention facility serving multiple properties. This approach is typically authorized within a context of promoting the siting and construction of more regional, as opposed to on-site, detention/ retention facilities. This objective is consistent with the intent of regional detention ordinances, which have proven effective as a vehicle to guide development patterns within a watershed, and as a tool to encourage comprehensive stormwater planning.

The problems most frequently encountered with fee-in-lieu construction involve cash flow and construction timing. The customary fee for a single property or development is rarely large enough to fund the construction of a regional facility. Therefore, either multiple developments must occur simultaneously or, more realistically, the project must be initially funded from alternative sources.

9. Plan Review and Inspection Fees

These fees are intended to recoup the expense of examining development plans to insure consistency with comprehensive or master plans, and to insure that design and construction standards are met. These fees are **not** designed to be primary revenue-generating sources. In theory, a detailed cost-accounting system can determine the actual costs of providing engineering review and field inspections/certifications. In practice, however, most drainage authorities apply a fee based on an average of their total costs. Four fee structures are commonly applied:

(a) Fees based on a flat rate for all projects reviewed or inspected, regardless of size or complexity.
(b) A variable or sliding-scale fee based on the size of development.
(c) A variable or sliding-scale fee based on the permitted construction cost of the development or project.
(d) A fee based on the fixed and variable costs to provide the review and inspection service. The fixed portion is usually a statistical estimation of the administrative costs to provide the service, while the variable portion reflects the actual time and materials required to perform the plan reviews and inspections.

One of the keys to a successful regulatory program is consistent enforcement of development/construction controls. By implementing a plan review/inspection charge based on the actual cost of providing the service (as in the case of code enforcement staffing), the program can both enhance the development/construction review process and avoid passing the costs of these direct services back to the general public.

10. Dedications

When new developments infringe on a floodplain the developer must consider the expense of meeting floodplain requirements. An alternative to structural improvements is to avoid the floodplain. The developer must compare the cost of structural improvements and potential revenues with the cost of not developing the floodplain. If the cost of development is excessive, the developer may choose to leave the floodplain alone and dedicate that area to a public entity (thereby providing valuable open space). Some jurisdictions promote the dedication of floodplain areas by granting density credits for land so dedicated, and by allowing cluster development on the remainder of the tract. It should be noted, however, that many jurisdictions still view floodplains as less desirable for recreation and open space, and resist the dedication of floodplain lands as the developer's open space requirement. Local governments should be aware that urbanization may accelerate erosion on "natural" drainageways, and that erosion control measures may be required in the future.

III. LEGAL

Legal considerations are often a determining factor behind drainage decisions. The engineer must appreciate drainage law and the limitations it may impose on design, and must think in terms of public safety and welfare and his professional liability throughout the course of a design effort. Designing and installing drainage systems that conflict with established drainage law or statutes, or that are inconsistent with good engineering practice, must be avoided because the engineer may be held liable for flood related damages that may occur. The information in this section is based on the Denver Urban Drainage Criteria Manual (Urban Drainage and Flood Control District 1984), Wright (1982), APWA (1981), and the Urban Land Institute (1975).

A. General Rules of Law

The "law" is derived from judicial decisions and statutory enactments. Over the years, two general rules of "law" have developed with respect to surface waters. More recently, judges have modified the original rules.

In the early years of rural America, the "common enemy" doctrine prevailed. Surface waters were considered a common enemy that each landowner could deal with as he pleased, by repelling, diverting or even retaining them. However, in some jurisdictions, judges applied a "civil law" rule which places a servitude upon the lower land in favor of the upper or dominant landowner to receive surface water in its natural way of draining. The passage of this water is not to be obstructed by the servient owner to the detriment of the dominant owner. This rule has been modified over the years to give some protection to servient owners and now follows the premise that "natural drainage conditions may be altered by an upper landowner provided the water is not sent down in a manner or quantity to do more harm than formerly."

The approach followed by judges today in resolving drainage disputes in many jurisdictions is called the "reasonable use" rule. Each property owner, including a municipality, can legally make use of his land, even though the flow of surface water is altered and causes some harm to others. However, liability attaches when his harmful interference with the flow of surface water is "unreasonable." Whether a landowner's use is unreasonable is determined by a nuisance-type balancing test involving three questions:

(a) Was there reasonable necessity for the actor to alter the drainage to make use of his land?
(b) Was the alteration done in a reasonable manner?
(c) Does the utility of the actor's conduct reasonably outweigh the gravity of harm to others?

The North Carolina Supreme Court in 1977 traced the evolution of common-law rule to the civil law, and then finally to the reasonable use rule by stating:

"It is no longer simply a matter of balancing the interests of individual landowners; the interests of society must be considered. On the whole the rigid solutions offered by the common enemy and civil law rules no longer provide an adequate vehicle by which drainage problems may be properly resolved."

As urbanization has taken place, drainage disputes have typically involved several parties, not just one upper owner and one lower owner. The "interests of society," referred to by the North Carolina Court, have put municipalities into the issues of drainage. Urban drainage problems and the engineering/legal solutions to them were identified by Shoemaker (1968). In 1969 the Colorado Legislature recognized the issue of urban drainage problems when it created the Urban Drainage & Flood Control District to address flooding problems in the Denver metropolitan area (Urban Drainage and Flood Control Act 1969). The Colorado General Assembly in its legislative declaration stated that:

"The necessity of this article results from the large population growth in the urban area included by this article within the district constituting a major portion of the state's population, from the numerous capital improvements and large amount of improved real property situated within such urban area, from the torrential storms occurring sporadically and intermittently in the urban area and other areas draining into such urban area, from the increasing danger of floods therein and the resultant risks to the property and to the health and safety of the persons within the urban area, . . . from the fragmentation and proliferation of powers, rights, privileges, and duties pertaining to water, flood control, and drainage within such urban area among a substantial number of public bodies, and from the resultant inabilities of such public bodies to acquire suitable capital improvements for the alleviation of such dangers and risks."

Many courts have recognized that:

(a) Since surface water flows downhill, don't try to stop it.
(b) He who casts more water downhill (by urbanization, for example) than would naturally flow, must provide for it.
(c) A municipality is generally treated like a private party in drainage matters.
(d) If you are not going to maintain it, do not build it.
(e) A municipality may be liable for issuing permits to a developer which increase flooding of downstream property.

And some legislatures have enacted laws that:

(a) Require runoff caused by developments in excess of historic flows be detained.
(b) Define "special benefits" for the purpose of determining special assessments to include "accepting the burden from specific property for discharging surface water into servient property in a manner or quantity greater than would naturally flow because the dominant owner made some of this property impermeable."
(c) Permit local governments to charge service fees to users of drainage facilities.
(d) Require or allow local governments to identify, designate, and regulate the 100-year floodplain.

B. Liability Issues

Potential liability is becoming of major concern to engineers, their employers, and their clients. The following discussion addresses some potential liability issues, however the engineer must rely on his own common sense, professional competence, and judgment to limit his exposure to liability. Expert legal advice should always be sought if there are questions or concerns.

1. Ordinary and Extraordinary Floods

Whether or not defendants were liable for their actions has often depended on whether the flood was determined to be "extraordinary" or "ordinary." The extraordinary flood was considered to be an act of

God and a defense against liability. Even early on, however, some courts did not recognize this and would not absolve a defendant when damages occurred due to an act of God, where the defendant's negligence was the proximate cause of injury. With the advances in hydrology and meteorology in recent years, courts are now able to base their decisions on scientific data presented to them by expert witnesses, with the result that it is virtually impossible to rely on the Act-of-God defense. For example, standard hydrologic techniques permit quantification of large floods such as the standard project flood (SPF) commonly used by the COE or the PMF, and dams in urban areas can and should be built to safely handle this extremely large event without failing. Dam design criteria are generally established by the states, and any such criteria must be carefully followed by a design engineer.

Other adjectives used by the courts are "unprecedented" and "unanticipated." In situations where the court may feel that the defendant's actions were reasonable and not negligent the "unprecedented flood" defense may still be used. One example is where a city maintained a natural waterway as a drainage ditch. A storm "in excess of 100 years" caused water to back up and damage plaintiff's property. The court found that the city had properly maintained the drainageway and held the city not liable because the storm was unprecedented. However, in another case where the flow was increased by upstream urbanization, and the city could have avoided the damage by simply replacing an inadequate culvert in a highway, the court held that the flood was not extraordinary but could have been anticipated.

Generally, an engineer should not assume that he will be able to use the "Act-of-God," "unprecedented," or "unanticipated" defense. Local flood or rainfall history may indicate flow peaks or rainfalls of comparable or greater magnitude than the event that is considered "unprecedented." When liability for flood damages becomes an issue it may be necessary for the engineer to demonstrate that he (1) practiced without negligence; (2) adopted the maximum design frequency specified by the regulation; (3) followed applicable drainage standards and criteria; (4) researched historic flood information; and (5) checked the design for events greater than the design flood.

2. Public Liability

A municipality is generally treated like a private party in drainage matters. Municipalities normally have no legal obligation to construct drainage improvements, unless public works such as schools or roads alter the drainage and require drainage improvements to prevent potential damage to others. Municipalities can be held liable, however, for negligent construction of drainage improvements as well as for their negligent maintenance and repair. In general, in the absence of negligence, a municipality will not be held liable for damages caused by overflow of its sewers or drains occasioned by extraordinary, unforeseeable rains or floods.

Municipalities can be held liable, however where they:

(a) Collect surface water and cause it to flow onto private property where it did not formerly flow.
(b) Divert surface water in such a way that it causes damage on private land where, but for the diversion, it would not go.
(c) Fill, dam, or otherwise divert a stream so that it flows onto the land of another.
(d) Fail to construct or have constructed drainage outfall facilities to carry developed flow from new subdivisions located in the upstream areas of a basin.

A municipality is also liable if it fails to provide a proper outlet for drainage improvements constructed to divert surface waters, or if it fails to exercise ordinary care in maintenance and repair of drainage improvements. Whether or not public entities have a "continuing duty" to upgrade drainage facilities in response to upstream development is not clear (in the law). The engineer should recognize, however, that design standards generally become stricter with time, and that facilities may need to be upgraded.

The courts have been reluctant to find liability when a municipality merely issues a permit to build, unless an ordinance or statute imposes a duty on a municipality to prevent or protect land from surface water drainage. The liability, if any, would be against the developer or builder. On the other hand, some courts are finding municipalities liable in situations where the municipality issued permits for development that caused injury to other property owners. An example is where a municipality granted a permit to develop an industrial complex that increased flooding on downstream property that had flooded previously. The municipality was held liable for granting the permit, had to pay damages, and was enjoined from issuing any more building permits.

There seems to be a trend toward imposing a greater burden of responsibility on municipalities for the drainage consequences of urban development. Municipalities should proceed with caution when issuing permits for development where the adverse impacts on drainage can be determined and foreseen.

3. Other Liability Considerations

Examples of questions or concerns of which the engineer should be aware are listed below. No opinions or answers are provided, but if these or similar concerns arise the engineer should seek legal advice.

(a) Should safety racks be installed at culvert entrances? At outlets?
(b) What constitutes a "dangerous" hydraulic structure from the standpoint of public safety?
(c) How much can and should the engineer disrupt an area's drainage "status quo" with new facilities? For example, a Federal District Court in Cheyenne, Wyoming found that a culvert with a capacity exceeding that required to pass the 100-year flood constituted an inadequate re-

placement structure for a far larger bridge, because reliance on the bridge had been established by upstream property owners. The bridge had been in place since 1910.

(d) To what extent are land developers, and their engineers, liable for downstream erosion, sedimentation and water quality problems?

(e) To what extent should buildings be allowed to encroach on floodplains downstream from dams?

Local case law should be searched to see if questions such as these have been litigated where the engineer is practicing, if there is a possibility that they could become an issue.

4. *Limiting Liability*

There are a number of common-sense rules that can be adopted to minimize potential liability. These include:

(a) Rigorous in-house review of all assumptions and calculations.

(b) Peer review by independent parties.

(c) Meticulous record-keeping, including calculations, telephone conversations, meeting notes, documentation of regulator/client decisions, contracts, etc.

(d) Step-by-step review and endorsement of design assumptions by review authorities.

(e) Searching for unusual and unforeseen problems that can arise, such as failure of upstream structures or changed geotechnical conditions.

(f) Avoidance of practices that "beg" litigation, such as failure to comply with contract provisions.

IV. REGULATORY CONCERNS

Several regulatory programs have significant impacts on urban stormwater management. Local governments generally enforce regulations by authority granted to them by state enabling legislation. There are some programs that are either required or enforced by the federal government, such as Section 404 of the Federal Water Pollution Control Act which includes the dredge and fill material permit program and the wetland protection program.

A. Floodplain Regulation

The purpose of floodplain regulation is to manage, not prevent, development within a defined floodplain (usually the 100-year floodplain) so as to preclude or mitigate future flood damages. Legal justification for floodplain regulation is the health, safety, and welfare of the public. Floodplain regulation should not preclude all uses of the floodplain, and should not be used to keep floodplains in an open undeveloped state.

With some exceptions, such as New Jersey, the authority to regulate floodplains is generally delegated by state legislatures to local govern-ment. Some states have minimum requirements; others do not. In a few states local governments must regulate floodplains or the state will do it for them.

A basic requirement for a successful floodplain regulation program is accurate delineation of the floodplain based on full development of the basin consistent with projected land uses. Detailed delineations are particularly important in urban and urbanizing areas because of the large number of residents, density of development, and quantity of drainage flows.

Floodplain regulations can take various shapes and forms, but own-ers/developers of floodplain lands have several basic choices. They can build outside of the floodplain and leave the floodplain alone. This alternative has a low capital cost, but less land is available for devel-opment. They can reclaim the floodplain by constructing a channel to contain flood flows. This alternative has a higher capital cost, but max-imizes the land available for development. The flood fringe can be filled, which reclaims some of the floodplain but generally not as much as channelization. Other approaches also are possible.

Floodplain regulation is most effective in undeveloped floodplains, where the opportunity to control future development is the greatest. In urbanized floodplains the potential for flood damage already exists. Regulation, however, can prevent increases in the damage potential, and can have a positive effect on damage reduction where older build-ings are being rehabilitated or the area is being redeveloped.

The National Flood Insurance Program (NFIP) has had a significant impact on floodplain regulation by local governments. This federal pro-gram makes flood insurance available at reasonable rates to individuals within communities that meet eligibility requirements by adopting and enforcing measures to reduce future flood risks to new construction in defined flood hazard areas (Federal Emergency Management Agency, October 1989). If local communities choose not to meet NFIP require-ments, FEMA will prepare flood hazard boundary maps for the com-munity, but will not make the federally subsidized insurance available, and will withhold federal financial assistance and support from the community for locations within identified flood hazard areas. For ex-ample, Small Business Administration, Veterans Administration, and Federal Housing Administration assistance, as well as other federal grants, loans, or guarantees, are prohibited within identified hazard areas unless the community participates in the program. Post-flood disaster aid is similarly reduced, which can be a serious loss to a com-munity that suffers a flooding disaster.

Community participation in the NFIP is voluntary, although some states require NFIP participation as a part of their flood management program. For all practical purposes, however, the regulation of flood-plains has become a matter of necessity for local governments rather than a matter of choice. Most local governments have taken the steps necessary to participate in the NFIP.

B. Section 404 Permits

Many engineers have heard (or asked) the question—"do we need a 404 permit for this project?" This refers to Section 404 of the Federal Water Pollution Control Act, as amended, which prohibits the discharge of dredged or fill material into the waters of the United States, including wetlands, unless a permit is obtained from the Corps of Engineers. Many urban drainage and flood control projects are impacted by 404 permit requirements since they deal with water and drainageways. It is therefore imperative that engineers be aware of 404 permit requirements and address the need for a 404 permit early in the planning or design process. The COE District Engineer should be consulted if any questions arise.

Some of the important definitions and requirements of the 404 permit process are discussed below. The COE has published summaries and brochures explaining the program. Specific requirements are set forth in the Code of Federal Regulations (33 CFR Parts 320 through 330 and 40 CFR Part 230).

The 404 permit program applies to "the waters of the United States," which are defined (33 CFR 328.3(a)) as:

(1) All waters which are currently used, or were used in the past, or may be susceptible to use in interstate or foreign commerce, including all waters which are subject to the ebb and flow of the tide.
(2) All interstate waters including interstate wetlands.
(3) All other waters such as interstate lakes, rivers, streams (including intermittent streams), mud flats, sand flats, wetlands, sloughs, prairie potholes, wet meadows, playa lakes, or natural ponds, the use, degradation or destruction of which could affect interstate or foreign commerce, and (a) which are or could be used by interstate or foreign travellers; (b) from which fish or shellfish can be taken and sold in interstate or foreign commerce; and (c) which are or could be used by industry engaged in interstate commerce.
(4) Tributaries of waters identified by items (1), (2) and (3) above.
(5) The territorial sea.
(6) Wetlands adjacent to waters defined above.

This definition obviously incorporates just about anything that is wet.

The definition of wetlands is also important. 40 CFR 230.41(a)(1) defines "wetlands" as those areas that are inundated or saturated by surface or groundwater at a frequency and duration sufficient to support, and that under normal circumstances does support, a prevalence of vegetation typically adapted for life in saturated soil conditions. Wetlands generally include swamps, marshes, bogs, and similar areas, though in arid regions they can also include washes and other areas that are not "wet" the year around.

The COE has limited the scope of the program somewhat by issuing nationwide permits for some activities which result in discharges into certain waters of the United States. If certain conditions are met, the specified action can take place without the need for an individual or regional permit. The nationwide permit has all the restrictions and conditions set forth, and little or no paperwork is involved. It takes a relatively short time (0–20 days) to initiate an individual project. For example, a nationwide permit authorizes discharges into waters that are located above the headwaters. The term "headwaters" means that point on a perennial stream above which the average annual flow is less than 5 cfs. On an intermittent stream the "headwaters" is that point where 5 cfs is equalled or exceeded 50% of the time. Maps of the headwaters have been published and copies may be obtained from the COE. The use of this nationwide permit is limited, however, and is subject to future modification. In cases where the project impact exceeds one acre, written coordination with the COE is required. Other nation-wide permits address storm drain lines, utility lines, and bank stabilization and maintenance activities.

Regional permits are a type of general permit and can be issued by a division or district engineer. A regional permit may require a case-by-case reporting and acknowledgment system. The regional permit will state what fill actions are allowed, what mitigation is necessary, how to get an individual project authorized, and how long it will take. The time required to initiate an individual project under a regional permit should be less than that for an individual permit. Projects that come under a regional permit must have minimal environmental impact, either separately or as a group.

An individual permit is for one action and the restrictions and conditions are tailored to the individual project. Extensive paperwork is involved and it usually takes 60 days or more to obtain. If there are any environmentally sensitive issues involved, or any objections to the work, it can take months or even years to obtain an individual 404 permit.

C. Erosion Control, Stormwater Detention, and Subdivision Ordinances and Codes

There are several areas of regulatory activity that are generally the responsibility of local government. They include erosion control ordinances, stormwater detention ordinances, and subdivision ordinances and codes. These activities are not influenced by federal programs in the same way that floodplain regulations are influenced by the National Flood Insurance Program. The authority which permits local governments to adopt and enforce such ordinances is derived from the state.

An important purpose of erosion control ordinances is to minimize the adverse effects of erosion and sedimentation from construction sites. The rapid conversion of land from natural or agricultural uses to urban uses may result in stripping the land of top soil, which can accelerate the processes of erosion and deposition. Construction-related sediment can be a source of pollution in downstream lakes, streams, ponds, and reservoirs and can result in large deposits on streets, and in drainage

channels, yards, etc. Detention ponds and small lakes can rapidly fill
with sediment, necessitating expensive dredging to return them to their
original effectiveness.

The greatest erosion potential occurs between the time when native
vegetation is removed from a construction site and when construction
is completed and restorative vegetation planted. Erosion control ordinances set forth practices, procedures, and objectives for the developer
and his contractor during the construction period, which include limiting the extent of native vegetation that is disturbed, limiting the time
during which construction takes place and the construction site is vulnerable to erosion, revegetating between phases of a construction project, and structural measures taken at the construction site such as
reducing the velocity of runoff or providing sediment traps. Cities
and/or counties commonly have erosion control manuals that provide
guidance for meeting local requirements. In arid or semi-arid areas, urbanization often results in reduced long-term erosion, compared to pre-
development conditions. Some useful references on erosion and
sediment control include Urban Land Institute (1978), U.S. Soil Conservation Service (1975), and U.S. Environmental Protection Agency
(1973).

Many local governments require that stormwater detention or retention basins be provided by developers. The purpose of such basins is
to limit increases in the rate of runoff from impervious surfaces. Most
detention and retention basins constructed by developers are in the
1–10 acre-foot capacity range. Larger detention facilities, generally built
by public-sector agencies, can be "regional" facilities which can take
the place of the smaller, randomly located developer-constructed
facilities.

Detention and retention basins are generally designed to control runoff from the more frequent storms in urban watersheds, but they must
be constructed to safely pass larger events. Design criteria vary considerably, and engineers must become familiar with local requirements.
Basins can be classified as wet or dry. Dry basins hold back water only
for short periods of time after storm events, whereas wet basins have
a permanent pool below the level needed to store storm waters. Detention and retention basins may have other beneficial uses such as
recreation, groundwater recharge, irrigation, industrial uses, water supply, sediment control, and pollution control. Local ordinances usually
set forth criteria for the design of detention and retention facilities and
require them to be constructed as a condition of development approval.

Detention and retention systems must not be considered substitutes
for a drainage system. Detention basins have their greatest impact
immediately downstream of the basin, but their effectiveness in reducing peak flows diminishes rapidly farther downstream. If downstream
facilities are to be designed on the assumption that basins will be built,
then a guarantee must be provided that all such facilities will in fact
be constructed as designed, will remain in place, and will be effectively
maintained. If this cannot be done, downstream facilities should be

designed as if the detention and retention facilities did not exist. Some ordinances will address the ownership and maintenance issue, and local governments will sometimes accept the ownership and maintenance responsibility. The transfer of maintenance responsibility to any group of private parties, such as homeowner's associations, is not recommended.

Subdivision ordinances guide the division of larger land parcels into smaller lots for development purposes. Most local governments of any size will have subdivision ordinances. The purpose of subdivision ordinances is to insure that development results in the provision of necessary basic facilities in a consistent manner. They control improvements such as roads, sewers, water, drainage facilities, recreation facilities, and dedication requirements. Floodplain regulations, erosion control ordinances, drainage ordinances, and detention ordinances may be included in a community's subdivision ordinance(s), or they may be separate from and in addition to those ordinances.

Building codes typically control building construction aspects such as the use of construction materials, but do not regulate the type or location of development. Building codes can include requirements that would tend to reduce flood damages. Examples are requiring suitable anchorage to prevent flotation of buildings during floods, requiring suitable locations for electrical outlets and mechanical equipment in flood-prone structures, restricting the use of materials that deteriorate when wet, and requiring adequate structural design to withstand effects of water pressure and flood velocities.

D. Stormwater Quality

Because of concerns over substantial pollution from nonpoint sources, the U.S. Environmental Protection Agency has undertaken an aggressive urban stormwater quality improvement program via the National Pollutant Discharge Elimination System (NPDES) permitting program under Sections 402 and 405 of the 1972 Federal Water Pollution Control Act (FWPCA) and 1987 Clean Water Act (CWA). Although the NPDES has traditionally focused on the reduction of pollutants from point sources (i.e. wastewater treatment facilities, industrial facilities), the current program will require permits for stormwater discharges.

Initially, this program has targeted larger American cities (those with population of over 100,000) and most categories of American industry. Smaller cities are required to obtain permits after October, 1992. Cities and industries are required to address such issues as: physical characteristics of storm drainage facilities; water quality sampling during "dry flow" and "wet weather" conditions to characterize runoff water quality; identifying "illicit discharges" and taking the steps necessary to remove these discharges from storm sewers; long-term mitigation planning and other activities. Traditionally, flood hazard reduction has been the driving force behind the implementation of storm drainage facilities. These regulations, however, in conjunction with a growing public awareness of the extent to which urban runoff degrades receiving water quality, have increased interest in planning and implementing systems that provide both runoff quantity **and** quality management.

V. REFERENCES

Choate, P. and Walter, S. (1983). *America in ruins*. Duke University Press, Durham, NC.

Denver Regional Council of Governments (1980). *Managing erosion and sedimentation from construction activities*. Denver, CO.

"Digest of Federal disaster assistance programs" (1989a). Federal Emergency Management Agency, Washington, D.C.

Hanks, E.H., Tarlock, A.D., and Hanks, J.L. (1974). *Cases and materials on environmental law and policy*. American Casebook Series, West Publishing Co., St. Paul, MN.

Meyers, C.J. and Tarlock, A.D. (1980). *Water resource management—A casebook in law and public policy*. The Foundation Press, Inc., Mineola, NY.

Public Works Historical Society (1988). *The flood control challenge: Past, present and future*. Rosen, H. and Reuss, M., eds. Chicago, IL.

Rosholt, J. and Piggott, S. (1986). *Financing and service charge alternatives for storm and surface water management*. URS Consultants.

Schaefer, G. (1989). *Investigation of user charge and fee systems for stormwater management*. Illinois Department of Transportation by the Northeastern Illinois Planning Commission, Springfield, IL.

Urban Drainage and Flood Control Act (1969). Colorado Revised Statutes 32-11-101 through 32-11-817 (as revised), Denver, CO.

Urban Drainage and Flood Control District (1984). *Urban storm drainage criteria manual*. Denver Regional Council of Governments, Denver, CO.

Urban Land Institute (1975). *Residential stormwater management: Objectives, principles and design considerations*. Published jointly with the American Society of Civil Engineers and the National Association of Home Builders, New York, NY.

Urban Land Institute (1978). *Residential erosion and sediment control*. Published jointly with the American Society of Civil Engineers and the National Association of Home Builders, New York, NY.

"Urban stormwater management" (1981). *APWA Special Report No. 49*, Chicago, IL.

U.S. Army Corps of Engineers (1990). *Nationwide permit summary*. Albuquerque, NM.

U.S. Army Corps of Engineers (1990) *Technical assistance and small project construction*. Omaha, NE.

U.S. Army Corps of Engineers (1985). "Regulatory program." *Pamphlet EP 1145-2-1.*

U.S. Department of Agriculture (1965). "Predicting rainfall-erosion losses from cropland east of the Rocky Mountains. *Agricultural Handbook No. 282,* Agricultural Research Service.

U.S. Department of Agriculture (1975). *Standards and specifications for soil erosion and sediment control in developing areas.* Soil Conservation Service.

U.S. Environmental Protection Agency (1973). "Comparative costs of erosion and sediment control, construction activities. *EPA-430/9-73-016,* Office of Water Program Activities, Washington, D.C.

U.S. Environmental Protection Agency (1973). Processes, procedures and methods to control pollution resulting from all construction activity. *EPA-430/9-73-007,* Washington, D.C.

U.S. Department of Housing and Urban Development (1989). *1989–1990 Programs of HUD,* Washington, D.C.

U.S. Soil Conservation Service (1984). "Small watershed projects." *Program Aid No. 1354,* U.S. Department of Agriculture, Washington, D.C.

Wright, R. (1982). "Legal aspects of drainage." *Urban storm drainage management.* Marcel Dekker, Inc., New York, N.Y.

Chapter 3

SURVEYS AND INVESTIGATIONS

I. INTRODUCTION

The purpose of surveys and investigations is to provide basic data for the replacement or upgrading of existing drainage facilities or for the design and construction of new storm drainage systems. Survey needs vary depending upon the stage of design, availability of reliable information, soil and groundwater conditions, requirements of agencies having jurisdiction over the project, and desired accuracy of hydraulic and water quality calculations. The engineer should approach surveys and investigations in a careful manner, and must learn to decide how much information is actually relevant and essential to the decisions that he will make. All subsequent work hinges on the accuracy, thoroughness and timeliness of field and office data.

Basic surveying theory and methods will not be discussed here. Rather, the emphasis is on the kinds of information and data typically required for urban stormwater management systems. Note that information should be collected both for existing and projected land use and regulatory conditions.

II. DEFINITIONS

The term "surveys" refers to the process of collecting and compiling information necessary to develop any given phase of the project. Surveys can occur either in the office or in the field. The office survey may include observations relating to general conditions affecting a project, such as historic, political, physical, fiscal, and others. The field survey may include the measurements necessary for the engineering design, as well as personal observations of the drainage area, present drainage facilities, geologic characteristics, and existing occupancy and improvements that merit particular attention. Extensive documentation of the

39

study area with photographs is encouraged during field surveys. The designer is urged to spend as much time as possible in the field prior to undertaking even the most rudimentary design, as there is no substitute for first hand observation, especially during rainfall events (U.S. Department of Transportation 1980). While the term "investigations" is often used interchangeably with "surveys," this manual uses it to mean the process of assimilation and analysis of data produced by surveys.

III. MONITORING AND DATA COLLECTION

On small catchments, and where detention/retention storage and water quality are not design issues, use of the rational method and/or published rainfall data may be acceptable (for further discussion, see Chapter 5). If this is not the case, however, the engineer may be faced with the problem of collecting the necessary and appropriate data on which to base plans and designs.

Most analysis for stormwater planning and design is now done using mathematical models on digital computers, and hydrologic data are needed to calibrate and run these models. As will be pointed out in later chapters, the return frequency of runoff from a given rainfall event will normally be different from the return frequency of the rainfall itself. The engineer must then decide whether to use synthetic rainfall data and to rely on his model to generate necessary runoff data, or to collect rainfall and runoff data to calibrate and verify his models and provide a design basis that more nearly approximates the local situation (Huber and Dickinson 1988).

Given the availability of rainfall data in computerized form in the data bases of the National Weather Service (also available commercially), it is difficult to justify not making at least a cursory analysis of rainfall data from a local (or the nearest) weather station to develop some understanding of the statistical aspects of the precipitation regime, and to identify significant historical rainfall events. It should also be possible to identify rainfall events that produce significant runoff, either by historical observation or by running a series of rainfall hyetographs through a simple runoff model. Since synthetic storms are not real rainfall events, but rather an aggregation of data from a number of discrete events, the engineer can then satisfy himself that at least the rainfall input he uses bears some relation to reality (ASCE 1986).

If rainfall and runoff data are to be collected, the statistical analysis referred to above will be invaluable in designing a monitoring program. For instance, if the analysis indicates that precipitation is evenly distributed throughout the year, data collected during any part of the year should be adequate. On the other hand, if precipitation is highly seasonal, then monitoring should be done during the critical season(s).

IV. INFORMATION REQUIRED FOR DESIGN

Depending on the type of storm drainage facilities envisioned and the stage of the design process, some or all of the following information will be required for proper design of drainage facilities for an area. In most cases, the information can be obtained from federal, state, or local governments, or through field inspection or instrument surveys.

A. Topographic Information

(a) City, county, USGS or other topographic mapping.
(b) Aerial photographs.
(c) Vegetation maps.
(d) Soil Maps.
(e) Property surveys and maps.
(f) Property ownership maps.
(g) Field investigations or surveys, to determine the following:
　　(1) Drainage basin boundaries to confirm interpretation of maps.
　　(2) Drainage basin areas and existing and projected land use characteristics (most communities have land use master plans).
　　(3) Typical overland flow paths, swales, channels, and major drainageways.
　　(4) Ground and drainageway slopes and lengths.
　　(5) Typical channel cross sections.
　　(6) Sites potentially suitable for detention storage.
　　(7) All relevant drainage and flood control facilities, such as culverts, bridges, drop structures, and utilities crossing channels.
　　(8) Properties that have actually sustained flood damage in the past.

The goal of field surveys is to confirm data obtained from maps and to assure that there are no unusual circumstances associated with the study area such as diversions out of the basin, or ponds that mapping does not show.

If adequate mapping is not available, the designer or his client will have to arrange for its preparation. Mapping needed for final design of local drainage systems typically requires a scale of 1 inch = 100 feet to 1 inch = 200 feet, with 1- or 2-foot contour intervals.

B. Survey and Boundary Data

(a) Land boundaries and corners.
(b) Bench marks.
(c) Aerial photographs and ground control.
(d) Existing streets, alleys, railroads, power lines, canals, schools, parks, and other physical features that will influence project feasibility and siting.
(e) Location of utilities.
(f) Existing rights-of-way and easements, along with their characteristics.

(g) Any other potential impediments to drainage system installation, such as community open space, protected habitat, etc. (a "site audit" for hazardous wastes or other contamination is often advisable).

C. Soils and Geologic Data

Excellent sources of soils and geologic data include U.S. Soil Conservation Service (SCS) soils reports, and files of state universities and departments of agriculture.

(a) Infiltration characteristics and permeability of basin soils.
(b) Other standard soil characteristics such as gradation, density, and classification.
(c) Bed and bank samples from drainageways and laboratory evaluations, if applicable.
(d) Soil strength properties for assessments of excavation stability and foundations for drainage structures.
(e) Data required to assess suitability of dam sites if detention ponds are envisioned.
(f) Groundwater elevations on a seasonal basis.
(g) Bedrock locations and characteristics, especially if foundations or trenches are involved.
(h) Other geotechnical characteristics required by either the hydraulic or structural designer.
(i) Local geologic maps or reports that could influence the project.
(j) Any other natural hazards that could affect the drainage system such as landslides or earthquakes.

D. Hydrologic and Hydraulic Data

This is a complex area, and the reader is advised to consult Chapters 5 and 6 for further information. The following list provides general guidance.

(a) Historical streamflow, sewer flow, and precipitation data.
(b) Local rainfall data, including records from local weather stations, intensity/duration/frequency curves, hyetographs, and design storm distributions.
(c) Channel and pipe characteristics including slope, roughness coefficients, vegetation, stability (erosivity), state of maintenance, including amount of debris, etc.
(d) Water "flow line" elevations required for hydraulic grade line analysis. Also sub-basin slopes for analysis of times of concentration and model input.
(e) Existing hydraulic structures including storm drains, inlets, culverts, channels, embankments, bridges, dams, ponds and other similar items.
(f) Existing intentional and inadvertent storage areas and associated flood routing characteristics.
(g) Water quality data.
(h) As much information on large historic floods in the vicinity of the study area as can be derived through library research, interviews with professionals and residents, and other sources.

E. Regulatory Data

One of the key ingredients of a successful design effort is regular, frank interaction between the designer and review authorities. Good communication results in well-conceived designs that benefit, and are well received by, the public (Edwards 1982). Ideas are frequently put forward that otherwise would not have influenced the design, and that help the designer and his client avoid surprises.

It is of particular importance for engineers to become familiar with local drainage goals, objectives, policies, and criteria (Texas Public Works Association 1986). Local codes and standards set the ground rules for designing storm drainage facilities, but they do not relieve the engineer of his responsibility to design a safe system (Wright undated). The engineer should consult with the department charged with administering local standards and codes to obtain pertinent information. Where these codes and regulations are absent or incomplete, the engineer should advise the regulatory agency of standards needed to protect the public interest. Examples of the sorts of information typically required can be found in Denver Urban Drainage and Flood Control District (1984), King County (1990), ASCE (1976), Florida Department of Environmental Regulation (1988), and U.S. Department of Transportation (1979), and can include:

(a) Zoning ordinances/maps.
(b) Floodplain zoning and requirements (is community participating in National Flood Insurance Program?).
(c) Subdivision regulations.
(d) Building and health codes.
(e) Water and sewer standards.
(f) Erosion, grading, water quality and environmental protection ordinances.
(g) Maps that show designated wetlands, wildlife habitat, receiving stream standards and classifications and other items related to environmental protection (wetlands are an especially important design consideration).
(h) Stormwater management policy and criteria documents, including representative stormwater master plans and final design drawings and specifications.
(i) Comprehensive land use planning reports and other information from local planning departments.
(j) EPA 208 plans (available from local government).
(k) Site specific or adjacent stormwater master plans.
(l) U.S. Soil Conservation Service PL-566 plans.
(m) Flood insurance studies and maps.
(n) Regional flood studies from ungauged basins, normally prepared by U.S. Geological Survey.
(o) Drainage reports and plans which detail specific functions and projected future changes for relevant upstream and downstream facilities or projects, especially if the designer intends to rely on these facilities for the proper functioning of his facilities (Rossmiller and Jones 1990).

V. FINANCIAL DATA

Collect information related to existing policies, obligations, or commitments that bear on the financing proposed for drainage facilities (to assure that adequate capital, operation and maintenance funds will be provided). Determine availability of federal, state, and local aid for projects. Determine local construction conditions that may affect project costs through review of bid prices, interviews with contractors, assessment of the state of the local economy, time of the year that the project will be built, etc.

VI. DATA MANAGEMENT

All of the data collected should be organized, reviewed for thoroughness and applicability, and disseminated to the owner, regulatory officials, members of the design team, and others as appropriate. Information should be arranged neatly and systematically to facilitate rapid retrieval.

The microcomputer allows even small engineering firms to manage large amounts of data with relative ease using a variety of commercially-available programs. These programs include both generalized data management software as well as specific data management packages tailored to the needs of engineering organizations. There are, in addition, many sources of computerized data. For instance, USGS daily streamflow records and NOAA daily weather summaries for the U.S. (and similar data for Canada) are available in microcomputer-compatible formats (floppy disks or CD/ROM). Successful stormwater master planning requires thorough and well-organized data, and the maintenance of a computerized data base or, at the very least, detailed project data files, is strongly recommended.

VII. REFERENCES

American Society of Civil Engineers (1976). "Guide for collection, analysis and use of urban stormwater data." *Proc. of an Engineering Foundation Conference,* ASCE, New York, N.Y.

American Society of Civil Engineers (1986). *Water Forum 86: World water issues in evolution.* vols I, II, ASCE, New York, N.Y.

American Society of Civil Engineers (1988). *Consulting engineering—A guide for the engagement of engineering services.* ASCE Manuals and Reports on Engineering Practice No. 45, ASCE, New York, N.Y.

Edwards, K.L. (1982). "Acceptance and/or resistance to detention basins." *Stormwater detention facilities, Proceedings of an Engineering Foundation Conference,* ASCE, New York, N.Y.

Livingston, E. et al. (1988). *The Florida development manual—A guide to sound land and water management.* vols 1, 2, State of Florida, Department of Environmental Regulations.

Gelonek, W. (1990). "Drainage law and the transportation engineer." Report for the California Department of Transportation, Sacramento, CA.

Huber, W.C. and Dickinson, R.E. (1988). "Storm water management model user's manual, Version 4." *EPA/600/3-88/001a* (NTIS PB88-236641/AS), Environmental Protection Agency, Athens, GA.

Kilpatrick, F.A. et al. (19??). "Development and testing of highway storm-sewer flow measurement and recording system," *Water Resources Investigations Report 85-4111,* U.S. Geological Survey, Reston, VA.

"Surface water design manual" (1990). Surface Water Management Division, Department of Public Works, King County, WA.

Rossmiller, R.L. and Jones, J. "Understanding and applying stormwater management techniques." Short Course notes from the Department of Engineering Professional Development, University of Wisconsin.

Texas Public Works Association (1986). "Guidelines for drainage design." *Proc. of a Conf. Sponsored by the Texas Public Works Assn.* (TPWA-002), Austin, TX.

Urban Drainage and Flood Control District (1984). *Urban storm drainage criteria manual.* rev. ed., Denver Regional Council Governments, Denver, CO.

U.S. Department of the Interior (1977). *Design of small dams.* U.S. Bureau of Reclamation, U.S. Government Printing Office, Washington, D.C.

U.S. Department of Transportation (1979). "Design of urban highway drainage—The state-of-the-art." *FHWA-TS-79-225,* Federal Highway Administration, Washington, D.C.

U.S. Department of Transportation (1980). *Hydrology for transportation engineers.* Federal Highway Administration, Washington, D.C.

Wright, R.M. (1979). "Oklahoma stormwater law." *Urban Drainage and Flood Control Criteria Manual and Handbook,* Stillwater, OK.

Chapter 4
DESIGN CONCEPTS AND MASTER PLANNING

I. INTRODUCTION

The conceptual design step is where objectives of the project are delineated in a manner consistent with stormwater control principles, water quality objectives, and local drainage policies. Design concepts provide a road map for preliminary and final design work. During the conceptual stage, the designer formulates the outline for and begins the "drainage master plan study" for the project, a report with accompanying drawings, photographs, maps and calculations, that discusses such subjects as the nature of the problem, hydrology, land use, alternatives evaluated, cost considerations, and environmental impacts. A master plan is normally prepared for both the client and review authorities (Sheaffer et al. 1982).

The total stormwater control system comprises a wide array of physical components, and includes overland flow paths, gullies, channels, streams, detention storage, floodplains, and larger downstream storage/treatment facilities and natural storage areas. The way in which natural and man-made components of the drainage system interrelate at the conceptual design level is the focus of this chapter.

II. PRINCIPLES

Experience has shown that the following general principles apply when planning for and designing urban storm drainage systems (King County 1990; Livingston et al. 1988; Urbonas and Roesner 1991; Schueler 1987; Urban Drainage and Flood Control District 1984)

(a) **Drainage is a Regional Phenomenon That Does Not Respect the Boundaries Between Government Jurisdictions or Between Public and Private**

Properties. This makes it necessary to formulate programs that include both public and private involvement. Overall coordination and master planning must be provided by the governmental units most directly involved, but drainage planning must be integrated on a regional level if optimum results are to be achieved. The ways in which proposed drainage systems fit existing regional systems must be quantified and discussed in the master plan.

(b) **Storm Drainage Is a Sub-System of the Total Urban Water Resource System**. Stormwater system planning and design must be compatible with comprehensive regional plans and should be coordinated particularly with planning for land use, open space and transportation. Erosion and sediment control, flood control, site grading criteria and regional water quality all closely interrelate with urban stormwater management. The master plan should normally address all of these considerations.

(c) **Every Urban Area has Two Drainage Systems, Whether or Not They Are Actually Planned For and Designed**. One is the **minor** or **primary** system, which is designed to provide public convenience and to accommodate relatively moderate frequent flows. The other is the **major** system, which carries more water and operates when the rate or volume of runoff exceeds the capacity of the minor system. Both systems should be carefully considered.

(d) **Runoff Routing Is a Space Allocation Problem**. The volume of water present at a given point in time in an urban region cannot be compressed or diminished. Channels and storm sewers serve both conveyance and storage functions. If adequate provision is not made for drainage space demands, stormwater runoff will conflict with other land uses, will result in damage, or will impair or even disrupt the functioning of other urban systems.

(e) **Planning and Design of Stormwater Drainage Systems Generally Should Not Be Based on the Premise That Problems Can Be Transferred From One Location to Another**. Urbanization tends to increase downstream peak flow by increasing runoff volumes and by increasing the speed of runoff. Stormwater runoff can be stored in detention reservoirs, which can reduce the downstream drainage capacity required.

(f) **An Urban Drainage Strategy Should Be a Multipurpose, Multimeans Effort**. The many competing demands placed upon space and resources within an urban region argue for a drainage management strategy that meets a number of objectives, including water quality enhancement, groundwater recharge, recreation, wildlife habitat, wetlands creation, protection of landmarks/amenities, control of erosion and sediment deposition, and creation of open spaces.

(g) **Design of the Stormwater Management System Should Consider the Features and Functions of the Natural Drainage System**. Every site contains natural features that may contribute to the management of stormwater under existing conditions. Existing features such as natural drainageways, depressions, wetlands, floodplains, permeable soils, and vegetation provide natural infiltration, help control the velocity of runoff, extend the time of concentration, filter sediments and other pollutants, and recycle nutrients. Each development plan should carefully map and identify the existing natural system. "Natural" engineering techniques can preserve and enhance the natural features and processes of a site and maximize post-development economic and environmental benefits, particularly in combination with open space and recreational uses. Good designs improve the effectiveness of natural systems, rather than negate, replace or ignore them.

(h) **In New Developments, Stormwater Flow Rates After Development Should Approximate Pre-Development Conditions, and Pollutant Loadings Should be Reduced.** Three interrelated concepts should be considered:

(1) The perviousness of the site should be maintained to the greatest extent possible.

(2) The rate of runoff should be slowed. Preference should be given to stormwater management systems which use practices that maintain vegetative and porous land cover. These systems will promote infiltration, filtering and slowing of the runoff. It should be noted that it may be difficult to restrict post-development **volumes**, and that existing storm water regulations instead require control of **peak flows** to pre-development levels. This may present no problems if the basin has a positive outfall to a stream or river. It can be a problem, however, for a small enclosed basin draining to a lake. Even if retention is provided, the lake might rise because of additional inflow via the shallow water table, and, more importantly, because the increased imperviousness reduces the area available for evapotranspiration. For such basins, the total water budget should be considered, not just peak flows.

(3) Pollution control is best accomplished by implementing a series of measures, which can include source control, minimization of directly connected impervious area (see also Chapter 12), and construction of on-site and regional facilities, to control both runoff and pollution.

(i) **The Stormwater Management System Should Be Designed, Beginning With the Outlet or Point of Outflow From the Project.** The downstream conveyance system should be evaluated to ensure that it has sufficient capacity to accept design discharges without adverse backwater or downstream impacts such as flooding, streambank erosion and sediment deposition.

(j) **The Stormwater Management System Should Receive Regular Maintenance.** Failure to provide proper maintenance reduces both the hydraulic capacity and pollutant removal efficiency of the system. The key to effective maintenance is the clear assignment of responsibilities to an established agency and a regular schedule of inspections to determine maintenance needs and to insure that required maintenance is done. Demonstrated past local maintenance performance should be the basis for the selection of specific design criteria.

III. DRAINAGE SYSTEMS

A. Natural Channels

Most of the time the natural system (hills, valleys, lakes, stream channels, floodplains, wetlands and coastal plains, etc.) appears to be in equilibrium. This "equilibrium" is relative, however, since changes due to erosion and deposition occur continually (Urban Land Institute 1987, 1975).

The size, shape and slope of a stream channel are functions of interrelated variables such as soil characteristics, lithology, width, depth, velocity, slope, sediment load, sizes of sediment and debris, hydraulic roughness and discharge. The stream channel accommodates itself to whatever discharge it receives. Increasing the flow alters the overall size, but the shape of the channel margins tends to remain constant. The cross-sections of most streams tend to be generally trapezoidal in straight reaches, but are asymmetric at curves or bends. They tend to become more rectangular as the stream becomes larger downstream, since width increases faster downstream than depth. Depth usually increases faster downstream than does velocity. Width usually increases in a more consistent manner than any other factor, roughly as the square root of the discharge; and mean velocity tends to decrease slightly downstream in most rivers (U.S. Bureau of Reclamation 1977).

In a sediment-laden stream, flow resistance is altered by the size of the bed particles, the form or configuration assumed by the particles on the channel bed, and the damping of turbulence by the sediment load. Channel width and depth increase, stream gradients generally flatten, and river bed particle sizes usually diminish as one moves downstream. A constant or gradually steepening stream gradient therefore would be associated with increasing bed particle sizes. A stable stream is one that reaches a slope that provides just the velocity required to transport the sediment load supplied from the drainage basin. Any change in the controlling factors will cause displacement of the equilibrium in a direction that will tend to absorb the effect of the change (Kolenkow et al. 1974; Linsley et al. 1982).

B. Effects of Urbanization

Urbanization disrupts the natural equilibrium of streams. Construction site erosion can result in order of magnitude increases in local sediment loads. The increased imperviousness and hydraulic efficiency of the urban flow paths can generate local (and sometimes quite large) increases in peak runoff rates. The effect on the stream increases with the percentage of the watershed that has undergone urbanization. Increases in water and sediment loads can be minimized by properly designing components of the urban drainage systems such as detention basins and on-site erosion control measures during construction phases. If such increases are not minimized, the natural streams will enlarge their channels to accommodate the increased loads by scouring their banks and beds, thus generating additional sediment loads (Urban Land Institute 1975, 1978; U.S. Soil Conservation Service 1973).

IV. BASIC CONCEPTUAL ELEMENTS

In some areas, urban stormwater management is being perceived and treated, at least in part, as a subsystem of a broader urban water resources management system. (This ideal, unfortunately, is not often seen, however, due to the "balkanized" (McPherson 1978) nature of the various components—water supply, waste disposal, flood control, urban drainage.) The advantage of overall urban water resources management is that urban drainage can be interrelated with such functions as groundwater management, water supply, waste disposal, slope stability control, aesthetic and recreational opportunity control, and others (King County, Washington 1990; Livingston et al. 1988; State of Delaware 1990).

As the capacity of the minor system is exceeded, streets or other surface channels (the major system) begin to carry excess flow. This condition does not constitute failure of the storm drainage system. This concept has implications for design and for analysis of total system costs, since the majority of costs are for small-diameter pipe and appurtenances in local neighborhoods. For the smaller, more frequent runoff events, the longer the runoff can be kept on the surface, the shorter and less costly the storm sewer system will be.

Denver, Colorado—A plaza that adjoins a major river. The lowest level of the plaza is inundated during the 2-year runoff event, and the 100-year event floods the entire plaza.

For the larger, less frequent storms, the designer should determine, at least in a general sense, the flow pathways, and related depths and velocities, of the major system. **This does not mean that a detailed analysis of the major system need be made in all cases, nor that existing systems, whose performance may be well understood, need always to be subject to this sort of analysis.** It is, rather, directed primarily at new developments, or to significant modifications to existing systems. The objective is to keep water out of buildings, and to assure that flow depths and velocities will not constitute a hazard (or at least to be able to define the hazard) to public convenience and safety.

Designs may combine portions of both the major and minor systems into a single system. One set of flow paths, both natural and man-made, can convey the runoff from both small and large storms without causing damage. It requires imagination, initiative, and ingenuity on the part of designers, developers, and local governments to incorporate such flow paths into the city in a way that is compatible with the urban setting. The result can be an overall system that costs less, is aesthetically pleasing, and is free from damage from all but catastrophic flood events (Urban Drainage and Flood Control District 1984).

V. PREVENTIVE AND CORRECTIVE ACTIONS

In existing urbanized areas, the designer will frequently find it necessary to develop a strategy based upon both preventive and corrective measures. Structural corrective actions affect and control the storm runoff and floodwaters directly, and can encompass such things as inlets, storm sewers, interceptor lines, channelized stream sections and reservoirs. There are also non-structural corrective actions which limit activities in the path of neighborhood storm runoff or in river floodplains. They include floodproofing and land use adjustments (Sheaffer et al. 1982).

A. Preventive Actions

Preventive actions available for reducing storm runoff and flood losses includes (Flood Insurance Administration 1981):

(a) Control of flood-prone land uses.
(b) Floodplain regulation.
(c) Flood-prone land acquisition.
(d) Subdivision regulations.
(e) Building code provisions.
(f) Control of water and sewer extensions.
(g) Flood-prone area information and education.
(h) Storm and flood forecasts and emergency measures.
(i) Measures to reduce the runoff rate.
(j) Measures to reduce erosion.
(k) Floodproofing (both preventive and corrective).

This slope (in the Pacific Northwest) should never have been developed. All of the homes shown were destroyed by a landslide which occurred shortly after this photograph was taken.

The designers of urban drainage systems must recognize that flood-prone lands and floodplains exist not only along rivers and streams, but also in headwater residential and commercial neighborhoods where storm drains are lacking or inadequate, where building floor elevations are too low, and where curbs and gutters are subject to overtopping.

By controlling the amount and type of economic and social growth in the floodplain, flood losses may be reduced and net benefits from suitable floodplain use increased (FEMA 1981a, 1981b). For instance, it may be found that it is more economical to turn a block of homes into a small park than to solve the storm drainage problem for an isolated area. Although this approach often requires a long period to realize its full effect, the community benefits can be significant. They include reduction in the exposure to risk, reduction in public costs for relief and rehabilitation, and decreased dependence on protective works. Public acquisition of properties at risk, with rent-back provisions that include flood insurance coverage requirements, can be used for long-term flood loss reduction at minimum public cost (FEMA 1981a, 1981b).

B. Delineation of Floodplains

The delineation of flood-prone areas is the first step in floodplain regulation and local drainage design. The runoff is sometimes calculated based upon the projected future development of the basins in accordance with long-term area-wide plans, and assuming storm sewers, surface drainage, floodplains, and watercourses in their existing condition. Adjacent properties that may discharge overland flow onto the study area should also be evaluated.

Acquisition of floodplain land may be a cost-effective means of avoiding unacceptable property damage. In addition, other benefits may be realized, such as the preservation of floodwater storage capacity and the provision of recreational open space.

C. Corrective Actions

Existing drainage, flood hazard, and water quality problems may require corrective actions, even under a basically non-structural management plan. These corrective actions fall into categories as follows:

    (a) Construction of storm sewers, stormwater storage, and water quality best management practices (BMPs).
    (b) Land use adjustments.
    (c) Channelization to enlarge streams.
    (d) Enlarging bridge and culvert openings.
    (e) Modifying bridge and culvert approach, entrance, and discharge transitions.
    (f) Designation of nonconforming uses.
    (g) Floodproofing of buildings.

Land use adjustments include rezoning, relocation of structures, programmed removal of incompatible structures, and purchase of floodplain properties (which may be leased back for temporary use). Urban renewal projects may be used to expedite adjustments in land use.

Rather than constructing new systems, it may sometimes be advantageous to retrofit existing systems to obtain increased capacity or to provide pollution control. Innovative retrofitting practices are continually being developed and improved, (Livingston 1986; Pisano 1990; Torno 1990), and the reader is urged to consult the current references.

It should be noted that, compared to construction in new areas, retrofitting can be expensive, principally because of such factors as land costs and disruption of traffic and/or utilities.

VI. STRUCTURAL COMPONENTS OF DRAINAGE SYSTEMS

The first sections of this chapter have emphasized the need to analyze drainage requirements in the context of the entire urban system, to relate the effects of local drainage solutions to their cumulative impact upon the larger drainage system, and to consider a wide range of non-structural and land-use measures in the design. Equal importance must be placed on the structural components of the drainage system. The following list of structural components and design considerations is not intended to be comprehensive, nor is the entire design process from start to finish explicitly summarized.

A. Major Drainageways

(a) Where are they now and where should they be after development?
(b) How far up into the basin should they extend?
(c) How will they behave in various floods?
(d) What are important geomorphologic characteristics?
(e) What channel improvements exist upstream and downstream?
(f) What does the floodplain look like presently, and can and/or should it be narrowed?
(g) How much land should be allocated for flood conveyance, and is this adequate? Does it fulfil community objectives?
(h) Are improvements required?
(i) Can/should multi-purpose uses for the channel be encouraged?
(j) Concrete-lined channels
    (1) Is there a limited right-of-way that constrains channel width?
    (2) Will flow be subcritical or supercritical?
(k) Grass-lined channels
    (1) Will a "soft," natural approach to channels fit better with the area?
    (2) Can velocities be adequately limited with drops?
    (3) Is there enough right-of-way for 4:1 or flatter side slopes?
(l) Box culverts
    (1) Box culverts can be used at streets.
    (2) Can internal pressure be controlled in long box culverts?
    (3) The capacity will lower if the culvert reaches full flow.
(m) Large pipes
    (1) Is an underground conduit preferable to an open channel?
    (2) Would an emergency open channel still be needed?
(n) Riprapped channels
    (1) Will a riprapped channel be aesthetically acceptable?
    (2) Stone riprap often is vandalized.
    (3) If wire gabions are used for drops, will the wire protective coating be eroded or corroded causing failure?

B. Streets

(a) What depth and velocity of flow is permitted in the gutters for the primary design runoff event? What depth for the major system runoff? Can velocities be maintained within safe limits?
(b) Can a minimum of 0.5% street slope be maintained? (note that there are existing developed urban areas in which 0.1 percent would be a luxury).
(c) How will flow across street crowns and intersections be handled?
(d) How much depth of flow over the street is appropriate for the major storm runoff, considering local conditions and classes of streets (local, collector, or arterial)?

(e) How much will future pavement overlays reduce the curb and gutter conveyance capacity?

C. Storm Sewers

(a) Will storm sewers be necessary?
(b) Will the storm sewer be designed for pressure flow or open channel flow?
(c) Are trash racks necessary for safety or would another safety feature serve just as well considering the problem of trash rack plugging?
(d) Cleaning of trash racks should be a design consideration, with adequate access for equipment during a major storm runoff.

D. Storm Inlets

(a) Where are curb opening inlets optimal? Where should grated inlets, sump inlets, combination inlets be used?
(b) How much flow will bypass inlets under various street grades?
(c) How much inlet capacity is required for full storm sewer capacity utilization?
(d) Should inlet capacity be limited to avoid over-charging the storm sewers in one location to the detriment of another area during the major storm runoff?
(e) The storm sewer design for the minor system design runoff should be compatible with the major runoff event surface flow.

E. Intersections

(a) Should street intersections be kept free from surface flows for the minor system design runoff? Should cross pans be used? What is the effect on traffic?
(b) How should tee intersections be handled at the foot of a steeply sloping street?

F. Flow Control Devices

(a) Energy dissipators are useful when changing from concrete-lined channels to grass-lined channels if velocities would otherwise be excessive, or anywhere else that hydraulic energy should be reduced.
(b) Hydraulic drops are necessary to keep channel velocities within design limits when the channel slope is excessive.
(c) Acceleration chutes are used when making a transition from slow-flow channels to higher velocity channels.
(d) Bends in an open channel need to be hydraulically analyzed for head loss, overtopping of the outside banks, and effectiveness.

G. Trash Racks (Safety Racks)

(a) Is a trash/safety rack essential?
(b) Will trash racks cause a system failure if they become plugged?
(c) What maximum and minimum velocities should be selected?

(d) Should detention/retention storage be used to reduce the required size of storm sewers?

(e) What construction problems are likely to be encountered and how will they affect costs?

(f) Where will the system discharge? Water quality may be an issue worthy of attention.

(g) Will downspouts/foundation drains be connected to the storm sewers? This has been common practice, primarily because the sewers provide a convenient means of disposal, but this conflicts with the notion of minimizing the impervious area directly connected to the storm sewers, reduces sewer capacity, and, in the case of foundation drains, can increase the risks of basement flooding. This issue should be carefully looked at when retrofitting existing systems.

H. Detention Facilities

(a) Is detention required? If yes, what are the appropriate regulations? Is capacity dictated by local ordinance or policy?

(b) What are the design discharge frequencies?

(c) What kinds of detention are appropriate? Is on-site ponding necessary, or should regional (large) ponds be relied upon?

(d) How will contemplated detention facilities address erosion/sediment control, runoff quality enhancement, creation of attractive and safe park areas, groundwater recharge, and other multi-use considerations?

(e) How will detention fit into the regional drainage system? For example, could on-site detention actually aggravate, rather than reduce, downstream peaks?

(f) If a dam is required, then standard dam design considerations and studies are necessary including geotechnical evaluations, adherence to state regulations for design flood spillway capacity, freeboard, hazard classification, etc.

I. Water Quality Mitigation Measures (Other Than Detention)

(a) What local requirements prevail (how much of which pollutants must be removed, and how frequently).

(b) What mitigation measures can be adopted, and how can they be optimized.

J. Other Special Structures

(a) How many special structures (outlet and inlet protection, flow splitters, multiple channel lining types, diversion boxes, etc.) will be necessary?

(b) What design considerations are associated with them?

(c) Can some be eliminated or modified?

It is characteristic of drainage system design problems (and water engineering in general) that the design considerations mentioned above cannot be reduced to simple rules. This is why imagination, experience and mature judgment play equally important roles in the conceptual design phase of successful urban drainage projects (see Chapter 9 for further details).

VII. RISK ANALYSIS

An essential step in the design of urban drainage and flood control systems is the selection of the recurrence frequencies (probabilities) of the runoff events for which the major and minor systems are to be designed, which in turn will determine the sizing of the various components of the systems. A clear understanding of the risks and costs associated with alternative drainage designs leads to better drainage systems and to wiser public and private investment. The process by which this understanding is achieved is risk analysis, the elements of which are described briefly herein. The additional cost of risk analysis (engineering analysis, regulator education, and negotiation) may restrict its application to larger projects.

There are no hard and fast rules regarding recurrence frequencies for design. The engineer must ascertain what local policy is regarding design return frequencies, and then ask if such standards are appropriate for the particular setting. Common sense and judgment on the part of the designer and local authorities should supersede uniform but arbitrary standards. Local regulators may accept (or at least carefully consider) proposed deviations based on principles of risk assessment.

A. Definitions

Risk is the expression of potential adverse consequences measured in terms of inconvenience, damage, safety, or even professional liability or political retribution. Risk analysis is the quantification of exposure, vulnerability and probability. Risk analysis involves the evaluation of alternative means to reduce risk and, finally, the determination of acceptable levels of risk (Earthquake Engineering Research Institute 1978).

In a risk analysis context the design runoff event is the event the drainage system must handle without permitting an unwanted consequence (unacceptable risk). It implies that uncertainty has been defined (the return period or the probability of the event), acceptable risk has been determined (no event smaller than the design event will exceed the capacity of the primary drainage system), and the vulnerability of the finite number of exposed improvements has been quantified and found consistent with the acceptable risk.

Because some of the criteria related to drainage design have come to be accepted as principles, there is confusion about their meaning and application. For example, the "100-year runoff event" is the event that has a probability of occurrence of 0.01 in any given year. It is often taken to mean that the event will occur only once in one hundred years, which although true on the average, may not be true for a particular 100-year period. Furthermore, within the context of risk analysis, the unwanted consequence is not the runoff but the damage which results. It is the uncertainty of this consequence which is of concern, and not the occurrence of the event.

The extent to which risk analysis is employed in the design and selection of a drainage system depends on the nature and extent of the unwanted consequences. For simple situations, mere recognition of risks, uncertainty and possible adverse consequences may be sufficient to permit informed judgments. When public health and safety are at stake, a more rigorous risk analysis should be undertaken. In any event, however, the elements of risk analysis should be systematically considered in formulating design parameters and guiding the decision-making process.

B. Methodology

The determination of risk-causing factors and unwanted consequences requires an assessment of the probability of occurrence and value of potential damage. Four determinations must be made. First, the critical events (i.e., the design events) must be defined and their magnitudes estimated. For drainage design, these events should be those that produce the unwanted consequences. Only the events that would cause damage need to be evaluated, and a probability of occurrence determined for each.

Next, the vulnerability of the exposure (the property that can be damaged) should be determined. The consequence, the third element, might range from public safety through inconvenience to severe damage. Finally, a value (such as dollars or time delays), for all exposed properties, is associated with each of the consequences.

For a given design, the probability of flooding, expected damage and system cost can be assessed. The resulting estimated probability, damage and cost give one point on the risk analysis curve. Similar analyses for alternative designs provide additional points. Information provided by the curve then allows assessment of risk avoidance or acceptance.

Alternative strategies should be pursued to the extent that their marginal costs of implementation are equal. This is to say that each additional dollar spent on avoidance should lower the risk an equivalent amount. The selected design should be the one that balances avoidance and acceptance at the point of acceptable cost and risk.

VIII. DESIGN ECONOMICS

While elements of design are important, information on the costs associated with stormwater management is also necessary to judge the feasibility of a program. Cost information is important to planning agencies and policy makers because the information can be used in setting user fees and comparing alternative basin-wide drainage strategies. Cost information is important to land developers because the information is basic to the determination of the feasibility of development.

Costs for stormwater control are not simple to quantify because of the uncertainties involved. The simplest estimation method consists of a table of mean values, one for each type of development. Such tables are usually applicable only for the jurisdiction for which they were developed. The mean value is the cost per acre of development and is limited to certain cost categories such as construction costs. Costs for such items as engineering, land, and annual operation and maintenance would be estimated separately, possibly by using either mean cost per acre values, or proportionality constants multiplied by the cost of construction.

There are empirical equations for predicting costs associated with other aspects of stormwater control such as erosion and sediment control. Mean cost estimates can be used for such erosion and sediment control methods as small sediment basins, interceptor berms, chemical control, and seeding with fertilizer and mulch (Schueler 1987).

Recognizing that many drainage systems include some form of stormwater runoff control, such as detention, statistical relationships for predicting the construction costs (per unit volume) of detention storage are useful. Similar methods are useful for estimating other costs including land costs, planning, design and supervision costs. Further discussion of economic evaluation is provided in Chapter 8.

IX. DRAINAGE MASTER PLANNING

Master planning is one of the most widely used and frequently misunderstood terms in drainage practice. There are few published definitions, but a master plan typically addresses such subjects as characterization of site development, grading plan, peak rates of runoff and volumes for various return frequencies, locations, criteria and sizes of detention ponds and conveyances, measures to enhance runoff quality, pertinent regulations and how the plan addresses them, and consistency with secondary objectives such as public recreation, aesthetics, protection of public safety, and groundwater recharge.

In its simplest form, a master plan may only identify the essential elements, alignments, and functions of a drainage system. Even at this conceptual level, the master plan should be based upon estimates of peak and total discharges for some selected runoff recurrence interval(s). These recurrence intervals, in turn, should be selected based on local standards and risk assessment, as discussed earlier in this chapter.

The next level of master planning should establish specific criteria consistent with acceptable risk, including design discharges and water surface profiles and elevations. Head losses at waterway crossings and other constructions or obstructions should be recognized in development of the water surface profiles. This level of master planning defines the ultimate drainage system components desired and provides information for their preliminary design and cost estimation. More importantly, it provides basic information useful to assess the feasibility and practicality of contemplated system components and permits local designs to be undertaken with reasonable assurance that they will be compatible with the ultimate overall system.

Where some components of a drainage system must be complete and in operation before other components may be wisely provided, the drainage master plan should identify essential component completion priorities.

A community cannot wisely allow development in a drainage basin, particularly when such development occurs over a long period of time and is not uniform throughout the basin, without first establishing all controlling drainage design parameters through drainage master planning. Failure to do so invites significant future flooding and drainage construction problems (Sheaffer et al. 1982).

X. REFERENCES

County of San Diego (1990). *Design Criteria—Standards for inundation of roads and time of concentration*, San Diego, CA.

Department of Stormwater Management (1991). *Storm drainage criteria manual.* Tulsa, OK.

Guy, H.P. and Jones, D.E. Jr. (1972). "Urban sedimentation—In perspective." *Journal of the Hydraulics Division*, 98 (HY12).

Johnson, W.K. (1978). *Physical and economic feasibility of nonstructural flood plain management measures.* Institute for Water Resources, U.S. Army Corps of Engineers, Fort Belvoir, VA.

Jones, J.E. (1983). "Water quality and institutional considerations associated with stormwater management and urban coastal environments." *Proc. of the Third Symp. on Coastal and Ocean Management*, ASCE, New York, NY.

Kolenkow, R.J. et al. (1974). *Physical geography today: A portrait of a planet.* CRM Books, Delmar, CA.

Linsley, R., Kohler, M., and Paulhus, J. (1982). *Hydrology for engineers.* 3rd ed., McGraw-Hill, New York, NY.

Livingston, E.H. (1986). "Stormwater regulatory program in Florida." *Urban runoff quality—Impact and quality enhancement technology, Proceedings of an Engineering Foundation Conference*, ASCE, New York, NY.

Livingston, E. et al. (1988). *The Florida development manual: A guide to sound land and water management*, State of Florida, Department of Environmental Regulation, Tallahassee, FL.

McPherson, M.B. (1978). "Urban runoff control planning, *EPA-600/9-78-035*, U.S. Environmental Protection Agency, Washington, D.C.

Pisano, W.C. (1990). "Inlet control concepts on catchbasins—U.S. experience," *Urban stormwater quality enhancement—Source control, retrofitting, and combined sewer technology, Proc. of an Engineering Foundation Conf.* ASCE, New York, NY.

St. Johns River Water Management District (1989). *Applicant's handbook—Management and storage of surface waters.* Palatka, FL.

Sheaffer, J.R. et al. (1982). *Urban storm water management.* Marcel Dekker, New York, NY.

Schueler, T. R. (1987). *Controlling urban runoff: A practical manual for planning and designing urban BMPs.* Department of Environmental Programs, Metropolitan Washington Council of Governments, Washington, D.C.

South Florida Water Management District (1990). "Management and Storage of Surface Waters." *Permit Information Manual Volume IV*, West Palm Beach, FL.

State of Delaware (1990). *Sediment and stormwater regulations.* Department of Natural Resources and Environmental Control, Wilmington, DE.

Surface Water Management Division (1990). *Surface water design manual.* Department of Public Works, King County, Washington.

Task Committee on Urban Sedimentation Problems (1975). "Urban sediment problems: A statement on scope, research, legislation, and education." *Journal of the Hydraulics Division*, 101 (HY4).

Torno, H.C., ed. (1990). "Urban stormwater quality enhancement—Source control, retrofitting, and combined sewer technology." *Proc. of an Engineering Foundation Conf.* ASCE, New York, NY.

Urban Drainage and Flood Control District (1984). *Urban storm drainage criteria manual*, rev. ed., Denver Regional Council of Governments, Denver, CO.

Urban Land Institute (1975). *Residential stormwater management: Objectives, principles and design considerations,* published jointly with the American Society of Civil Engineers and the National Association of Home Builders, New York, NY.

Urban Land Institute. (1978). *Residential erosion and sediment control: objectives, principles and design considerations,* published jointly with the American Society of Civil Engineers and The National Association of Home Builders, New York, NY.

Urbonas, B. and Roesner, L. (1991). "Hydrologic design for urban drainage and flood control." *The McGraw Hill Handbook of Hydrology*, D.R. Maidment, ed. New York, NY.

U.S. Bureau of Reclamation (1977). *Design of small dams.* U.S. Government Printing Office, Washington, D.C.

U.S. Federal Emergency Management Agency (1981). "Design guidelines for flood damage reduction." FEMA Publication, Washington, D.C.

U.S. Federal Emergency Management Agency (1981). "Multi-government management of floodplains and small watersheds." FEMA Publication, Washington, D.C.

Vanoni, V.A., ed. (1975). "Sedimentation engineering." *ASCE Manuals and Reports on Engineering Practice No. 54*, ASCE, New York, NY.

Wiggins, J.H. (1978). "Generalized description of the concept of risk, acceptable risk/balanced risk, socioeconomic aspect of risk taking/decision making," *EERI Short Course on Seismic Risk Analysis*, University of Southern California Earthquake Engineering Research Institute, El Cerrito, CA.

U.S. Federal Emergency Management Agency (1987). "Integrated emergency management system, mitigation program development guidance." FEMA Publication, Washington, D.C.

U.S. Federal Emergency Management Agency (1986). "A unified national program for floodplain management." FEMA Publication, Washington, D.C.

U.S. Federal Flood Insurance Administration (1981). *Evaluation of the economic, social and environmental effects of floodplain regulations.* U.S. Government Printing Office, Washington, D.C.

U.S. Soil Conservation Service (1973). *SCS national engineering handbook.* U.S. Department of Agriculture, Washington, D.C.

Chapter 5

HYDROLOGY AND INTRODUCTION TO WATER QUALITY

I. INTRODUCTION

A. Effect Of Urbanization On Streamflows

Figure 5.1 (Schueler 1987) depicts typical changes in watershed hydrology that can be expected as a result of urbanization. Under undeveloped conditions, losses (abstractions) such as evapotranspiration, canopy interception and soil infiltration tend to be large. Under developed conditions, the increase in the amount of impervious surface area (streets, roofs, parking lots, driveways, and sidewalks) in the watershed increase surface runoff. Wet-weather stream discharges generally increase, while dry-weather discharges may either decrease or increase. Pavements, gutters, and storm sewers convey runoff more rapidly than do natural surfaces. Development-related straightening, cleaning, and lining of natural channels increase flow velocities.

The typical impact of urbanization on the runoff hydrograph is shown qualitatively in Figure 5.2. The post-development hydrograph differs from the pre-development hydrograph in three important ways: (1) the total runoff volume is greater, (2) the runoff occurs more rapidly, and (3) the peak discharge is greater. The increase in runoff volume results from the decrease in infiltration and depression storage. The shortened time base results from the greater flow velocities in the drainage system. The increase in peak discharge is the inevitable consequence of a larger runoff volume occurring over a shorter time. This increase in peak discharge for any storm means a related high discharge occurs more frequently. **Urbanization has a greater impact on frequent events than on rare events.** Because urbanization tends to reduce both shallow subsurface flow and groundwater recharge, dry-weather stream discharges may decline. However, in arid regions, the intensive lawn irrigation that often accompanies urbanization may increase dry-weather flows.

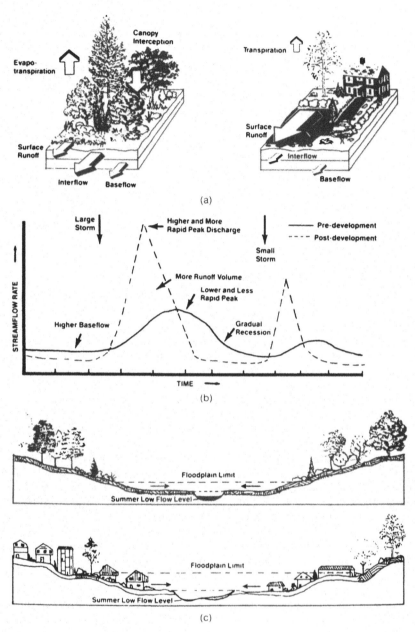

Figure 5.1—Changes in watershed hydrology as a result of urbanization.

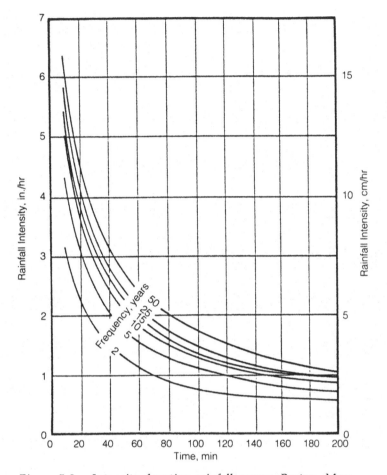

Figure 5.2—Intensity-duration rainfall curves, Boston, Mass.

This chapter describes some commonly used methods for computing peak discharges and runoff hydrographs for individual rainfall events. Modelling the long-term hydrologic and water quality response of urban watersheds to continuous climatological inputs is addressed in Chapter 7.

B. Quality Of Urban Runoff

Storm runoff may contain significant concentrations of sediment and other substances classified as pollutants. Classes of pollutants typically monitored in stormwater quality studies include suspended solids, heavy metals, nutrients, organics, oxygen-demanding substances, and bacteria. The urban transportation system can be a major source of pollutants. Exhaust emissions contribute heavy metals and hydrocarbons. Some snow and ice control measures contribute sand and salts. Construction activity can be a major source of soil sediments. Other sources of pollutants include particulate emissions from heating systems and industrial processes, fertilizers and pesticides washed off lawns, and lawn clippings, used crankcase oil, washoff from commercial establishments such as filling stations, and solvents dumped in the drainage system.

The effects of urban runoff on receiving waters depend on the hydrology, chemistry, and beneficial uses of the receiving body, as well as the quantity and quality of the discharges. This manual will describe general classes of receiving water impacts that can occur. Because receiving-water impacts are highly site-specific, local evaluations, such as instream biosurveys, sampling of sediment quality, and measurements of discharge quantity and quality, can provide the best information on those impacts. Receiving water impacts are caused by a combination of physical and chemical effects. More frequent occurrences of high discharges may cause or intensify channel erosion problems, disrupting the riparian habitat both where the erosion occurs and where the additional sediment is deposited downstream. Development-related changes in water quality and dry-weather flows may also alter the riparian habitat.

Later sections of this chapter provide a brief introduction to sources, concentrations, and loadings of pollutants in urban stormwater. Chapters 10 and 11 address receiving water impacts and mitigation measures.

II. QUANTITY OF STORMWATER

A. Overview

Whenever the rainfall exceeds the interception by vegetation and infiltration into the soil, water accumulates on the catchment surface. This water flows overland for some distance, filling surface depressions along the way, before concentrating in small channels such as gutters and swales. These small channels convey the surface runoff to underground storm sewers or larger natural or man-made drainage channels. These storm sewers and larger channels comprise the stormwater transport system. The hydrologic methods described in this chapter are useful for computing runoff hydrographs and peak discharges on small urban catchments. These catchment flows comprise the inflows to the stormwater transport system. The hydraulics of flow in the stormwater transport system are covered in Chapter 6.

To compute and interpret stormwater flows, the engineer must perform the following basic steps:

(a) Define the locations at which flows are required and the corresponding tributary areas.

(b) Determine the present and projected land uses and other physical characteristics for each catchment.

(c) Specify the return periods for which runoff hydrographs or peak discharges are to be computed.

(d) Select appropriate hydrologic procedures for computing runoff hydrographs or peak discharges.

(e) Construct a design storm hyetograph based on local rainfall characteristics.

(f) Compute rainfall abstractions (losses) and thereby rainfall excess.

(g) Compute runoff hydrographs or peak discharges.

(h) Assess the reasonableness of the computed flows.

(i) Evaluate the significance of the computed flows.

(j) Evaluate the significance of any data available on historic floods.

B. Design Points, Catchments, and Return Periods

The first step in surface runoff analysis is identifying the points at which flows must be determined, and their corresponding catchments. For computational purposes, it is often advantageous to divide a catchment into sub-catchments with more or less homogeneous physical characteristics. The outlet of each sub-catchment is termed a design point. Pipes and channels have design points wherever significant quantities of additional flow are introduced. Sub-catchments are linked together computationally by means of a hydraulic or hydrologic routing procedure. Catchment discretization is generally based on drainage patterns, surface slopes, and land use patterns. Sub-catchments may vary in size from a few acres to tens of acres. Areas of catchments and sub-catchments are usually determined from topographic maps, maps showing existing underground storm drains, and grading plans.

Existing and projected drainage patterns, land uses, and physical characteristics for each sub-catchment should be estimated carefully. Current conditions can be determined from field inspection, topographic maps, soils maps, drainage-system maps, aerial photographs, and other sources. The engineer should strive to develop a thorough understanding of actual drainage conditions in the catchment. Detailed inspection helps identify such features as depressions that do not contribute runoff, grading that disrupts natural flow paths, and underground storm drains that cut across surface drainage divides. Most communities have long-range master plans from which future conditions can be estimated, and most regulatory agencies require that stormwater flows be computed on the basis of maximum projected development in the catchment. This helps assure the long-term adequacy of the improvements (it should be noted here that master plans—even "ultimate" plans—can change, often without adequate consideration of the hydrologic consequences).

A key physical characteristic of an urban catchment is the amount of impervious surface area. Impervious surfaces typically comprise between 10–30% of the total surface area in low-density residential districts, between 30–60% in high-density residential districts, and between 80–100% in central business districts. The impervious area within a catchment should be further divided into two categories: directly connected impervious area and non-connected impervious area. Directly connected impervious surfaces drain directly to storm sewers or drainage channels. Examples of impervious surfaces that are usually directly connected include parking lots, streets with curbs and gutters, and roofs with drains leading directly to storm sewers. Non-connected impervious surfaces are impervious surfaces that drain onto pervious surfaces, such as roofs that drain onto lawns.

Peak discharges on small urban catchments may be determined largely by the runoff from directly connected impervious surfaces, particularly for the more frequent storms. It is therefore important that the directly connected impervious area in a catchment be estimated with reasonable accuracy. Driveways, sidewalks, playgrounds, roofs, and streets without curbs and gutters might or might not be directly connected, depending on site-specific conditions. When analyzing an existing drainage system, impervious surfaces should be inspected in the field to determine whether or not they are directly connected.

Selection of design return periods for the minor (primary) and major drainage systems is discussed in Chapter 8. Regardless of the design basis, it is recommended that performance of the drainage system be examined for a range of return periods. A comparison of results for the various return periods may indicate the need for a different design basis.

C. Methods For Computing Stormwater Flows

There are two basic approaches to computing stormwater flows. The first approach relates runoff to rainfall through a proportionality factor and yields only a peak discharge. This approach, termed the rational method, has been used by engineers since the nineteenth century. The second approach starts with a rainfall hyetograph, accounts for abstractions and temporary detention in transit, and yields a discharge hydrograph. Most modern methods use the hydrograph approach.

Traditionally, design discharges for street inlets and storm sewers have been computed using the rational method, although hydrograph methods also can be used for these purposes. The primary attraction of the rational method has been its simplicity. However, now that computerized procedures for hydrograph generation are readily available to all engineers, computational simplicity no longer need be the primary consideration. Experience has shown that the rational method can provide satisfactory estimates of peak discharge on most small catchments. For larger catchments, storage and timing effects can become significant, and therefore one of the hydrograph methods should be used. The rational method cannot be used to design a detention storage facility, because a complete hydrograph is needed. Various methods have been devised to form pseudo-hydrographs based on the rational formula, but their reliability is uncertain.

Sections D–F present some concepts and methods that are useful for developing design hydrographs for storm drainage facilities. The rational method is covered in Section G. More information on these topics can be found in the references provided, and in textbooks on engineering hydrology (e.g. Chow, Maidment, and Mays 1988; Viessman, Lewis, and Knapp 1989; McCuen 1989; Ponce 1989; Linsley, Kohler, and Paulhus 1982; and Hjelmfelt and Cassidy 1975).

D. Design Rainfall

The specification of a rainfall event, sometimes called a "design storm," as a design criteria is widely used in engineering practice. Despite this widespread use, however, the subject of design storms is controversial. Much of the controversy stems from the lack of realistic and accurate definitions of design storms, and confused thinking about their application. The main criticisms of design storms arise from the practice of assigning a particular frequency to a design storm, neglect of antecedent catchment conditions, and design on the basis of the return frequency of rainfall rather than runoff.

The use of a design storm requires minimal resources, and this, as well as the lack, in the past, of well-defined and inexpensive alternatives, has contributed to the popularity of this approach. Although the design storm must reflect required levels of protection, the local climate, and catchment conditions, it need not be scientifically rigorous. It is probably more important to define the storm and the range of its applicability fairly precisely to ensure safe, economical, and standardized usage.

Two types of design storms are recognized—synthetic and actual (historic) design storms. The former are derived by synthesis and generalization of a large number of actual storms. The latter are events that have occurred in the past, and which may be well documented in terms of their impacts on the drainage system. Most of the following discussion concentrates on synthetic storms.

In catchments with low imperviousness, the pervious segments will control the generation of runoff peaks, and neglect of antecedent conditions may no longer be acceptable. As catchment area increases, there may be more opportunities to incorporate runoff storage into the design, and spatial effects may become more significant. The application of conventional design storms to the design of drainage systems incorporating storage should be avoided, though many engineers use them for this purpose due to the perceived lack of practical alternatives (see also McPherson 1969, 1978 and Adams and Howard 1986 for a comprehensive discussion of limitations of the design storm approach).

1. Synthetic Design Storms

Synthetic design storms (which in fact bear little or no resemblance to real rainfall at the site in question) are derived from synthesis and generalization of point-rainfall data for a large number of actual storms. These design storms are normally defined by their duration, total rainfall depth, temporal pattern, spatial characteristics (average spatial distribution, storm movement, and spatial development and decay) and some measure of antecedent rainfall. The total rainfall depth is normally selected so that the depth-duration combination has some specified return period. It should be noted that any frequency associated with the "storm" (rainfall) is likely to be different from the frequency associated with peak runoff flow, runoff volume, or pollutant loading which results from that rainfall.

Design Storm Return Period   Ideally, the return period should be selected on the basis of economic efficiency. In practice, however, economic efficiency is typically replaced by the concept of level of protection. The selection of this level of protection (or return period), which actually refers to the exceedance probability of the design storm, rather than the probability of failure of the drainage system, is largely based on local experience. Typical return periods used in the United States and Canada, are given in Table 5.1, although longer return periods are sometimes used.

Storm Duration   Design storm duration is an important parameter that defines the rainfall depth or intensity for a given frequency, and therefore affects the resulting runoff peak. In selecting a duration, one should consider both the hydrologic time-response characteristics of the catchment and the typical durations of intense storms in the region. The design storm duration that produces the maximum runoff peak depends on the catchment time constant, traditionally defined as the time of concentration. The time of concentration is commonly expressed as the travel time (properly the **wave** travel time—see Section G. 4) from the most remote point in the catchment to the point under design. Such a definition ignores the relative runoff-producing capabilities of pervious and impervious areas, and possible variations in rainfall intensity. Nevertheless, the current practice is to select the design storm duration

TABLE 5.1:   Typical Design Storm Frequencies

Land Use (1)	Design Storm Return Period (Frequency) (2)
Minor Drainage Systems	
Residential	2– 5 years
High value general commercial area	2–10 years
Airports (terminals, roads, aprons)	2–10 years
High value downtown business areas	5–10 years
Major Drainage System Elements	up to 100 years

as one equal to or longer than the time of concentration for the catchment (or some minimum value when the time of concentration is short). Intense rainfalls of short durations (30 minutes or less) usually occur within longer-duration storms rather than as isolated events. It is common practice (Packman and Kidd 1980) to compute discharges for several design storms with different durations, and then base the design on the storm which produces the maximum discharge. **Note that the storm durations discussed above may not be suitable for storage design.**

2. *Rainfall Depth*

The total storm rainfall depth at a point, for a given rainfall duration and frequency, is given by the local climate. Rainfall depths for various durations and frequencies are published in maps of precipitation and in the reports of governmental agencies that collect precipitation data. Rainfall depths can be further processed and converted into rainfall intensities (intensity = depth/duration), which are then presented in rainfall intensity-duration-frequency (IDF) curves. Such curves are particularly useful in storm drainage design because many computational procedures require rainfall input in the form of intensities.

Although some municipalities collect and process their own rainfall data, the drainage design is more often based on rainfall data compiled and processed by national agencies. In the United States, the best source of rainfall data are the data banks of the National Climatic Data Center, and reports (rainfall frequency atlases) of the National Weather Service. Besides rainfall data, it is possible to obtain maps containing isolines of constant rainfall depth for specific durations and return periods (U.S. Weather Bureau 1958). Technical Paper 40 (Hershfield 1961) contains maps of the entire United States showing rainfall depths for durations from 30 minutes–24 hrs and return periods from 1–100 years. More-detailed maps for eleven western states are found in NOAA Atlas 2 (Miller et al. 1973). This atlas, issued in eleven volumes, contains maps for durations of 6 and 24 hours and return periods from 2–100 years. For the eastern and central U.S., a report by Frederick et al. (1977) provides maps for durations of 5, 15, and 60 minutes and return periods of 2 and 100 years. All three of these reports provide procedures for interpolating, and in some cases extrapolating, to durations and return periods not included in the reports.

Rainfall data from these reports can be abstracted, converted into intensities, and presented in the form of IDF curves. Similarly, the total storm rainfall depth can be determined from such maps for a chosen storm duration and return period.

It should be noted here that, since the last century, consistent underestimation of rainfall (due to wind effects upon rain gauges) has been recognized (National Weather Service 1963). Despite this, published rainfall intensity/duration/frequency data have been developed from these measurements, without adjustments for wind effects. This can be a particular problem in areas where convective thunderstorms (with attendant high winds) account for rainfalls of the highest intensity.

The rainfall intensity-duration curves for individual frequencies also are sometimes approximated by mathematical expressions in one of the following forms:

$$i = \frac{a}{(t_d + c)^b} \tag{5-1}$$

or

$$i = \frac{a}{t_d^b + c} \tag{5-2}$$

in which i is the average intensity for duration t_d, and a, b, and c are fitting constants. The values of the constants vary with location and return period. One should not apply an equation of this type outside the range of the data to which it was fitted. Figure 5.2 shows a family of curves of this type.

Recognizing that the precipitation data in maps were subject to some interpolation and smoothing, it may be justifiable to develop IDF curves directly from local rain-gage records if these records are sufficiently long and reliable. An example of this type of analysis for a single rain gage is presented by Chow, Maidment, and Mays (1988). Wenzel (1982) also describes procedures for performing such analyses.

3. Temporal Distribution

The temporal distribution of rainfall within the design storm is an important factor that affects the runoff volume, and the magnitude and timing of the peak discharge. Realistic estimates of temporal distributions are best obtained by analysis of local rainfall data from recording gauge networks. Such an analysis may have to be done for several widely varying storm durations to cover various types of storms and to produce distributions for various design problems. Where such analyses cannot be justified, the designer must adopt one of the existing distributions. Note that different distributions may apply to different climatic regions of the country. A brief discussion of several well-known distributions follows.

Three approaches are commonly used to distribute rainfall within a design storm. The first approach uses an average temporal distribution derived from hyetographs of actual storms. The second approach uses a simple temporal pattern (e.g., triangular) fitted to local storm data by the method of moments. The third approach uses a temporal pattern derived from the local depth-duration-frequency relationship. Two excellent sources of information on the three approaches are ASCE (1983) and Harremöes (1983).

Average temporal patterns can be developed from local point-rainfall data if such data are available in short time intervals (15 min. or less). Such an analysis should consider only major storms, and storms should be grouped according to duration to account for different storm types. If an average temporal pattern is applied outside the region for which it was developed, caution and judgment must be used in interpreting the results.

Huff (1967) developed four average temporal patterns for heavy storms in east-central Illinois. Historical storms were divided into four groups according to the relative timing of the peak intensity, and average temporal patterns were developed for each group. The short-duration high-intensity storms that typically govern the designs of storm-drainage facilities were particularly common in the first-quartile group. Consequently, Terstriep and Stall (1974) recommend the first-quartile median distribution for urban design storms in Illinois. Huff's procedure has also been applied in Canada to derive average temporal patterns for one-hour and twelve-hour durations (Hogg 1980). These durations were selected to provide samples of both local convective storms and large-scale cyclonic storms. Hershfield (1962) presents average temporal patterns for longer-duration storms (6–24 hours).

An alternative approach is to use a simple idealized rainfall distribution fitted to local storm data by the method of moments. In this procedure, the fitting parameters of the idealized rainfall distribution are expressed in terms of the dimensionless moments of the distribution. The dimensionless moments of the idealized distribution are then set equal to the mean values of these dimensionless moments for major historical storms. The number of moments required equals the number of fitting parameters.

A simple triangular rainfall distribution, as shown in Figure 5.3, is often used in computing runoff hydrographs. A single fitting parameter, the time to the peak intensity, determines the shape of the distribution. This parameter, t_p, is related to the first moment of the hyetograph (the time to the centroid), $\bar{t}$, by the equation:

$$\frac{t_p}{t_d} = 3\frac{\bar{t}}{t_d} - 1 \qquad (5\text{-}3)$$

where t_d is the storm duration. Yen and Chow (1980) first used the method of moments to develop non-dimensional triangular hyetographs for various locations, durations, depths, and seasons. Times to peak on these triangular hyetographs range from 32–51% of the storm duration. However, most of the storms included in this study produced relatively small amounts of rainfall. McEnroe (1986) found much shorter times to peak by considering only major storms (in this case, storms that produced one-hour rainfalls with return periods of two years or greater). The following example illustrates the development of a triangular design storm.

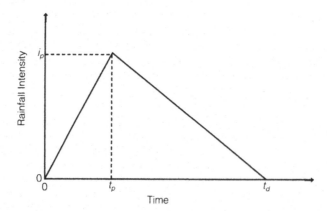

Figure 5.3—Triangular design storm.

Example 5–1: A 2-hour, 10-year triangular design storm is to be developed for Dodge City, Kansas. The 2-hour 10-year rainfall depth for this location is 2.80 inches. The mean value of t/t_d for intense storms in this region is 0.38 (McEnroe 1986).

Solution: Using Equation (5-3), one finds that the time to the peak rainfall intensity, t_p, is only 14 percent of the storm duration, or 17 minutes. The average rainfall intensity during the storm is 1.40 inches, the depth divided by the duration. The peak intensity, i_p is 2.80 inches per hour, twice the average intensity. These dimensions completely define the triangular design storm.

Two-parameter rainfall distributions permit one to adjust the peakedness of the storm as well as the timing of the peak intensity. A flexible two-parameter beta-function distribution has been proposed by Voorhees and Wenzel (1984). A two-parameter distribution used in Canada consists of a linear rise followed by an exponential decay, or vice versa (Marsalek and Watt 1983).

The third general approach for distributing rainfall within a design storm makes use of the local intensity-duration relationship for the design return period. This approach is based on the assumption that the maximum rainfall for any duration less than or equal to the total storm duration should have the same return period. For example, a 10-year three-hour design storm of this type would contain the 10-year rainfall depths for all durations from the shortest time interval considered (perhaps 5 minutes) up to three hours. These rainfalls are generally skewed. This distribution can be readily derived from the local IDF curves and the analysis of skewness of actual storms. A study of the conditional probabilities of intense rainfalls of different durations (Frederick and Tracey 1977) indicates that this assumption is very conservative, particularly for longer storms.

Kiefer and Chu (1957) used this approach in the development of the so-called Chicago storm. Although the Chicago storm distribution is widely used in practice, a word of caution is in order. Recent extensive analyses in Canada (Hogg 1980) indicate that the Chicago-type distribution is totally inappropriate for some Canadian climates, and, for the bulk of the country, is not among the most probable distributions. A recently proposed modification of the Chicago distribution attempts to reduce the excessive sharpness of the storm hyetograph by averaging the storm segment that contains the intensity peak. The modified storm profile that results no longer contains all maximum rainfalls for some short durations.

The United States Soil Conservation Service (SCS) used this assumption in developing 24-hour design-rainfall distributions for four geographical regions in the United States (SCS 1986). The ratio of peak intensity to average intensity is higher for these SCS 24-hour design distributions than for most historical storms of this approximate duration.

One simple method for developing a design storm from intensity-duration-frequency data is termed the alternating-block method. This method is illustrated in the following example.

Example 5-2:   Use the alternating-block method to develop a design storm with a 10-year return period and a 60-minute duration for St. Louis, Missouri. Use a 5-minute time interval. The 10-year intensity-duration relationship for St. Louis is described by the equation $i = 104.7 / (t_d^{0.89} + 9.4)$ where i is the average rainfall intensity in inches per hour and t_d is the duration in minutes (Wenzel 1982).

Solution: First, average rainfall intensities are computed for durations from 5–60 minutes in 5-minute increments. These values are shown in column 2 of Table 5.2. Cumulative depth, shown in column 3, is the product of duration and average intensity. The incremental depths in column 4 are the differences between successive values of cumulative depth. These values are the 5-minute rainfall amounts, arranged in descending order. These rainfall amounts are expressed as average intensities in column 2. The period of highest intensity is assumed to occur just before the midpoint of the 60-minute storm duration. The other 5-minute blocks of rainfall are arranged in descending order alternately to the right and left of the largest block, as shown in Figure 5.4. With this arrangement, the maximum rainfall for any duration from 5–60 minutes has a 10-year return period.

For practical applications, it is recommended to use one of the existing standard distributions approved by the client. Where the designer must develop the temporal distribution, in the absence of comprehensive evaluations and comparisons, it is recommended to use the simpler ones, such as the triangular or combined linear/exponential distribu-

TABLE 5.2:   Development of Alternating-Block Design Storm

Duration min. (1)	Average Intensity in./hr (2)	Cumulative Depth in. (3)	Incremental Depth in. (4)	Incremental Intensity in./hr (5)
5	7.68	0.640	0.640	7.68
10	6.09	1.015	0.375	4.49
15	5.09	1.272	0.257	3.09
20	4.39	1.463	0.191	2.31
25	3.88	1.617	0.154	1.82
30	3.48	1.740	0.123	1.49
35	3.16	1.843	0.103	1.25
40	2.90	1.933	0.090	1.07
45	2.68	2.010	0.077	0.93
50	2.50	2.083	0.073	0.82
55	2.34	2.145	0.062	0.73
60	2.20	2.200	0.055	0.66

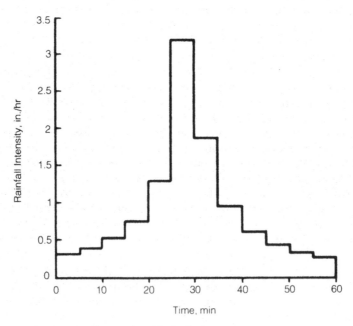

Figure 5.4—Alternating-block design storm.

tions. The fitting of these conceptual models to actual rainfall data is done by the method of moments. The goodness of fit has to be further tested. The selected distribution is then applied to the total rainfall to produce the storm hyetograph.

4. Spatial Distribution

Storm spatial characteristics are important for larger catchments. In general, the larger the catchment and the shorter the rainfall duration, the less uniformly the rainfall is distributed over the catchment. For any specified return period and duration, the average rainfall depth over an area is less than the point rainfall depth. The ratio of the areal average rainfall with a specified duration and return period to the point rainfall with the same duration and return period is termed the areal reduction factor. The areal reduction factor can be estimated using Figure 5.5, developed by the National Weather Service (Miller et al. 1973).

Once the areal average rainfall has been determined, this rainfall is normally assumed to be distributed uniformly over the catchment. Various models have been developed for simulating the effects of storm dynamics on the temporal and spatial distribution of rainfall over a catchment. Storm direction and movement can have marked effects, particularly in areas with predominating weather patterns, and are particularly relevant to the case of operation and/or control of a large system of combined sewers (James and Drake 1980).

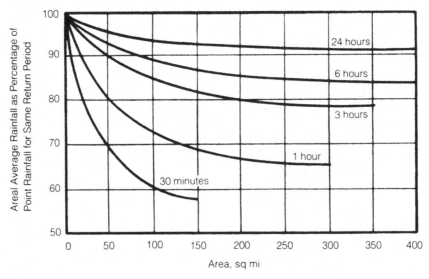

Figure 5.5—*Ratio of areal average rainfall to point rainfall for a constant return period.*

Although temporal distributions are often expressed as continuous functions, for actual use in runoff models they must be discretized into time intervals generally coinciding with the computational time step.

In light of the uncertainties inherent in the selection of other design inputs, and the general unavailability of the detailed rainfall data needed for calibration, modelling of storm dynamics would appear to be an unnecessary refinement at this time.

5. Other Methods for Design Hydrology

Another technique for arriving at design rainfall is to analyze statistically the long-term rainfall record for a station in or near the catchment, to determine the return frequencies of actual rainfall events. Appropriate events can then be used as inputs to runoff calculations. Alternatively, **actual** storms that produce rainfall equivalent to the "design" storm can be used. The graphics capabilities of modern computers provide a quick and easy means of making those comparisons (also see the following section on Infiltration). Marsalek (1978), Wenzel (1982), and Bedient and Huber (1988) should be consulted for further information on actual storms.

Frequency analysis of synthetic runoff records generated by continuous simulation is another alternative to the use of design storms (McPherson 1978). It is particularly useful when flow volumes are of concern (as when designing detention/retention storage). Despite the fact that synthetic runoff records may not be entirely "accurate" unless the continuous simulation model has been calibrated using local data, the method allows the selection of **actual** rainfall events that have produced **runoff** of some specified return frequency.

E. Rainfall Abstractions

The physical processes of interception of rainfall by vegetation, infiltration of water into the soil surface, and storage of water in surface depressions are commonly termed rainfall abstractions. Although these three processes are physically complex, some simplified modelling procedures have been found acceptable in urban areas. Evaporation is generally insignificant during the short-duration storms of concern in storm drainage design. The portion of the rainfall that is not abstracted by interception, infiltration, or depression storage is termed the excess rainfall.

1. Interception

The amount of rainfall intercepted by vegetation depends on vegetation type, growth stage, wind speed, and rainfall intensity and duration. Interception is averaged over the surface area and expressed as a depth. The data of Horton (1919) and others show that the interception storage capacity of vegetation can range from less than 0.01–0.5 inches. A typical value for grass turf is 0.05 inches.

2. Infiltration

The infiltration process plays the dominant role in determining storm runoff from pervious surfaces. The primary factors affecting rainfall infiltration are the initial soil wetness, the temporal distribution of rainfall intensity, the soil type, the vegetative cover, and special considerations such as whether the soil surface is crusted or frozen. At any instant, the maximum rate at which the soil surface can absorb water is termed the potential infiltration rate. At the beginning of a storm, all rainfall reaching the soil surface is absorbed into the soil. However, the potential infiltration rate decreases as the total infiltrated depth increases. This is caused by a reduction in the hydraulic gradient at the surface, and in some cases also by surface sealing, crusting, or other factors. When the rate at which rainfall reaches the soil surface exceeds the potential infiltration rate, water ponds on the soil surface and runoff begins.

Several approximate methods are widely used by engineers to estimate infiltration losses for storm rainfall. These include the equations of Green and Ampt, Horton, and the U.S. Soil Conservation Service (SCS), which are discussed below. Other approximate methods sometimes used in watershed modelling include the Holtan equation and the exponential loss-rate function of the U. S. Army Corps of Engineers' Hydrologic Engineering Center (HEC). More sophisticated methods of infiltration analysis are based on numerical solution of the complete equations governing unsaturated vertical flow in soils. These methods are almost never used in engineering practice on the watershed scale because of their computational complexity and because of the difficulties associated with obtaining the necessary inputs.

3. Green-Ampt Equation

The Green-Ampt equation is an approximate infiltration equation based on Darcy's law (Green and Ampt 1911). Infiltrated water is assumed to move downward through the soil with an abrupt wetting front separating the wetted and unwetted zones. At any instant the potential infiltration rate, f_p, is given by the equation:

$$f_p = K\left[1 + \frac{\Psi_f(\phi - \theta_i)}{F}\right] \qquad (5\text{-}4)$$

where K is the hydraulic conductivity of the transmission zone, Ψ_f is the capillary suction head at the wetting front, ϕ is the porosity of the soil (the volumetric water content at saturation), θ_i is the initial volumetric water content of the soil, and F is the cumulative infiltration (the total amount of infiltration that has occurred since rainfall began). Ponding begins when the potential infiltration rate equals the rainfall rate. The cumulative infiltration at any time during ponding is given by the equation:

$$F - F_p - \Psi_f(xH - \theta_i)\, ln\left[\frac{F + \Psi_f(\phi - \theta_i)}{F_p + \Psi_f(\phi - \theta_i)}\right] = K(t - t_p) \qquad (5\text{-}5)$$

where t_p is the time at the onset of ponding, F_p is the cumulative infiltration at time t_p, and F is the cumulative infiltration at time t.

The Green-Ampt infiltration equation has several advantages over other commonly-used infiltration equations. Some of these advantages are its physical basis, its explicit consideration of initial soil water content, and its direct applicability to conditions of unsteady rainfall. The values of the Green-Ampt parameters can be estimated from physical soil characteristics and knowledge of the initial moisture content (for θ_i). Mein and Larsen (1973), present the easiest and most often cited method of using the Green-Ampt equation.

The values of the Green-Ampt parameters for a particular soil can be estimated in several different ways. Procedures have been developed for estimating the values of the parameters K and Ψf based on the soil's porosity, saturated hydraulic conductivity, and water retention characteristics (Bouwer, 1966; Brakensiek and Onstad, 1977). Alternatively, the Green-Ampt equation can be fitted to in-situ infiltrometer data (Brakensiek and Onstad, 1977). Table 5.3 presents some average values of the Green-Ampt parameters sorted according to soil texture class. These values are based on statistics for porosity, saturated hydraulic conductivity, and water-retention parameters for some 5000 soil horizons compiled by the U.S. Agricultural Research Service (Rawls et al. 1983). Actual values of these parameters vary widely within a soil texture class. The values in Table 5.3 should not be used for bare soils

TABLE 5.3:   Average Values of Green-Ampt Parameters by USDA Soil-Texture Class (Rawls et al. 1983)

USDA Soil-Texture Class (1)	Hydraulic Conductivity K_1 in/hr (2)	Wetting-Front Suction Head ψ_{f1} in (3)	Porosity in³/in³ (4)	Water Retained @ Field Capacity in³/in³ (5)	Water Retained @ Wilting Point in³/in³ (6)
Sand	4.74	1.93	0.437	0.062	0.024
Loamy Sand	1.18	2.40	0.437	0.105	0.047
Sandy Loam	0.43	4.33	0.453	0.190	0.085
Loam	0.13	3.50	0.463	0.232	0.116
Silt Loam	0.26	6.69	0.501	0.284	0.135
Sandy Clay Loam	0.06	8.66	0.398	0.244	0.136
Clay Loam	0.04	8.27	0.464	0.310	0.187
Silty Clay Loam	0.04	10.63	0.471	0.342	0.210
Sandy Clay	0.02	9.45	0.430	0.321	0.221
Silty Clay	0.02	11.42	0.479	0.371	0.251
Clay	0.01	12.60	0.475	0.378	0.265

with crusted surfaces. This table also provides some typical values of water content at field capacity (pore-water pressure of −1/3 bar, a typical value after prolonged gravity drainage) and wilting point (pore-water pressure of −15 bars, roughly the limiting value for transpiration) to guide the selection of an initial water content. Table 5.3 provides a reasonable basis for engineering estimates of infiltration on the watershed scale.

The following example shows how the Green-Ampt equation is used to compute infiltration losses for constant-intensity rainfall. The case of variable-intensity rainfall is covered in detail by Chow et al. (1988).

Example 5-3: A 30-minute rainfall with a constant intensity of 5 cm/hr occurs on a typical silt-loam soil that is initially at field capacity. The objective is to determine the time to the onset of ponding and the total amount of runoff produced.

Solution: The values of the Green-Ampt parameters and the water content at field capacity for this soil are assumed to be the mean values for soils in this texture class, from Table 5.2: $K = 0.65$ cm/hr, $\Psi_f = 17$ cm, $\phi = 0.501$, and $\theta_i = 0.284$. The cumulative infiltration at the onset of ponding, F_p, is found using equation (5-4) by setting f_p, the potential infiltration rate, equal to the rainfall rate. This yields a value of 0.55 cm for F_p, indicating that ponding begins when 0.55 cm of infiltration has occurred. Until ponding begins, the actual infiltration rate equals the rainfall rate, so the time to the onset of ponding, t_p, is 0.11 hours, or 7 minutes. The cumulative infiltration at any time during ponding can be found using equation (5-5). Setting t equal to 30 minutes, the rainfall duration, and solving for F by trial, one obtains a total infiltration of 1.67 cm. The total runoff is 0.83 cm, the difference between the 2.50 cm of rainfall and the 1.67 cm of infiltration.

Note that rainfall interception is neglected in this example for the sake of simplicity. In practice, one should make reasonable estimates of these losses.

4. Horton Equation

The Horton infiltration equation (Horton, 1939 and 1940) is an empirical three-parameter equation that describes the decrease in infiltration rate over time when water is ponded on a soil surface. The Horton equation is:

$$f = f_c + (f_o - f_c)e^{-kt} \tag{5-6}$$

where f is the infiltration rate, f_o is the maximum or initial infiltration rate, f_c is the minimum or ultimate infiltration rate that is approached

t is the elapsed time since infiltration began, and k is a constant that reflects how rapidly the infiltration rate decreases with time.

Equation (5-6) is not directly applicable to storm rainfall applications. The potential infiltration rate will not actually decrease as rapidly as indicated by this equation unless the rainfall intensity always exceeds the infiltration rate. This problem is sometimes sidestepped by considering the potential infiltration rate to be a function of cumulative infiltration rather than elapsed time (see Viessman et al. 1989, for an excellent discussion). This form of the Horton equation is used in the U.S. Environmental Protection Agency's Storm Water Management Model (SWMM). The relationship between potential infiltration rate and cumulative infiltration is assumed to be identical to the relationship between actual infiltration rate and cumulative infiltration under ponded conditions. This assumption leads to a rather awkward procedure for applying the Horton infiltration equation to steady and unsteady rainfalls. This procedure is presented in detail by Chow et al. (1988).

Since the Horton equation is strictly empirical, the values of its three parameters should be determined by fitting the equation to field or laboratory infiltrometer data. These fitted values are sensitive to the antecedent wetness of the soil. No reliable procedure exists for estimating the values of the Horton parameters from the physical characteristics of a soil, or for adjusting these values to account for initial soil wetness or vegetative cover.

5. Soil Conservation Service Equation

The well known rainfall-runoff relationship of the U.S. Soil Conservation Service (SCS 1972) is often misused as an infiltration equation. The SCS procedure, often referred to as the curve-number procedure, was originally developed to provide an estimate of total runoff volume for a storm event based on total storm rainfall, antecedent watershed conditions, an SCS soil classification, and surface cover conditions. Derived from storms of 24-hour or less duration on small to very small rural watersheds, this empirical equation does not consider the effect of rainfall intensity on runoff volume. Differentiating the SCS rainfall-runoff equation with respect to time yields a rate equation. This rate equation indicates that the infiltration rate is directly proportional to the rainfall intensity, a relationship that is unacceptable from a physical standpoint. **For this reason, use of the SCS rainfall-runoff relationship as an infiltration equation is discouraged.**

6. Depression Storage

Depression storage refers to water that accumulates in surface depressions during a storm. This water eventually evaporates or infiltrates following the storm. Like infiltration, depression storage is expressed as an average depth over the catchment area. Typical depths of depression storage are 0.05 to 0.10 inches for impervious surfaces such as pavements and roofs (Tholin and Kiefer 1960), 0.10–0.20 inches for lawns (Hicks 1944), 0.2 inches for pasture, and 0.3 inches for forest litter. These typical values are for surfaces with moderate slopes. Typical values would be larger for flat slopes and smaller for steep slopes. The manual for the SWMM is a good source of European depression storage data (Huber and Dickinson 1988).

F. Runoff Hydrographs

Many different procedures are used to compute runoff hydrographs for storm events on small urban catchments. Most of these procedures make use of synthetic unit hydrographs, time-area curves, or kinematic-wave analysis. This section discusses these three general methods, plus the nonlinear-reservoir procedure used in the SWMM. A more detailed treatment of each of these methods can be found in the references provided, and in standard textbooks on engineering hydrology.

1. Synthetic Unit Hydrograph

A unit hydrograph is a hydrograph resulting from an excess rainfall of unit depth with a specified duration and a fixed spatial and temporal pattern. Hydrographs for complex historical and design storms can be computed using the unit hydrograph. Unit hydrograph theory assumes that a linear transfer function relates the excess rainfall on the catchment to the stream discharge at the catchment outlet. This assumption is often a reasonable and useful approximation. Unit hydrographs should be developed from actual streamflow and rainfall data if such data are available. Unit hydrograph theory and procedures, including derivation of unit hydrographs from actual streamflow and rainfall data, are discussed in detail in engineering hydrology texts.

In the analysis and design of storm drainage facilities, developing unit hydrographs from field data is often impossible or impractical. Instead, a synthetic unit hydrograph procedure is used. Synthetic unit hydrograph procedures relate unit hydrograph characteristics to watershed and channel characteristics. Snyder (1938) developed the first synthetic unit hydrograph procedure using data for Appalachian Highland watersheds with areas of 10 mi^2 to 10,000 mi^2. Snyder's basic procedure has been applied by others to different regions and watershed types.

One adaptation of Snyder's procedure is a 10-minute unit hydrograph for urban areas developed by Espey and Altman (1978). The Espey 10-minute unit hydrograph was developed from data for 41 watersheds throughout the United States with areas ranging from 10 acres to 15 mi^2 and impervious fractions ranging from 2–100 percent. The dimensions of the Espey 10-minute unit hydrograph depend upon five watershed characteristics: the drainage area, the percentage of the total surface area that is impervious, the length of the main drainage channel from the study point to the watershed divide, the average slope of the main channel, and the average Manning n value of the main channel. The following parameters were selected to describe the shape of the hydrograph (see also Figure 5.6):

T_R = Time of rise, minutes.
Q = Peak discharge, cfs.
T_B = Time base, minutes.
W_{50} = Time, in minutes, between the two points on the unit hydrograph at which the discharge is half the peak discharge.
W_{75} = Time, in minutes, between the two points on the hydrograph at which the peak discharge is three-fourths of the peak discharge.

The Colorado Urban Hydrograph Procedure (CUHP) (Urban Drainage and Flood Control District 1984) is another adaptation of Snyder's procedure based on data for Colorado urban watersheds ranging in size from 100–2000 acres. The dimensions of the CUHP hydrograph depend upon the unit excess-rainfall duration and five watershed characteristics: the drainage area, the percentage of the total surface area that is im-

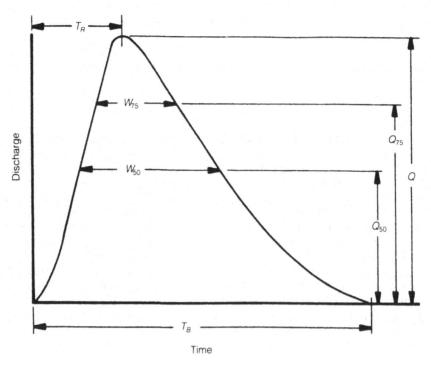

Figure 5.6—Definition of unit hydrograph parameters.

pervious, the length of the main drainage channel from the study point to the watershed divide, the distance along the main drainage channel from the study point to a point adjacent to the centroid of the watershed, and a weighted-average slope of the main drainage channel.

The dimensionless synthetic unit hydrographs of the U.S. Soil Conservation Service are also often applied to urban watersheds. These two dimensionless synthetic hydrographs, one triangular and the other curvilinear, are shown in Figure 5.7. To convert one of the dimensionless hydrographs to dimensional form, one needs values for the peak discharge, Q_p, and the time to peak, t_p. The time to peak is given by the formula:

$$t_p = 0.5t_r + 0.6t_c \tag{5-7}$$

where t_r is the unit excess-rainfall duration and t_c is the time of concentration for the watershed. Estimation of the time of concentration is discussed in Section G. The SCS recommends that t_r not exceed two-tenths of t_c. The time base, t_b, equals $2.67t_p$ for the triangular unit hydrograph and $5t_p$ for the curvilinear unit hydrograph. Both t_p and t_b should be rounded to the nearest whole multiple of t_r. From its simple ge-

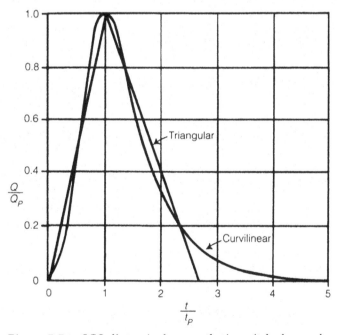

Figure 5.7—SCS dimensionless synthetic unit hydrographs.

ometry, the peak discharge on the triangular unit hydrograph is given by the formula:

$$Q_p = \frac{2C\,D\,A}{t_b} \tag{5-8}$$

where A is the drainage area, D is the unit depth of excess rainfall, and C is a units-conversion constant. When D equals one inch and Q_p is in cfs, A is in acres, and t_p is in minutes, C has a value of 60.5. When D equals one centimeter and Q_p is in m³/s, A is in hectares and t_b is in minutes, C has a value of 1.67. The values of t_p and Q_p computed using Equations (5-7) and (5-8) can also be applied to the curvilinear dimensionless unit hydrograph.

Example 5-4: Use the SCS procedure to develop a triangular synthetic unit hydrograph for a 250-acre watershed with a time of concentration of 30 minutes.

Solution: Because the unit excess-rainfall duration should not exceed two-tenths of the time of concentration, a 5-minute unit hydrograph is selected. A peak discharge and time to peak are needed to define the unit hydrograph. Equation (5-7) gives a time to peak of 20.5 minutes. The corresponding time base is 54.7 minutes. After rounding to the nearest whole multiple of t_r, t_p equals 20 minutes and t_b equals 55 minutes. From equation (5-8), the peak discharge is 550 cfs. Using these values of t_p and Q_p, the dimensionless unit hydrograph of Figure 5.7 can be converted to dimensional form.

2. Time-Area Curves

A particularly simple type of synthetic unit hydrograph is the time-area histogram. The time-area routing method is used in the Illinois Urban Drainage Area Simulator (ILLUDAS) computer program (Terstriep and Stall 1974). This method neglects all storage effects in the catchment. This is a reasonable approximation in many, but not all, urban catchments. The time-area routing method is a simplified version of Clark's method for watershed routing (Clark 1945). In Clark's method, the hydrograph developed using the time-area histogram is routed through a hypothetical linear reservoir to account for the effects of storage in the catchment.

Development of a time-area histogram requires a map of the catchment showing isochrones, lines of constant travel time to the catchment outlet. These isochrones are constructed for whole multiples of the time increment used in the infiltration and routing calculations. Travel-time should be the wave travel time (see discussion in Section II G, rational method), though estimates have often been based on velocities for uniform flow computed using Manning's equation. The incremental areas between adjacent isochrones are determined and plotted against travel time in the form of a histogram. This histogram represents the synthetic unit hydrograph for the catchment. For catchments that contain both pervious and impervious surfaces, it is advantageous to develop two separate time-area relationships: one for the directly connected impervious area and another for the remaining area. An advantage of the time-area method is its ability to account for unusual catchment shapes and nonuniform spatial distributions of excess rainfall.

3. Kinematic Wave

The kinematic-wave method is a hydraulic method for routing runoff across planar surfaces and through small channels and pipes. The kinematic-wave formulation couples the continuity equation with a simplified form of the momentum equation that includes only the bottom-slope and friction-slope terms. This is usually a valid approximation provided that backwater effects are not present. In modelling overland flow over a pervious surface, the problem formulation also includes an infiltration equation.

The kinematic-wave equations are solved numerically by a finite-difference method or by the method of characteristics. Analytical solutions are available for some special cases, such as for overland flow with constant rates of rainfall and infiltration (Eagleson 1970). The widely used HEC-1 flood hydrograph program of the Army Corps of Engineers contains options for kinematic-wave routing of overland and channel flows. This application of kinematic-wave procedures is discussed in the HEC-1 User's Manual (Hydrologic Engineering Center 1985) and a related background document (Hydrologic Engineering Center 1979). The kinematic wave method for urban stormwater simulation is also used in the U.S. Geological Survey's Distributed Routing Rainfall Runoff Model (DR3M) (Alley and Smith 1982), and in the SWMM Transport Block.

Kinematic-wave theory provides a useful formula for the time of concentration for one-dimensional overland flow on a planar surface. This time of concentration is defined as the time required for equilibrium discharge (outflow = inflow) to become established at the point of interest. For a constant excess rainfall rate, this time of concentration, t_c, is given by the formula:

$$t_c = C \frac{n^{0.6} L^{0.6}}{i_e^{0.4} S^{0.3}} \tag{5-9}$$

in which L is the distance from the upper end of the plane to the point of interest, n is the Manning resistance coefficient, i_e is the excess-rainfall rate, S is the dimensionless slope of the surface, and C is a constant that depends on the units of the other variables. For t_c in minutes, i_e in in./hr, and L in feet, C equals 0.938. For t_c in minutes, i_e in mm/hr, and L in meters, C equals 6.99. This equation is for turbulent flow and for use of Manning's equation (for a more general equation, see Eagleson (1970)). Estimates of Manning's n for overland flow are shown in Table 5.4. It should be noted that the kinematic wave velocity for turbulent overland flow using Manning's equation is 5/3 of the water velocity. Thus a t_c found by conventional methods (based on water parcel travel time) could be adjusted by multiplying by 5/3.

4. Nonlinear Reservoir

In the nonlinear reservoir method, the catchment is conceptualized as a very shallow reservoir. The discharge from this hypothetical reservoir is assumed to be a nonlinear function of the depth of water in the reservoir. The SWMM uses a nonlinear reservoir approach to compute surface runoff hydrographs. The SWMM procedure is described here (Huber and Dickinson 1988).

Figure 5.8 shows the catchment conceptualized as a reservoir with rainfall as inflow, and infiltration and surface discharge as outflows. The depth y represents the average depth of surface runoff, and the depth y_d represents the average depression storage in the catchment. The continuity relationship for this system is:

$$A \frac{dy}{dt} = A(i - f) - Q \tag{5-10}$$

where A is the catchment area, i is the rainfall intensity, f is the infiltration rate, and Q is the discharge at the catchment outlet. The model assumes uniform overland flow at the catchment outlet at a depth equal

TABLE 5.4:   Estimates of Manning's n for Overland Flow

Surface Type (1)	Manning's n (2)	Range (3)
Concrete/Asphalt**	0.011	0.01–0.013
Bare Sand**	0.01	0.01–0.016
Bare Clay—Loam (eroded)**	0.02	0.012–0.033
Gravelled Surface**	0.02	0.012–0.03
Packed Clay*	0.03	
Short Grass Prairie**	0.15	0.10–0.20
Light Turf*	0.20	
Lawns*	0.25	0.20–0.30
Dense Turf*	0.35	
Pasture*	0.35	0.30–0.40
Dense Shrubbery and Forest Litter*	0.40	
Bluegrass Sod**	0.45	0.39–0.63

*From Crawford and Linsley (1966)—obtained by calibration of Stanford Watershed Model.
**From Engman (1986) by kinematic wave and storage analysis of measured rainfall-runoff data.

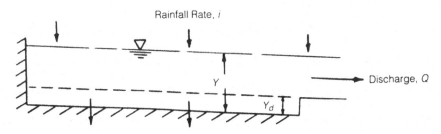

Figure 5.8—Definition sketch for nonlinear-reservoir model.

to the difference between y and y_d. Based on the Manning friction relationship, the catchment discharge, Q, is given by:

$$Q = \frac{CW}{n} (y - y_d)^{5/3} S^{1/2} \qquad (5\text{-}11)$$

where C is a constant, W is a representative width for the catchment, n is an average value of the Manning roughness coefficient for the catchment, and S is an average surface slope. The constant C has a value of 1.49 when y and y_d are in feet and Q is in ft³/s, and a value of 1 when y and y_d are in meters and Q is in m³/s. Substituting equation (5-11) into equation (5-10) yields a nonlinear differential equation for y. A simple finite difference form of the equation is used to solve for the depth y at the end of each time step. This equation is:

$$\frac{y_2 - y_1}{\Delta t} = \bar{i} - \bar{f} - \frac{CWS^{1/2}}{An} \left(\frac{y_1 + y_2}{2} - y_d \right)^{5/3} \qquad (5\text{-}12)$$

where Δt is the time-step increment, y_1 is the depth at the beginning of the time step, y_2 is the depth at the end of the time step, and $\bar{i}$ and $\bar{f}$ are the average rainfall and infiltration rates over the time step.

For each time step, three separate calculations are performed. First, an infiltration equation is used to compute the average potential infiltration rate over the time step (in the SWMM program, the user selects either the Green-Ampt or the Horton infiltration equation), then equation (5-12) is solved iteratively for y_2, and, finally, equation (5-11) yields the corresponding discharge.

Unlike the synthetic unit hydrograph and time-area methods, which use excess rainfall as input, the nonlinear-reservoir method couples the processes of infiltration and surface runoff. The nonlinear-reservoir model assumes that infiltration occurs at the potential rate over the entire surface area whenever the ponded depth is non-zero. The excess-rainfall models, on the other hand, entirely neglect infiltration of ponded water. This difference becomes important following cessation of rainfall, or whenever the rainfall intensity drops below the potential infiltration rate. In reality, infiltration does continue for some time after rainfall ceases, but the area over which infiltration continues to occur decreases rapidly with time. Thus the runoff that occurs after rainfall ceases would tend to be underestimated by the nonlinear-reservoir methods and overestimated by the excess-rainfall methods, though the difference will depend on the degree of discretization used, since a more detailed schematization can partially account for the phenomenon of decreasing area of infiltration.

G. Rational Method For Peak Discharge

In the rational method, known as the Lloyd-Davies method in the United Kingdom, peak discharge is related to rainfall intensity by the formula:

$$Q = CiA \qquad\qquad (5\text{-}13)$$

in which Q is the peak discharge, in cfs; C is a non-dimensional runoff coefficient; i is the average rainfall intensity, in in./hr, over a duration equal to the time of concentration for the contributing area; and A is the contributing area, in acres (1 acre-inch/hr = 1.008 cfs).

The rational method is based on the following assumptions:

(a) The peak discharge at any point is directly proportional to the average rainfall intensity during the time of concentration to that point.
(b) The return period of the peak discharge is the same as the return period of the average rainfall intensity.
(c) The time of concentration is the travel time from the most remote point in the contributing area to the point under consideration. This assumption applies to the point most remote in time, not necessarily in distance.
(d) The contributing area can be the entire drainage area upstream of the design point or some subset of this area, such as only the directly connected impervious portion of the drainage area.

1. Limitations of the Rational Method

Experience has shown that **when applied properly**, the rational method can provide satisfactory estimates for **peak** discharges on small catchments where storage effects are insignificant. The rational method is not recommended for drainage areas much larger than 100–200 acres, for any catchment where ponding of stormwater in the catchment might affect peak discharge, or where the design and operation of large (and hence more costly) drainage facilities is to be undertaken, particularly if they involve storage.

It may be possible to use the rational method in some cases where ponding affects peak discharge, by adjusting the runoff coefficient to compensate for the ponding. While procedures have been developed for the use of the rational method in detention basin design (see Chapter 11), care should be exercised since the rational method does not produce a discharge hydrograph (Adams and Howard 1986). Procedures that use the rational method as a basis for constructing pseudo-hydrographs are not recommended.

2. Runoff Coefficient

The runoff coefficient, C, accounts for the integrated effects of rainfall interception, infiltration, depression storage, and temporary storage in transit on the peak rate of runoff. When estimating a value for the runoff coefficient, the engineer should consider the roles played by these hydrologic processes. The runoff coefficient depends on rainfall intensity and duration as well as the catchment characteristics. The greater the rainfall depth, the lesser the relative effect of rainfall abstractions on the peak discharge, and therefore the greater the runoff coefficient.

Decades of practical experience with the rational formula have led to some accepted ranges of values for the runoff coefficient. Table 5.5 shows normal ranges of values for several types of surfaces. Table 5.6 shows some typical ranges for the composite runoff coefficient for various urban land uses. In practice, the composite runoff coefficient for a heterogeneous catchment should be computed as the area-weighted average of the runoff coefficients for the different types of surfaces in the catchment.

The ranges of values in Tables 5.5 and 5.6 are typical for return periods of 2–10 years. Higher values are appropriate for longer return

TABLE 5.5: Normal Range of Runoff Coefficients*

Character of Surface (1)	Runoff Coefficients (2)
Pavement	
Asphalt and Concrete	0.70 to 0.95
Brick	0.70 to 0.85
Roofs	0.75 to 0.95
Lawns, Sandy Soil	
Flat (2 percent)	0.05 to 0.10
Average (2 to 7 percent)	0.10 to 0.15
Steep (>7 percent)	0.15 to 0.20
Lawns, Heavy Soil	
Flat (2 percent)	0.13 to 0.17
Average (2 to 7 percent)	0.18 to 0.22
Steep (>7 percent)	0.25 to 0.35

*The range of "C" values presented are typical for return periods of 2–10 years. Higher values are appropriate for larger design storms.

TABLE 5.6:   Typical Composite Runoff
Coefficients, by Land Use*

Description of Area (1)	Runoff Coefficients (2)
Business	
Downtown	0.70 to 0.95
Neighborhood	0.50 to 0.70
Residential	
Single Family	0.30 to 0.50
Multi-units, detached	0.40 to 0.60
Multi-units, attached	0.60 to 0.75
Residential (suburban)	0.25 to 0.40
Apartment	0.50 to 0.70
Industrial	
Light	0.50 to 0.80
Heavy	0.60 to 0.90
Parks, cemeteries	0.10 to 0.25
Playgrounds	0.20 to 0.35
Railroad yards	0.20 to 0.35
Unimproved	0.10 to 0.30

*The ranges of "C" values presented are typical for return
periods of 2–10 years. Higher values are appropriate for
larger design storms.

periods, because infiltration and other losses then have a proportionally
smaller effect on runoff.

3. Rainfall Intensity

The rainfall intensity, i, in the rational formula represents the average
rainfall intensity over a duration equal to the time of concentration for
the catchment. This combination of average intensity and duration must
have a return period equal to the desired return period of the peak
discharge. This rainfall intensity therefore depends on the following:
(1) the desired return period of the peak discharge, (2) the local rainfall
depth-duration relationship for this return period, and (3) the time of
concentration for the catchment. Sources of rainfall depth-duration-
frequency data were discussed in Section D.

4. Time of Concentration

The time of concentration is the travel time of a **wave** from the most
hydraulically remote point in the contributing area to the point under
study (it should be noted that many references define time of concentration as the travel time of a parcel of water to move down the catchment). This can be considered the sum of an overland-flow time and
times of travel in street gutters, roadside swales, storm sewers, drainage

channels, small streams, and other drainageways. The major factors affecting time of concentration for overland flow are maximum flow distance, surface slope, surface roughness, rainfall intensity, and infiltration rate. The wave speed in a gutter, pipe, or channel at a particular discharge can be calculated using equation (5-14):

$$c = Vt \pm \sqrt{\frac{gA}{B}} \qquad (5\text{-}14)$$

where

V = water velocity
g = gravitational acceleration
A = cross-sectional area of flow
B = top width of flow

For subcritical flow, the wave of interest is in the downstream direction, for which the plus sign is used. Here, c is obviously $> V$, and, once again, the t_c based on wave speed is less than t_c based on water velocity. In the design of storm drainage systems, the time of concentration is often considered the sum of the time of travel to an inlet and the time of travel in the storm sewer or drainage channel. The inlet time usually includes both overland flow time and flow time in gutters or roadside swales.

For overland flow, wave speed is usually given by the kinematic wave equation, and the time of concentration for overland flow by kinematic wave theory is given by equation (5-9). Since the t_c equation (5-9) involves the use of rainfall excess, i_e, which is not a constant, the calculation of time of concentration requires iteration between the IDF curve and the kinematic wave equation for t_c until t_c = rainfall duration. Excellent discussions of this issue can be found in Bedient and Huber (1988) and Eagleson (1970). It should be noted that any of the popular microcomputer spreadsheet programs would be very useful in such calculations.

Many other equations, mostly empirical, have been proposed for estimating times of concentration for urban watersheds. For watersheds with varied land use and significant channel-flow time, no single equation can be expected to provide an accurate estimate of time of concentration. In such cases, the recommended procedure is to break the flow path into segments that are somewhat homogeneous and then use a velocity-based equation to estimate the flow time in each segment. It is common practice to assume an arbitrary minimum time of concentration that has been observed to range, depending on local practice, from 5–20 minutes.

In any method for calculating times of concentration, uncertainties about the actual overland flow path, roughness, slope, and temporal and spatial rainfall variations can overwhelm any assumptions about physics. Equation (5-9) has the advantages of using the correct physical concept and resulting in a shorter t_c (and hence a more conservative design).

5. Contributing Area

The contributing area in the rational formula can be either the entire drainage area upstream of the point under study or some part of this area. The rational formula might yield a larger peak discharge for a subarea than for the entire drainage area if the subarea has either a larger runoff coefficient or a much shorter time of concentration than the total area. As a result, the rational formula should be applied in two ways: (1) using the entire drainage area, and (2) using only a portion of the most densely developed hydraulically connected area to see which portion of the tributary watershed governs.

Some fundamental points regarding the proper application of the rational method are illustrated in the following simple example.

Example 5-5: Use the rational formula to determine design discharges for the two inlets and two pipes shown in Figure 5.9. The design return period is 5 years. The local 5-year rainfall intensity duration relationship is fitted by the equation $i_e = 51.4 / (t_d + 7.8)^{0.75}$, where i_e is the average rainfall intensity in in./hr and t_d is the duration in minutes. Sub-catchment A has an area of 2.5 acres, 40 percent of which is impervious. Sub-catchment B has an area of 4.0 acres, 15 percent of which is impervious. The runoff coefficients are estimated to be 0.9 for the impervious surfaces and 0.2 for the pervious surfaces.

T_c is calculated by iterating between Equation 5-9 and the intensity-duration relationship above, and the results are shown in the following table:

T_d	i_e	$i_e^{0.4}$	Catchment A		Catchment B	
			$T_{c(imp)}$	$T_{c(per)}$	$T_{c(imp)}$	$T_{c(per)}$
5	7.59	2.24	9.95	17.23	9.95	25.36
10	5.93	2.03	10.98	19.01	10.98	27.99
15	4.92	1.89	11.79	20.42	11.79	30.06
20	4.24	1.78	12.52	21.69	12.52	31.92
25	3.75	1.69	—	22.84	13.18	34.68
30	3.37	1.62	—	—	—	35.07
35	3.07	1.56	—	—	—	36.42
40	2.82	1.51	—	—	—	37.62

We can see that, for Catchment A, the T_c is between 10–15 minutes for the impervious area, and 20–25 minutes for the pervious area. Similarly, for Catchment B, T_c is between 10–15 minutes (impervious) and 35 and 40 minutes (pervious). Interpolating, we determine that:

	Catchment A		Catchment B	
	Imp	Per	Imp	Per
T_c	11	22	11	37
i_e	5.69	4.03	5.69	2.96

(a) Inlet 1 and Pipe 1—Considering the entire Catchment A, the contributing area is 2.5 acres, the time of concentration is 22 minutes, the corresponding average rainfall intensity is 4.03 in./hr, and the area-weighted average runoff coefficient is 0.48. The rational formula yields a peak discharge of 4.89 cfs for this case.

Considering only the impervious portion of the catchment, the contributing area is 1.0 acres, the time of concentration is 11 minutes, the corresponding average rainfall intensity is 5.69 in/hr, and the runoff coefficient is 0.9. The rational formula yields a peak discharge of 5.1 cfs for this case. Therefore the design discharge is 5.1 cfs, controlled by the runoff from the impervious surfaces.

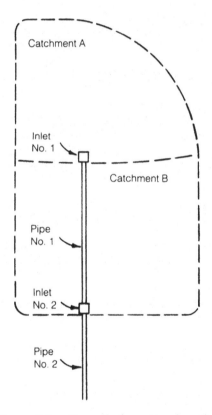

Figure 5.9—Example for rational method.

(b) Inlet 2—Considering the entire Catchment B, the contributing area is 4.0 acres, the time of concentration is 37 minutes, the corresponding average rainfall intensity is 2.96 in./hr, and the area-weighted average runoff coefficient is 0.30. The rational formula yields a peak discharge of 3.6 cfs for this case.

Considering only the impervious portion of the catchment, the contributing area is 0.6 acres, the time of concentration is 11 minutes, the corresponding average rainfall intensity is 5.69 in/hr, and the runoff coefficient is 0.9. The rational formula yields a peak discharge of 3.1 cfs for this case. Therefore the design discharge is 3.6 cfs, controlled by the runoff from the entire basin.

(c) Pipe 2—Consider first the case where the entire 6.5-acre area tributary to Pipe 2 is contributing runoff. To determine the time of concentration for this area, one would first compute the travel time in Pipe 1 using a hydraulic formula. Assume here that this travel time is 2 minutes. The travel time from the most remote point in Catchment A is 13 minutes, the sum of the inlet time for the pervious portion of this catchment and the travel time in Pipe 1. The travel time from the most remote point in Catchment B is 37 minutes, the inlet time for the entire catchment. The time of concentration for the entire 6.5-acre area is 37 minutes, the longest travel time to the point under study. The corresponding rainfall intensity is 2.96 in./hr, and the area-weighted average runoff coefficient is 0.37. The rational formula yields a peak discharge of 7.11 cfs for this case.

Next consider the case where only the impervious portions of Catchments A and B are contributing runoff. The contributing area is 1.6 acres. The time of concentration is 13 minutes, the sum of the inlet time for the impervious portion of Catchment A and the travel time in Pipe 1. The corresponding average rainfall intensity is 5.28 in./hr, and the runoff coefficient is 0.9. The rational formula yields a peak discharge of 7.6 cfs for this case. Therefore the design discharge is 7.6 cfs, controlled by the runoff from the impervious area.

H. Assessing the Reasonableness of Computed Flows

After computing hydrographs or peak discharges for the desired return periods, the engineer must assess the reasonableness of the results. This assessment is generally based on the engineer's own experience and the experience of other engineers in previous flood studies on hydrologically similar watersheds. Results for equal return periods on watersheds with similar physical characteristics in the same vicinity can be compared in terms of runoff rates per unit surface area. For example, in the Denver area, engineers assume that the 100-year runoff rate is about 1 cfs per acre for undeveloped land whenever the watershed is approximately one square mile in size. Any result that deviated significantly (i.e., more than minus 25% or plus 50%) from this value would merit reappraisal. Approximate unit rates of runoff and flood envelope curves may be available from an agency of local or state government. In comparing unit runoff rates for catchments of different sizes, one should remember that the peak discharge per unit surface area tends to decrease as catchment area increases because of storage effects and spatial variability of rainfall.

The reasonableness of a flow estimate can also be checked by comparison with the results of another, usually more approximate, hydrologic procedure. Peak discharges may differ significantly (a variation of 25% or more is common), but other hydraulic characteristics such as the total volume and time to peak should be comparable. Substantial differences may indicate an error in applying one of the procedures.

If hydrographs or peak discharges are generated by more than one method, the engineer must decide which results to use as a basis for design. This decision requires thoughtful appraisal, and the engineer should have a sound basis for his ultimate decision. Each procedure should be carefully considered with regard to its origin, assumptions, limitations, applicability to local conditions, and the reliability of the needed input data. If, based on these considerations, one procedure appears to be more appropriate than the others, the results of this procedure should be used as the basis for design. If no one procedure is clearly most appropriate, one could reasonably opt for the results of the procedure most often used by local engineers. If one procedure produces flows that are significantly lower than alternative procedures, these flows should not be used as a basis for design unless this decision can be justified convincingly.

I. Interpretation of Computed Flows

After hydrographs or peak discharges have been computed for various return periods, the engineer must select the design return periods for the minor (primary) and major drainage systems. The local government usually specifies the minimum return period to be used in designing the minor system. However, it may be desirable to examine the additional cost of designing for a higher return period. In some instances the frequency with which the capacity of the minor system is exceeded can be reduced greatly for a rather nominal increase in cost. In designing the major drainage system, the engineer should keep in mind that the major system design runoff event probably does not represent the worst possible case. Where a potential for catastrophic losses exists, one should attempt to evaluate the consequences of larger flood.

J. Significance of Historic Flood Data

The engineer should determine whether any information on historic floods is available for the catchment under study or for any hydrologically similar catchment nearby. Information on urban floods can often be obtained from local offices of the U.S. Geological Survey (Water Resources Division), or by interviewing long-time residents of the community. Municipal public works personnel also may be able to provide useful information. If the approximate date of a historic flood is known, a newspaper account often can be found in the local public library. Rainfall data should be sought in the records of the National Weather Service. Peak discharges can often be estimated roughly from high-water marks.

It is usually not possible to make a reliable estimate of the return period of a historic flood on an ungauged stream. Therefore historic flood data are generally of limited value in assessing the reasonableness of computed flows. However, if such data are available, the engineer should attempt to determine how the storm drainage system would perform under these extreme conditions. The results of such an analysis may suggest ways in which the major drainage system may be modified to improve public safety during extreme flooding.

III. QUALITY OF STORM SEWER DISCHARGES

Sources of pollutants that affect the quality of discharges from separate storm sewer systems can be described in terms of two major classes: non-storm water sources; and runoff related sources. This section describes these two classes of pollutant sources as well as water quality impacts associated with discharges from separate storm sewers. Discharges from combined sewer overflows can differ significantly, and are discussed in more detail in a WPCF Manual of Practice (1989).

A. Pollutant Sources: Non-Storm Water Sources

Although separate storm sewers are primarily designed to remove runoff from storm events, materials other than storm water find their way into, and are ultimately discharged from, separate storm sewers. A wide range of pollutants can be associated with non-storm water discharges, including pathogens, metals, nutrients, oil and grease, metals, phenols, and solvents. Removal of non-storm water sources of pollutants often provides opportunities for dramatic improvement in the quality of discharges from separate storm sewers.

Major classes of non-storm water discharges to storm sewers include:

1. Illicit or Cross Connections

Illicit connections, also referred to as cross connections, to separate storm water sewers are physically connected conveyances that carry untreated wastewaters other than storm water. Illicit connections take a variety of forms, including improper connections of residential sewer service lines or sumps, cross-connections with sanitary sewers, improper connections of industrial sewer lines, and the improper disposal of wastes to floor drains or outdoor drains connected to separate storm sewers. For many of these connections, there is a mistaken belief that materials are going to a sanitary sewer or some other type of treatment facility.

Illicit connections to separate storm sewers can occur in a number of ways. In older sections of cities with separate storm sewers, illicit connections can often be traced to the initial development of the storm sewer system. Illicit connections can occur in newer systems where flows in sanitary sewers have grown to exceed the hydraulic capacity of sanitary sewers, and formal connections or overflow devices have been installed or where holes are punched into the sanitary sewer to relieve the sanitary sewer of high flows. Discharges from malfunctioning sanitary sewage pumping stations can also be directed towards storm sewers. Incomplete separation of combined sewers can result in significant numbers of cross-connections between the sanitary sewer system and the storm sewer system.

2. Interactions with Sewage Systems

Sewage exfiltration out of a sanitary sewer collection system can result from aged sanitary sewers, poorly constructed manholes and joints, and main breaks. Sewage from a leaky sanitary system can flow to a storm sewer or contaminate ground water supplies.

3. Improper Disposal

Discharges from separate storm sewers can be contaminated by materials that are improperly disposed directly to a catchbasin or are improperly disposed to the ground and drained or washed into a storm sewer. These materials include used oil, household toxic materials, radiator fluids, litter and cement-truck washings. Direct improper disposal to separate storm sewers occurs in part because much of the public believes that disposal of materials to street catchbasins is an environmentally sound practice, rather than a conduit to a receiving water. In some areas, the use of used oil for road oiling is an accepted practice.

4. Spills

Spilled materials often have a high potential for entering man-made drainage systems. A wide variety of materials may spill during transportation, transfer, use, and storage.

5. Malfunctioning Septic Tanks

In rural and suburban areas served by septic tanks, malfunctioning septic systems can contribute pollutants to separate storm sewers. Surface malfunctions are caused by clogged or impermeable soils, or when stopped up or collapsed pipes force untreated wastewater to the surface. These discharges have high bacteria, nitrate, and nutrient levels and can contain a variety of household chemicals. One type of improper remedy to a surface malfunction is to install a pipe or trench over soil absorption systems to route overflow away from the septic system, resulting in direct discharges to drainage ditches, empty lots, or surface waters.

6. Infiltration of Contaminated Ground Water

Many separate storm sewers are subject to infiltration from surrounding groundwater. Usually this infiltrated water is not contaminated and poses no direct pollutant threat to the receiving water. However, separate storm sewers may serve as a conduit to surface waters for ground water contaminated by industrial or other sources.

B. Pollutant Sources: Runoff Related Sources

The type and concentration of pollutants in runoff discharged from separate storm sewer systems will depend on the types of land use activities occurring in the area served by the storm sewer system. The quality of runoff from several major types of land uses are discussed below:

1. Runoff from Residential and Commercial Areas

Pollutants in runoff from residential and commercial lands has been addressed in a number of studies, including the Nationwide Urban Runoff Program (NURP). The NURP program evaluated data collected between 1978 and 1983 from 81 sites in 22 of the 28 cities funded by the program. Of the 81 sites selected, 39 were completely or primarily residential, 14 were commercial, 20 were mixed commercial and residential, and 8 were runoff from open space in urban areas. The NURP study provides insight on what can be considered background levels of pollutants in runoff from residential and commercial land uses as sites were carefully selected so they were not impacted by pollutant contributions from construction sites, industrial activities, or illicit connections. Data from several sites had to be eliminated from the study because of elevated pollutant loads associated with these sources.

The majority of samples collected in the NURP study were analyzed for seven conventional pollutants and three metals. A summary of the concentrations of these parameters is provided in Table 5.7.

One way to evaluate the NURP data is to compare annual pollutant loads in runoff from commercial and residential areas with discharges from sewage treatment plants providing secondary treatment. Such a comparison indicates that on an annual loading basis: suspended solids in runoff are around an order of magnitude or more greater; chemical oxygen demand (COD) in runoff is comparable in magnitude; and nutrients in runoff are around an order of magnitude less than annual loadings of discharges from sewage treatment plants providing secondary treatment. When analyzing annual pollutant loadings, it is im-

TABLE 5.7:   Water Quality Characteristics of Runoff from Residential
and Commercial Areas

Constituent (1)	Average Residential or Commercial Site Concentration (2)		Weighted Mean Residential or Commercial Site Concentration (3)		NURP Recommendations for Load Estimates (4)
TSS	239	mg/l	180	mg/l	180–548 mg/l
BOD	12	mg/l	12	mg/l	12–19 mg/l
COD	94	mg/l	82	mg/l	82–178 mg/l
Total P	0.5	mg/l	0.42	mg/l	0.42–0.88 mg/l
Sol. P	0.15	mg/l	0.15	mg/l	0.15–0.28 mg/l
TKN	2.3	mg/l	1.9	mg/l	1.90–4.18 mg/l
NO2 + 3 − N	1.4	mg/l	0.86	mg/l	0.86–2.2 mg/l
Total Cu	53	ug/l	43	ug/l	43–118 ug/l
Total Pb	238	ug/l	182	ug/l	182–443 ug/l
Total Zn	353	ug/l	202	ug/l	202–633 ug/l

Developed from results of the nationwide urban runoff program (EPA 1983).

portant to recognize that discharges of runoff are highly intermittent, and short-term loadings associated with individual events could be high and may have shock loading effects on receiving waters.

Additionally, 121 samples at 61 sites were analyzed for 120 of the pollutants EPA classifies as priority pollutants. Heavy metals were by far the most prevalent priority pollutant found in the study, with 10 metals detected in discharges at frequencies of greater than 10%, and copper, lead, and zinc each found in at least 91% of the samples. Sixty-three of the 106 organics measured were detected, with concentrations of organic pollutants in discharges exceeding water quality criteria less frequently than with heavy metals. The NURP found fecal coliforms in runoff from residential and commercial lands at concentrations approaching dilute sewage at a number of sites. Seventeen sites analyzed fecal coliform levels, with fecal coliform counts in runoff typically in the tens of thousands per 100 ml during warm weather conditions (average NURP site concentrations were 27,000 counts/100 ml), with lower concentrations during colder weather (average NURP site values were 1,000 counts/100 ml for cold weather). Other pollutants were either not considered in the NURP study (e.g. oil and grease, floatables, chlorides, non-polar pesticides, asbestos, etc.) or were found less frequently.

When considering relatively large commercial and residential drainage basins, the most important factors influencing pollutant loadings are usually the degree of imperviousness and the amount of precipitation. The NURP concluded that, for general planning purposes, the concentrations of pollutants in runoff from different large residential and commercial areas can be assumed to be roughly equivalent, but the degree of imperviousness plays an important role in determining pollutant loads. Central business districts, which have a very high degree of imperviousness, will usually have the highest pollutant loadings per unit area. Commercial land uses can also have high degrees of imperviousness. The degree of imperviousness of residential lands is generally significantly lower than commercial land uses, and depends on the type of housing provided and the resulting density.

Seasonal variations are also important considerations in many parts of the country. The Milwaukee NURP project showed that for northern climates, concentrations of total suspended solids and associated metals were highest in winter, with spring pollutant concentrations following closely. High winter concentrations were attributed to the release of pollutants accumulated in snow. In addition, chloride concentrations will be highest in the winter in areas were road salting is performed. In areas with long dry seasons, accumulated pollutants result in peak levels of heavy metal and oil and grease in runoff from early rain season storm events. Peak concentration levels can be on the order of three or four times greater than the concentrations in discharges occurring later in the season. Areas with high intensity rainfalls, such as the southwest and parts of Texas, can have high rates of erosion from unprotected soils.

2. Runoff from Construction Sites

The amount of sediment in storm water discharges from construction sites can vary considerably, depending on whether effective management practices have been implemented at the site. Uncontrolled or inadequately controlled construction site sediment loads have been reported to be on the order of 35–45 tons/acre/year. Sediment runoff rates from construction sites are typically 10–20 times those from agricultural lands, and typically 1,000–2,000 times that of forest lands. Over a short period of time, construction sites can contribute more sediment to streams than was previously deposited over several decades.

3. Runoff from Industrial Lands

Discharges from separate storm sewers serving industrial lands may contain a larger number of toxic constitutes at higher concentrations. In general, a greater variety and larger amounts of toxic materials are used, produced, stored or transported in industrial areas. Material management practices, and atmospheric deposition can contribute to significant levels of toxic constituents in runoff. Many industrial areas have a high potential for illicit connections, spills, leaks, and other sources that may contribute a wide variety of pollutants to discharges from separate storm sewer systems. In addition, many heavy industrial areas are highly impervious, which results in high volumes of runoff with high pollutant loads.

4. Runoff from Roads and Highways

Pollutant concentrations in runoff from roads and highways are generally higher than those found in typical runoff from residential and commercial areas. Traffic-related pollutants come from leakage of oil, fuel oil, hydraulic fluids, coolants, incomplete combustion of fuel, clutch and brake lining wear, particulate exhaust emissions and debris from vehicles. Rust, dirt, metals, litter, plastic, and glass are pollutants from weathering and wear.

Research indicates that the median concentrations of pollutants in urban road and highway runoff are typically three times higher than pollutant concentrations in runoff from roads in rural areas. Higher pollutant concentrations in urban areas were attributed to higher traffic volumes and more atmospheric deposition of pollutants.

Road maintenance activities, including right-of-way grass mowing, vegetation control, road repair, snow removal, and road de-icing activities, can significantly impact the pollutants in runoff. Spraying of herbicides and growth regulators has become an increasingly popular method to control vegetation along roadsides. De-icing salts can be major sources of sodium and chlorides in storm water runoff from roads and highways. De-icing salts can also be a source of toxic metals (i.e., lead, nickel, chromium) and cyanide (used as an anticaking agent to prevent the granulated salt from solidifying).

5. Estimation of Pollutant Loads

A number of models have been developed for estimating pollutant loads in discharges from separate storm sewers. A detailed description of quality and quantity models is provided in Chapter 7 of this manual. As discussed previously, pollutants in discharges from separate storm sewers can come from a variety of sources, making pollutant loading estimates without discharge-specific pollutant concentration data difficult. Typically, where discharge-specific data is not available, models assume pollutants concentrations in discharges can be approximated by the pollutant concentrations associated with runoff from lands used for residential and commercial activities which are the predominate land uses in most urbanized areas, typically occupying between 55 to 85 percent of the total area. Such an assumption can be used to provide initial planning estimates of the background level of pollutants expected in discharges from separate storm sewers. However, the resulting estimates should be presented with the caveats that they do not account for pollutants from many sources such as industrial runoff, illicit connections, construction site runoff, improper dumping, spills, etc, and that pollutants in separate storm sewer discharges exhibit considerable variability both from event to event and from place to place. Bearing this in mind, two relatively simple methods that may be appropriate for providing initial planning level estimates of pollutants are discussed below:

Concentration Times Flow   Pollutant loads associated with discharges from separate storm sewer systems can be estimated by multiplying the estimated volume of the discharge over a given time period (e.g. event or annual) by an estimated representative concentration for the event(s) to be modelled. Hydraulic models that can be used to estimate discharge volumes are discussed in more detail in Chapter 7 of this manual. Table 5.7 summarized NURP data of the event mean concentrations for 10 pollutants in runoff from residential and commercial land uses.

Regression Formulas   The United States Geological Survey (USGS) (Driver and Tasker 1988) has developed regression equations for estimating pollutant loads in urban runoff. The regression equations are based on the NURP data base supplemented by USGS data for 2813 storms at 173 urban stations in 30 metropolitan areas throughout the United States. Equations are provided for estimating pollutant loads from individual storms, and mean seasonal or mean annual pollutant loads for eleven constituents: chemical oxygen demand, suspended solids, dissolved solids, total nitrogen, total Kjeldahl nitrogen, total phosphorus, dissolved phosphorus, cadmium, copper, lead, and zinc.

C. Water Quality Impacts of Storm Water Discharges

A National Water Quality Inventory, (EPA 1988) indicates that discharges from municipal separate storm sewers are a major source of surface water use impairment. The Inventory, which is based on reports from the States, indicates that of the rivers, lakes, and estuaries that were assessed by States, roughly 70–75% are supporting the uses for which they are designated. Of those waters that are experiencing use impairment, separate storm sewers are the leading cause of impairment for 7% of rivers and streams, 41% of estuaries, 8% of lakes, 20% of coastal areas and 35% of the Great Lakes shoreline. The National Water Quality Inventory generally only addresses larger receiving water bodies and does not address major portions of the natural drainage system of most watersheds, such as smaller feeder streams and wetlands where impacts from separate storm sewers are often the most dramatic. A number of other major sources cited in the report can, to varying degrees, discharge through urban drainage systems including construction, land disposal of wastes, and resource extraction. With continued urban expansion, further degradation of surface waters from separate storm sewers remains a concern.

The effects of discharges from separate storm sewers on receiving waters depend on the hydrology and chemistry of the receiving body as well as the quantity and quality of the discharge. This manual will describe classes of general receiving water impacts that can occur; however, local evaluations such as instream biosurveys, sampling of sediment quality, and measuring discharge quantity and quality, can provide the most accurate assessment of impacts to a given receiving water.

1. Parameters Associated with Impacts

Siltation/Sedimentation Siltation from sediment pollutant loads can cause a broad range of interrelated impacts in receiving waters including:

(a) Loss of benthic habitat—Channel scour and bank erosion can result in habitat destruction. Suspended solids may be deposited as sediment bars or sediment blankets in pools and other areas and may smother benthic organisms, including the eggs and immature forms of free-swimming organisms.
(b) Reduced water storage capacity—Increased sediment loads reduce water storage capacity in reservoirs (in urban areas, these loads result largely from construction activities). Sediment loads can also reduce stream depths, decreasing their retention and conveyance capacity, which can then result in increased flooding.
(c) Impaired oxygen exchange—Increased turbidity levels may impair the ability of aquatic organisms to obtain dissolved oxygen from the water by interfering with the gill movements and associated water circulation.
(d) Decreased light penetration—The depth of light penetration into surface waters is sharply diminished by turbidity. As a result, photosynthetic activity and food sources are reduced. Loss of submerged aquatic vegetation may also remove habitat for juvenile fish and shellfish.
(e) Increased water treatment costs—Sediments can increase the costs of treating potable water supplies. Inadequate sediment removal may limit the effectiveness of chlorination.

In addition, many of the pollutants associated with runoff may become chemically or physically bound with sediment particles and accumulate in the bottom of receiving waters. As a result, many of the pollutants in urban runoff are trapped in bottom sediments and are thereby immobilized. It is possible that these accumulated pollutants may remobilize and contaminate the overlying water column and benthic biota, and may even enter the food chain. However, much research is still needed to define the physiochemical and biological processes and activities that may be related to these pollutants. Oxygen-demanding pollutants in sediment deposits can create oxygen deficits during and after storm water discharge events. Variable flows in receiving waters can resuspend sediments. The repetitive process of deposition, resuspension and redeposition of sediments can result in pollutants associated with sediments taking a long time to pass through receiving waters.

Organic Enrichment/Oxygen Demand Aquatic organisms such as fish and water-dwelling insects require minimum levels of dissolved oxygen (DO). Excessive oxygen—demanding pollutants can lead to periods of oxygen sag which may cause fish kills or create anoxic conditions accompanied by foul-smelling odors. Oxygen levels in receiving waters can be lowered by the decomposition of organic matter by microorganisms, by the chemical oxidation of material, or by aquatic vegetation that uses more oxygen at night than it produces.

The two parameters most commonly used to describe the oxygen demand of pollutants are the five-day biological oxygen demand (BOD5) and chemical oxygen demand (COD). Of the two, COD is more accurate for the purpose of comparing the oxygen demand of storm water discharges to the oxygen demand of other types of discharges. The BOD5 test underestimates the true oxygen demand of storm water because heavy metals in the storm water slow the bacterial action used in the test.

The impacts of oxygen-demanding pollutants can be more dramatic in shallow, slow-moving waters due to limited aeration and the tendency of these pollutants to accumulate in bottom sediments.

Pathogens   Pathogens are disease-causing organisms including viruses and some bacteria. Water-borne pathogens can be transmitted to man or animals through direct contact recreation, drinking water supplies or through eating contaminated shellfish. Pathogens may enter separate storm sewers from leaking sanitary sewers, cross-connections with sanitary sewers, malfunctioning septic tanks and animal wastes.

Toxicity (Metals, Toxic Organics, Pesticides, Inorganics, and Oil and Grease)   A wide range of chemicals can exhibit toxicity, including metals, organics, pesticides, inorganic pollutants, and oil and grease. Toxic impacts can be classified in terms of acute and chronic effects.

Many of the toxic metals and other toxic constituents in storm water discharges are attached to suspended solids in the discharge, and may accumulate in the bottom sediments of receiving waters where they may persist for long periods of time. Toxics concentrated in bottom sediments can cause adverse impacts on benthic organisms, may become resuspended during high flows resulting from other large storm events, or may dissolve into the water as parameters such as Ph and dissolved oxygen change. Accumulated pollutants in bottom sediments may also adversely affect fish during periods of continuous low flow.

Nutrients   Excessive nutrients over-stimulate the growth of aquatic plants which may result in low oxygen levels, accelerated eutrophication, unsightly conditions, interference with navigation, interference with treatment processes, and disagreeable tastes and odors. Eutrophic conditions are evidenced by surface algal scums, reduced water clarity, odors, and dense algal growth on shallow water substrates. Algal blooms block light from submerged aquatic vegetation which may remove habitat for juvenile fish and shellfish. After blooms or at the end of a growing season, the decomposition of dead vegetation may result in reduced oxygen levels, leading to fish kills and mass mortality of benthic organisms.

Excessive nutrients usually have more adverse affects in surface water bodies with slow flushing rates, such as slow moving rivers, lakes and estuaries. Nutrients delivered during storm events may settle out, and later be solubilized or resuspended.

Temperature Increased temperature can have detrimental effects on fish and other aquatic life during various stages of their life cycle. Water holds less oxygen as it gets warmer, which may affect habitat and make the water more susceptible to oxygen-demanding pollutants.

Floatables, Plastics Large amounts of litter and plastics are flushed from storm sewers. Litter and other floatables can degrade aesthetic quality and diminish the attractiveness of receiving waters and lower the value of adjacent property. Litter can impact the operating effectiveness of the drainage system and related facilities such as detention ponds. Economic losses caused by the aesthetic degradation of recreational areas such as beaches have been found to be significant. Plastic debris present hazards to wildlife.

2. Assessing Impacts

Assessing the impact (if any) of discharges from separate storm sewers on receiving waters is difficult for a number of reasons, including:

(a) Water quality impacts may be caused by a combination of the quality and quantity of discharges, or a combination of sources.
(b) Water quality impacts are site-specific.
(c) Water quality impacts may occur considerable distances downstream from discharge locations.
(d) Water quality standards may be difficult to relate to actual impacts.

Impacts: Quality and Quantity Impacts on receiving waters associated with storm water discharges can be discussed in terms of three general classes: 1) short-term changes in water quality; 2) long-term water quality impacts; and 3) physical impacts. Use impairment of receiving waters often is caused by a combination of all three types of impacts. Physical impacts and short-term water quality changes are generally more critical than long-term water quality impacts for receiving waters with relatively short residence times (such as smaller streams and rivers). Receiving waters with long residence times (lakes, estuaries) are generally more sensitive to long-term water quality changes, although certain physical changes, such as loss of reservoir capacity due to siltation, can be important.

Short-term changes in water quality occur during and shortly after storm events. Examples include periodic dissolved oxygen depressions due to oxidation of pollutants, short-term increases in the receiving water concentrations of one or more toxic pollutants, high bacteria levels, and peak acidity. These conditions can result in fish kills, loss of submerged macrophytes, and other temporary impairment of uses.

Long-term water quality impacts are caused by the cumulative effects associated with repeated storm water discharges. These impacts often result from pollutants from a number of different types of sources. Examples of the long-term water quality impacts that runoff can cause or contribute to include: depressed dissolved oxygen caused by the oxygen demanding pollutants in bottom sediments; biological accumulation of toxics as a result of uptake by organisms in the food chain; chronic toxicity to organisms subject to repeated exposures of toxic pollutants; destruction of benthic habitat; loss of storage capacity in receiving waters; and increased lake eutrophication. Long-term water quality impacts also include impacts caused by pollutants attached to suspended solids that settle in receiving waters and by nutrients that enter receiving water systems with long retention times. In both cases, water quality impacts are related to the increased residence times of pollutants in receiving waters. Long-term water quality impacts of pollutants from storm water discharges may be manifested during critical periods other than during storm events, such as during low stream flow conditions, or during sensitive life cycle stages of organisms. When evaluating long-term impacts, the cumulative and relative effects of seasonal and long-term pollutant loadings from all relevant sources (e.g., storm water, sewage treatment plants, industrial discharges, nonpoint sources, atmospheric deposition, in-place pollutants, etc.) should be considered.

Physical impacts due to erosional effects of high stream velocities that occur after the natural hydraulic cycle has been altered can deteriorate fish habitat. These changes are often accompanied by the installation of engineered structures such as concrete walls or underground culverts which may further degrade the habitat and aesthetic values of the receiving water. In addition, where ground water recharge has been limited by the placement of impervious structures on the land, dry weather base flows may be lowered to the detriment of the receiving water.

The Bellevue, Washington NURP project provides an example where water quality impacts were caused by a combination of factors. The project involved the study of a stream that received runoff from residential lands. The pollutant concentrations in the runoff to the stream had significantly lower concentrations than the average NURP site. Runoff to the stream did not appear to cause short term acute toxicity problems. However, massive fish kills were observed on several occasions during the study and were attributed to the illegal dumping of toxic materials into storm drains. In addition, the study recognized that potential long term problems may be associated with settleable solids, lead, and zinc which may have silted up spawning beds and introduced high concentrations of potentially toxic materials directly to the sediments. Up to three-fourths of the fish in the receiving stream had damaged gills and respiratory anomalies. Further, benthic organisms, which are sensitive indicators to environmental degradation, were rarely found in the receiving stream.

Site Specific Considerations   Factors to consider when evaluating the pollutants in discharges from separate storm sewers include:

(a) Size of area served by separate storm sewers.
(b) Rainfall patterns and amounts and seasonal effects.
(c) Potential for non-storm water discharges to the separate storm sewer system.
(d) Receiving water characteristics.
(e) Interaction with other pollutant sources such as discharges from sewage treatment plants and combined sewer overflows.
(f) Potential for development and/or renovation.
(g) Potential for pollutants in storm water associated with industrial activity.

IV. REFERENCES

Abraham, C., Lyon, T.C. and Schulze, K.W. (1976). "Selection of a design storm for use with simulation models." *Nat. Symp. on Urban Hydrology, Hydraulics, and Sediment Control*, University of Kentucky, Lexington, KY.

Adams, B.J., and Howard, C.D.D. (1986). "Design storm pathology." *Can. Water Res. J.*, 11(3).

Alley, W.M., and Smith, P.E. (1982). "Distributed routing rainfall runoff model—Version II." *U.S. Geological Survey Open File Report 82-344*, Reston, VA.

American Society of Civil Engineers. (1983). *Annotated bibliography of design storms.* Urban Water Resources Research Council, ASCE, New York, NY.

Bedient, P.B., and Huber, W.C. (1988). *Hydrology and floodplain analysis.* Addison-Wesley, New York, NY.

Bouwer, H. (1966). "Rapid field measurement of air-entry value and hydraulic conductivity of soil as significant parameters in flow system analysis." *Water Resources Research*, 2, 729–738.

Brakensiek, D.L., and Onstad, C. (1977). "Parameter estimation of the Green and Ampt infilitration equation." *Water Resources Research*, 13(6), 1009–1012.

Chow, V.T., Maidment, D.R., and Mays, L.R. (1988). *Applied hydrology*, McGraw-Hill, New York, NY.

Clark, C.O. (1945). "Storage and the unit hydrograph," *Trans. of the American Society of Civil Engineers*, ASCE, 110, 1419–1446.

Crawford, N.H., and Linsley, R.K. (1966). "Digital simulation in hydrology: Stanford watershed model IV." *Technical Report No. 39*, Civil Engineering Department, Stanford University, Palo Alto, CA.

Driscoll, E.D. (1986). "Log-normality of point and non-point source pollutant concentrations." *Proc. of the stormwater and water quality model users group mtg.*, EPA-600/9-86/023 (NTIS PB 87-117438/AS), U.S. Environmental Protection Agency, Athens, GA.

Driver, N.E., and Tasker, G.D. (1988). "Techniques for estimation of storm-runoff loads, volumes, and selected constituent concentrations in urban watersheds in the United States." *Open-File Report 88-191*, U.S. Geological Survey, Denver, CO.

Eagleson, P. (1970). *Dynamic hydrology*. McGraw-Hill, New York, NY.

Engman, E.T. (1986). "Roughness coefficients for routing surface runoff." *Journal of the Irrigation and Drainage Division*, ASCE, 112(12).

Espey, W.H., and Altman, D.G. (1978). "Nomographs for ten-minute unit hydrographs for small urban watersheds." *Addendum 3 of Urban Runoff Control Planning, Report EPA-600/9-78-035*, U.S. Environmental Protection Agency, Washington, D.C.

Frederick, R.H., Myers, V.A., and Auciello, E.P. (1977). "Five-to-60-minute precipitation frequency for the eastern and central United States." *NOAA Technical Memorandum NWS HYDRO-34*, U.S. National Weather Service, Silver Springs, MD.

Green, W.H., and Ampt, G.A. (1911). "Studies on soil physics, Part I.—The flow of air and water through soils." *J. Agric. Science*, 4(1).

Harremöes, P., ed. (1983). "Rainfall as the basis for urban runoff design and analysis." *Proc. of the Seminar*, Pergamon Press, New York, NY.

Hershfield, D.M. (1961). "Rainfall frequency atlas of the United States for durations from 30 minutes to 24 hours and return periods of 1 to 100 years." *Technical Paper No., 40*, U.S. National Weather Service, Silver Springs, MD.

Hershfield, D.M. (1962). "Extreme rainfall relationships." *Journal of the Hydraulics Division*, 88(HY6).

Hicks, W.I. (1944). "A method of computing urban runoff." *Transactions of the American Society of Civil Engineers*, 109, ASCE, 1217.

Hjelmfelt, A.J., Jr., and Cassidy, J.J. (1975). *Hydrology for engineers and planners*. Iowa State University Press, Ames, IA.

Hogg, W.D. (1980). "Time distribution of short storm rainfall in Canada." *Proc. of the Canadian Hydrology Symp. '80*, National Research Council, Ottawa, Canada, 52–63.

Horton, R.E. (1919). "Rainfall interception." *Monthly Weather Review*, 47, 603–623.

Horton, R.E. (1939). "Analysis of runoff plot experiments with varying infiltration capacity," *Trans., American Geophysical Union, Part IV*, 693–694, Washington, D.C.

Horton, R.E. (1940). "An approach toward a physical interpretation of infiltration capacity." *Proc., Soil Science Society of America*, 5, 399–417.

Huber, W.C., and Dickinson, R.E. (1988). Storm water management model, version 4, User's manual. EPA-600/3-88/001a (*NTIS PB88-236641/AS*), U.S. Environmental Protection Agency, Athens, GA.

Huff, F.A. (1967). "Time distribution of rainfall in heavy storms," *Water Resources Research*, 3(4), 1007–1019.

Hydrologic Engineering Center. (1985). *HEC-1 flood hydrograph package: User's manual*, U.S. Army Corps of Engineers, Davis, CA.

Hydrologic Engineering Center. (1979). "Introduction and application of kinematic wave routing techniques using HEC-1." *Training Document No. 10*, U.S. Army Corps of Engineers, Davis, CA.

James, W., and Drake, J.J. (1980). "Kinematic design storms incorporating spatial and time averaging." *Proc. of the Stormwater Management Model Users Group Mtg.*, EPA-600/9-80/064 (NTIS PB 81-173858), U.S. Environmental Protection Agency, Athens, GA.

James, W., and Shtifter, Z. (1981). "Implications of storm dynamics on design storm inputs." *Proc. of the Stormwater and Water Quality Model Users Group Mtg.*, USEPA and the Ontario Ministry of the Environment, Department of Civil Engineering, McMaster University, Hamilton, Ontario, Canada.

Kiefer, C.J., and Chu, H.H. (1957). "Synthetic storm pattern for drainage design." *Journal of the Hydraulics Division*, 83(HY4).

Linsley, R.K., Jr., Kohler, M.A., and Paulhus, J.L.H. (1982). *Hydrology for engineers*. 3rd ed., McGraw-Hill, New York, NY.

Marsalek, J. (1978). "Research on the design storm concept." *Technical Memo 33*, ASCE, New York, NY.

Marsalek, J., and Watt, W.E. (1984). "Design storms for urban drainage design." *Canadian Journal of Civil Engineering*, 11, 574–584, Ottawa, Canada.

McCuen, R.H. (1989). *Hydrologic analysis and design*. Prentice-Hall, Englewood Cliffs, NJ.

McEnroe, B.M. (1986). "Characteristics of intense storms in Kansas." *World Water Issues in Evolution*, ASCE, New York, NY, 933–940.

McPherson, M.B. (1969). "Some notes on the rational method of storm drain design." *ASCE Urban Water Resources Research Program Technical Memorandum No. 6 (NTIS PB-184701)*, ASCE, New York, NY.

McPherson, M.B. (1978). "Urban runoff control planning." *EPA-600/9-78-035*, U.S. Environmental Protection Agency, Washington, D.C.

Mein, R.C., and Larsen, C.L. (1973). "Modeling infiltration during steady rain." *Water Resources Research*, 9(2).

Miller, J.F., Frederick, R.H., and Tracey, R.J. (1973). *Precipitation frequency atlas of the western United States*. NOAA Atlas 2, U.S. National Weather Service, Silver Springs, MD.

Packman, J.C., and Kidd, C.H.R. (1980). "A logical approach to the design storm concept." *Water Resources Research*, 16(6).

Ponce, V.M. (1989). *Engineering hydrology: Principles and practices*. Prentice-Hall, Englewood Cliffs, NJ.

Rawls, W.J., Brakensiek, D.L., and Miller, N. (1983). "Green-ampt infiltration parameters from soils data." *Journal of Hydraulic Engineering*, 109(1), 1316–1320.

Rovey, E.W., and Woolhiser, D.A. (1977). "Urban storm runoff model," *Journal of the Hydraulics Division*, 103(HY11).

Schueler, T.R. (1987). *Controlling urban runoff: A practical manual for planning and designing urban BMPs.* Metropolitan Washington Council of Governments, Washington, D.C.

Snyder, F.F. (1938). "Synthetic unit graphs." *Trans., American Geophysical Union,* 19, 447–454.

Tasker, G.D., and Driver, N.E. (1988). "Nationwide regression models for predicting urban runoff water quality at unmonitored sites." *Water Resources Bulletin,* 24(5).

Terstriep, M.L., and Stall, J.B. (1974). "The Illinois urban drainage area simulator, ILLUDAS." *Bulletin 58,* Illinois State Water Survey, Urbana, IL.

Tholin, A.L., and Keifer, C.J. (1960). "The hydrology of urban runoff." *Transactions, ASCE,* 125, 1308.

Urban Drainage and Flood Control District. (1984). *Urban storm drainage criteria manual.* rev. ed., Denver Regional Council of Governments, Denver, CO.

U.S. Environmental Protection Agency. (1983). "Results of the nationwide urban runoff program (NURP)." vol. 1, (*NTIS PB 84-185552*), Water Planning Division, Washington, D.C.

U.S. Environmental Protection Agency. (1988). *National water quality inventory, 1988 Report to Congress.* Washington, D.C.

U.S. Soil Conservation Service. (1972). *National engineering handbook.* Springfield, VA.

U.S. Soil Conservation Service. (1986). "Urban hydrology for small watersheds." *Technical Release 55, NTIS PB87-101580,* 2nd ed., U.S. Department of Agriculture, Springfield, VA.

U.S. Weather Bureau. (1958). *Key to meteorological records documents No. 3.081— Excessive precipitation techniques.* Washington, D.C.

Viessman, W., Jr., Lewis, G.L., and Knapp, J.W. (1989). *Introduction to hydrology.* 3rd ed., Harper and Row, New York, NY.

Voorhees, M.L., and Wenzel, H.G., Jr. (1989). "Urban design storm sensitivity and reliability." *Journal of Hydrology,* 68.

Watt, W.E. et al. (1985). "A 1-hour urban design storm for Canada." *Proc. of the Conf. on Stormwater and Water Quality Management Modeling,* E.M. James and W. James, eds., Computational Hydraulics, Inc., Guelph, Ontario, Canada.

Wenzel, H.G., Jr. (1982). "Rainfall for urban stormwater design." *Urban Stormwater Hydrology,* American Geophysical Union Water Resources Monograph No. 7, AGU, Washington, D.C.

Yen, B.C., and Chow, V.T. (1980). "Design hyetographs for small drainage structures." *Journal of the Hydraulics Division,* 106(HY6).

Chapter 6
STORM DRAINAGE HYDRAULICS

I. INTRODUCTION

The purpose of this chapter is to discuss the basic hydraulic principles inherent in the design of a storm drainage system (Figure 6.1). The basic elements covered include street inlets, storm sewers, natural and man-made channels, culverts, detention basins, and outlet structures. The primary design objective for each of these elements is usually to provide a discharge capacity sufficient to convey the design flow at velocities that are self-cleansing without being destructive. In the case of storm sewers and man-made channels, the designer may be asked to make additional calculations involving hydraulic grade line or water surface profile, total energy grade line, normal and critical depth, and location of hydraulic jumps, surcharge, or out-of-bank flows.

The design and analysis of drainage system facilities proceed according to fundamental principles of water movement in open channels, closed conduits, and other special hydraulic structures. This chapter treats water as an incompressible Newtonian fluid, and emphasizes the following concepts and principles: flow classification; mass conservation; momentum conservation; total and specific energy; friction losses and minor losses; hydraulic jump; water surface profiles; special hydraulic structures; and flood routing by both hydraulic and hydrologic methods. Pump hydraulics are not discussed here because of the ample treatment provided in standard hydraulic texts.

II. FLOW CLASSIFICATION

Identifying the type of flow at the outset of a particular design problem is essential because the design equations themselves are often developed for specific flow classes. Flow in a storm drainage system can be classified in a number of ways, depending on the particular

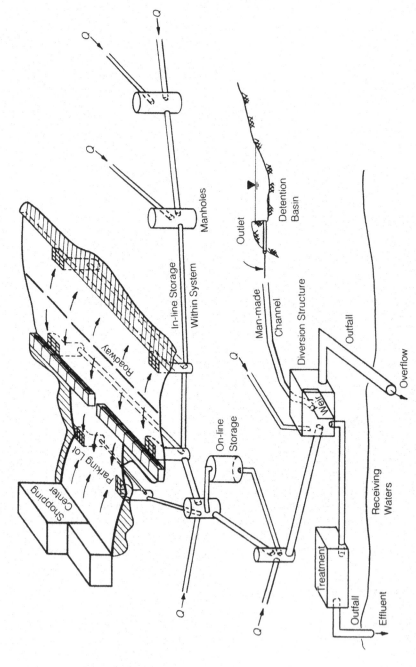

Figure 6.1—Principal hydraulic elements in urban storm drainage system.

circumstances. For example it might be classified as laminar vs. turbulent, steady vs. unsteady, uniform vs. nonuniform, gradually varied vs. rapidly varied, open channel vs. closed conduit, or free surface vs. pressure flow. Flow is thus classified according to certain properties that change with time and distance along the flow element. The brief outline below summarizes the traditional flow types encountered frequently in a storm sewer system:

A. Steady vs. Unsteady Flow

At a single location in a conduit or channel there may be changes in depth, velocity, or discharge that occur with time as the result of hydrograph inflows at an upstream inlet or perhaps tidal discharges at a lower location. Variations in depth, velocity, or discharge with time at a single point cause the flow to be unsteady. Steady flow on the other hand requires that discharge, depth, and velocity are constant with time, as in spring-fed baseflow in a small stream or interceptor. The flow is considered to be quasi-steady if depth, velocity, and discharge are changing very slowly over time, such as the sanitary flow in a combined sewer during dry weather. The flow in storm sewer systems will be unsteady during periods of rainfall-runoff. This unsteadiness is often ignored in the design of small municipal storm sewers and subdivision drainage systems based on peak flows only (note that it cannot be ignored in the design of complex networks involving looped sewers, diversion structures, and combined sewer overflow facilities).

B. Uniform vs. Nonuniform Flow

In uniform flow, depth, discharge, and velocity are constant with distance along the channel or conduit. Channel slope, energy or friction slope, and water surface slope are all equal. The depth of uniform flow is called the "normal" depth. Uniform flow can occur as non-pressure or free surface flow in a prismatic conduit flowing partially full. It also can occur in a storm sewer flowing full if minor losses due to contractions, expansions, and bends are negligible. Uniform flow equations such as the Manning or Hazen-Williams are still applicable, provided the slope of the energy grade line is used and the equations are applied in their proper flow ranges—Manning in the rough flow range and Hazen-Williams in the smooth flow range.

Nonuniform flow by contrast is characterized by changing depth and velocity with distance along the channel or sewer. Sometimes these depth and velocity changes take place over considerable lengths and the nonuniform flow is called gradually varied. Uniform flow equations are often applied to very short distance intervals within a gradually varied flow. Examples of gradually varied flow include sheet flow on paved surfaces, gutter flow, and flow in storm drains and floodways. Rapidly varied flow, on the other hand, produces abrupt changes in depth and velocity over very short distances, as in the case of flow over an emergency spillway, through a hydraulic jump, or beneath a sluice gate. Rapidly varied flow usually involves wave phenomena which preclude the use of uniform flow formulas. Nonuniform flow can also be unsteady, as in the passage of a runoff peak or flood wave through a storm sewer or man-made channel and in fact most storm drainage flow can be classified as unsteady and nonuniform during periods of heavy runoff.

C. Open Channel vs. Closed Conduit Flow

Open channel flow has a free surface as in a natural stream, swale, or man-made channel. The slope of the channel will be classified as hydraulically mild, critical, or steep depending on whether this slope is less than, equal to, or greater than the critical slope computed for the channel on the basis of its critical depth, discharge, and roughness. The slope could of course also be horizontal or adverse. A closed conduit may flow full or partially full, depending on whether the runoff event is larger or smaller than the design value. Most storm sewers will be sized to flow full at the design discharge, although in certain instances where ground elevations are sufficient, a limited surcharge above the pipe crown may be permitted. Full flow in a conduit is confined without a free surface and is sometimes referred to as pipe flow or pressure flow. Gravity forces still govern, but the additional pressure head of any surcharge above the pipe crown must be taken into account. Closed conduits flowing partially full are analyzed as open channels. A procedure for computing normal depth and backwater profiles in partially full circular storm sewers has been described by Christensen (1984).

D. Laminar vs. Turbulent Flow

The flow in a pipe or channel may also be classified according to the nature of the flow streamlines and velocity distribution within the flow section, depending on whether viscous or inertia forces predominate. Laminar flow occurs in water supply conduits where viscous forces are predominant and the Reynolds number is less than about 500 (computed with hydraulic radius). Laminar flow may develop in overland sheet flow in the early stages of runoff on the rising side of the hydrograph well below the peak discharge. Storm sewer and open channel flow on the other hand tends to be turbulent, at least during periods of peak flow, and will have Reynolds numbers (computed from hydraulic radius) exceeding 500. Within the turbulent regime there are sub-classes designated as the smooth range, the transition range, and the rough range based on the thickness of the viscous sub-layer compared to the roughness size. These sub-classes are important since certain uniform flow formulas, such as the Manning equation, are considered more appropriate in the rough range, while the Hazen-Williams and Colebrook-White equations are better suited for the smooth and transition ranges, respectively. In most design situations, the flow is turbulent, and the Manning equation can be used.

E. Subcritical vs. Supercritical Flow

The flow in open channels and closed conduits may also be classified according to the level of energy contained in the flow itself as represented by the Froude number. The subcritical range has Froude numbers less than 1.0 and is characterized by low velocities and high depths found typically on hydraulically mild slopes. Supercritical flow has a Froude number greater than 1.0 and is characterized by high velocities and low depths developed in a hydraulically steep channel or pipe. Critical flow occurs when the Froude number equals 1.0 and the actual depth is equal to the critical depth for the flow element. The classification of the flow according to subcritical, critical, or supercritical conditions is important for two reasons related to design of storm drainage facilities. First, the location of hydraulic jumps where the flow passes through an abrupt transition from supercritical to subcritical should always be determined by the designer so that the associated energy loss and depth increase can be accommodated in the designed system. Secondly, the location of a critical depth section in a channel or storm sewer is important because that section serves as a control from which water surface profile calculations can proceed. It also marks a point in the sewer or channel where a unique relation between depth and discharge exists and therefore constitutes an ideal flow monitoring location. More details about critical depth and water surface profile calculations are provided later in this chapter.

A systematic classification of flow types has been developed by Christensen (1985) and is shown in Figure 6.2. Having outlined all the principal flow types, it is fair to state that stormwater runoff usually will be classified as: unsteady, nonuniform, closed conduit flowing partially full, turbulent, and subcritical. Many departures from this general rule can of course occur. The analysis of complex unsteady, nonuniform flow problems sometimes can be accomplished satisfactorily by using less rigorous methods based on steady, uniform flow approximations. The purpose of the remaining sections in this chapter is to define fundamental flow relations and to provide examples illustrating the application of these principles to typical problems of storm drainage design and analysis.

III. CONSERVATION OF MASS

The principle of mass conservation states that the difference between the mass inflow and mass outflow rates over a time interval must equal the change in mass storage over the same time interval. Since stormwater is treated as incompressible, the mass rates are expressed as

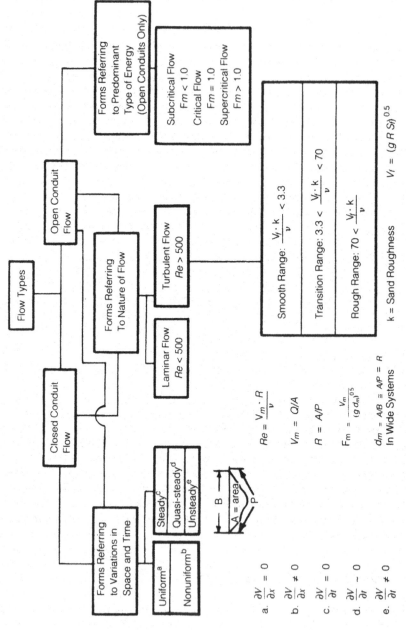

Figure 6.2—Schematic classification of flow types.

volume rates over time. The general mass conservation equation, otherwise known as the continuity equation, is in any consistent unit system:

$$\frac{\Delta S}{\Delta t} = \Sigma \bar{I} - \Sigma \bar{O} \tag{6-1}$$

in which ΔS is storage change, Δt is the time interval, $\bar{I}$ is the average rate of inflow over the time interval, and $\bar{O}$ is the average outflow rate. For the system in Figure 6.3, the continuity equation written for the control volume represented by the manhole, would be:

$$\frac{\Delta S}{\Delta t} = \frac{A_s \times \Delta H}{t_1 - t_0} = Q_A + Q_U + Q_S + Q_D \tag{6-2}$$

where A_s is the cross-sectional area of the manhole and H is the depth change over the time interval $t_1 - {}_0$. If ΔS were zero, as it would be under steady flow, then the continuity equation for the system reduces to: $Q_D = Q_S + Q_A + Q_U$. Thus the continuity equation requires that the algebraic sum of inflows to a junction is equal to the sum of outflows from that junction when there is no storage change. Similarly, the continuity equation in pipe A assuming full flow is: $Q_A = A_1 V_1 = A_2 V_2$, where the subscripts 1 and 2 refer to two sections normal to the flow, A is the cross-sectional area, and V is mean velocity in the cross-section. Again, Figure 6.3 illustrates this situation.

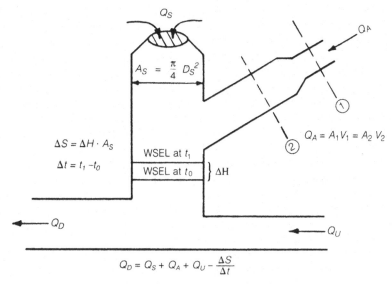

Figure 6.3—Continuity principle applied to unsteady and non-uniform flow.

During flood conditions in an open channel or storm sewer, flow is both unsteady and nonuniform, and the discharge into any control volume will not equal the discharge out of the control volume. If Q is replaced with its equivalent, AV, and the time change of storage within the control volume is taken into account, the partial differential form of the equation of continuity becomes (without loss or gain of water along the length of the control volume):

$$V\frac{\partial A}{\partial X} + A\frac{\partial V}{\partial X} + b\frac{\partial y}{\partial t} = 0 \qquad (6\text{-}3)$$

in which B is the water surface width, Y is the depth of water, X is in the direction of flow, and t is time. Figure 6.4 gives a definition sketch for the control volume under nonuniform unsteady flow in an open channel.

Users of computer models that simulate the routing of stormwater flows through conduits must be careful to confirm that continuity is being maintained by the routing procedure. Approximations have to be made in any dynamic routing algorithm, and under extreme conditions the model may over or underestimate routed flow. Depending on computational structure, some models may be prone to numerical

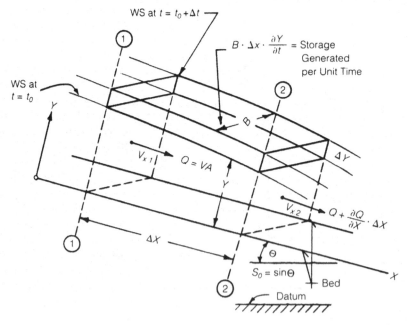

Figure 6.4—Continuity equation for non-uniform and unsteady flow in an open channel.

truncation or round off error. A check of cumulative inflow volume versus cumulative outflow volume will provide a rough check for the presence of continuity errors and indicate the significance of numerical truncation errors.

IV. CONSERVATION OF MOMENTUM

The momentum conservation equation is derived from Newton's second law which states that the time rate of change in linear momentum of a fluid mass equals the sum of external forces acting on that mass. For steady, uniform flow there are no velocity changes with time or distance and hence the net sum of external forces must be zero. For the nonuniform flow shown in Figure 6.4 the momentum conservation principle requires that the sum of external forces acting on the control volume in the X-direction between sections 1 and 2 be equal to the momentum flux or time rate of change in linear momentum, also in the X-direction. For the channel shown in Figure 6.4, the forces in the X-direction include gravity, friction, and hydrostatic pressure imbalance. Other forces due to contraction/expansion, wind shear, and rainfall momentum transfer can also be included depending on the particular application. For the case of steady, nonuniform flow in the open channel element of Figure 6.4, the momentum equation is shown in (6-4):

$$\Sigma F_x = \frac{\gamma Q}{g} (V_{x2} - V_{x1}) \qquad (6\text{-}4)$$

in which ΣF_x is the summation of the X components of the external forces acting on the fluid body, γ is the specific weight, Q is the discharge, V_x is the spatial mean velocity in the x-direction at cross-sections 1 and 2, and g is the acceleration due to gravity.

This form of the momentum equation is adequate for steady flow in storm sewers and man-made channels which have simple prismatic cross-sections. The assumption of the velocity being approximately equal to the mean velocity across the channel is usually sufficiently accurate. However, in natural channels with flood plains and in complex man-made channels, this assumption can be inaccurate. In these cases, a momentum correction factor, β, must be applied to each of the velocity terms in Equation (6-4). The equation is then written as:

$$\Sigma F_x = \frac{\gamma Q}{g} (\beta_2 V_{x2} - \beta_1 V_{x1}) \qquad (6\text{-}5)$$

where

$$\beta = \frac{\int V^2 dA}{V_m^2 A}$$

Under actual rainfall-runoff conditions, the flow in man-made channels and storm drains is unsteady and nonuniform. Referring to Figure 6.4, the force balance equation becomes:

$$\Sigma F_x = \gamma A S_o X - A S_f \Delta X - A \left(\frac{\partial Y}{\partial X}\right) \Delta X \qquad (6\text{-}5a)$$

where S_o is the channel or pipe slope defined as $\sin\theta$, and S_f is the energy or friction slope. The terms on the right hand side of equation (6-5a) represent weight, friction, and hydrostatic pressure forces, respectively acting in the X-direction. Letting $\Delta X - > \partial X$, $\beta = 1.0$, and setting ΣF_x in equation (6-5a) equal to the rate of change of momentum in the main fluid mass entering from upstream in the element shown in Figure 6.4, the momentum equation can be written in simplified form:

$$\frac{\partial V}{\partial V} + V\left(\frac{\partial V}{\partial X}\right) + g\left(\frac{\partial Y}{\partial X}\right) = g(S_o - S_f) \qquad (6\text{-}5b)$$

where forces due to local losses, wind shear, and lateral inflows have been neglected. Equations (6-3) and (6-5b), together, form the Saint-Venant or gradually varied unsteady flow equations, after Barre de Saint-Venant who first developed them for unsteady flow in an open channel in 1871. It should be noted that equation 6-5b can be rewritten in terms of the major flow classes discussed earlier in this chapter as (Henderson 1966):

$$S_f = S_o - \left(\frac{\partial Y}{\partial X}\right) - \frac{V}{g}\left(\frac{\partial V}{\partial X}\right) - \frac{1}{g}\left(\frac{\partial V}{\partial t}\right) \qquad (6\text{-}5c)$$

steady
uniform flow

steady nonuniform flow

unsteady nonuniform flow

Equations (6-3) and (6-5b) are called the nonconservation form of the gradually varied unsteady flow equations because they are written in terms of velocity, V. For reasons of numerical stability, the conservation form, written in terms of discharge, Q, is often used (as in the extended transport hydraulic routing routine (EXTRAN) in the EPA SWMM program). Numerical solution of the full dynamic or gradually varied unsteady flow equation is important when analyzing complex sewer networks where pipe looping and various diversion/outfall structures can create significant backwater effects. These act to reduce the hydraulic capacity of sewer elements leading to increased flooding and surcharge in heavily developed urban watersheds.

The solution of the full equations (6-3) and (6-5b) and the kinematic approximation for sewer and channel routing is discussed in a later section of this chapter. It is important to note that equation (6-5b) can also be developed from energy conservation principles, as well as from Newton's second law as shown above. The energy equation is introduced below because of its importance in developing the energy grade line and water surface profiles in steady flow sewer/channel problems.

V. CONSERVATION OF ENERGY

The total energy in a moving fluid at any point is the sum of potential energy, pressure energy, and kinetic energy. The energy conservation principle states that, in an ideal fluid without external energy sources and sinks, this total energy sum does not change along the flow element, but that only its distribution among the individual energy components changes. Thus, total fluid energy, although transformed, is always conserved. In the field of applied hydraulics, we must account not only for these principal energy components, but also for other energy gains and losses by boundary friction, form losses, pumps, turbines, hydraulic jumps, and energy dissipators. The total energy at a point on a streamline in Figure 6.5 (a) or (b) is equal to:

$$H = z + y + \frac{p}{\gamma} + \frac{V^2}{2g} \tag{6-6}$$

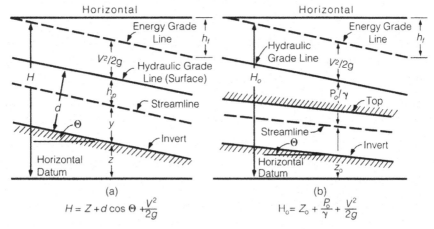

Figure 6.5—Definition of terms for total energy in an open channel and closed conduit.

in which H is the total energy, z is the difference between the elevation of a point and some arbitrary horizontal datum (such as mean sea level), y is measured normal to the invert and is the difference in elevation between this point and the elevation of some streamline, p is the pressure on the streamline, γ is the fluid unit weight, v is the velocity of the streamline, and g is the acceleration due to gravity. The invert slope, θ, has been neglected in equation 6-6 ($\cos \theta$ is assumed to equal unity).

Since H is energy per unit weight, it has units of length and is often referred to as the total head or total energy in whatever length unit is used. The term p/γ is the pressure energy per unit weight and is known as the pressure head, h_p. Likewise, $V^2/2g$ is the kinetic energy per unit weight and is known as the velocity head, h_v.

Since z and y can be referenced to any streamline, it is sometimes convenient in closed conduit flow to let z be the difference in elevation between the centerline of the conduit and the elevation of the arbitrary datum and let y be zero. Therefore, for pressure flow, equation 6-6 becomes:

$$H = z + h_p + \frac{V^2}{2g} \qquad (6\text{-}7)$$

where z is now the vertical distance to the conduit centerline.

For the open channel in Figure 6.5 (a) it is convenient to let z be the difference between the elevation of the channel invert and the elevation of the arbitrary datum, and to let d be the depth. The term p/γ is then zero. Therefore, for open channel flow, equation 6.6 becomes:

$$H = z + d\cos\theta + \left(\frac{V^2}{2g}\right) \qquad (6\text{-}8)$$

for a plane section normal to the streamlines and with hydrostatic pressure distribution. The velocity V is the spatial average velocity in the section obtained as Q/A. As indicated above, the depth of flow is always measured normal to the channel bed and equals $d \cos\theta$ for the sketch in Figure 6.5 (a).

As in the earlier discussion for momentum, equation (6-8) applies to steady flow in storm sewers and man-made channels that have simple prismatic cross-sections and where any local velocity is approximately equal to the spatial average velocity in that section. In natural channels with floodplains and in complex man-made channels, the velocity head based on the average velocity V must be multiplied by a velocity distribution factor, α, to account for the variation in velocity across the cross-section. Equation 6-8 then becomes:

$$H = z + d\cos\theta + \alpha \left(\frac{V^2}{2g}\right) \qquad (6\text{-}9)$$

Referring to Figure 6.5, the energy equation can now be written as:

$$z_1 + d_1 \cos\theta + \alpha \left(\frac{V_1^2}{2g}\right) = z_2 + d_2 \cos\theta + \alpha \left(\frac{V^2}{2g}\right) + H_L \qquad (6\text{-}10)$$

in which H_L is the sum of all the head losses between two sections 1 and 2 normal to the flow. These losses include major and minor losses. The major loss is that due to the friction of the water against the sides of the channel or pipe. The minor losses are called form losses and can include entrance, contraction, expansion, junction, exit, bend, and man-hole losses. Major and minor losses will be discussed in more detail later.

A. Hydraulic and Energy Grade Lines

Two very useful concepts in flow analysis are the hydraulic and energy grade lines. These are shown in Figure 6.5 (a) and 6.5 (b) for open channel and pressure flow, respectively. For the case of pressure flow, the hydraulic grade line (HGL) is the piezometric surface, i.e. the height to which water will rise in a piezometer. It is often referred to as the piezometric head line (PHL). The energy grade line (EGL) is the line showing the total energy of the flow above some arbitrary horizontal datum. The slope of the EGL is called the energy slope or the friction slope and is designated S_f. The vertical difference between the HGL and the EGL is the velocity head.

For open channel flow, the HGL is equal to the water surface. As shown in Figure 6.5 (a), the sum of y and p/γ, the potential energy above the channel invert, is equal to d $\cos\theta$. If the slope of the channel invert is small, so that $\cos\theta$ can be assumed to be 1.00, then the potential energy above the invert can be assumed to be equal to the vertical depth of flow. This assumption is made for the remainder of this chapter, unless noted otherwise. Further, the velocity distribution coefficient is assumed equal to 1.00, unless otherwise noted.

B. Specific Energy

When the arbitrary horizontal datum is taken as the channel or pipe invert for open channel flow, Equation 6-8 becomes:

$$E = d + \left(\frac{V^2}{2g}\right) \qquad (6\text{-}11)$$

in which E is known as the specific energy per unit weight of fluid, usually expressed in feet. The specific energy curve giving E as a function of depth, d, has many uses and is constructed as shown in Example 6-1.

Example 6-1:   Determine the specific energy curve for a rectangular channel 10 feet wide conveying a flow of 250 cfs. The calculations are shown in Table 6.1. Columns 1 and 5 are plotted to form a curve similar to that in Figure 6.6.

The depth corresponding to minimum specific energy is defined as critical depth, d_c. The region of flow above critical depth is known as the subcritical flow or tranquil zone. The region below critical depth is known as the supercritical or shooting flow zone. A vertical line through any specific energy greater than minimum specific energy will cut the curve at two locations, at a depth less than critical depth and at a depth greater than critical depth. These two depths are known as alternate depths.

One other aspect of the specific energy is worth noting. The nose of the curve near critical depth can be either somewhat pointed or flat depending on the cross-sectional shape. If it tends to be flat for a given situation, then the water surface will be unstable when depths are within 10–15% of critical depth because a slight change in energy may impart large fluctuations in either of the alternate depths.

C. Froude Number

When flow is at critical depth, specific energy equals E_{min}, the depth equals d_c, and the Froude number, F, is equal to one. If F is less than

TABLE 6.1.   Example Calculation of Specific Energy in a Rectangular channel, width = 10 feet, Q = 250 cfs and α = 1.

Depth (ft.) (1)	Area (sq.ft.) (2)	Velocity (fps) (3)	$V^2/2g$ (ft.) (4)	E (ft.) (5)
0.0	0.0	—	—	—
0.5	5.0	50.00	38.87	39.37
1.0	10.0	25.00	9.72	10.72
1.5	15.0	16.67	4.32	5.82
2.0	20.0	12.50	2.43	4.43
2.5	25.0	10.00	1.55	4.05
2.69 (d_c)	-----	------	------(E_{min})	4.03
3.0	30.0	8.33	1.08	4.08
3.5	35.0	7.14	0.79	4.29
4.0	40.0	6.25	0.61	4.61
4.5	45.0	5.56	0.48	4.98
5.0	50.0	5.00	0.39	5.39
5.5	55.0	4.54	0.32	5.82
6.0	60.0	4.17	0.27	6.27
6.5	65.0	3.85	0.23	6.73
7.0	70.0	3.57	0.20	7.20
7.5	75.0	3.33	0.17	7.67
8.0	80.0	3.12	0.15	8.15

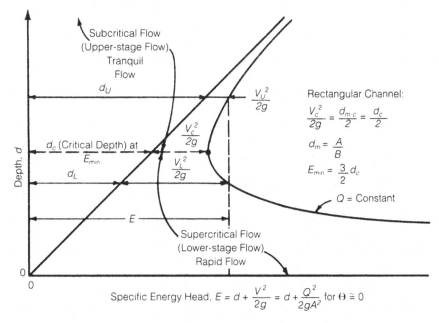

Figure 6.6—Specific energy curve for rectangular channels.

one, then the flow is subcritical. If F is greater than one, the flow is supercritical. The Froude number is defined as:

$$F = \frac{V}{(gd_m)^{1/2}} \qquad (6\text{-}12)$$

in which V is the spatial average velocity in the cross-section, g is the acceleration due to gravity, and d_m is the hydraulic mean depth, determined from:

$$d_m = \frac{A}{T} \qquad (6\text{-}13)$$

in which A is the cross-sectional area and T is the top width of flow.

The primary significance of the Froude number in urban drainage design is that it represents the ratio of the average flow velocity to the propagation velocity of a small gravity wave and thus determines the direction of water surface profile computations. A gravity wave will be propagated in both the upstream and downstream directions in subcritical flow since the wave velocity is greater than the flow velocity, but can only be propagated downstream in supercritical flow. Consequently, water surface profile computations always begin at a control section and proceed upstream when the control depth is greater than d_c and downstream when the control depth is less than d_c. The Froude number also is used to detect the occurrence of a hydraulic jump forming in a channel or conduit.

D. Critical Depth

As noted above, critical depth is defined as the depth at the point of minimum specific energy for constant discharge. Critical depth is a function of discharge, size of channel, and shape of channel. Its value can be calculated in one of three ways: (1) Equation 6-14 for an irregular channel of any cross-section; (2) from tables contained in French (1985) for rectangular and trapezoidal channels; and (3) from nomographs contained in FHWA (1981). Equation 6-14 is a general expression for critical depth in a channel of any cross-sectional shape:

$$\alpha \left(\frac{Q^2}{g} \right) = \frac{A^3}{T} \tag{6-14}$$

in which Q is the discharge, g is the acceleration due to gravity, A is the cross-sectional area of flow, and T is the top width of flow.

Example 6-2:   Use the above three methods to calculate critical depth for the rectangular channel described in Example 6-1, with $Q = 250$ cfs, $b = 10$ feet, and $= 1$.

Method 1: Equation (6-14)

$$\frac{Q^2}{g} = \frac{A^3}{T} = \frac{b^3 d_c^3}{b} = b^2 d_c^3$$

$$(250)^2/32.15 = (10)^2 \, (d_c)^3$$

$$d_c = (62,500/3215)^{1/3}$$

$$= 2.69 \text{ feet}$$

Alternatively, for a rectangular channel, Equation 6-14 reduces to:

$$d_c = \left(\frac{g^2}{g} \right)^{1/3} \tag{6-15}$$

where q = discharge per unit width in cfs/ft. Solving (6-15) yields d_c = 2.69 feet.

Method 2: Table of K_c' values for trapezoidal channels.

$$K'_c = \frac{Q}{b^{5/2}}$$

$$= 250/316 = 0.790$$

Table 6.2 is taken from Brater and King (1976), and for a rectangular channel, the value of $K_c' = 0.790$ lies between d_c/b equal to 0.26 and 0.27. Linear interpolation produces:

$$\frac{d_c}{b} = 0.269$$

$$d_c = 0.269 \times 10 = 2.69 \text{ feet}$$

Method 3: Nomograph (FHWA, 1961)
Figure 6.7 is a nomograph from FHWA (1961), which includes other nomographs for circular and trapezoidal channels. From Figure 6.7, for a discharge of 250 cfs, critical depth is equal to 2.7 feet. Figure 6.7 also can be used to determine the normal depth as noted in the following section.

Figure 6.8 is a nomograph for critical depth and velocity in circular conduits. The nomograph solutions in Figures 6.7 and 6.8 are presented to illustrate the ease of graphical solution to problems of computing critical and normal depth. Programmable calculators and microcomputers can also be used to obtain iterative numerical solutions. Refer to Croley (1979) and Smith (1986) for details of computational programs pertinent to storm drainage hydraulics.

VI. NORMAL DEPTH

Normal depth is defined simply as the depth of uniform flow under constant discharge. Recall that in uniform flow the losses due to boundary friction are just balanced by the gravity component in the direction of flow. In other words, friction and gravity forces in the direction of flow are equal but act in opposite directions. At normal depth the slope of the invert, the slope of the HGL, and the slope of the EGL are numerically equal and parallel to each other.

Normal depth is a function of discharge, size of channel, shape of channel, slope of channel, and frictional resistance to flow. Its value can be calculated in one of three ways: (1) by the Manning equation (6-16) and Equation (6-17); (2) by tables contained in French (1985); and (3) by nomographs contained in FHWA (1961). Equation (6-16) is Manning's Equation:

$$V = \frac{1.49}{n} R^{2/3} S_f^{1/2} \tag{6-16a}$$

TABLE 6.2. Values of K_c' in the Formula $Q = K_c' b^{5/2}$ for Critical Depth in Trapezoidal Channels (Brater and King 1976).

$\dfrac{D_c}{b}$	Side slopes of channel, ratio of horizontal to vertical									
	Vertical	¼–1	½–1	¾–1	1–1	1½–1	2–1	2½–1	3–1	4–1
.01	.0057	.0057	.0057	.0057	.0057	.0057	.0057	.0057	.0058	.0058
.02	.0160	.0161	.0161	.0162	.0162	.0163	.0164	.0165	.0165	.0167
.03	.0295	.0296	.0297	.0298	.0299	.0302	.0304	.0306	.0309	.0314
.04	.0454	.0456	.0458	.0461	.0463	.0468	.0473	.0478	.0483	.0493
.05	.0634	.0638	.0642	.0646	.0650	.0659	.0668	.0677	.0686	.0704
.06	.0833	.0840	.0846	.0853	.0859	.0873	.0887	.0902	.0916	.0946
.07	.1050	.1060	.1069	.1079	.1089	.1109	.1130	.1151	.1173	.1218
.08	.1283	.1296	.1310	.1323	.1337	.1366	.1395	.1426	.1456	.1520
.09	.1531	.1549	.1567	.1585	.1604	.1643	.1683	.1724	.1766	.1852
.10	.1793	.1816	.1840	.1864	.1889	.1940	.1992	.2046	.2101	.2214
.11	.2069	.2098	.2128	.2159	.2191	.2256	.2323	.2392	.2463	.2607
.12	.2357	.2394	.2431	.2470	.2509	.2591	.2676	2762	.2851	.3032
.13	.2658	.2702	.2748	.2796	.2844	.2945	.3049	.3156	.3265	.3488
.14	.2971	.3024	.3079	.3137	.3196	.3318	.3444	.3574	.3706	.3975
.15	.3295	.3358	.3424	.3493	.3563	.3710	.3861	.4015	.4173	.4495
.16	.363	.370	.378	.386	.395	.412	.430	.448	.467	.505
.17	.397	.406	.415	.425	.435	.455	.476	.497	.519	.563
.18	.433	.443	.454	.465	.476	.499	.524	.549	.547	.625
.19	.470	.481	.493	.506	.519	.546	.574	.603	.632	.691
.20	.507	.520	.534	.549	.564	.594	.626	.659	.692	.760
.21	.546	.561	.576	.593	.610	.645	.681	.718	.755	.832
.22	.585	.602	.620	.638	.657	.697	.737	.779	.822	.908
.23	.626	.644	.664	.685	.706	.751	.796	.843	.891	.988
.24	.667	.688	.710	.733	.757	.806	.858	.910	.963	1.071
.25	.709	.732	.757	.782	.809	.864	.921	.979	1.038	1.158
.26	.752	.777	.805	.833	.862	.923	.986	1.051	1.116	1.248
.27	.796	.824	.854	.885	.918	.985	1.054	1.125	1.197	1.343
.28	.840	.871	.904	.938	.974	1.048	1.124	1.202	1.281	1.441
.29	.886	.919	.955	.993	1.032	1.113	1.197	1.283	1.368	1.543
.30	.932	.969	1.008	1.049	1.092	1.180	1.272	1.365	1.458	1.649
.31	.979	1.019	1.062	1.107	1.153	1.249	1.349	1.450	1.552	1.759
.32	1.027	1.070	1.116	1.165	1.216	1.320	1.428	1.537	1.648	1.873
.33	1.075	1.122	1.172	1.225	1.280	1.393	1.510	1.628	1.748	1.991
.34	1.124	1.175	1.229	1.286	1.345	1.468	1.594	1.722	1.851	2.113
.35	1.174	1.229	1.287	1.349	1.413	1.545	1.680	1.818	1.958	2.240
.36	1.225	1.283	1.346	1.413	1.481	1.623	1.769	1.917	2.067	2.370
.37	1.276	1.339	1.407	1.478	1.552	1.704	1.860	2.019	2.180	2.505
.38	1.328	1.395	1.468	1.544	1.623	1.786	1.954	2.124	2.296	2.644
.39	1.381	1.453	1.530	1.612	1.697	1.871	2.050	2.232	2.416	2.787
.40	1.435	1.511	1.594	1.681	1.771	1.958	2.149	2.343	2.539	2.935
.41	1.489	1.570	1.658	1.752	1.848	2.046	2.250	2.457	2.665	3.087
.42	1.544	1.630	1.724	1.823	1.926	2.137	2.353	2.573	2.795	3.243
.43	1.599	1.691	1.791	1.896	2.005	2.229	2.460	2.693	2.929	3.404
.44	1.655	1.752	1.859	1.970	2.086	2.324	2.568	2.816	3.066	3.570
.45	1.712	1.815	1.928	2.046	2.168	2.421	2.679	2.942	3.206	3.740

Q = discharge D_c = critical depth b = bottom width of channel

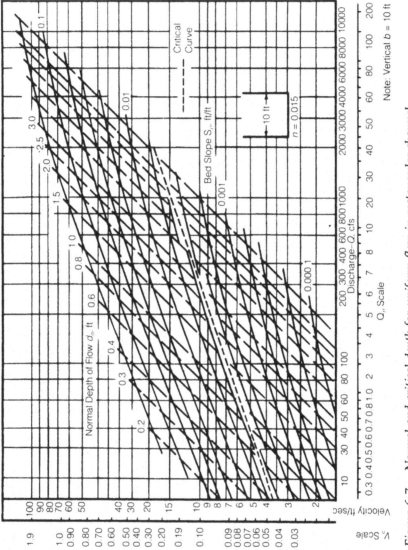

Figure 6.7 — Normal and critical depth for uniform flow in rectangular channel.

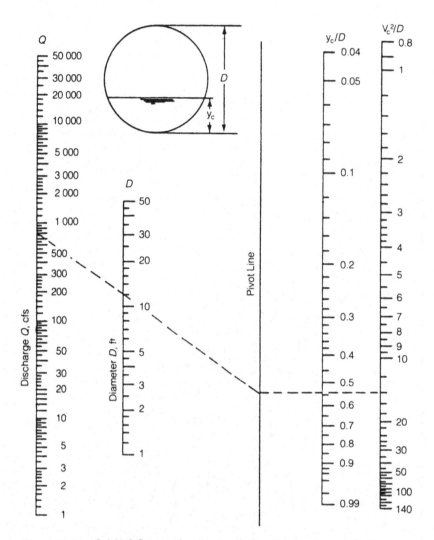

Figure 6.8—Critical flow and critical velocity in circular conduits.

in which V is the average velocity of flow in feet per second, R is the hydraulic radius in feet, S_f is the friction slope in feet per foot, and n is Manning's roughness coefficient. The hydraulic radius, R, in Manning's Equation is A/WP, where A is the cross-sectional area of flow in square feet and WP is the wetted perimeter in feet (the perimeter of the channel wetted by water, excluding the free surface width).

In the SI system of units, Manning's Equation is:

$$V = \frac{1}{n} R^{2/3} S_f^{1/2} \qquad (6\text{-}16b)$$

where V is in meters per second and R is in meters. It should be noted that Mannings n in SI units $= 0.6730n$ in English units of measure.

As noted, S_f is the friction slope, or the slope of the EGL. When S_f is set equal to the slope of the channel, S_o, the resulting depth of flow calculated from Manning's equation is the normal depth. Since $Q = AV$, Manning's Equation can be written as:

$$Q = \frac{1.49}{n} AR^{2/3}S_o^{1/2} \tag{6-17}$$

Equation 6-17 sometimes is written as:

$$q = k(S_f^{1/2}) \tag{6-18}$$

where

$$K = \frac{1.49}{n} AR^{2/3} \tag{6-19}$$

and is called the conveyance of the channel section.

Example 6-3: Use the three methods shown in the previous section to calculate normal depth for the flow situation described in Example 6-1 where $Q = 250$ cfs in a 10-foot wide rectangular channel. Assume the slope of the channel is 0.2 percent and Manning's n is 0.014.

Method 1
Manning's equation is:

$$Q = \frac{1.49}{n} AR^{2/3}S_o^{1/2}$$

$AR^{2/3}$ is a function of the geometric properties of the channel. Solving for $AR^{2/3}$:

$$A^{2/3} = \frac{Qn}{1.49S_o^{1/2}}$$

$$= 250 \times \frac{0.014}{1.49(0.002)^{1/2}}$$

$$= 52.53$$

Using the relationships $A = bd$, $WP = b + 2d$, and $R = A/WP$, determine the value of normal depth by trial and error as shown in Table 6.3.

TABLE 6.3. Calculations for Normal Depth Using Manning's Equation.

Depth (ft.) (1)	Area (sq. ft.) (2)	WP (feet) (3)	R (ft.) (4)	$R^{2/3}$ (5)	$AR^{2/3}$ (6)
5.00	50.0	20.00	2.50	1.84	92.1
4.00	40.0	18.00	2.22	1.70	68.1
3.50	35.0	17.00	2.06	1.62	56.6
3.40	34.0	16.80	2.02	1.50	54.4
3.31	33.1	16.62	1.99	1.58	52.4
3.32	33.2	16.64	2.00	1.58	52.6

Normal depth = 3.32 feet = d_n.

Method 2
Table 6.4 is a portion of Table 7.11 from Brater and King (1976).

$$K'_c = \frac{Qn}{b^{8/3}S^{1/2}}$$

$$= 250 \times \frac{0.014}{10^{8/3}(0.002)^{1/2}}$$

$$= 250 \times \frac{0.014}{464} \times 0.0447$$

$$= 0.1686$$

From Table 6.4, for a rectangular channel, the value of $K' = 0.1686$ lies between d_n/b equal to 0.33 and 0.34. Linear interpolation gives:

$$d_n/b = 0.332$$

$$d_n = 0.332 \times 10 = 3.32 \text{ feet}$$

Method 3
Figure 6.7 is a nomograph from FHWA (1961). Other nomographs for d_n are presented there for circular and trapezoidal channels. Note that Figure 6.7 is set for a value of n equal to 0.012. Therefore, in this case, we must use the Qn scale. For a discharge of 250 cfs and n equal to 0.014, Qn is equal to 3.5. Using this value and a channel slope of 0.002, normal depth is equal to 3.3 feet.

TABLE 6.4. Values of K' in formula $Q = \dfrac{k'}{n} b^{2/3}s^{1/2}$ for trapezoidal channels (Brater and King 1976)

$\dfrac{D}{b}$	Side slopes of channel, ratio of horizontal to vertical									
	Vertical	¼–1	½–1	¾–1	1–1	1½–1	2–1	2½–1	3–1	4–1
.01	.00068	.00068	.00069	.00069	.00069	.00069	.00069	.00069	.00070	.00070
.02	.00213	.00215	.00216	.00217	.00218	.00220	.00221	.00222	.00223	.00225
.03	.00414	.00419	.00423	.00426	.00428	.00433	.00436	.00439	.00443	.00449
.04	.00660	.00670	.00679	.00685	.00691	.00700	.00708	.00716	.00723	.00736
.05	.00946	.00964	.00979	.00991	.01002	.01019	.01033	.01047	.01060	.01086
.06	.0127	.0130	.0132	.0134	.0136	.0138	.0141	.0143	.0145	.0150
.07	.0162	.0166	.0170	.0173	.0175	.0180	.0183	.0187	.0190	.0197
.08	.0200	.0206	.0211	.0215	.0219	.0225	.0231	.0236	.0240	.0250
.09	.0241	.0249	.0256	.0262	.0267	.0275	.0282	.0289	.0296	.0310
.10	.0284	.0294	.0304	.0311	.0318	.0329	.0339	.0348	.0358	.0376
.11	.0329	.0343	.0354	.0364	.0373	.0387	.0400	.0413	.0424	.0448
.12	.0376	.0393	.0408	.0420	.0431	.0450	.0466	.0482	.0497	.0527
.13	.0425	.0446	.0464	.0480	.0493	.0516	.0537	.0556	.0575	.0613
.14	.0476	.0502	.0524	.0542	.0559	.0587	.0612	.0636	.0659	.0706
.15	.0528	.0559	.0585	.0608	.0627	.0662	.0692	.0721	.0749	.0805
.16	.0582	.0619	.0650	.0676	.0700	.0740	.0777	.0811	.0845	.0912
.17	.0638	.0680	.0716	.0748	.0775	.0823	.0866	.0907	.0947	.1026
.18	.0695	.0744	.0786	.0822	.0854	.0910	.0960	.1008	.1055	.1148
.19	.0753	.0809	.0857	.0899	.0936	.1001	.1059	.1115	.1169	.1277
.20	.0812	.0876	.0931	.0979	.1021	.1096	.1163	.1227	.1290	.1414
.21	.0873	.0945	.101	.106	.111	.120	.127	.135	.142	.156
.22	.0934	.1015	.109	.115	.120	.130	.139	.147	.155	.171
.23	.0997	.1087	.117	.124	.130	.141	.150	.160	.169	.187
.24	.1061	.1161	.125	.133	.140	.152	.163	.173	.184	.204
.25	.1125	.1236	.133	.142	.150	.163	.176	.188	.199	.222
.26	.119	.131	.142	.152	.160	.175	.189	.202	.215	.241
.27	.126	.139	.151	.162	.171	.188	.203	.218	.232	.260
.28	.132	.147	.160	.172	.182	.201	.217	.234	.249	.281
.29	.139	.155	.170	.182	.194	.214	.232	.250	.268	.302
.30	.146	.163	.179	.193	.205	.228	.248	.267	.287	.324
.31	.153	.172	.189	.204	.218	.242	.264	.285	.306	.347
.32	.160	.180	.199	.215	.230	.256	.281	.304	.327	.371
.33	.167	.189	.209	.227	.243	.271	.298	.323	.348	.396
.34	.174	.198	.219	.238	.256	.287	.316	.343	.370	.423
.35	.181	.207	.230	.251	.269	.303	.334	.363	.392	.450
.36	.189	.216	.241	.263	.283	.319	.353	.385	.416	.478
.37	.196	.225	.252	.275	.297	.336	.372	.406	.440	.507
.38	.203	.234	.263	.288	.312	.353	.392	.429	.465	.537
.39	.211	.244	.274	.301	.326	.371	.413	.452	.491	.568
.40	.218	.253	.286	.315	.341	.389	.434	.476	.518	.600
.41	.226	.263	.297	.328	.357	.408	.456	.501	.546	.633
.42	.233	.273	.309	.342	.373	.427	.478	.526	.574	.668
.43	.241	.283	.321	.357	.389	.447	.501	.553	.603	.703
.44	.248	.293	.334	.371	.405	.467	.525	.580	.633	.740
.45	.256	.303	.346	.386	.422	.488	.549	.607	.664	.777

D = depth of water b = bottom width of channel

VII. WATER SURFACE PROFILES

As noted in the first section of this chapter, open channel flow can be classified as uniform or nonuniform. Uniform flow is characterized by normal depth computed from Manning's equation. Nonuniform flow, on the other hand, will be either gradually or rapidly varied. Gradually varied flow can be analyzed by making the following assumptions:

(a) The slope of the EGL at a cross-section in a channel is the same as it would be for uniform flow at that section with the same velocity and hydraulic radius.
(b) The vertical curvature of the streamlines is small enough for the pressure in the flow to be hydrostatic in plane sections normal to the bed.
(c) The channel is prismatic (of constant shape with straight alignment or negligible curvature and constant bed slope).
(d) The momentum and energy correction factors are constant along the length of channel or pipe.

Rapidly varied flow by contrast does not conform to the above conditions. It occurs in abrupt transitions such as flow over weirs, under gates, and through junctions. The hydraulic jump is perhaps the best example of rapidly varied flow of interest in urban drainage design. Rapidly varied flow has the following characteristics:

(a) The local curvature of streamlines in the flow is so pronounced that the pressure distribution of the flow is not hydrostatic.
(b) A rapid variation in flow velocity and depth occurs within a relatively short reach of the channel.
(c) Separation zones, eddies, and rollers may occur and complicate flow patterns or distort the velocity distribution.
(d) The energy and momentum correction factors vary greatly and may be difficult to determine.

The identification of particular water surface profiles in steady gradually varied flow is determined by the relative magnitudes of the actual flow depth, the normal depth, the critical depth, and the location of control sections. Water surface profiles in gradually varied flow are characterized first by one of the following types that describe the slope of the channel.

$$
\begin{array}{lll}
\text{M—Mild} & : & d_n > d_c \\
\text{C—Critical} & : & d_n = d_c \\
\text{S—Steep} & : & d_n < d_c \\
\text{H—Horizontal} & : & d_n \text{ does not exist} \\
\text{A—Adverse} & : & d_n \text{ does not exist}
\end{array}
$$

Water surface profiles are then characterized by the location of the actual depth of flow in relation to normal and critical depth, according to the following zones:

Zone 1—depth of flow is greater than d_n and d_c
Zone 2—depth of flow is between d_n and d_c
Zone 3—depth of flow is less than d_n and d_c

Thus, a water surface profile might be classified as M-1 (mild slope, zone 1), S-2 (steep slope, zone 2) etc. **Preliminary identification of the water surface profile type is an important first step in actual computation of water surface elevations in a channel or major storm sewer.**
A water surface profile is computed by the following procedure.

(a) Assemble the basic hydraulic data for each channel segment, including slope, width, side slopes, roughness, and discharge.
(b) Establish the location of "control" sections where flow is a unique function of depth, such as at a natural falls or the crest of a spillway.
(c) Determine d_n and d_c for each channel segment and plot on rough profile.
(d) Locate all the control sections on the profile and sketch the possible water surface profiles. Recall that transition from supercritical to subcritical flow is only possible in a hydraulic jump.
(e) Compute the actual water surface elevations by applying the energy equation to individual sections of the channel above or below the control.

The actual calculation of water surface profile is made by direct step or standard step methods, as discussed later in this chapter. Both methods proceed from a control section. A control is any section in a channel where the depth of flow is known, such as critical depth, depth upstream of a culvert, depth of flow over a weir or dam, or depth of flow under a gate. If the flow above the control is subcritical, then the water surface profile calculation will proceed in the upstream direction from the control. If flow below the control is supercritical, then the calculations proceed downstream. This latter situation is illustrated in Figure 6.9(a), while the former occurs in Figure 6.9(b). Most water surface profiles take the form of a draw-down, Figure 6.9(a), or a backwater curve, Figure 6.9(b).
Figure 6.10 depicts three water surface profiles for gradually varied flow on mild and steep slopes. Other profiles are discussed in the following paragraphs. A more complete listing of water surface profile types with detailed discussion can be found in basic hydraulic texts (Chow 1959, Brater and King 1976, Rouse 1961). One of the most common water surface profiles is the M-1 profile shown in Figure 6.11. Here water in a channel on a mild slope flows into a pool. The pool could be the water surface upstream of a dam, a culvert, a bridge, or at the upstream end of a detention basin. If the channel is on a hydraulically steep slope instead of a mild slope, the water surface profile will show a hydraulic jump. A hydraulic jump occurs whenever flow passes through critical depth enroute from a hydraulically steep to a hydraulically mild slope.

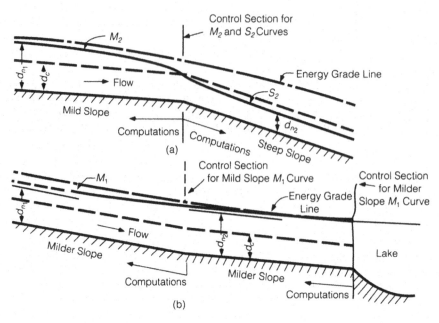

Figure 6.9—Drawdown and backwater profiles in gradually varied flows.

VIII. HYDRAULIC JUMP

The hydraulic jump is a rapidly varied flow phenomenon in which flow in a channel changes abruptly from supercritical flow at a relatively shallow depth (less than d_c) to subcritical flow at a greater depth (greater than d_c). The depth before the jump is called the initial depth, while the depth after the jump is known as the sequent depth. The situation is illustrated in Figure 6.12.

The hydraulic jump may be employed as a device for the dissipation of excess energy, as where a steep sewer enters a larger sewer at a junction. In stormwater projects, the hydraulic jump may be used to consume excess energy and avoid scour of earthen channels. Thus, the analysis of hydraulic jumps usually has three objectives. First, the location of the jump is important because of the potential for unexpected surcharge or channel scour. This can be determined by searching for pipe/channel elements where the flow is supercritical upstream and subcritical downstream. Once this is determined, it is important to compute the two depths, d_1 and d_2, which are the initial and sequent depths, respectively. Third, the energy loss H_j dissipated by the jump

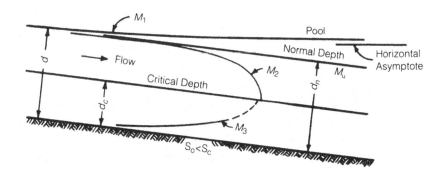

M_u—Uniform Flow
M_1—Backwater from Resevoir or from Channel of Milder Slope ($d > d_n$);
M_2—Drawdown, as from Change of Channel of Mild Slope to Steep Slope ($d_n > d > d_c$); and
M_3—Flow Under Gate on Mild Slope, or Upstream Profile Before Hydraulic Jump on Mild Slope ($d < d_c$).

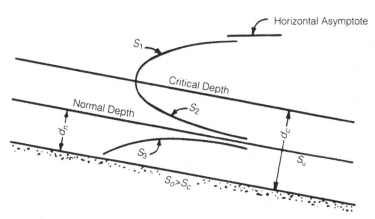

S_u—Uniform Flow
S_1—Downstream Profile After Hydraulic Jump on Steep Slope ($d > d_c$);
S_2—Drawdown, as from Mild to Steep Slope or Steep Slope to Steeper Slope ($d_c > d > d_n$); and
S_3—Flow Under Gate on Steep Slope, or Change from Steep Slope to Less Steep Slope ($d < d_n$).

Figure 6.10—Water surface profiles for gradually varied flow on (a) mild slope and (b) steep slope.

is often an important design consideration. The pertinent depth equation for a rectangular section is:

$$\frac{d_2}{d_2} = 0.5[(1 + 8F_1^2)^{1/2} - 1] \qquad (6\text{-}20)$$

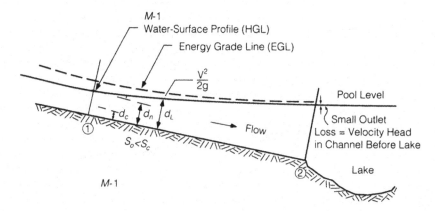

Figure 6.11—Water surface profile in flow from a channel to a pool on a mild slope.

in which F_1 is the Froude number at the upstream section. The energy lost in the jump, H_j, is obtained by subtracting the specific energy at section 2 in Figure 6.12 from that at section 1.

$$H_j = E_1 - E_2 = \frac{(d_2 - d_1)^3}{(4d_1d_2)} \qquad (6\text{-}21)$$

Equations similar to (6-20) and (6-21) are readily derived from energy conservation principles for other section geometries.

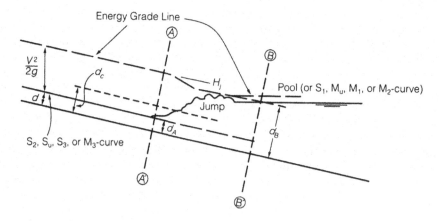

Figure 6.12—Water surface profile for hydraulic jump.

IX. FRICTION LOSSES

The major loss in a channel or pipe is the friction or boundary shear loss. The head loss due to friction in a pipe or channel is computed from the general definition:

$$H_f = Lx\overline{S_f} \qquad (6\text{-}22)$$

in which H_f is the head loss due to friction, L is the length of channel or conduit, $\overline{S_f}$ is the average friction slope for the length L, and the subscripts 1 and 2 refer to the ends of the reach. The average friction slope $\overline{S_f}$ can be computed as the simple mean $(S_{f1} + S_{f2})/2$, or it can be evaluated from the mean depth or conveyance for the entire reach length. The friction slope between sections 1 and 2 at a single point along a pipe or channel is computed from one of several so-called friction formulas developed for uniform and gradually varied flow. The present practice is to use the Hazen-Williams formula for closed conduit or pressure flow when flow is in the hydraulically smooth range, and to use the Manning's equation in open channel and pipe flow when flow is in the transition and hydraulically rough range. The Hazen-Williams formula may be used when:

$$\frac{V_f(\kappa)}{\upsilon} < 3 \qquad (6\text{-}23)$$

and

$$\frac{V_m(R)}{\upsilon} > 30{,}000 \qquad (6\text{-}24)$$

where V_f equals the shear velocity $(g\ R\ S_f)^{0.5}$, κ is equivalent sand roughness, υ is the kinematic viscosity, V_m is the mean velocity, R is the hydraulic radius. For flow with Reynolds numbers less than 30,000, other formulas such as the Blasius formula may be used in the hydraulically smooth range. Christensen suggests the Manning Equation is valid when:

$$\frac{V_f(\kappa)}{\upsilon} > 3 \qquad (6\text{-}25)$$

and

$$5 < \frac{R}{K} < 300 \qquad (6\text{-}26)$$

According to French (1985) fully rough flow and Manning's Equation apply when:

$$n^6\sqrt{RS_f} \geq 1.9 \times 10^{-13} \tag{6-27}$$

where n is the Manning roughness coefficient. As a matter of interest consider a 24-inch concrete sewer flowing full at a 1% slope with $n = 0.012$. The pipe carries a discharge of 25 cfs. If we say that $S_f = S_o$ and that κ, the equivalent sand roughness, equals 0.01 feet for precast concrete pipe the following results are obtained with $= 1.217 \times 10^{-5}$ ft²/sec for water at 60°F:

$$\frac{V_f(\kappa)}{v} = \frac{0.40 \times 0.01}{1.217 \times 10^{-5}} = 329$$

$$\frac{V_m(R)}{v} = \frac{7.96 \times 0.5}{1.217 \times 10^{-5}} = 326,941$$

$$\frac{R}{K} = \frac{0.50}{0.01} = 50$$

$$n^6\sqrt{RS_f} = (0.012)^6(0.5 \times 0.01)^{0.5} = 50$$

Comparing these parameter values with Equations 6-23 through 6-27 it seems we are clearly in the fully rough zone where the Manning Equation should be valid. As a practical matter, we would continue the analysis of friction head loss in this particular pipe by the Manning equation. A brief discussion of the two principal friction loss formulas follows.

A. Hazen-Williams Formula

The Hazen-Williams formula for smooth flow in a pipe is:

$$V = 1.3HxC_{HW}R^{0.63}S_f^{0.54} \tag{6-28a}$$

where V is the mean flow velocity in feet per second, C_{HW} is the Hazen-Williams coefficient, and all other terms are as defined previously.

The Hazen-Williams formula in SI units is:

$$V = 0.85C_{HW}R^{0.63}S_f^{0.54} \tag{6-28b}$$

where V is in meters per second.

The following are values of C_{HW} suggested by Brater and King (1976) for pipes carrying water. The C_{HW} values for new, smooth pipes generally are taken to be from 130 to 140. To estimate friction losses for future conditions, lower values of C_{HW} are used to allow for reductions in flow capacity resulting from the factors listed above. For smooth concrete and cement-lined pipes, a C_{HW} value of 100–120 commonly is used for future conditions. A wide variety of values published in table form are available (Williams and Hazen 1945). For the example cited above, the friction slope computed by the Hazen-Williams equation, assuming $C_{HW} = 120$, is:

$$V = \frac{Q}{A} = \frac{25}{3.1416} = 7.96 \text{ ft/sec}$$

From Equation (6-28):

$$Sf = \left(\frac{V}{1.318 C_{HW} R^{0.63}} \right)^{1.852}$$

$$S_f = \left(\frac{7.96}{1.318 \times 120 \times 0.5^{0.63}} \right)^{1.852} \tag{6-29}$$

$$= 0.0088 \text{ ft/ft}$$

The head loss, H_f, in 100 feet of precast concrete storm sewer would be 0.88 feet.

B. Darcy-Weisbach Equation

The Darcy-Weisbach Equation, also developed primarily for flow in pipes, is:

$$H_f = f \frac{L}{d_o} \frac{v^2}{2g} \tag{6-30}$$

where:

f = friction factor
L = length of pipe, feet
d_o = diameter of pipe, feet

C. Manning Equation

The Manning Equation 6-16 is used widely in analyzing uniform and gradually varied flow in pipes and open channels. Typical values of "n" for both closed conduits and open channels are listed in Table 6.5. More complete listings of n values are found in Chow (1959) and FHWA (1961). The indirect effect of depth of flow or the height of vegetal cover on the roughness coefficient has been investigated in grass-lined chan-

TABLE 6.5. Values of Manning Coefficient for Various Materials (ASCE 1982).

Conduit Material (1)	Manning n (2)[a]
Closed conduits	
Asbestos-cement pipe	0.011–0.015
Brick	0.013–0.017
Cast iron pipe	
Cement-lined & seal coated	0.011–0.015
Concrete (monolithic)	
Smooth forms	0.012–0.014
Rough forms	0.015–0.017
Concrete pipe	0.011–0.015
Corrugated-metal pipe	
(½-in. × 2½-in. corrugations)	
Plain	0.022–0.026
Paved invert	0.018–0.022
Spun asphalt lined	0.011–0.015
Plastic pipe (smooth)	0.011–0.015
Vitrified clay	
Pipes	0.011–0.015
Liner plates	0.013–0.017
Open channels	
Lined channels	
a. Asphalt	0.013–0.017
b. Brick	0.012–0.018
c. Concrete	0.011–0.020
d. Rubble or riprap	0.020–0.035
e. Vegetal	0.030–0.40[b]
Excavated or dredged	
Earth, straight and uniform	0.020–0.030
Earth, winding, fairly uniform	0.025–0.040
Rock	0.030–0.045
Unmaintained	0.050–0.14
Natural channels (minor streams,	
top width at flood stage < 100 ft)	
Fairly regular section	0.03–0.07
Irregular section with pools	0.04–0.10

[a]Dimensional units contained in numerical term in formula.
[b]See References 2, 5, 16. (Vanes with depth and velocity.)
Note: 1 in. = 2.54 cm; 1 ft = 0.305 m.

nels (Chow 1959; FHWA 1961; SCS 1954) and also for natural channels and floodplains (Chow 1959). Equation 6-16 can be rewritten as:

$$S_f = \left(\frac{Q_n}{1.49AR^{2/3}}\right)^2 \qquad (6\text{-}31a)$$

or, in SI units:

$$S_f = \left(\frac{Q_n}{AR^{2/3}}\right)^2 \qquad (6\text{-}31b)$$

and

$$S_f = \frac{29.1n^2}{R^{4/3}}\left(\frac{V^2}{2g}\right) \qquad (6\text{-}32a)$$

or, in SI units:

$$S_f = \frac{19.85n^2}{R^{4/3}}\left(\frac{V^2}{2g}\right) \qquad (6\text{-}32b)$$

For the 24-inch concrete sewer discussed in the preceding section, the friction slope by Manning's equation would be computed from Equation 6-31 as:

$$S_f = \left(\frac{25 \times 0.012}{1.49 \times 3.1416 \times 0.5^{2/3}}\right)^2 = 0.0104 \text{ feet/foot}$$

or by (6-32) as:

$$S_f = \frac{29.1 \times (0.012)^2}{0.5^{4/3}}\left(\frac{7.96^2}{64.4}\right) = 0.0104 \text{ feet/foot}$$

This produces a friction head loss, H_f, of 1.04 feet for 100 feet of sewer, which is about 16% higher than H_f computed from the Hazen-Williams equation. This difference is attributed to the difference between the Manning n value and C_{HW} in the Hazen-Williams formula.

In view of the importance of analyzing normal depth and other hydraulic properties in partially-full circular conduits, a hydraulic element chart is presented in Figure 6.13. The ratios Q/Q_f, V/V_f, A/A_f, R/R_f, and n/n_f are called the hydraulic elements. A symbol without subscript represents the value of a variable when the conduit is flowing partially full and the subscript "f" represents values for the full conduit with the same slope of the energy grade line. Figure 6.13 is a hydraulic-

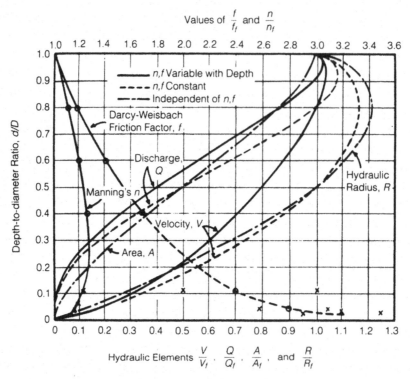

Figure 6.13—Hydraulic elements graph for circular sewers.

elements graph for circular conduits with uniform roughness through-out the surface area. These curves are shown in tabular form in Table 6.6. Geometric relationships for various types of flow sections are listed in Table 6.7. Note the apparent variation in Manning's n with depth in Figure 6.13. This problem is avoided in an alternative method for normal depth proposed by Christensen (1984) for flow in partially full conduits.

X. MINOR LOSSES

In addition to the friction loss along a channel or conduit, there is usually a local loss of energy associated with any sudden change due to transitions, junctions, bends, entrances, exits, obstructions, and con-trol devices, such as orifices and gates. These losses occur over a rel-atively short distance and are usually represented by a steep slope or sudden drop in the energy grade line.

TABLE 6.6. Tabular Values of Hydraulic Elements of Pipes (n constant) (ASCE 1972).

$\dfrac{d}{D}$	$\dfrac{A}{D^2}$	$\dfrac{Q_n}{D^{8/3}S^{1/2}}$	$\dfrac{Q_c}{D^{5/2}}$	$\dfrac{d}{D}$	$\dfrac{A}{D^2}$	$\dfrac{Q_n}{D^{8/3}S^{1/2}}$	$\dfrac{Q_c}{D^{5/2}}$
0.01	0.0013	0.00007	0.0006	0.51	0.4027	0.239	1.4494
0.02	0.0037	0.00031	0.0025	0.52	0.4127	0.247	1.5041
0.03	0.0069	0.00074	0.0055	0.53	0.4227	0.255	1.5598
0.04	0.0105	0.00138	0.0098	0.54	0.4327	0.263	1.6166
0.05	0.0147	0.00222	0.0153	0.55	0.4426	0.271	1.6741
0.06	0.0192	0.00328	0.0220	0.56	0.4526	0.279	1.7328
0.07	0.0242	0.00455	0.0298	0.57	0.4625	0.287	1.7924
0.08	0.0294	0.00604	0.0389	0.58	0.4724	0.295	1.8531
0.09	0.0350	0.00775	0.0491	0.59	0.4822	0.303	1.9147
0.10	0.0409	0.00967	0.0605	0.60	0.4920	0.311	1.9773
0.11	0.0470	0.01181	0.0731	0.61	0.5018	0.319	2.0410
0.12	0.0534	0.01417	0.0868	0.62	0.5115	0.327	2.1058
0.13	0.0600	0.01674	0.1016	0.63	0.5212	0.335	2.1717
0.14	0.0668	0.01952	0.1176	0.64	0.5308	0.343	2.2886
0.15	0.0739	0.0225	0.1347	0.65	0.5404	0.350	2.3068
0.16	0.0811	0.0257	0.1530	0.66	0.5499	0.358	2.3760
0.17	0.0885	0.0291	0.1724	0.67	0.5594	0.366	2.4465
0.18	0.0961	0.0327	0.1928	0.68	0.5687	0.373	2.5182
0.19	0.1039	0.0365	0.2144	0.69	0.5780	0.380	2.5912
0.20	0.1118	0.0406	0.2371	0.70	0.5872	0.388	2.6656
0.21	0.1199	0.0448	0.2609	0.71	0.5964	0.395	2.7416
0.22	0.1281	0.0492	0.2857	0.72	0.6054	0.402	2.8188
0.23	0.1365	0.0537	0.3116	0.73	0.6143	0.409	2.8977
0.24	0.1449	0.0585	0.3386	0.74	0.6231	0.416	2.9783
0.25	0.1535	0.0634	0.3667	0.75	0.6319	0.422	3.0606
0.26	0.1623	0.0686	0.3957	0.76	0.6405	0.429	3.1450
0.27	0.1711	0.0739	0.4259	0.77	0.6489	0.435	3.2314
0.28	0.1800	0.0793	0.4571	0.78	0.6573	0.441	3.3200
0.29	0.1890	0.0849	0.4893	0.79	0.6655	0.447	3.4111
0.30	0.1982	0.0907	0.5226	0.80	0.6736	0.453	3.5051
0.31	0.2074	0.0966	0.5969	0.81	0.6815	0.458	3.6020
0.32	0.2167	0.1027	0.5921	0.82	0.6893	0.463	3.7021
0.33	0.2260	0.1089	0.6284	0.83	0.6969	0.468	3.8062
0.34	0.2355	0.1153	0.6657	0.84	0.7043	0.473	3.9144
0.35	0.2450	0.1218	0.7040	0.85	0.7115	0.477	4.0276
0.36	0.2546	0.1284	0.7433	0.86	0.7186	0.481	4.1466
0.37	0.2642	0.1351	0.7836	0.87	0.7254	0.485	4.2722
0.38	0.2739	0.1420	0.8249	0.88	0.7320	0.488	4.4057
0.39	0.2836	0.1490	0.8672	0.89	0.7384	0.491	4.5486

TABLE 6.6.   Continued

$\dfrac{d}{D}$	$\dfrac{A}{D^2}$	$\dfrac{Q_n}{D^{8/3}S^{1/2}}$	$\dfrac{Q_c}{D^{5/2}}$	$\dfrac{d}{D}$	$\dfrac{A}{D^2}$	$\dfrac{Q_n}{D^{8/3}S^{1/2}}$	$\dfrac{Q_c}{D^{5/2}}$
0.40	0.2934	0.1561	0.9104	0.90	0.7445	0.494	4.7033
0.41	0.3032	0.1633	0.9546	0.91	0.7504	0.496	4.8724
0.42	0.3130	0.1705	0.9997	0.92	0.7560	0.497	5.0602
0.43	0.3229	0.1779	1.0459	0.93	0.7612	0.498	5.2727
0.44	0.3328	0.1854	1.0929	0.94	0.7662	0.498	5.5182
0.45	0.3428	0.1929	0.1410	0.95	0.7707	0.498	5.8119
0.46	0.3527	0.201	1.1900	0.96	0.7749	0.496	6.1785
0.47	0.3627	0.208	1.2400	0.97	0.7785	0.494	6.6695
0.48	0.3727	0.216	1.2908	0.98	0.7817	0.489	7.4063
0.49	0.3827	0.224	1.3427	0.99	0.7841	0.483	8.8261
0.50	0.3927	0.232	1.3956	1.00	0.7854	0.463	——

In long conduits where $L/D >> 1000$ these local losses are usually very small in comparison to the friction losses and the minor losses can be neglected. However, if the channel or conduit is very short and/or there are a number of manholes, changes in direction, junctions, or changes in pipe size then the sum of these losses can exceed the friction loss.

In terms of calculations, the loss is expressed either as a coefficient times the velocity head or as a coefficient times the difference in velocity heads, depending on the type of loss involved. This is usually written:

$$H_L = K_c \left(\frac{V^2}{2g} \right) \tag{6-33}$$

in which H_L is the minor head loss, K_c is a loss coefficient dependent on the type of loss, and $V^2/2g$ is the velocity head.

The loss coefficient and the form of the equation are different depending on the type of loss, whether flow is open channel or pressure flow, and at times whether flow is subcritical or supercritical. Full discussion and values of coefficients are given in several references (Chow 1959; Brater and King 1976; Rouse 1961; Hendrickson 1964; USBR 1977; FHWA 1978; FHWA 1985; NBS 1938; Bowers 1950). The following are useful minor head loss formulas for hydraulic structures commonly found in storm sewer systems and open channels.

A. Transition Losses

A transition is a location where a conduit or channel changes size. The change in cross-sectional area results in a change in velocity, which

means a loss of head. The energy losses in contraction or expansion can be expressed in terms of the kinetic energy at the two ends of the transition as shown in Equations (6-34) and (6-35). K_c (contraction) and K_e (expansion) are the transition loss coefficients. Table 6.8 gives some typical values. V_1 is the upstream velocity above the transition and V_2 is the downstream velocity below the transition.
Contraction:

$$H_c = K_c \left(\frac{V_2^2}{2g} - \frac{V_1^2}{2g} \right) \text{ for } V_2 > V_1 \tag{6-34}$$

Expansion:

$$H_e = K_e \left(\frac{V_1^2}{2g} - \frac{V_2^2}{2g} \right) \text{ for } V_1 > V_2 \tag{6-35}$$

Henderson (1969) recommends using $H_t = K_t (V_1 - V_2)^2/2g$ (where H_t = transition loss coefficient) in place of 6-34 and 6-35, but indicates the two methods give very similar results when $1.5 < V_1/V_2) < 2.5$. For pipes with pressure flow, the loss coefficients listed in Tables 6.9 and 6.10 can be used in conjunction with Equation 6-33 for sudden and gradual expansions, respectively. For sudden contractions in pipes with pressure flow, the loss coefficients listed in Table 6.11 can be used in conjunction with Equation 6-33, in which K_c is replaced by K_3. For values of K_2 and K_3 outside the ranges found in Tables 6.10 and 6.11, see Daily and Harleman (1966). As will be noted from these tables and from the discussion on junctions below, the designer is advised to design transitions carefully to minimize losses.

B. Entrance Losses

Entrance losses to box culverts and pipes of various materials can be estimated by using the entrance loss coefficients listed in Table 6.12 in conjunction with Equation 6-33. See also the detailed discussion in Chapter 8.

C. Manhole and Junction Losses

Junctions are locations where two or more pipes join together to form another pipe or channel. They represent another critical point in a storm drainage system that must be designed as a transition through which the flow is changing direction.

Multiple pipes or channels coming together at a junction should flow together smoothly to avoid high head losses. Items that promote turbulent flow and high losses include a large angle between the two (>60°), a large vertical difference between the two (greater than 6 inches between the two inverts), and absence of a semicircular channel or

TABLE 6.7. Geometric Relationships for Various Flow Sections.

Section	Area (A)	Wetted Perimeter (P)	Hydraulic Radius (R)	Top Width (T)
Rectangular	by	$b + 2y$	$\dfrac{by}{b + 2y}$	T
Trapezoidal	$(a + my)y$	$a + 2y\sqrt{1 + m^2}$	$\dfrac{(a + my)y}{a + 2y\sqrt{1 + m^2}}$	$a + 2my$

Shape	Area			
Triangular	my^2	$2y\sqrt{1+m^2}$	$\dfrac{my}{2\sqrt{1+m^2}}$	$2my$
Circular $\alpha = 2\sin^{-1}\left[\dfrac{2\sqrt{y(D-y)}}{D}\right]$	$(\alpha - \sin\alpha)D^2/8$	$\dfrac{\alpha D}{2}$	$\dfrac{D}{4}\left[1 - \dfrac{\sin\alpha}{\alpha}\right]$	$(\sin\tfrac{1}{2}\alpha)D$ or $2\sqrt{y(D-y)}$

TABLE 6.8.   Storm Sewer Energy Loss Coefficient. (expansion, contraction) (Linsley and Franzini 1964)

(a) Expansion (K_e)

θ^*	$\dfrac{D_2}{D_1} = 3$	$\dfrac{D_2}{D_1} = 1.5$
10	0.17	0.17
20	0.40	0.40
45	0.86	1.06
60	1.02	1.21
90	1.06	1.14
120	1.04	1.07
180	1.00	1.00

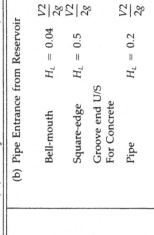

*The angle θ is the angle in degrees between the sides of the tapering section.

(b) Pipe Entrance from Reservoir

Bell-mouth	$H_L = 0.04$	$\dfrac{V2}{2g}$
Square-edge	$H_L = 0.5$	$\dfrac{V2}{2g}$
Groove end U/S For Concrete		
Pipe	$H_L = 0.2$	$\dfrac{V2}{2g}$

(c) Contractions (K_c)

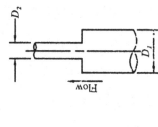
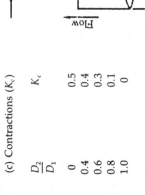

$\dfrac{D_2}{D_1}$	K_c
0	0.5
0.4	0.4
0.6	0.3
0.8	0.1
1.0	0

TABLE 6.9. Values of K_e for Determining Loss of Head Due to Sudden Enlargement in Pipes, from the Formula $H_2 = K_2(V_1^2/2g)$ (AISI 1985)

d_2/d_1 = ratio of larger pipe to smaller pipe

V_1 = velocity in smaller pipe

$\dfrac{d_2}{d_1}$	Velocity, V_1, in feet per second												
	2	3	4	5	6	7	8	10	12	15	20	30	40
1.2	.11	.10	.10	.10	.10	.10	.10	.09	.09	.09	.09	.09	.08
1.4	.26	.26	.25	.24	.24	.24	.24	.23	.23	.22	.22	.21	.20
1.6	.40	.39	.38	.37	.37	.36	.36	.35	.35	.34	.33	.32	.32
1.8	.51	.49	.48	.47	.47	.46	.46	.45	.44	.43	.42	.41	.40
2.0	.60	.58	.56	.55	.55	.54	.53	.52	.52	.51	.50	.48	.47
2.5	.74	.72	.70	.69	.68	.67	.66	.65	.64	.63	.62	.60	.58
3.0	.83	.80	.78	.77	.76	.75	.74	.73	.72	.70	.69	.67	.65
4.0	.92	.89	.87	.85	.84	.83	.82	.80	.79	.78	.76	.74	.72
5.0	.96	.93	.91	.89	.88	.87	.86	.84	.83	.82	.80	.77	.75
10.0	1.00	.99	.96	.95	.93	.92	.91	.89	.88	.86	.84	.82	.80
∞	1.00	1.00	.98	.96	.95	.94	.93	.91	.90	.88	.86	.83	.81

TABLE 6.10.   Values of K_2 for Determining Loss of Head Due to Gradual Enlargement in Pipes from the Formula $H_2 = K_2(V_1^2/2g)$ (AISI 1985)

d_2/d_1 = ratio of diameter of larger pipe to diameter of smaller pipe. Angle of cone is twice the angle between the axis of the cone and its side.

$\dfrac{d_2}{d_1}$	Angle of cone													
	2°	4°	6°	8°	10°	15°	20°	25°	30°	35°	40°	45°	50°	60°
1.1	.01	.01	.01	.02	.03	.05	.10	.13	.16	.18	.19	.20	.21	.23
1.2	.02	.02	.02	.03	.04	.09	.16	.21	.25	.29	.31	.33	.35	.37
1.4	.02	.03	.03	.04	.06	.12	.23	.30	.36	.41	.44	.47	.50	.53
1.6	.03	.03	.04	.05	.07	.14	.26	.35	.42	.47	.51	.54	.57	.61
1.8	.03	.04	.04	.05	.07	.15	.28	.37	.44	.50	.54	.58	.61	.65
2.0	.03	.04	.04	.05	.07	.16	.29	.38	.46	.52	.56	.60	.63	.68
2.5	.03	.04	.04	.05	.08	.16	.30	.39	.48	.54	.58	.62	.65	.70
3.0	.03	.04	.04	.05	.08	.16	.31	.40	.48	.55	.59	.63	.66	.71
∞	.03	.04	.05	.06	.08	.16	.31	.40	.49	.56	.60	.64	.67	.72

TABLE 6.11. Values of K_3 for Determining Loss of Head Due to Sudden Contraction from the Formula $H_3 = K_3(V_2^2/2g)$ (AISI 1985)

d_2/d_1 = ratio of larger to smaller diameter

V_2 = velocity in smaller pipe

$\dfrac{d_2}{d_1}$	Velocity, V_2, in feet per second												
	2	3	4	5	6	7	8	10	12	15	20	30	40
1.1	.03	.04	.04	.04	.04	.04	.04	.04	.04	.04	.05	.05	.06
1.2	.07	.07	.07	.07	.07	.07	.07	.08	.08	.08	.09	.10	.11
1.4	.17	.17	.17	.17	.17	.17	.17	.18	.18	.18	.18	.19	.20
1.6	.26	.26	.26	.26	.26	.26	.26	.26	.26	.25	.25	.25	.24
1.8	.34	.34	.34	.34	.34	.34	.33	.33	.32	.32	.31	.29	.27
2.0	.38	.38	.37	.37	.37	.37	.36	.36	.35	.34	.33	.31	.29
2.2	.40	.40	.40	.39	.39	.39	.39	.38	.37	.37	.35	.33	.30
2.5	.42	.42	.42	.41	.41	.41	.40	.40	.39	.38	.37	.34	.31
3.0	.44	.44	.44	.43	.43	.43	.42	.42	.41	.40	.39	.36	.33
4.0	.47	.46	.46	.46	.45	.45	.45	.44	.43	.42	.41	.37	.34
5.0	.48	.48	.47	.47	.47	.46	.46	.45	.45	.44	.42	.38	.35
10.0	.49	.48	.48	.48	.48	.47	.47	.46	.46	.45	.43	.40	.36
∞	.49	.49	.48	.48	.48	.47	.47	.47	.46	.45	.44	.41	.38

TABLE 6.12.   Entrance Loss Coefficents for Culverts (FHWA 1985)
Outlet Control, Full or Partly Full Entrance Head Loss.

$$H_e = k_e \left(\frac{V^2}{2g} \right)$$

Type of *Structure and Design of Entrance*	Coefficent k_e
Pipe, Concrete	
Projecting from fill, socket end (groove-end)	0.2
Projecting from fill, sq. cut end	0.5
Headwall or headwall and wingwalls	
Socket end of pipe (groove-end)	0.2
Square-edge	0.5
Rounded (radius = 1/12D)	0.2
Mitered to conform to fill slope	0.7
*End-section conforming to fill slope	0.5
Beveled edges, 33.7° or 45° levels	0.2
Side- or slope-tapered inlet	0.2
Pine, or Pipe-Arch, Corrugated Metal	
Projecting from fill (no headwal)	0.9
Headwall or headwall and wingwalls square-edge	0.5
Mitered to conform to fill slope, paved or unpaved slope	0.7
*End-section conforming to fill slope	0.5
Beveled edges, 33.7° or 45° bevels	0.2
Side- or slope-tapered inlet	0.2
Box, Reinforced Concrete	
Headwall parallel to embankment (no wingwalls)	
Square-edged on 3 edges	0.5
Rounded on 3 edges to radius of 1/12 barrel dimension, or beveled edges on 3 sides	0.2
Wingwalls at 30° to 75° to barrel	
Square-edged at crown	0.4
Crown edge rounded to radius of 1/2 barrel dimension, or beveled top edge	0.2
Wingwall at 10° to 25° to barrel	
Square-edged at crown	0.5
Wingwalls parallel (extension of sides)	
Square-edged at crown	0.7
Side- or slope-tapered inlet	0.2

*Note: "End-section conforming to fill slope," made of either metal or concrete, are the sections commonly available from manufacturers. From limited hydraulic tests they are equivalent in operation to a headwall in both *inlet* and *outlet* control. Some end sections, incorporating a *closed* taper in their design have a superior hydraulic performance.

benching at the bottom of the junction box in the case of the pipes. Special problems arise when smaller pipes join a larger one at a junction.

Losses at sewer junction manholes can typically account for 20–30% of total head losses, though wide variances are possible. In extreme cases, junction manholes can account for much higher percentages of losses. These losses can be minimized by careful design and construction. For a complete discussion, see Marsalek (1985, 1986, 1987).

In a straight-through manhole where there is no change in pipe size, the minor loss can be estimated by:

$$H_m = 0.05 \left(\frac{V^2}{2g}\right) \qquad (6\text{-}36)$$

Junction losses for other configurations in closed conduits can be estimated from the equations shown in Figures 6.14 and 6.15 (City of Austin 1987), and Figure 6.16 (AISI 1985) (see also discussion of losses at manholes in Chapter 8).

D. Bend Losses

Bend losses in open channels can be estimated by using the bend loss coefficients listed in Table 6.13 in conjunction with Equation (6-33). If the ratio of the radius of the bend, r, to the width of the channel, b, is equal to or greater than 3.0, the loss is negligible.

Bend losses in closed conduits can be estimated by using the bend loss coefficients shown in Figures 6.15 (in conjunction with Equation 6-33) and 6.16. These values are for the high Reynolds numbers usually encountered in hydraulic engineering practice.

XI. CALCULATION OF WATER SURFACE PROFILES

The prediction of drawdown and backwater curves such as those shown in Figure 6.9 is essential for the design of storm sewers and open channels. In the case of drawdown curves, it is sometimes possible to achieve a substantial savings in cost by reducing the wall height or by reducing the size of the conduit, thereby also avoiding overhead structures.

Two stepwise calculation methods are available: the Direct Step Method, which yields the **location** of chosen depths; and the Standard Step Method giving **depths** at selected locations. Both methods provide information necessary for plotting the nonuniform water surface profile. The Direct Step Method is recommended for hand calculations under prismatic channel/pipe conditions since it does not involve tedious iterative operations. The Standard Step Method on the other hand, is iterative and is best handled by a computer.

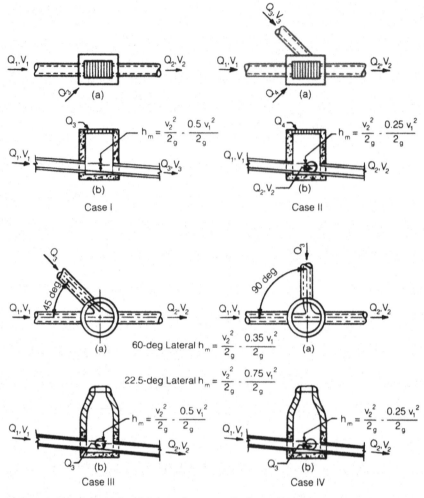

Figure 6.14—Minor head losses due to turbulence at structures.

As noted in previous sections, all calculations of water surface profiles must begin at a control section where the depth is known. They will proceed in the direction of flow if the flow is supercritical, or in the upstream direction if the flow is subcritical. Numerous references are available for step-wise and analytical calculations (Chow 1959; Von Seggern 1950; Kiefer and Chu 1955; Chow 1955) as well as for graphical water surface profile computations (Chow 1959; Thomas 1934).

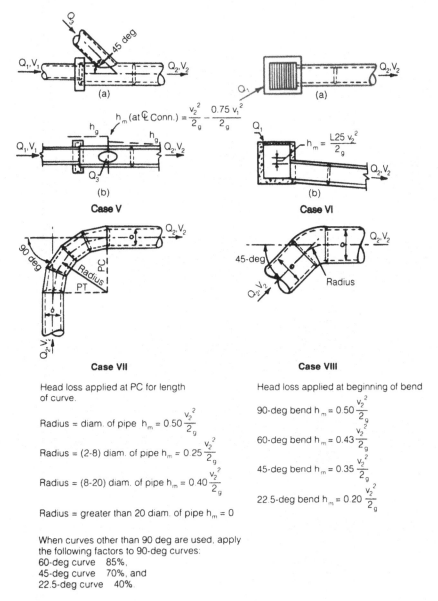

Case V

$$h_m \text{ (at } \mathcal{C} \text{ Conn.)} = \frac{v_2^2}{2_g} - \frac{0.75 \, v_1^2}{2_g}$$

Case VI

$$h_m = \frac{L25 \, v_2^2}{2_g}$$

Case VII

Head loss applied at PC for length of curve.

Radius = diam. of pipe $h_m = 0.50 \dfrac{v_2^2}{2_g}$

Radius = (2-8) diam. of pipe $h_m = 0.25 \dfrac{v_2^2}{2_g}$

Radius = (8-20) diam. of pipe $h_m = 0.40 \dfrac{v_2^2}{2_g}$

Radius = greater than 20 diam. of pipe $h_m = 0$

When curves other than 90 deg are used, apply the following factors to 90-deg curves:
60-deg curve   85%,
45-deg curve   70%, and
22.5-deg curve   40%.

Case VIII

Head loss applied at beginning of bend

90-deg bend $h_m = 0.50 \dfrac{v_2^2}{2_g}$

60-deg bend $h_m = 0.43 \dfrac{v_2^2}{2_g}$

45-deg bend $h_m = 0.35 \dfrac{v_2^2}{2_g}$

22.5-deg bend $h_m = 0.20 \dfrac{v_2^2}{2_g}$

Figure 6.15—Minor head losses due to turbulence at structures.

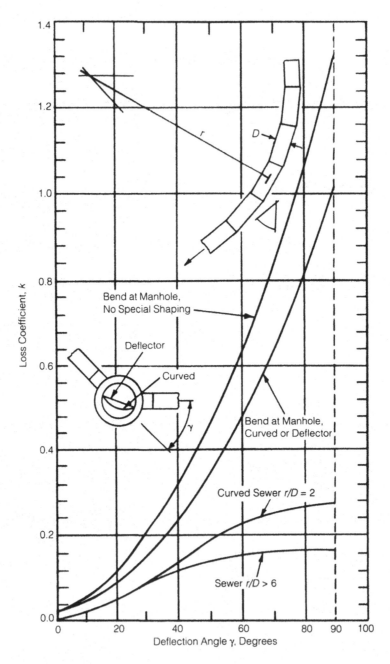

Figure 6.16—Sewer bend loss coefficient.

TABLE 6.13. Bend Loss Coefficients in Open Channels.

r/b (1)	k_b (2)
2.5	0.02
2.0	0.07
1.5	0.12
1.0	0.25

The Direct Step Method for calculating the length of prismatic channel or conduit between cross-sections where the water surface elevation is known is based on the following equation:

$$L = \frac{E_2 - E_1}{S_o - \overline{S}_f} = \frac{(d + h_v)_2 - (d + h_v)_1}{S_o - \overline{S}_f} \qquad (6\text{-}37)$$

where

L = length between cross-sections.
E = sum of depth and velocity heads (specific energy).
d = depth.
h_v = velocity head.
$\overline{S}_f$ = average friction slope between the two cross-sections.
S_o = bed slope of the channel or conduit.

Examples of both the direct-step and standard-step methods are illustrated below.

Example 6-4: Direct Step Method Calculation of M-1 Backwater Curve An 8-foot diameter circular storm sewer is laid at a slope of 0.001 ft per ft and is conveying a flow of 80 cfs. At its confluence with an open channel, the depth of flow is 8.0 ft. Assume a roughness coefficient of 0.013 when flowing full. Critical depth is 2.20 ft and normal depth is 3.24 ft. Determine the length required to reach normal depth. The calculations are shown in Table 6.14, adapted from ASCE (1972).

Example 6-5: Standard Step Calculation of S-2 Drawdown Curve Flow from a channel on a mild slope enters a 36-inch circular pipe which is laid on a slope of 0.02 ft per ft. Determine the water surface profile downstream from the pipe entrance for a discharge of 40 cfs. Assume a roughness coefficient of 0.013, which is constant with depth. Critical depth is 2.06 feet and normal depth is 1.36 ft. Since $d_c > d_n$, the uniform flow is supercritical, so calculations should start at critical depth and proceed downstream.

TABLE 6.14.   Calculation of M-1 Backwater Curve by Direct-Step Method (ASCE 1972).

d (ft) (1)	d/D (2)	Q/Q_f (3)	Q_f (cfs) (4)	V_f (fps) (5)	V/V_f (6)	V (fps) (7)	h_v (ft) (8)	$d + h_v$ (ft) (9)	S (10)	S_e (11)	$S_e - S_o$ (12)	$\Delta(d + h_v)$ (ft) (13)	ΔL (ft) (14)
8.00	1.00	1.00	80	1.59	1.00	1.59	0.04	8.04	0.85×10^{-4}	1.1×10^{-4}	-8.9×10^{-4}	-1.98	2,220
6.00	0.75	0.79	101	2.01	0.98	1.97	0.06	6.06	1.35×10^{-4}	1.7×10^{-4}	-8.3×10^{-4}	-0.97	1,170
5.00	0.62	0.60	133	2.64	0.90	2.38	0.09	5.09	2.1×10^{-4}	3.4×10^{-4}	-6.6×10^{-4}	-0.93	1,410
4.00	0.50	0.40	200	3.97	0.80	3.18	0.16	4.16	4.8×10^{-4}	6.0×10^{-4}	-4.0×10^{-4}	-0.36	900
3.60	0.45	0.33	242	4.80	0.75	3.60	0.20	3.80	7.2×10^{-4}	8.6×10^{-4}	-1.4×10^{-4}	-0.29	2,080
3.24	0.40	0.28	290	5.77	0.72	4.16	0.27	3.51	10.0×10^{-4}				

$$L = 7,780$$

Explanation:

Col. 1: Assumed depths between initial depth of 3.24 ft and terminal depth of 8.00 ft.

Col. 2: Col. 1 ÷ D (diameter of sewer).

Col. 3: From Fig. 6-13 for d/D in Col. 2.

Col. 4: 80 cfs ÷ Col. 3.

Col. 5: Col. 4 ÷ 50.3 sq ft.

Col. 6: From Fig. 6-13 for d/D in Col. 2.

Col. 7: Col. 5 × Col. 6.

Col. 8: Velocity head for Col. 7.

Col. 9: Col. 1 + Col. 8.

Col. 10: S from Manning equation (6-17) for $D = 8$ ft, Q_f of Col. 4, and $n = 0.013$.

Col. 11: Arithmetic mean of successive pairs in Col. 10.

Col. 12: Col. 11 − S_o.

Col. 13: Difference between successive amounts in Col. 9.

Col. 14: Col. 13 ÷ Col. 12.

Note 1: To obtain accuracy, the assumed depths should be chosen such that differences between successive values of velocities shown in Col. 7 will be less than 10 percent; to make Table 6.14 concise, this limit was not met.

Note 2. Ft × 0.3048 = m; cfs × 1.7 = cu m/min.

Arbitrary lengths of reach and assumed depths of flow are selected and S_f is obtained by averaging successive S_f values. The friction loss is applied to the previous H and the result is compared with the assumed value. If they are sufficiently close, the calculation proceeds to the next station. If not, a new trial depth of flow is selected for the current station and the process repeated until the assumed and computed water surface elevations agree, within some specified tolerance. The calculations are shown in Table 6.15, adapted from ASCE (1972).

XII. SPECIAL HYDRAULIC STRUCTURES

A. Storm Sewer Inlets

Storm sewer inlets are the means by which urban runoff enters the sewer system. A storm sewer system is usually designed on the assumption of full-flowing pipes, often with little regard for how surface runoff is delivered to it. In fact, the storm sewer inlet is an important element of the design in its own right. As shown in Figure 6.17, stormwater inlets can take many forms, but usually are classified as curb inlets, gutter inlets, or slotted drains. No one inlet is best suited for all conditions. The hydraulics of flow into an inlet are based on principles of weir and orifice flow, modified by laboratory and field observation of entrance losses under controlled conditions.

Curb inlets are installed along street sections having curbs and gutters to intercept stormwater runoff and to allow its passage into a storm sewer. Inlets can be located at low points (sumps), directly upstream from street intersections, and at intermediate locations as well. The spacing of these intermediate curb inlets depends on several criteria but is usually controlled by rate of flow and the permissible water spread toward the street crown. The type of road is also important since the greater the speed and volume of traffic, the greater the potential for hydroplaning. On the other hand, it is also considered acceptable practice to allow some periodic and temporary flooding of low volume streets.

Given the maximum allowable width of street flooding, the designer can compute the allowable rate of runoff in the curb section represented as a flat triangular channel by:

$$Q = 0.56 \left(\frac{Z}{n}\right) d^{8/3} S^{1/2} \qquad (6\text{-}38a)$$

or, in SI units:

$$Q = 0.38 \left(\frac{Z}{n}\right) d^{8/3} S^{1/2} \qquad (6\text{-}38b)$$

TABLE 6.15.   Calculation of S–2 Drawdown Curve by Standard-Step Method (ASCE 1982).

Q (cfs) (1)	$Qn/D^{8/3}$ (2)	Sta. (3)	Elev Water Surface (ft) (4)	Elev Invert (ft) (5)	d (ft) (6)	d/D (7)	A (sq ft) (8)	V (fps) (9)	$V^2/2g$ (ft) (10)	Assume Elev H (ft) (11)	$Qn/D^{8/3}S^{1/2}$ for d/D (12)	S_f (13)	Avg S_f (14)	H_f (15)	Computed Elev H (16)
40	.02778	0 + 00	102.06	100.00	2.06	0.687	5.18	7.72	0.93	102.99	0.378	.0054	—	—	102.99
—	—	0 + 10	101.54	99.80	1.74	0.580	4.25	9.41	1.37	102.91	0.295	.0089	.0072	0.07	102.92
—	—	0 + 30	100.99	99.40	1.59	0.530	3.80	10.53	1.72	102.71	0.255	.0119	.0104	0.21	102.71
—	—	0 + 60	100.29	98.80	1.49	0.497	3.50	11.43	2.03	102.32	0.230	.0146	.0133	0.40	102.31
—	—	1 + 50	98.39	97.00	1.39	0.463	3.20	12.50	2.43	100.82	0.203	.0187	.0167	1.50	100.81
—	—	2 + 50	96.37	95.00	1.37	0.457	3.14	12.74	2.52	98.89	0.199	.0195	.0191	1.91	98.90

Explanation:

Col. 2: Useful constant for reach calculations.

Col. 3: Stationing arbitrarily established to define backwater curve.

Col. 4: First line of calculation is known elevation at point of control, remaining lines assumed elevations.

Col. 6: Depth of flow, Col. 4 − Col. 5.

Col. 8: Area of channel for depth in Col. 6, from Table 6.6 A/D^2 values for d/D in Col. 7.

Col. 9: Q/A, Col. 1 ÷ Col. 8.

Col. 10: $V^2/2g$ from Col. 9.

Col. 11: Col. 4 + Col. 10.

Col. 12: From Table 6.6 $Qn/D^{8/3}S^{1/2}$ for d/D in Col. 7.

Col. 13: (Col. 2 ÷ Col. 12)².

Col. 14: Average friction slope of adjacent stations (no entry on first line of reach).

Col. 15: Length between stations × Col. 14.

Col. 16: First line for actual elevation of energy head at point of control. Remaining lines are values at prior station plus or minus Col. 15. Col. 16 should be in approximate agreement with Col. 11 before proceeding.

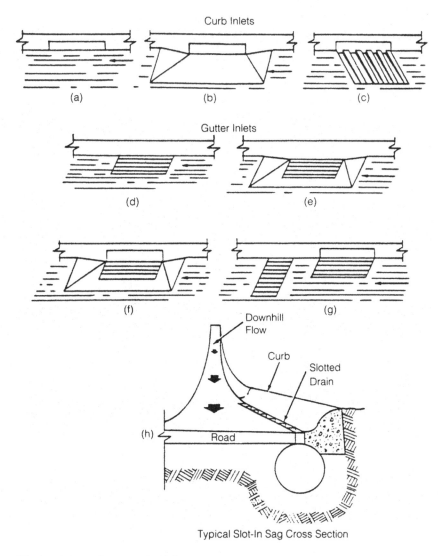

Curb Inlets

(a)   (b)   (c)

Gutter Inlets

(d)   (e)

(f)   (g)

Downhill
Flow

Curb

Slotted
Drain

(h)   Road

Typical Slot-In Sag Cross Section

Figure 6.17—Stormwater inlets.

Equation (6-38) is derived from Manning's equation where Z is the reciprocal of the street cross slope ($1/S_x$), n is Manning's roughness coefficient, S is the longitudinal slope of the street, and d is the depth of flow at the curb, and Zd (used in the nomograph) is the width of water spread. A nomograph is presented in Figure 6.18 for Equation 6-38. Note that the Manning equation was modified in deriving equation (6-38) because the hydraulic radius does not adequately describe the

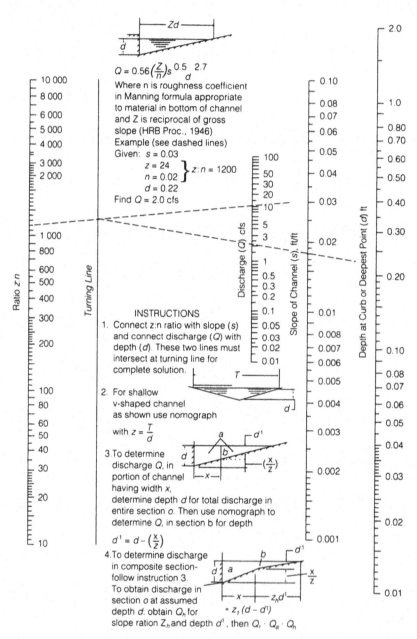

Figure 6.18—Nomograph for flow in triangular channels.

gutter cross-section, particularly where the top width Zd may be more than 40 times the depth at the curb.

If the actual runoff exceeds the allowable flooding, then the inlets need to be redesigned, placed closer together, or made larger. The capacity of each individual inlet depends on whether the inlet is located on a continuous grade or in a sump. The capacity also depends on the reduction factor used for blockage, the interception ratio for the total flow, whether the inlet throat is depressed, and whether deflectors are used.

Actual capacity is determined from nomographs or equations found in state highway department manuals, textbooks, and in some regional specifications. Two very good references are the Denver Regional Urban Storm Drainage Criteria Manual (1969) and the Federal Highway Administration report on Drainage of Highway Pavements (1984).

As a practical matter, curb inlets are notoriously inefficient and there never can be too many. The final design should show roughly one 4-foot wide inlet for every 3 cfs on a street with longitudinal slope of 2% or less.

Grate inlets are flush-mounted inlets installed along drainageways and streets for the purpose of intercepting storm water and directing it into the underground storm drainage system. Grate inlets can be sized large enough to drain an area quickly, which in turn will reduce infiltration into sanitary sewers and minimize freeze-thaw cycle effects under nearby pavements. Conversely, grate inlets can be undersized to reduce outflow and intentionally back-up storm water into a detention basin which will reduce peak rates of flow downstream.

Grate inlets function best when located in a sump. For depths of water not exceeding 0.4 feet, the weir equation is used:

$$Q = C_w \, P \, d^{3/2} \qquad (6\text{-}39)$$

where $C_w = 3.0$ (1.66 in SI units), P is the perimeter of the grate opening assuming no bars, and d is the depth of flow above the grate. If the grate is adjacent to a curb, that side of the grate is not counted in the perimeter. For depths of flow exceeding 1.4 feet, the orifice equation is used:

$$Q = C_o \, A \, (2gh)^{1/2} \qquad (6\text{-}40)$$

where $C_o = 0.6$, A is the total area of opening, g is the acceleration due to gravity, and h is the head above the center of the orifice. For depths greater than 0.4 feet but less than 1.4 feet, the capacity will be somewhere between those calculated by the weir and orifice equations.

The minimum length of clear opening parallel to the direction of flow necessary to allow the flow to fall through the opening and clear the downstream end of the bars can be estimated by the equation:

$$L = 0.5 \ V(t + d)^{1/2} \qquad\qquad (6\text{-}41a)$$

where L is the minimum length of clear opening in feet, V is the average velocity of the approach water, d is the approaching water depth in feet, and t is the grate thickness in feet.

In SI units:

$$L = 0.91 \ V(t + d)^{1/2} \qquad\qquad (6\text{-}41b)$$

Again, reference is made to FHWA (1984) for detailed capacity calculations.

B. Culverts

Culverts are defined as short conduits sized to pass a stream or drainage swale under a roadway or railway embankment without overtopping. Culverts can be circular, elliptical, arched, rectangular, or square in cross-section. They are frequently mounted in a headwall with an improved entrance. The headwall may have multiple barrels if allowable headwater is limited. Under heavy runoff conditions, the culvert inlet can restrict the amount of water passing into the culvert, and the culvert is said to flow under inlet control. The culvert capacity is given by the orifice equation, as modified by a projecting or non-projecting entrance.

In cases where the culvert is long and there is tailwater submergence, the culvert can flow under outlet control. Here the capacity is dependent on the culvert length, slope, roughness, and entrance and exit loss, as well as the tailwater depth. The hydraulics of flow through culverts under outlet control are based on the energy equation, plus the Manning equation to represent the friction slope. Culvert design charts for both inlet and outlet control are abundant and are commonly used in an iterative design process. Design criteria include size and cross-section of the culvert, culvert type, amount of headwater available, and the details of the entrance. The best known charts are those by the Federal Highway Administration (1985). The hydraulic design of culverts is discussed further in Chapter 8.

C. Energy Dissipators

Energy dissipators are used in a storm drainage system to reduce flow velocity (kinematic energy) and thereby reduce pipe scouring or stream bank and bed erosion. Depending on the type of channel or pipe lining, flow velocities may become excessive well below the critical velocity.

Open channel or non-pressurized flow velocities in a conduit or channel can be estimated by Manning's equation. For culverts under inlet control (partially full) velocities can be estimated from Manning's equation or by dividing the rate of flow by the cross-sectional area under outlet control (full flow).

Allowable flow velocities vary with conditions. For example, according to the Virginia Erosion and Sediment Control Handbook, (Virginia Soil and Water Conservation Commission 1982), permissible velocities for earth-lined channels range from 2.5 feet per second for loamy sand and sandy loam soils to 6.0 feet per second for coarse gravel, shale, and hard pan. Permissible velocities for grass-lined channels range from 2.5 feet per second for red fescue and red top grasses to 6.0 feet per second for Bermuda grass on soils that are not highly erodible. According to the Virginia Department of Highways and Transportation Drainage Manual, the allowable velocity in riprapped channels ranges from 12 feet per second for 50–150 pound stone to about 25 feet per second for 3–10-ton stone.

There are countless types of energy dissipators, including riprap, poured in place concrete, drop overs (which reduce channel slopes), and manufactured devices. Even grass linings on channels and natural undergrowth can be considered as energy dissipators since they act to retard flow velocity, though their effectiveness is limited under high depth and velocity conditions.

The effectiveness of a dissipator varies by type and magnitude of storm. However, it should be emphasized that certain practices produce poor energy dissipation and should be avoided. For example, straw bales, silt fences, and small stone will not only prove worthless, but they will have to be cleaned up somewhere downstream following a severe storm.

D. Drop Structures

Drop structures are used in open channels as a means of reducing channel gradient, and thereby channel velocity. Thus, drops can be considered a form of energy dissipator.

Reduced channel velocity means less capacity for a given cross-section. Other disadvantages of a vertical drop include the turbulence and erosive effect of the falling water on the drop structure, necessitating high maintenance. Because of these disadvantages, vertical drops are not recommended and sloping drops are used instead. As with vertical drops, stream banks in the vicinity of slope drops must be protected from erosion and scouring. The hydraulics of drop structures are complex, and reference is made to various publications by the USDA Soil Conservation Service for design specifications.

E. Outlet Structures

The purpose of an outlet structure in a storm drainage system is to deliver storm water that has been collected in a conduit, channel, or detention basin to a downstream receiving channel or body of water. Four factors influence the design of outlet structures, of which adequate capacity is the first and most obvious. A constricting outlet structure inadequate to carry the design flow will cause flooding and possible damage upstream. On the other hand, an oversized outlet structure is a waste of funds since peak flows will pass through unattenuated. In cases such as detention basins, only a predetermined amount of flow through an outlet can be permitted. Usually multiple stages will be required to maintain outflows at pre-development levels across the full range of design flows.

Second, non-erosive velocities must be maintained, recognizing that inadequate velocity will produce sediment deposits within the drainage way. This of course will cause increased maintenance and/or reduced capacities, which together can combine to increase the potential for flooding and related property damage.

XIII. ROUTING

With recent emphasis on watershed-wide control of stormwater run-off in many states, the essence of stormwater management planning has become the analysis of individual subarea runoff impacts at a downstream point of interest under changing land use in the watershed. This concept requires that runoff control facilities be sized and located in such a way that not only local runoff control is achieved, but also that flow targets at downstream points are met. Thus, drainage engineers must be able to generate subarea hydrographs as well as route these discharge hydrographs through the system to identify those subareas best suited for various runoff control facilities. The flow routing operation is critical to the analysis and to the identification of sewer/channel elements susceptible to surcharge and street flooding.

Routing of unsteady discharge is essential to the analysis of subarea timing effects and also to the design of storm drainage facilities, such as detention basins and pumping stations. Inputs to the routing sequence will be outflow hydrographs from individual subareas. The routing operation consists of computing the movement and attenuation of an inflow hydrograph as it passes through the storm sewer system, or, under surcharge conditions, through both the storm sewer and overland system. Flow routing may involve anything from a simple time shift based on time-of-travel to solution of the Saint-Venant or gradually varied unsteady flow equations, taking into account backwater and pressure flow conditions. In both cases, we are computing the discharge hydrograph (or stage hydrograph) at the downstream end of a pipe, channel, or detention basin and accounting mathematically for the effects of storage on flow through the element.

A. Types of Routing Methods

There are numerous methods for routing flows through a sewer, channel or storage basin in an urban drainage system. All methods can be classified as either hydraulic (based on Saint-Venant equations) or hydrologic (based on mass continuity plus a storage-outflow relation). The intent here is to identify those most commonly used and provide references for the detailed computational procedures. The simplest routing method is the time-of-travel shift whereby the discharge or even the entire inflow hydrograph is translated without attenuation to the downstream end of the flow element.

The travel time is defined simply as the element length divided by the full-flow velocity estimated by the Manning or the Hazen-Williams equation. This has become a satisfactory routing technique for relatively simple systems without significant backwater effects created by pipe looping, diversion structures, or submerged outfalls. It is particularly well-suited to the separate storm drainage system where storage is insufficient to cause significant peak reduction.

Flow routing in an open channel on the other hand can have substantial peak attenuation because of the storage in overbank areas during flood stage. This should be treated by one of the hydrologic routing methods such as the Muskingum method or the kinematic wave technique to simulate important properties of the flood wave. Neither of these methods has the ability to represent backwater conditions emanating from a downstream location, however.

The Muskingum method is a well-documented hydrologic procedure for channel routing and can be found in any standard reference on engineering hydrology. It simply combines the equation of mass continuity with a two-parameter storage-inflow-outflow relation for the channel reach to obtain a value of the outflow hydrograph at the end of a specified time step. The routing calculations continue until the entire inflow hydrograph has been routed through the reach.

The kinematic wave routing method was developed originally for overland flow on a plane surface and has also been applied to flow routing in pipes and channels. The method derives from the assumption that gravity and friction forces dominate all other terms in the momentum equation for gradually varied unsteady flow and, therefore, that the bed slope of the channel or pipe equals the friction slope. This allows use of a uniform flow equation in combination with mass conservation to obtain a solution for the outflow hydrograph leaving a pipe or channel segment.

Theoretically, the kinematic wave method cannot attenuate the flood wave unless kinematic shock is incorporated in the computations. Nor can this method be used when severe backwater conditions are present, since it is based on tracking disturbances that move only downstream. Despite these disadvantages, the kinematic wave routing method is widely used.

A modified form that permits attenuation of the peak has been developed by the Soil Conservation Service (1983) in its hydrologic model TR-20. Similarly, the kinematic wave method is used to simulate urban drainage systems in HEC-1, developed by the U.S. Army Corps of

Engineers Hydrologic Engineering Center. It also serves as the principal routing technique in the urban drainage model known as DR3M, developed by the U.S. Geological Survey Urban Studies Program (Alley and Smith 1982), and is used in the Transport block of SWMM. The reader is referred to a text on kinematic wave modeling by Stephenson and Meadows (1986) for a full discussion of numerical methods and comparisons between time-shift, Muskingum, and kinematic wave routing methods.

Routing of inflow hydrographs through flood control reservoirs and detention basins is performed by storage-indication or modified Puls methods. These two are computationally equivalent and involve the construction of an elevation-storage-outflow relationship which can be used with the mass continuity equation to obtain the outflow hydrograph. An example of modified Puls routing is presented later in this chapter.

The routing of combined sewer flows in a complex network involving looped pipes, diversion structures, and submerged outfalls usually requires solving the one-dimensional unsteady, gradually varied flow equations presented previously in this chapter. Under these conditions, the flow is controlled by acceleration terms, in addition to gravity and friction forces, and consequently the kinematic wave method is no longer applicable.

The methods available for routing combined sewer flows are generally referred to as full dynamic methods and require the use of complex finite difference schemes working at small time intervals for computational stability. While numerous computer models are available, the best known in the United States is the Extended Transport model (EXTRAN) in the SWMM program. EXTRAN and other hydraulic transport models in its class have the capability to represent complex backwater conditions caused by flow diversion structures, sewer looping, tidal outfalls, flow reversals, and pressure or surcharge flow.

EXTRAN can also handle a variety of sewer configurations including parallel pipes and trapezoidal channels. EXTRAN uses an explicit form of the Saint-Venant equations particularly sensitive to time step size and prone to numerical instability. In general, all models in this class, while affording complete representation of all elements in a complex sewer system, require considerable hydraulic modeling and computing expertise. Further information on hydraulic and hydrologic routing models is contained in Chapter 7.

B. Detention Basin Routing By Modified Puls Method

1. Basic Equations

The Modified Puls method consists of repetitive solutions of the continuity equation and is based on the assumptions that (1) the reservoir water surface remains horizontal; and (2) the outflow from the reservoir is a unique function of storage. This method is sometimes called the storage-indication method. The methods are numerically equivalent and can be applied to major flood control projects as well as to small detention basins.

The continuity equation may be expressed as:

$$\bar{I} - \bar{O} = \frac{\Delta S}{\Delta t} \qquad (6\text{-}42)$$

where

$\bar{I}$ = mean inflow into reservoir during routing period Δt,
$\bar{O}$ = mean outflow from reservoir during routing period Δt,
ΔS = change in reservoir storage during routing period Δt.

Equation 6-42 may be expressed by:

$$\frac{I_1 + I_2}{2} - \frac{O_1 + O_2}{2} = \frac{S_2 - S_1}{\Delta t} \qquad (6\text{-}43)$$

where subscripts 1 and 2 denote the beginning and end, respectively, of a routing period, Δt.

The assumption implicit in Equation 6-43 is that the discharge varies linearly with time during a routing period Δt. This assumption must be borne in mind when selecting a routing period.

Equation 6-43 may be re-stated as follows:

$$(I_1 + I_2) + \left(\frac{2S_i}{\Delta t} - O_1\right) = \frac{2S_2}{\Delta t} + O_2 \qquad (6\text{-}44)$$

In Equation 6-44, all terms on the left-hand side are known from preceding routing computations. The terms on the right-hand side involving S_2 and O_2 are unknown and must be determined by storage routing.

2. General Routing/Design Procedure for Stormwater Detention Facilities

Subtract the pre-development runoff volume from the post-development volume to arrive at an estimation of trial detention basin size. The difference represents the approximate storage requirement. Then refer to storage-elevation data for the site to estimate the headwater available. Next, refer to culvert design charts by FHWA for inlet/outlet control to determine the approximate size of outlet required to pass maximum allowable outflow at a headwater depth corresponding to the storage requirement. A full routing analysis of the trial basin/outlet is based on the following steps in the Modified Puls method:

(a) Compute pre-development hydrograph from which the maximum basin outflow is determined.

(b) Compute post-development hydrograph which is to be routed through the proposed detention basin.

(c) Compute a table and/or curve of water depth versus storage (a function of basin geometry); water depth is measured above the basin floor or lowest outflow pipe invert elevation.

(d) Compute a table and/or curve of water depth versus outflow, recognizing that stage-discharge relationships are a function of the particular outlet structure and whether it is flowing under inlet or outlet control.

(e) Select a routing period Δt such that there are 5 or 6 points on the rising side of the inflow hydrograph, one of which coincides with the inflow peak.

(f) Construct a graph of $[(2S/\Delta t) + O]$ versus O where
    S = storage volume, cfs-hours
    Δt = routing period, hours
    O = outflow rate, cfs

(g) A routing procedure is now initiated using a tabular method for the solution of Modified Puls equation in equation (6-44).

(h) Compare the maximum outflow rate with the allowable rate of discharge from the drainage area.

(i) Adjust size, shape, and/or outlet structure if the maximum routed outflow is greater than the allowable.

(j) Repeat the entire routing procedure for alternative designs.

Example 6-6: Single-Stage Detention Basin Routing and Design   The example shown in Figure 6.19 (Kibler 1988) is a 150-acre watershed located in central Pennsylvania. The developed condition in Figure 6.19 includes a 53-acre site in the middle of the watershed which will be developed as a detached multi-family town house project. In addition, the area to the right will be developed as single family homes. The pre-development land use was corn/hay agricultural in the town house area. The detention basin adjacent to the roadway embankment is to be designed as a single-stage structure which will control the Q_{25} post-development runoff peak at the pre-development level. The pre- and post-development hydrographs were computed by the tabular hydrograph method in SCS Technical Release 55 (SCS 1986) and are shown in Figure 6.20 (Kibler 1988).

Based on the TR55 tabular hydrographs in Figure 6.21, the storage volume is estimated at approximately 6 acre-feet from Fig. 6.1, TR55 (SCS 1986). The pond depth and headwater at this storage level will be about 8.0 feet from Figure 6.21. With 8.0 feet of head, a 48-inch CMP culvert will discharge 125 cfs under inlet control, as shown in Figure 6.22. Since this is close to the target outflow of 113 cfs, try a single 48-inch CMP outlet. The following steps constitute the detention routing procedure (FHWA 1985):

(a) The pre- and post-development hydrographs are given in Figure 6.21. The problem is to size the detention basin and outlet works to reduce the post-development flood peak of 186 cfs to the pre-development level of 113 cfs.

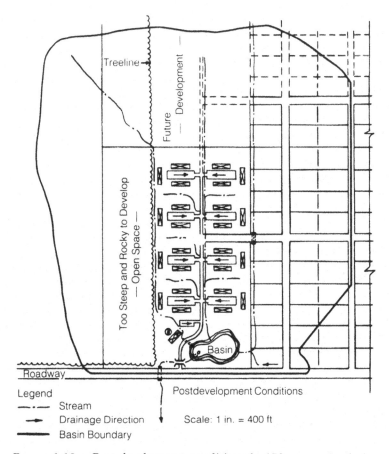

Figure 6.19—Post-development conditions in 150-acre watershed used in detention basin routing example.

(b) The storage-elevation curve for the proposed site of the detention facility is shown in Figure 6.21. The outlet invert is assumed to be at elevation 1060.0 feet for dry-pond alternative.

(c) The hydraulic performance curve for a 48-inch CMP culvert from FHWA HDS No. 5 (1985) with projecting inlet flowing under inlet control is shown in Figure 6.22. This is an assumed size and flow condition.

(d) Select the routing interval Δt such that there are 5 or 6 points on the rising limb of the inflow hydrograph, one of which coincides with the inflow peak. From Figure 6.21, $\Delta t = 10$ min.

(e) Construct table and graph of $[(2S/\Delta t) + O]$ versus outflow using Figures 6.21 and 6.22 with $\Delta t = 10$ minutes. These are shown in Table 6.16 and Figure 6.23.

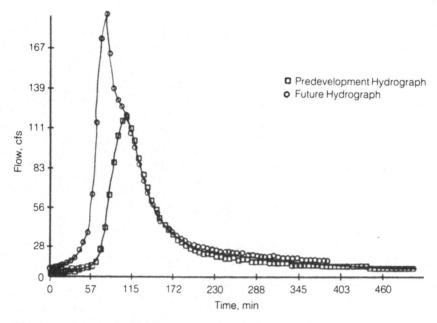

Figure 6.20—Q_{25} hydrographs by SCS tabular method for existing and developed conditions in 150-acre example watershed.

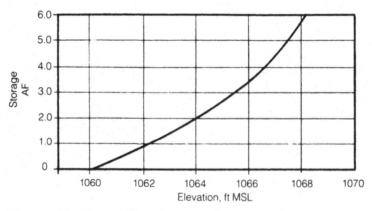

Figure 6.21—Storage-elevation curve for proposed detention site.

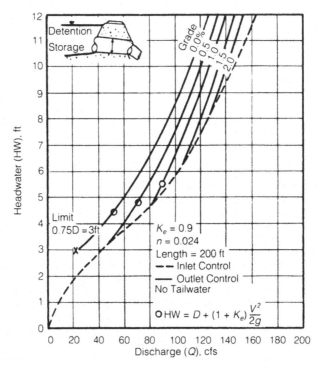

Figure 6.22—Discharge performance curves for 48-inch CMP culvert.

TABLE 6.16. $2S/\Delta t$ for 48-inch CM Pipe Culvert with $\Delta t = 10$ Minutes

Elev (1)	Storage, AF (2)	Storage, cfs-hrs (3)	O cfs (4)	$2S/\Delta t + O$ (5)
1060	0.00	0.0	0	0
1061	0.40	4.8	6	63
1062	1.00	12.0	21	165
1063	1.50	18.0	40	256
1064	2.00	24.0	65	353
1065	2.60	31.2	87	461
1066	3.40	40.8	103	593
1067	4.20	50.4	115	720
1068	5.40	64.8	125	903
1069	7.20	86.4	135	1172
1070	10.00	120.0	147	1587

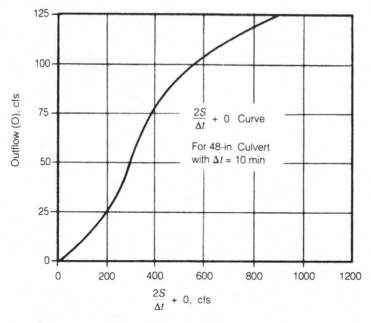

Figure 6.23—Plot of outflow vs 2s/Δt + o for 48-inch culvert.

(f) Construct a routing table for solution of equation (6-43), using the post-development inflow hydrograph in Figure 6.20 and the $2S/\Delta t$ plot in Figure 6.23. The completed routing table is shown in Table 6.17.

(g) Compare the maximum outflow rate, cfs, with the allowable rate of 113 cfs. Since this 48-inch CMP design results in a peak outflow rate of 113 cfs it is probably acceptable without further modification. However, other options should be investigated as necessary. Note that TR55 produces a storage estimate which is about 38% higher than that by Modified Puls routing. The TR55 approximate routing procedure is very useful for preliminary sizing purposes, but should always be followed by a full routing analysis to determine design adequacy.

3. Multi-stage Detention Facilities

Because of the need to control multiple storms with return periods usually between 2–50 years, the engineer is faced with the challenge of sizing multiple outlets in a detention facility. These outlets take the form of various orifices, weirs, and drop inlets which will control detention basin releases at pre-development levels in the watershed. Multiple openings are usually mounted in a riser box that discharges to an outlet pipe or culvert. The design of the total structure is complicated by: (1) submergence of the inlets by water in the riser; and (2) shifting control between inlet and outlet control in the outfall pipe. An inter-

TABLE 6.17.   Routing table and detention storage analysis for 48-inch pipe.

Time Hrs:min (1)	I_1 cfs (2)	$I_2 + I_2$ cfs (3)	$2S/\Delta t - O$ cfs (4)	$2S/\Delta t + O$ cfs (5)	O_2 cfs (6)
11:00	7	15	0	0	0
11:10	8	18	13	15	1
11:20	10	22	25	31	3
11:30	12	30	39	47	4
11:40	18	46	57	69	6
11:50	28	87	81	103	11
12:00	59	207	126	168	21
12:10	148	334	217	333	58
12:20	186	320	359	551	96
12:30	134	256	463	679	108
12:40	122	232	495	719	112
12:50	110	201	501	727	113
13:00	91	163	480	702	111
13:10	72	129	433	643	105
13:20	57	103	370	562	96
13:30	46	84	305	473	84
13:40	38	71	255	389	67
13:50	33	62	222	326	52
14:00	29	55	202	284	41
14:10	26	49	187	257	35
14:20	23	44	74	136	31
14:30	21	41	64	118	27
14:40	20	39	57	105	24
14:50	19	37	48	96	22
15:00	18	33	43	85	21

Notes: (1) Max. storage occurs at max. outflow at time 12:50. Since $2S/\Delta t + O = 727$ cfs at this time, $S = 51.17$ cfs-hrs $= 4.3$ AF. From storage-elevation curve, this produces a maximum depth of 6.8 feet at elevation 1066.8.

(2) The TR55 (1986) estimate of required storage is based on the following calculations:

$Q_o/Q_t = 113/186 = 0.61$

$V_s/V_r = 0.23$ from Fig. 6.1 TR55

V_s (AF) $= (0.23 \times 2.07$ inches $\times 150$ acres$)/12$

$= 5.96$ AF (about 38% high)

active computer solution is usually required. Chamberlain (1985) has developed a micro-computer program capable of analyzing up to 9 stages with 10 different outlet types including: circular and rectangular orifices, V-notch weir, proportional weir, rectangular weir, perforated riser, drop inlet with grate; emergency spillway; and circular discharge pipe. This multi-stage program has been documented by Kibler and Seybert (1989).

XIV. REFERENCES

Alley, W.M. and Smith, P.E. (1982). "Distributed routing rainfall-runoff model—Version II, computer program documentation, user's manual." *Open-file Report 82–344*, U.S. Geological Survey, Washington, D.C.

American Iron and Steel Institute. (1985). *Modern sewer design*. AISI, Washington, D.C.

American Society of Civil Engineers. (1969). *"Design and construction of sanitary and storm sewers."* ASCE Manual of Practice No. 37, WPCF Manual of Practice No. 9, ASCE, New York, NY.

American Society of Civil Engineers. (1982). *"Gravity sanitary sewer design and construction."* ASCE Manual of Practice No. 60, WPCF Manual of Practice No. RD-5, ASCE, New York, NY.

Bowers, C.E. (1950). "Studies of open channel hydraulics, Part V: Hydraulic model studies for Whiting Naval Air Station, Milton, Fla." *Tech. Paper No. 6, Series B*, St. Anthony Falls Hydraulics Laboratory, University of Minnesota, Minneapolis.

Brater, E.F. and King, H.W. (1976). *Handbook of hydraulics*. 6th edition, McGraw-Hill, New York, NY.

Chamberlain, A.S. (1986). *A multiple stage detention basin analysis and design model*, Master's thesis, The Pennsylvania State University, University Park, PA.

Chow, V.T. (1955). "Integrating the equation of gradually varied flow." *Journal of the Hydraulics Division*, 81 (HY1).

Chow, V.T. (1959). *Open channel hydraulics*. McGraw-Hill, New York, NY.

Christensen, B.A. (1984). "Design of partially filled circular pipes: A rational approach to storm sewer analysis." *Proc intl symp on urban hydrology, hydraulics and sediment control*, University of Kentucky, Lexington, KY.

Christensen, B.A. (1985). *Hydraulics: A Compendium of Formulas, Diagrams, Programs, and Solved Problems for Use in Design and Analysis of Hydraulic Engineering Problems*. University of Florida, Gainesville, FL.

City of Austin. (1987). *Austin drainage criteria manual*. 2nd ed., Watershed Management Division, Austin, TX.

Croley, T.E. III. (1979). *Hydrologic and hydraulic computations on small programmable calculators*. Iowa Institute of Hydraulic Research, University of Iowa, Iowa, City, IA.

Daily, J.W. and Harleman, D.R.F. (1966). *Fluid dynamics*. Addison-Wesley Publishers, Reading, MA.

"Design charts for open-channel flow." (1961). *Hydraulic design series No. 3*, Federal Highway Administration, U.S. Department of Transportation, Washington, D.C.

"Hydraulics of bridge waterways." (1978). *Hydraulic design series No. 1*, Federal Highway Administration, U.S. Department of Transportation, Washington, D.C.

"Drainage of highway pavements." (1984). *Hydraulic engineering no. 12*, Federal Highway Administration, Washington, D.C.

"*Handbook of channel design for soil and water conservation.*" (1954). U.S. Government Printing Office, Washington, D.C.

"Hydraulic design of highway culverts." (1985). Federal Highway Administration, *Hydraulic design series no. 5, Report no. FHWA-IP-85-15*, Washington, D.C.

French, R.H. (1985). *Open-channel hydraulics*. McGraw-Hill, New York, NY.

Henderson, F.M. (1966) *Open-channel flow*. MacMillan Co., New York, NY.

Hendrickson, J.G. Jr. (1964). *Hydraulics of culverts*. American Concrete Pipe Association, Chicago, IL.

Kibler, D.F. (1988). "Computational Methods in Stormwater Management." *Proc of Conf*, Pennsylvania State University, University Park, PA.

Kibler, D.F. and Seybert, T.A. (1989). *Penn State urban hydrology model users manual*. Pennsylvania State University, University Park, PA.

Kiefer, C.J. (1955). "Backwater function by numerical integration." *Transactions of the American Society of Civil Engineers*, 120, 429, New York, NY.

Linsley, R.K. and Franzini, J.B. (1964). *Water Resources Engineering*. McGraw-Hill, New York, NY.

"Little Lehigh Creek Watershed Storm Water Management Plan." (1988). high-Northampton Counties Joint Planning Commission. Allentown, PA.

Marsalek, J. (1985). "Head losses at selected sewer manholes." *APWA Special Report No. 52*, American Public Works Association, Chicago, IL.

Marsalek, J. (1986). "Hydraulically efficient junction manholes." *APWA Reporter*, 53 (12).

Marsalek, J. (1987). "Improving flow in junction manholes." *Civil Engineering*, 57 (1).

National Bureau of Standards. (1938). "Pressure losses for fluid flow in 90-degree bends." *Journal of Research*, 21, Paper RP1110.

Rouse, H. (1961). *Fluid Mechanics for Hydraulic Engineers*. Dover Publications Inc., New York, NY.

Smith, P.D. (1986). *BASIC hydraulics*. Butterworth Scientific Press, London, England.

Soil Conservation Service. (1983). "Computer program for project formulation—hydrology." *SCS Technical Release 20 (TR-20)*, Washington, D.C.

Stephenson, D. and Meadows, M.E. (1986). *Kinematic hydrology and modeling*. Elsevier Science Publishing Co., Inc., New York, NY.

Thomas, H.A. (1934). *Hydraulics of flood movements in rivers*. Carnegie Institute of Technology, Pittsburgh, PA.

Urban Drainage and Flood Control District. (1984). *Urban storm drainage criteria manual*. rev. ed., Denver Regional Council of Governments, Denver, CO.

U.S. Army Corps of Engineers. (1973). *HEC-1, Flood hydrograph package.* Hydrologic Engineering Center, Davis, CA.

U.S. Bureau of Reclamation. (1977). *Design of small dams.* U.S. Government Printing Office, Washington, D.C.

"Urban hydrology for small watersheds." (1986). *Technical Release 55, NTIS PB87-101598,USDA/SCS,* Washington, D.C.

Virginia Department of Highways and Transportation. (1980). *Drainage manual.* Richmond, VA.

Virginia Soil and Water Conservation Commission. (1982). *Virginia erosion and sediment control handbook.* 2nd ed., Richmond, VA.

Von Seggern, M.E. (1950). "Integrating the equation of non-uniform flow." *Trans of the American Society of Civil Engineers,* 115, 71, New York, NY.

Williams, G.S. and Hazen, A. (1945). *Hydraulic tables.* 3rd ed., John Wiley and Sons, Inc., New York, NY.

Chapter 7

COMPUTER MODELING

I. INTRODUCTION

Computer modeling became an integral part of storm drainage planning and design in the mid-1970s. Several federal agencies undertook major software developments, including The Hydrologic Engineering Center (HEC) of the U.S. Army Corps of Engineers with the Storage, Treatment and Overflow Runoff Model (STORM) and the HEC-1 and HEC-2 series; the U.S. Soil Conservation Service with TR-20 and WSP2; and the U.S. Environmental Protection Agency, who sponsored development of the Storm Water Management Model (SWMM). These were soon supplemented by a plethora of proprietary models, many of which were simply variants on the originals. The proliferation of microcomputers in the 1980s has made it possible for virtually every engineer to use state-of-the-art analytical technology for purposes ranging from analysis of individual pipes to comprehensive stormwater management plans for entire cities.

In addition to the simulation of hydrologic and hydraulic processes, models can have other uses. They can provide a quantitative means to test alternatives and controls before implementation of expensive measures in the field. If a model has been calibrated and verified at a minimum one site, it may be used to simulate non-monitored conditions and to extrapolate results to similar ungauged sites. Models may be used to extend time series of flows, stages, and quality parameters beyond the duration of measurements, from which statistical performance measures then may be derived. They may also be used for design optimization and real-time control.

This chapter presents modeling guidelines and describes some frequently-used operational (see Section IV for definition of "operational") models that are readily available for various kinds of analyses related to stormwater management. No attempt will be made to describe all available models, nor will details of individual models be presented.

II. PROBLEM IDENTIFICATION

While models may be useful for the solution of many stormwater problems, the following types of problems have been found to be especially amenable to analysis using computer modeling.

A. Drainage and Flooding

The traditional urban drainage design problem of avoidance of street and surface flooding is commonly analyzed by models. Only a few models can analyze the general problem of surcharging, the condition in which the hydraulic grade line is above the sewer crown. Conditions during surcharging are hydraulically similar to pressurized conditions found in a water distribution network and often are modeled in a similar manner (see Chapter 6). Models can be used to simulate flows in both the major and minor systems, as well as the effect of possible inlet controls on combined sewers (Wisner 1982; Wisner et al. 1986).

B. Detention/Retention Storage

Many of the storm drainage models currently available can simulate the hydraulics of detention facilities. Simulation of the behavior of such facilities for removal of pollutants in urban runoff is in its infancy, due largely to the lack of scientifically controlled studies on the pollutant removal efficiency of these devices.

C. Sedimentation

Sedimentation in storm sewers can constitute both a quantity and quality problem. Excessive sedimentation can diminish the hydraulic capacity of the system and result in upstream flooding. Sediment deposited during small storms can flush during larger storm events. Simulation of the complex scour and deposition cycle of sediment found in sewers remains at a low stage of development. However, some empirical modeling and attempts at more deterministic modeling will be discussed later.

D. Water Quality

Studies and projects involving urban stormwater runoff quality can relate to many problems. In a narrower sense, a water quality study may address a particular issue, such as bacterial contamination of a beach, release of oxygen demanding material into a stream or river, unacceptable aesthetics of an open channel receiving urban runoff, eutrophication of a lake, contamination of basements from surcharged sewers due to wet-weather flooding, etc. In some instances, local or state regulations may prescribe a nominal "solution" without recourse to any water quality analysis as such (see Section IIIB).

Simulation of stormwater impacts on receiving water quality involves modeling of both quantity and quality. Superimposed on almost any water quality modeling effort is the need to analyze controls and abatement strategies. The considerable uncertainty in quality modeling makes the effort especially difficult.

III. URBAN MODELING OBJECTIVES AND CONSIDERATIONS

If a problem does require modeling, particular modeling objectives will probably result. Models may be used for objectives such as the following:

(a) Characterize the urban runoff as to temporal and spatial flow distributions, concentration/load ranges, etc.
(b) Provide input to a receiving water quality analysis, e.g., drive a receiving water quality model.
(c) Determine effects, magnitudes, locations, combinations, etc. of control options.
(d) Perform frequency analysis on hydrologic or quality parameters, e.g., to determine return periods of concentrations/loads.
(e) Provide input to economic analyses.

A. Planning, Analysis/Design, and Operation

Another way of looking at modeling objectives is to consider that there are several levels of the process—planning, analysis/design and operation—that can use modeling.

Planning involves a comparison of general design and/or flood mitigation strategies and may include optimization and risk assessment. At the planning level, the relative effectiveness of alternative drainage and flood control practices may be assessed and economic trade-offs evaluated. Modeling is likely to be somewhat less detailed in an effort to screen several alternative strategies. Continuous simulation (discussed below) can be useful at this level to determine relative flooding frequency as affected by alternative stormwater management programs, selection of hydrologic events for detailed design and assessment of the reliability of a proposed design, and economic optimization.

At the analysis/design level, the detailed analysis of an existing system, proposed system, or system improvements is investigated. Examples include analysis of alternative surface drainage patterns, location of detention storage facilities, and alternative runoff transport systems (e.g. swales vs. pipes). Design models must be capable of realistic simulation of hydrologic, hydraulic and, possibly, water quality phenomena.

Operational controls are devices that function during a storm such as variable weirs, pumps, and gates. These devices are most often found in combined sewer systems and are used to minimize combined sewer overflows rather than provide flood protection. More information on combined sewer system analysis can be found in Chapter 10.

B. When Should A Model Be Used?

Modeling may not always be necessary to address the problems identified earlier or the objectives listed above. Since the process of implementing a model, collecting the necessary data, calibrating and verifying the model, and assessing the results can be very time consuming and expensive, it is important to understand that problems may be amenable to a simpler form of analysis. For example, field investigation of a flooding problem may reveal culverts choked by debris and/or sediment, or overgrown channels. Simple maintenance may solve this problem. In another case, field inspection may reveal that a single channel or road crossing is undersized. Upsizing this one structure to the capacity of the channel immediately upstream may solve the problem so that it is not necessary to model the entire upstream watershed. Modeling may also be unnecessary to implement required control measures.

By the same token, it should not be assumed that every water quality problem requires a water quality modeling effort. Some problems may be strictly hydraulic in nature, e.g., the basement flooding problem. That is, the solution may often reside primarily in a hydrologic or hydraulic analysis in which the concentration or load of pollutants is irrelevant.

For example, the State of Florida requires the capture of the first one-half inch of runoff for water quality control for certain size developments. Storage requirements for this case are simply calculated as one-half acre-inch per acre of development. In other cases the State requires that the runoff from the first inch of rainfall be captured for water quality control. For this case, a runoff model may be required to calculate the runoff volume produced by a one-inch rainfall.

While modeling generally yields more information, simpler methods may provide sufficient information for developing a control strategy. **In general, the simplest method that provides the desired analysis should be used.** The risk of using a more complex (and presumably "better") model is that it requires more expertise, data, support, etc. to use and understand, with a consequent higher probability of misapplication.

If modeling is necessary, it still may not be necessary to simulate quality processes since most control strategies are based on hydrologic or hydraulic considerations. Quality processes are very difficult to simulate accurately and generally incorporate many heuristic procedures that require extensive calibration (Huber 1985). If abatement strategies can be developed without the simulation of water quality parameters, the overall modeling program will be greatly simplified.

Computer models allow some types of analysis (such as frequency analysis) to be performed that could rarely be performed otherwise, since periods of runoff or quality measurements in urban areas are seldom very long. It should always be borne in mind, however, that use of measured data is usually preferable to use of simulated data, particularly for objectives (a) and (b), in which accurate concentration values are needed. Modeling is **not** a good substitute for data collection, especially for water quality parameters. Although modeling is generally cheaper than data collection, the uncertainties involved, especially in water quality simulation, mandate the collection of data for model calibration and verification.

Models sometimes may be used to extrapolate beyond the measured data record. It is important to recognize, however, that models do not extend data, but rather generate mathematically simulated numbers that should never be assumed to be the same as data collected in the field.

Careful consideration should be given when using models to provide input to receiving water quality analyses. The first urban runoff quality model (SWMM) inadvertently overemphasized the concept of simulation of detailed intra-storm quality variations, e.g., production of a "pollutograph" (concentration vs. time) at 5 or 10 minute intervals during a storm for input to a receiving water quality model. The fact is, however, that the quality response of most receiving waters is insensitive to such short-term variations, as illustrated in Table 7.1. In most instances, the total storm load will suffice to determine the receiving water response, eliminating the need to calibrate against detailed pollutographs. Instead, only the total storm loads need be matched, a much easier task. Simulation of short time increment changes in concentrations and loads is generally necessary only for analysis of control options, such as storage or high-rate treatment, whose efficiency may depend on the transient behavior of the quality constituents.

Any consideration of water quality modeling means that some additional data will be required for model input. As described later, such

TABLE 7.1. Required Temporal Detail for Receiving Water Analysis
(After Driscoll 1979; and Hydroscience 1979)

Type of Receiving Water (1)	Key Constituents (2)	Response Time (3)
Lakes, Bays	Nutrients	Weeks—Years
Estuaries	Nutrients, OD*	Days—Weeks
	Bacteria	
Large Rivers	OD, Nitrogen	Days
Streams	OD, Nitrogen	Hours—Days
	Bacteria	Hours
Ponds	OD, Nutrients	Hours—Weeks
Beaches	Bacteria	Hours

*OD = oxygen demand, e.g., BOD, that affects dissolved oxygen.

requirements may be as simple as a constant concentration or much more complex. Data may be obtained from existing studies or their acquisition may require extensive field monitoring. For some conceptualizations of the urban quality cycle, e.g., buildup and washoff, it may not be routinely possible to physically measure fundamental input parameters, and such parameters will only be obtained through model calibration. Involvement in acquisition of quality data, be it through literature reviews or field surveys, profoundly escalates the level of effort required for the study. Details on data requirements for urban areas will be deferred until modeling techniques are described.

IV. MODEL DEFINITION

In the broadest context, a model can be defined as any organized procedure for the analysis of a problem. With such a definition, almost any traditional technique could be included for discussion, from the rational method to unit hydrographs. However, this chapter treats a model in the popular sense of a computer program designed to analyze one or many problems encountered in storm drainage systems. The program may well incorporate traditional procedures, but will also include extensive routines for data management, including input and output procedures, and possibly including graphics and statistical capabilities. More specifically, this chapter discusses only those models that are "operational," i.e. defined as satisfying the following three criteria:

(a) An operational model must have documentation. This must include a user's manual that describes input data requirements, output to be expected, and computer requirements. In addition, the theory and numerical procedures used in the model must be explained so that the user will understand the basis for the model predictions. Documentation is the characteristic that most often distinguishes a model that can be accessed and used by others from the many computerized procedures described in the literature.
(b) An operational model must have support. This may be provided by the original model developer, the commercial vendor for the model or a sponsoring government agency. The outstanding hydrologic example is the Corps of Engineers, Hydrologic Engineering Center, in Davis, California which maintains a staff specifically for the purpose of model development, maintenance, and updating. Support means that the user can obtain answers, by telephone or written correspondence, to problems that arise during model implementation and use.
(c) An operational model should have been widely used by other than just the model developer. Regardless of its technical virtues, a procedure described in a single journal article or report with no experience or "review" by the engineering community is a weak candidate for use by a third party. Furthermore, user feedback is an invaluable means for identifying model limitations and "bugs," and initiating improvements and corrections to a model. Of course, no model will meet this

third criterion initially, and the prospective user must decide on the relative merits of new options versus older ones. Several models satisfying these criteria will be discussed later.

V. OVERVIEW OF AVAILABLE MODELING OPTIONS

A. Introduction

Several classification schemes can be developed for models, to differentiate the type and versatility of various models, e.g., deterministic versus stochastic, transient versus steady-state, lumped versus distributed, number of dimensions, quality and/or simulation, etc. These classifications are relatively unimportant when considering simulation of drainage systems because nearly all models are transient and schematize the whole system by linking lumped or one-dimensional schematizations of individual components such as catchments or pipes. Thus, for drainage system simulation there are only a few major distinguishing features among models.

Quantity modeling is relatively well understood. Many models can convert rainfall into runoff and perform flow routing; the user can make a selection on the basis of method used, computer supported, options included, etc. A reasonably accurate prediction of a runoff hydrograph will result if the modeler knows just three input parameters: the catchment imperviousness (or directly connected or hydraulically effective imperviousness), the watershed area and the rainfall hyetograph. Given these three parameters, hydrograph volumes and peaks may be predicted within "ball park" accuracy even before calibration.

Quality modeling is quite different. In a review of quality modeling methodologies, Huber (1985) concluded that prediction of absolute values of concentrations of quality parameters was not possible without calibration and verification data. That is, first-cut modeling attempts may differ from "true" values by orders of magnitude for concentrations and loads. About all that might be safely concluded from quality modeling without measured calibration and verification data is the **relative** effect of control strategies, although even these might be open to question depending upon the assumptions incorporated into the model. The result is that simulation of quality parameters should be performed only when necessary, and only when requisite calibration and verification data are available.

Modeling of runoff water quality parameters or development of water quality input data are often a prelude to receiving water quality modeling. A discussion of receiving water quality models is beyond the scope of this chapter. However, reviews are provided by Hinson and Basta (1982), EPA (1983a), Barnwell (1984), Thomann and Mueller (1987) and Ambrose et al. (1988). Corps of Engineers models are summarized briefly in a brochure (U.S. Army Corps of Engineers 1987).

B. Continuous Versus Single-Event Simulation

The first major hydrologic model was the Stanford Watershed Model (Crawford and Linsley 1966). It simulated the long-term response (e.g., over several years) of a watershed to precipitation input. It was the first continuous model, and its hydrologic components may be found in the current HSPF model (Johanson et al. 1984). Subsequent models typically were single-event models that converted an individual rainfall hyetograph into an individual runoff hydrograph, perhaps with quality processes superimposed. Single-event models characteristically employ a short-time step (e.g. 5–15 min.) and a detailed catchment and drainage network schematization. These models are often used for detailed design. Continuous models use a longer time step (usually 1 hour to correspond to available recorded hourly rainfall data) and generally employ a coarse schematization of the catchment and drainage network to avoid excessive computer run times.

A distinct advantage of continuous simulation is its ability to provide antecedent conditions as an implicit component of the modeling. That is, continuous records of soil moisture, water table elevation, surface storage, surface pollutant loads, dry-weather activities, etc. may all be maintained, eliminating the vexing questions of initial conditions inherent in single-event modeling. Critical design conditions fall out naturally from a continuous analysis. If necessary, these conditions can later be simulated in more detail in a single-event model, since antecedent conditions can be obtained from the earlier continuous simulation. Advantages such as these are summarized by Linsley and Crawford (1974), and James and Robinson (1982), who discuss implications of continuous simulation on design of detention storage.

Continuous simulation is most useful for planning and optimization of the preliminary design; single-event simulation may then be subsequently used for detailed design and analysis. Individual models may perform both functions by varying the level of schematization and the time step. Output from continuous simulation may be analyzed as a time series and thus reveal more information than statistical methods (Goforth et al. 1983), although greater effort is generally required.

Single-event simulation has the advantage of well-defined data sets for calibration and verification and the ability to incorporate details of the sewer system that are usually too time-consuming or expensive to simulate in the continuous mode. In order to reap the benefits of continuous modeling at less cost and effort, combined approaches are sometimes used in which event simulation is performed on a number of storms and a frequency analysis performed on the subset of continuous meteorological input. Examples are provided by Walesh et al. (1979), Walesh and Snyder (1979) and Murphy et al. (1982).

C. Modeling Options

Several modeling options exist for simulation of quantity and quality in urban storm and combined sewer systems. These have been reviewed by Huber (1985, 1986) and range from simple to involved, although some "simple" methods, e.g., the EPA statistical methods, can incorporate quite sophisticated concepts. The principal methods available to the contemporary engineer are outlined generically below, in a rough order of complexity. Before choosing one of the methods, there are a number of selection criteria that must be considered. They include:

(a) Have a clear statement of project objectives. Verify the need for quality modeling. (Perhaps the objectives can be satisfied without quality modeling.)
(b) Use the simplest model that will satisfy the project objectives. Often a screening model, e.g., regression or statistical, can determine whether more complex simulation models are needed.
(c) To the extent possible, use a quality prediction method consistent with available data. This would often rule against buildup-washoff formulations, although these might still be useful for detailed simulation, especially if calibration data exist.
(d) Only predict the quality parameters of interest and only over a suitable time scale. That is, storm event loads and event mean concentrations (EMCs, the average concentration measured over the entire storm event) will usually represent the most detailed prediction requirement, and seasonal or annual loads will sometimes be all that are required. Do not attempt to simulate intra-storm variations in quality unless it is necessary.
(e) Perform a sensitivity analysis on the selected model and familiarize yourself with the model characteristics.
(f) If possible, calibrate and verify the model results. Use one set of data for calibration and another independent set for verification. If no such data exist for the application site, perhaps they exist for a similar catchment nearby.

The following methods are commonly used for runoff quality prediction. It should be noted that these methods all involve quantity and quality, and some are suitable for simulating quantity alone.

1. Constant Concentration or Unit Loads

As its name implies, constant concentration means that all runoff is assumed to have the same, constant concentration at all times for a given pollutant. At its very simplest, an annual runoff volume can be multiplied by a concentration to produce an annual runoff load. However, this option may be coupled with a hydrologic model, wherein loads (product of concentration and flow) will vary if the model produces variable flows. This option may be quite useful because it may be used with any hydrologic or hydraulic model to produce loads, merely by multiplying by the constant concentration. For instance, the highly sophisticated SWMM EXTRAN Block may be used for hydraulic analysis of a sewer system, prediction of overflows and diversions to receiving waters, etc., yet it performs no quality simulation as such. In many instances, it may be most important to get the volume and timing of such overflows and diversions correctly, and simply estimate loads by multiplying by a concentration.

An obvious question is what (constant) concentration to use? Early (pre-1977) concentration and other data are summarized in publications such as Manning et al. (1977) and Lager et al. (1977). The more recent EPA NURP studies (U.S. EPA 1983) have produced a large and invaluable data base from which to select numbers, but the 30-city coverage of NURP will most often not include a site representative of the area under study. Nonetheless, a large data base does exist from which to review concentrations. Another option is to use measured values from the study area. This might be done from a limited sampling program. However, the NURP study conclusively demonstrated the variation that exists in EMCs at a site, within a city, and within a region or the country as a whole. Thus, while use of a constant concentration may produce **load** variations, EMC variations will not be replicated. These variations may be important in the study of control options and receiving water responses.

Unit loads are perhaps an even simpler concept. These consist of values of mass per area per time, typically lb/ac-yr or kg/ha-yr, for various pollutants, although other normalizations such as lb/curb-mile are sometimes encountered. Annual (or other time unit) loads are thus produced upon multiplication by the contributing area. Such loadings are obviously highly site-specific and depend upon both demographic and hydrologic factors. They must be based on an average or "typical" runoff volume and cannot vary from year to year, but they can conveniently be subject to reduction by best management practices (BMPs) if the BMP effect is known. Although early EPA references provide some information for various land uses (U.S. EPA 1973; U.S. EPA 1976a; McElroy et al. 1976), unit loading rates are exceedingly variable and difficult to transpose from one area to another. Constant concentrations can sometimes be used for this purpose, since mg/l x 0.2265 = lb/ac per inch of runoff. Thus, if a concentration estimate is available, the annual loading rate, for example, may be calculated by multiplying by the inches per year of runoff. Finally, the Universal Soil Loss Equation (Wischmeier and Smith 1958; Heaney et al. 1975)) was developed to estimate tons per acre per year of sediment loss from land surfaces. If a pollutant may be considered as a fraction ("potency factor") of suspended solids concentration or load, this offers another option for prediction of annual loads. Lager et al. (1977), Manning et al. (1977) and Zison (1980) provide summaries of such values.

2. Spreadsheets

Microcomputer spreadsheet software, e.g., Lotus, Quattro, Excel, is now ubiquitous in engineering practice. Very extensive and highly sophisticated engineering analysis is routinely implemented on spreadsheets, and water quality simulation is no exception. The spreadsheet most definitely may be used to automate and extend the concept of the constant concentration or unit load idea. In the usual manifestation of this spreadsheet application, runoff volumes are calculated very simplistically, usually using a runoff coefficient times a rainfall depth. The coefficient may vary according to land use, or an SCS procedure may be used, but the hydrology is inherently simplistic in the spreadsheet predictions. The runoff volume is then multiplied by a constant concentration to predict runoff loads. Alternatively, unit loads are input directly and then multiplied by corresponding land use areas. The advantage of the spreadsheet is that a mixture of land uses (with varying concentrations or loads) may easily be simulated, and an overall load and flow-weighted concentration obtained from the study area (Walker et al. 1989). The study area itself may range from a single catchment to an entire urban area, and "delivery ratios" can be added to simulate loss of pollutants along drainage pathways between the simulated land use and the receiving waters. The relative contributions of different land uses may be easily identified, and handy spreadsheet graphics tools used for display of the results.

As an enhancement, control options may be simulated by application of a constant removal fraction for an assumed BMP. Although spreadsheet computations can be amazingly complex, BMP simulation is rarely more complicated than a simple removal fraction because anything further would require simulation of the dynamics of the removal device (e.g., a wet detention pond), which is usually beyond the scope of the hydrologic component of the spreadsheet model. Nonetheless, if simple BMP removal fractions can be believed, the spreadsheet can easily be used to estimate the effectiveness of control options. Loads with and without controls can be estimated and problem areas, by contributing basin and land use, can be determined. Since most engineers are familiar with spreadsheets, such models can be developed in-house in a logical manner.

The spreadsheet approach is best suited to estimation of long-term loads, such as annual or seasonal, because very simple prediction methods generally perform better over a long averaging time and poorly at the level of a single storm event. Hence, although the spreadsheet could be used at the microscale (at or within a storm event) it is most often applied for much longer time periods. It is harder to obtain the variation of predicted loads and concentrations using the spreadsheet method because this can ordinarily only be done by varying the input concentrations or rainfall values. A Monte Carlo simulation may be attempted (i.e., systematic variation of all input parameters according to an assumed frequency distribution) if the number of such parameters is not too large. These results may then be used to estimate the range and/or frequency distribution of predicted loads and concentrations.

In a generic sense, the spreadsheet idea may be used in methods programmed in other languages, e.g., Fortran. For example, comprehensive assessments of coastal zone pollution from urban areas are made by NOAA (1987) by assembling land use data with different runoff coefficients, predicting daily and seasonal runoff volumes from daily rainfall, and predicting seasonal pollutant loads using constant concentrations. Although the demographic data base and use of magnetic tapes may dictate use of mainframes, the computational concept is still that of a spreadsheet.

Again, the question arises of what concentrations or unit loads to use, this time potentially for multiple land uses and subareas. And again, the NURP data base will usually be the first one to turn to, with the possibility of local monitoring to augment it.

3. Statistical Approaches

Uncertainties in deterministic or physically-based methods for water quality modeling have led to the use of statistical approaches as a data-based alternative to uncertain model predictions. These methods basically ignore the process conceptualizations employed by physically-based models and substitute generalizations about flow, pollutant concentration and load means, standard deviations, and probability distributions based on observations gathered at many sites.[1] The statistical approach leads to prediction of the probability distribution of event mean concentrations for the site in question. The key problem is how to determine the statistical properties of EMCs at ungauged sites, not a dissimilar problem from the use of physically-based models. Another problem with statistical approaches is studying the effect of control options or changes to the physical system. Nonetheless, the statistical approach has been used in several important applications.

Statistical analyses of various kinds are also incorporated into most evaluations of measured data. For instance, water quality data may be averaged over time intervals from the start of a storm event to identify first flush phenomena. Similarly, the concentration time series may reflect the storm hyetograph and hydrograph. Regression methods are often applied in order to extrapolate observed data to new areas (Driver and Tasker 1988, Tasker and Driver 1988), subject to the usual caveats about regression analysis.

The so-called "EPA Statistical Method" is somewhat generic and until recently was not implemented in any off-the-shelf model or even very well in any single report (Hydroscience 1979; U.S. EPA 1983). A new FHWA study (Driscoll et al. 1989) partially remedies this situation. The concept is straightforward, namely that of a derived frequency distribution for EMCs. This idea has been used extensively for urban runoff

---

[1]It should be noted that it is also possible to use a physically-based continuous simulation model to generate a time series of flows, pollutant concentrations, or loads, which can then be analyzed statistically.

quantity (e.g., Howard 1976; Loganathan and Delleur 1984; Zukovs et al. 1986) but not as much for quality predictions.

The EPA Statistical Method is based on the fact that EMCs are not constant but tend to exhibit a lognormal frequency distribution. When coupled with an assumed distribution of runoff volumes (also lognormal), the distribution of runoff loads may be derived. When coupled again to the distribution of streamflow, an approximate (lognormal) probability distribution of in-stream concentrations may be derived (Di Toro 1984)—a very useful result, although assumptions and limitations of the method have been pointed out by Novotny (1985) and Roesner and Dendrou (1985). Further analytical methods have been developed to account for storage and treatment (Di Toro and Small 1979; Small and Di Toro 1979). The method was used as the primary screening tool in the EPA NURP studies (U.S. EPA 1983) and has also been adapted to combined sewer overflows (Driscoll 1981) and highway-related runoff (Driscoll et al. 1989). This latter publication is one of the best for a concise explanation of the procedure and assumptions and includes spreadsheet software for easy implementation of the method.

A primary assumption is that EMCs are distributed lognormally at a site and across a selection of sites. The concentrations may thus be characterized by their median value and by their coefficient of variation (CV = standard deviation divided by the mean). There is little doubt that the lognormality assumption is good (Driscoll 1986), but similar to the spreadsheet approach, the method is then usually combined with weak hydrologic assumptions, e.g., prediction of runoff using a runoff coefficient (the accuracy of a runoff coefficient increases as urbanization and imperviousness increase.) However, since many streams of concern in an urban area consist primarily of stormwater runoff during wet weather, the ability to predict the distribution of EMCs is very useful for assessment of levels of exceedance of water quality standards. The effect of BMPs can again be estimated crudely through constant removal fractions that lower the EMC median, but it is harder to determine the effect on the coefficient of variation. Overall, the method has been very successfully applied as a screening tool.

Input to the method as implemented for the FHWA (Driscoll et al. 1989) includes statistical properties of rainfall (mean and CV of storm event depth, duration, intensity and interevent time), area, and runoff coefficient for the hydrologic component, plus EMC median and CV for the pollutant. Generalized rainfall statistics have already been calculated for many locations in the U.S. Otherwise, the EPA SYNOP model (U.S. EPA 1976b; Hydroscience 1979; U.S. EPA 1983; Woodward-Clyde 1989) must be run on long-term hourly rainfall records. If receiving water impacts are to be evaluated, the mean and CV of the streamflow are required plus the upstream concentration. A Vollenweider-type lake impact analysis is also provided based on phosphorus loadings.

As with the first two methods discussed, the choice of median concentration may be difficult, and the Statistical Method requires a coefficient of variation as well. Fortunately, from NURP and highway studies, CV values for most urban runoff pollutants are fairly consistent, and a value of 0.75 is typical. If local and/or NURP data are not available or inappropriate, local monitoring may be required, as in virtually every quality prediction method. The estimation of the whole EMC frequency distribution for a pollutant is a definite advantage of the Statistical Method over some applications of constant concentration and simple spreadsheet approaches. Frequency analyses of water quantity and quality parameters may also be performed on the output of continuous simulation models such as HSPF, SWMM, and STORM. The derived distribution approach of the Statistical Method avoids the considerable effort required for continuous simulation at the expense of simplifying assumptions that may or may not reflect the prototype situation adequately.

4. Regression—Rating Curve Approaches

With the completion of the NURP studies in 1983, there are measurements of rainfall, runoff and water quality at well over 100 sites in over 30 cities. Some regression analysis has been performed to try to relate loads and EMCs to catchment, demographic and hydrologic characteristics (e.g., McElroy et al. 1976; Miller et al. 1978; Brown 1984), the most recent of which is the work by the USGS (Tasker and Driver 1988; Driver and Tasker 1988), described briefly below. Regression approaches have also been used to estimate dry-weather pollutant deposition in combined sewers (Pisano and Queiroz 1977), a task at which no model is very successful. What are termed "rating curves" herein are just a special form of regression analysis, in which concentration and/or loads are related to flow rates and/or volumes. This is an obvious exercise attempted at most monitoring sites and has a historical basis in sediment discharge rating curves developed as a function of flow rate in natural river channels.

A rating curve approach is most often performed using total storm event load and runoff volume although intra-storm variations can sometimes be simulated in this manner as well (e.g., Huber and Dickinson 1988). It is usually observed (Huber 1980; U.S. EPA 1983; Driscoll et al. 1989) that concentration (EMC) is poorly or not correlated with runoff flow or volume, implying that a constant concentration assumption is adequate. Since the load is the product of concentration and flow, load is usually well correlated with flow regardless of whether or not concentration correlates well. This manifestation of spurious correlation (Bensen 1965) is often ignored in urban runoff studies. If load is proportional to flow to the first power (i.e. linear), then the constant concentration assumption holds; if not, some relationship of concentration with flow is implied. Rating curve results can be used by themselves for load and EMC estimates and can be incorporated into some models (e.g., SWMM, HSPF).

Rainfall, runoff and quality data were assembled for 98 urban stations in 30 cities (NURP and other) in the U.S. for multiple regression analysis by the USGS (Driver and Tasker 1988; Tasker and Driver 1988). Thirty-four multiple regression models (mostly log-linear) of storm runoff constituent loads and storm runoff volumes were developed, and 31 models of storm runoff EMCs were developed. Regional and seasonal effects were also considered. The two most significant explanatory variables were total storm rainfall and total contributing drainage area. Impervious area, land use, and mean annual climatic characteristics also were significant explanatory variables in some of the models. Models for estimating loads of dissolved solids, total nitrogen, and total ammonia plus organic nitrogen (TKN) generally were the most accurate, whereas models for suspended solids were the least accurate. The most accurate models were those for the more arid Western United States, and the least accurate models were those for areas that had large mean annual rainfall.

These USGS equations represent the best generalized regression equations currently available for urban runoff quality prediction. Note that such equations do not require preliminary estimates of EMCs or local quality monitoring data except for the very useful exercise of verification of the regression predictions. Regression equations only predict the mean and do not provide the frequency distribution of predicted variable, a disadvantage compared to the statistical approach (the USGS documentation describes procedures for calculation of statistical error bounds, however). Finally, regression approaches, including rating curves, are notoriously difficult to apply beyond the original data set from which the relationships were derived. That is, they are subject to very large potential errors when used to extrapolate to different conditions. Thus, the usual caveats about use of regression relationships continue to hold when applied to prediction of urban runoff quality.

5. Buildup and Washoff

In the late 1960s, a Chicago study by the American Public Works Association (1969) demonstrated the (assumed linear) buildup of "dust and dirt" and associated pollutants on urban street surfaces. During a similar time frame, Sartor and Boyd (1972) also demonstrated buildup mechanisms on the surface as well as an exponential washoff of pollutants during rainfall events. These concepts were incorporated into the original SWMM model, as well as into the STORM, USGS, and HSPF models to a greater or lesser degree (Huber 1985). "Buildup" is a term that represents all of the complex spectrum of dry-weather processes that occur between storms, including deposition, wind erosion, street cleaning, etc. The idea is simply that all such processes lead to an accumulation of solids and other pollutants that are then "washed off" during storm events.

Although ostensibly physically based, models that include buildup and washoff mechanisms really employ conceptual algorithms because the true physics is related to principles of sediment transport and erosion that are poorly understood in this framework. Furthermore, the inherent heterogeneity of urban surfaces leads to use of average buildup and washoff parameters that may vary significantly from what may occur in an isolated street gutter, for example. Thus, except in rare instances of measurements of accumulations of surface solids, the use of buildup and washoff formulations inevitably results in a calibration exercise against measured end-of-pipe quality data. It then holds that in the absence of such data, inaccurate predictions can be expected.

Different models offer different options for conceptual buildup and washoff mechanisms, with SWMM having the greatest flexibility. In fact, with calibration, good agreement can be produced between predicted and measured concentrations and loads with such models, including intra-storm variations that cannot be duplicated with most of the methods discussed earlier. (When a rating curve is used in SWMM instead of buildup and washoff, it is also possible to simulate intra-storm variations in concentration and load.) A survey of linear buildup rates for many pollutants by Manning et al. (1977) is probably the best source of generalized buildup data, and some information is available in the literature to aid in selection of washoff coefficients (Huber 1985; Huber and Dickinson 1988). However, such first estimates may not even get the user in the ball park (i.e., quality—not quantity—predictions may be off by more than an order of magnitude); the only way to be sure is to use local monitoring data for calibration and verification. Thus, as for most of the other quality prediction options discussed herein, the buildup-washoff model may provide adequate comparisons of control measures, ranking of loads, etc., but cannot be used for prediction of absolute values of concentrations and loads, e.g., to drive a receiving water quality model, without adequate calibration and verification data. Since buildup and washoff are somewhat appealing conceptually, it is somewhat easier to simulate potential control measures such as street cleaning and surface infiltration using these mechanisms than with, say, a constant concentration or rating curve method. In the relatively unusual instance in which intra-storm variations in concentration and load must be simulated, as opposed to total storm event EMC or load, buildup and washoff also offer the most flexibility. This is sometimes important for the design of storage facilities in which first-flush mechanisms may be influential.

As mentioned above, generalized data for buildup and washoff are sparse (Manning et al. 1977) and such measurements are almost never conducted as part of a routine monitoring program. For buildup, normalized loadings, e.g., mass/day-area or mass/day per curb-length, or just mass/day, are required, along with an assumed functional form for buildup vs. time, e.g., linear, exponential, Michaelis-Menton, etc. For washoff, the relationship of washoff (mass/time) vs. runoff rate must be assumed, usually in the form of a power equation. When end-of-pipe concentration and load data are all that are available, all buildup and washoff coefficients end up being calibration parameters.

6. Related Mechanisms

In the discussion above, washoff is assumed proportional to the runoff rate, as for sediment transport. Erosion from pervious areas may instead be proportional to the rainfall rate. HSPF does the best job of including this mechanism in its algorithms for erosion of sediment from pervious areas. SWMM includes a weaker algorithm based on the Universal Soil Loss Equation (Wischmeier and Smith 1958; Heaney et al. 1975).

Many pollutants, particularly metals and organics, are adsorbed onto solid particles and are transported in particulate form. The ability of a model to include "potency factors" (HSPF) or "pollutant fractions" (SWMM) enhances the ability to estimate the concentration or load of one constituent as a fraction of that of another, e.g., solids (Zison 1980).

The groundwater contribution to flow in urban areas can be important in areas with unlined and open channel drainage. Of the urban models discussed, HSPF far and away has the most complex mechanisms for simulation of subsurface water quality processes in both the saturated and unsaturated zones (there are rarely, if ever, adequate local data to calibrate all HSPF parameters, and the user is forced to accept default values). Although SWMM includes subsurface flow routing, the quality of subsurface water can only be approximated at present using a constant concentration.

The precipitation load may be input in some models (SWMM, HSPF), usually as a constant concentration. Point source and dry-weather flow (baseflow) loads and concentrations can also be input to SWMM, STORM and HSPF to simulate background conditions. Other quality sources of potential importance include catchbasins (SWMM) and snowmelt (SWMM, STORM, HSPF).

Scour and deposition within the sewer system can be very important in combined sewer systems and some separate storm sewer systems. The state of the art in simulation of such processes is poor (Huber 1985), though SWMM offers a crude but calibratable attempt.

In the last analysis, the most complicated model may not be the most accurate one when estimating runoff quantity and quality. It is the experience and understanding of the engineer that is most important when generating runoff quantity and quality calculations.

VI. COMPUTER REQUIREMENTS

Virtually every engineer has access to a microcomputer, and most major hydrologic models have been modified to run on these systems. This model availability, coupled with advances in data acquisition and management and the extraordinary graphical capabilities of microcomputers, means that the most advanced technology and largest data bases are now accessible to almost anyone. However, there are still advantages and disadvantages of both mainframe and microcomputing. Micros are "personal" and permit immediate access and "turnaround" (the time taken to run a program). Use of micros is inherently interactive, which encourages inspection and evaluation of data and results. Other microcomputer software, especially spreadsheets, makes it easy to perform simple calculations without the need of more complex models at all and without the need for programming (although spreadsheet manipulation is essentially the same thing as programming). Disadvantages of micros include execution time for long programs, and lengthy waiting periods for input and output of large data sets.

Mainframes have the advantages of speed, including fast peripherals such as high-speed printers and tape drives. Thus, a program that may require several hours to execute on a micro may execute in seconds on a mainframe.

Micros continue to increase in their performance, and already will satisfy most engineering needs. Most of the models to be discussed in this chapter are available in both mainframe and microcomputer versions.

VII. STEPS IN MODELING

A. Data Requirements

Before a model is even considered, some data will be available to indicate that there is a problem. For instance, receiving water samples may indicate a bacteriological problem, inspection may reveal sedimentation, or high-water marks may be seen on basement walls. Receiving water evaluations may include common parameters such as DO, oxygen demand, nutrients, and bacteria as well as a bio-assay or other biological assessment of receiving water quality. Such observations constitute a data set in and of themselves, and usually indicate the direction for subsequent data collection. At every stage of the preliminary analysis, one must ask if measured data can resolve the problem. If so, there is no need to model.

If modeling is required, there are three types of required data; model input data, calibration data, and verification data. Input data consist simply of the required parameters to run the model, and typically include rainfall information, area, imperviousness, runoff coefficient and other quantity prediction parameters, plus quality prediction parameters such as constant concentration, constituent median and CV, regression relationships, buildup and washoff parameters, etc.. Calibration is the process of parameter adjustment to obtain a match between predicted and measured output. Verification holds the parameters constant and tests the calibration on an independent data set.

Thus, calibration and verification data consist of the very elements the model predicts, such as hydrographs or pollutographs (a plot of concentration versus time), whereas input data may resemble physical characteristics of the system. Some input data for some models are mostly conceptual, such as a time of concentration, and cannot be readily measured. Calibration is used to estimate the value of these parameters, and verification is used to test the validity of the estimate.

Data sets which can be used for calibration and verification exist (e.g., Huber et al. 1982; Driver et al. 1985; Noel et al. 1987) but seldom for the site of interest. If the project objectives absolutely require such data (e.g., if a model must be calibrated to drive a receiving water quality model), then expensive local monitoring may be necessary.

B. Basic Input Data

All models require some form of input data. For quantity simulation, these data include catchment areas, imperviousness, slopes, roughnesses, etc.; channel and conduit linkages, shapes, sizes, slopes, roughnesses; invert and ground elevations; characteristics of hydraulic structures or controls such as weirs, orifices and pumps; depth-area-volume-outflow relationships for storage units; information on downstream hydraulic controls, such as river stages or tidal elevations. Since the overall system is driven by rainfall, suitable rainfall hyetographs must be found (see below), as well as base flow, if any, in the drainage channels.

Not all models need all these data; some use very simple methods to convert rainfall into runoff, but greater model complexity generally means greater model input requirements. A critical factor in successful hydraulic modeling of older drainage systems is an accurate survey to determine invert elevations and conduit or channel condition. These are seldom the same as shown on as-built plans because of settlement, deterioration, and modifications to the system. This is typically an expensive component of data preparation.

Dry-weather flow concentrations must be measured to simulate the mix of stormwater and base flow that occurs during a storm, as well as characteristics of solids if scour and deposition are to be simulated.

C. Rainfall Input

Rainfall is the driving force for all hydrologic simulation models. If adequate measured rainfall is not available, a good calibration between measured and predicted hydrographs cannot be expected. Even though the Stanford Watershed Model introduced the concept of continuous simulation to hydrologic modeling, most models have evolved from single-event formulations, for which a single hyetograph is used to simulate runoff. For calibration purposes, measured rainfall must be input to produce output for comparison with measured hydrographs. However for design purposes, synthetic design storms have mostly replaced historic rainfall hyetographs as input to models (Keifer and Chu 1957; USDA 1971; Arnell 1982) because of the ease with which they may be constructed. The perils of single design storms are many, as discussed by McPherson (1978), not the least of which is the necessity for choosing antecedent conditions and the assumption that the return period of the rainfall input will be the same as the return period of the runoff (or pollutant) output.

Continuous simulation or statistical methods offer alternatives for selection of design rainfall. For example, a selection of historic storms can be made from a continuous simulation on the basis of the return period of the runoff or quality parameter of interest, e.g., peak flow, maximum runoff volume, maximum stage, peak runoff load, peak runoff concentration (Huber et al. 1986b). These events, with their antecedent conditions for runoff and quality, can then be analyzed in more detail in a single-event mode. Rainfall is variable in space as well as in time; models that accept multiple hyetographs can simulate storm motion and spatial variation that can strongly affect runoff hydrographs (Surkan 1974; James and Shtifter 1981). Many considerations related to the selection of rainfall input for modeling are provided in proceedings edited by Harremöes (1983).

D. Sensitivity Analysis

Before attempting to calibrate and verify a model, the user should be familiar with its capabilities and nuances. Some models have very few parameters to adjust, simplifying the calibration process, but others may have 30 or more. Which ones affect the output the most? Since this is seldom documented adequately, the user should perform a sensitivity analysis (with hypothetical data if necessary), varying key parameters by known percentages and inspecting the change in output. In this way, it will be far easier to know which parameters should be changed during the calibration process. For instance, in urban areas, most models are highly sensitive to imperviousness but only weakly sensitive to soil infiltration parameters. First runs with any model should deal with a very simple configuration for which the result is known, e.g., steady rain on an impervious surface, and build up gradually to more complex and realistic systems. In this way, the user can exercise good judgment regarding the validity and reasonableness of the results.

E. Calibration

Model calibration consists of adjusting model parameters (e.g., imperviousness, roughness) until the predicted output agrees with measured observations. For example, the predicted hydrograph or pollutograph may be adjusted to agree with the measured hydrograph or pollutograph. For most models, calibration will be performed using observed storm events. How many storms are required cannot be answered exactly, but 3–6 are desirable. The calibration process should be performed simultaneously for all available storms in order to produce a robust calibration (Maalel and Huber 1984). In this instance, the single set of calibration parameters will result in less-than-perfect fits for any single storm but better for all storms together, and presumably better for further predictions.

Calibration tends to be subjective. For example, how well does one hydrograph match another? When several storms are used, it is customary to plot predicted versus measured peaks and predicted versus measured volumes, seeking to produce points that fall on the 45-degree line indicating perfect agreement. Deviations from the line of perfect fit are one measure of the goodness of fit; hydrographs can (and should) also be compared visually. James and Burges (1982) furnish many practical guidelines for calibration, testing, and quantification of errors. Thomann (1982) provides guidelines for assessing goodness of fit, and Reckhow et al. (1990) have expanded upon Thomann's work.

During the calibration process, care must be taken to make sure that the physical parameters are not adjusted outside their reasonable range to achieve a "calibration." For example, if the Manning roughness coefficient for a concrete pipe has to be set at 0.10 to achieve calibration, most likely there is an error in the input data of some other variable such as pipe slope or model conceptualization.

Calibration usually provides the only means for determining values for input parameters related to water quality, such as buildup rates and washoff coefficients. Although limited measurements of surface constituents have been conducted (see Manning et al. 1977 and Terstriep et al. 1982), such data are generally useful only for a first parameter estimate. Quality concentrations and loads are so difficult to predict that calibration data provide almost the only means for parameter estimation (Huber 1985).

F. Verification

Ideally, an equal number of storms should be used for verification as for calibration; however, 1–3 often seems to be the pragmatic limit of the number of storms that can be afforded for this purpose. Goodness of fit may be assessed similarly to the method used for calibration. In the not unlikely event that the verification is poor, an improved calibration can be attempted. This is sometimes performed using a different grouping of storms for calibration and verification than was used during the first attempt.

G. Uncertainty Analysis

Uncertainty analysis is rapidly moving from state-of-the-art to state-of-practice. Uncertainty analysis can be used to compute expected output variability as a function of ill-defined input parameters. Additionally, this technique can serve as a means of quantifying model acceptability (does the range of probable model outcomes intersect the range of observations).

Uncertainty analysis has been performed via first-order approximations and Monte Carlo simulation techniques. First-order uncertainty analysis provides estimates of the mean and standard deviation for the dependent variable (model output) as functions of the mean and standard deviation associated with each independent variable. While this technique is relatively fast computationally, it does require evaluation of all partial derivatives for the dependent variable with respect to each independent variable. Additionally, the assumptions inherent in first-order uncertainty analysis (system linearity, and type of output distribution) can cause erroneous estimations of true model output behavior.

Monte Carlo analysis is relatively simple to program, and does not suffer from the aforementioned assumptions. Unfortunately, this technique can be rather computer-intensive. Monte Carlo analysis involves the random selection of input parameter values from prescribed statistical distributions followed by model execution. The process is repeated until an adequate description of model output has been obtained. The required number of Monte Carlo iterations is a function of the questions being asked (mean response, 95% confidence intervals, complete model output probability density function) and the level of accuracy needed. Simply stated, more iterations will result in a better definition of model output variability. The user must, through prior experimentation, ascertain the appropriate number of iterations required.

Uncertainty analysis can be particularly useful in evaluating the relationship between field sampling and modeling. Hypothetical sampling scenarios can be tested to understand the expected uncertainty in model output. If the level of output variability is too large, the sampling strategy can be increased until an acceptable level of model output uncertainty is achieved. Finally, uncertainty analysis can also be used to quantify model acceptability (expansion of goodness-of-fit testing). Warwick and Wilson (1990) used uncertainty analysis, applied to the STORM model, to determine if the use of default areal pollution accumulation rates were acceptable for the site under investigation. The use of these default values was rejected because the range of probable STORM output did not encompass field observations of pollutant runoff concentrations. The use of areal accumulation rates taken from a nearby location could not be rejected based upon this test, and was thereby viewed as acceptable for future predictions.

H. Production Runs

Following the calibration and verification processes, the model is ready for application to the problem. During this phase, just as earlier, all model parameters and results should be double-checked for reasonableness. Continuity checks built into a model often aid in checking results so that an unrealistic gain or loss of water (or pollutants) can be noticed. Violations of continuity sometimes indicate numerical problems, especially with sophisticated hydraulic routing models.

VIII. MODELS

A. Published Reviews

Several publications provide reviews of available models. Recent reviews that consider surface runoff quantity/quality models include Huber and Heaney (1982), Kibler (1982), Whipple et al. (1983), Barnwell (1984, 1987), Huber (1985, 1986), Bedient and Huber (1988), Viessman et al. (1989), WPCF (1989), and Donigian and Huber (1990). Renard et al. (1982) review water resources models, some of which are useful in urban hydrology. HEC models are described in detail by Feldman (1981). Descriptions of EPA nonpoint source water quality models are provided by Ambrose et al. (1988) and Ambrose and Barnwell (1989).

B. Models To Be Reviewed

There are so many potential models that might be used for analysis of storm drainage systems that some screening is in order. For this chapter, the most significant criterion is that the model be operational, as defined earlier, and have a fairly large user base. That still leaves potentially many models, at least if rainfall-runoff and other quantity-only models are to be included. Modeling and software development is highly dynamic, and the following model descriptions may be quite different in the future.

Nine of the models reviewed below are listed in Table 7.2, which characterizes the models in terms of various criteria. The other model, HEC-2, is reviewed because of its importance in simulating hydraulic processes, though it, like HEC-1, has no quality modeling capabilities.

1. DR3M-QUAL

U.S. Geological Survey has updated earlier model development by Dawdy et al. (1972) into the Distributed Routing Rainfall Runoff Model (DR3M), including quality, designed specifically for urban hydrology (Alley and Smith 1982a, 1982b). Runoff generation and subsequent routing use the kinematic wave method, and parameter estimation assistance is included in the model. Quality is simulated using buildup and washoff functions, with settling of solids in storage units dependent upon a particle size distribution. The model has been used in some of the EPA Nationwide Urban Runoff Program (NURP) studies that were conducted by the USGS (U.S. EPA 1983; Alley 1986). The FORTRAN version of the model is available from the U. S. Geological Survey's

TABLE 7.2.   Comparison of Model Attributes.

Attribute (1)	Model: DR3M-QUAL (2)	HSPF (3)	ILLUDAS (4)	Penn State (5)	Statistical (6)	STORM (7)	SWMM (8)	TR55 (9)	HEC-1 (10)
Sponsoring agency	USGS	EPA	Ill. State Water Survey	OWRT and City of Phil.[a]	EPA	HEC	EPA	SCS	HEC
Simulation type[b]	C,SE	C,SE	SE	SE	N/A	C	C,SE	SE	SE
No. pollutants	4	10	None[c]	None	Any	6	10	None	None
Rainfall/runoff analysis	Y	Y	Y	Y	N	Y	Y	Y	Y
Sewer system flow routing	Y	Y	Y	N	N	N	Y	Y	Y
Full, dynamic flow routing equations	N	N	N	N	N/A	N	Y[d]	N	N
Surcharge	Y[e]	N	Y[e]	N	N/A	N	Y[d]	N	N
Regulators, overflow structures, e.g., weirs, orifices, etc.	N	N	N	N	N/A	Y	Y	N	N
Special solids routines	Y	Y	N/A	N/A	N	N	Y	N/A	N/A

Storage analysis	Y	Y	Y	Y	Y[f]	Y	Y	Y	Y
Treatment analysis	Y	Y	N/A	N/A	Y[f]	Y	Y	N/A	N/A
Suitable for planning (P), design (D)[g]	P,D	P,D	D	D	P	P	P,D	D	D
Available on microcomputer	N	Y	Y	Y	Y[h]	N	Y	Y	Y
Data and personnel requirements	Medium	High	Low	Low	Medium	Low	High	Medium	Medium
Overall model complexity[j]	Medium	High	Low	Low	Medium	Medium	High	Low	High

[a]Currently supported by Penn State University.

[b]Y = yes, N = no, N/A = not applicable, C = continuous simulation, SE = single event simulation.

[c]Undocumented quality routines added during applications.

[d]Full dynamic equations and surcharge calculations only in Extran Block of SWMM.

[e]Surcharge simulated by storing excess inflow at upstream end of pipe. Pressure flow not simulated.

[f]Storage and treatment analyzed analytically. See references in Section 3.7.9.

[g]See Section 3.3.

[h]See Driscoll et al. 1989.

[i]General requirements for model installation, familiarization, data[j] requirements, etc. To be interpreted only very generally.

[j]Reflection of general size and overall model capabilities. Note that complex models may still be used to simulate very simple systems with attendant minimal data requirements.

National Center in Reston, Virginia. No microcomputer version is available.

2. HSPF

The Hydrologic Simulation Program—Fortran (HSPF) was developed from hydrologic routines that began with the Stanford Watershed Model (Crawford and Linsley 1966) and nonpoint source water quality routines that were included in such models as the EPA Nonpoint Source Model (NPS), (Donigian and Crawford 1976) and the Agricultural Runoff Model (ARM) (Donigian and Davis 1978). The user's manual (Johanson et al. 1984) includes information on all hydrologic and water quality routines, including the "IMPLND" (impervious land) segment for use in urban areas. Additional guidelines for application are provided by Donigian et al. (1984). The model has special provisions for management of time series that result from the continuous simulation.

HSPF includes subsurface water balance and quality routing and contains the most comprehensive pollutant kinetics of any of the models discussed. The model is maintained by, and mainframe and microcomputer versions are available from, the U. S. Environmental Protection Agency, Athens, Georgia.

3. ILLUDAS

The Illinois Urban Drainage Area Simulator (ILLUDAS) (Terstriep and Stall 1974) evolved from the British Road Research Laboratory Model (Watkins 1962; Stall and Terstriep 1972). The model uses time-area methods to generate hydrographs from the directly connected paved area and from the pervious area. For pervious areas, the Horton infiltration equation is used to generate typical infiltration rates based on input of the soil's SCS hydrologic group category. A design routine is included that will re-size pipes of insufficient hydraulic capacity. User-provided stage/discharge/storage relationships are used to provide detention facilities anywhere in the system. Plots of calculated and observed hydrographs may be produced. Its simplicity and metric option have given ILLUDAS widespread use. Although quality is not formally included in the model, it has been added for special applications (Noel and Terstriep 1982).

4. Penn State

The Penn State Urban Runoff Model (PSURM) (Aron 1987) was originally developed in cooperation with the City of Philadelphia for drainage analysis, and has been applied to combined sewers in that city. Nonlinear reservoir routing is used for generation of the runoff hydrograph, coupled with a user-defined lag for routing within the sewer system. A hydraulic design capability helps to size pipes. No quality routines are included. The model has seen considerable use in the northeastern United States (Kibler and Aron 1980, and Kibler et al. 1981). The PSURM is available in mainframe and microcomputer versions.

5. Statistical

The statistical approach is not a simulation model in the conventional sense, but rather a combined sequence of analysis of rainfall, runoff, and quality data coupled with analytical solutions (see extended discussion in Section V.C.3. of this chapter). The best explanation and implementation of the statistical method is in an adaptation developed for the U. S. Federal Highway Administration (Driscoll et al. 1989). This publication is one of the best for a concise explanation of the procedure and assumptions and includes spreadsheet software for easy implementation of the method. A useful component is the SYNOP ("synoptic precipitation") model for statistical analysis of hourly rainfall time series from magnetic tapes and floppy disks supplied by the National Climatic Data Center of NOAA at Asheville, NC.

6. STORM

The first significant use of continuous simulation in urban hydrology came with the Storage, Treatment, Overflow, Runoff Model (STORM), a program whose development was funded by the Corps of Engineers, Hydrologic Engineering Center (HEC 1977; Roesner et al. 1974). Early applications included the San Francisco master plan for combined sewer overflow (CSO) pollution abatement (McPherson 1974). The support of the HEC led to the wide use of STORM for planning purposes, especially for evaluation of the trade-off between treatment and storage as CSO control options (e.g., Heaney et al. 1977). Statistics of long-term runoff and quality time series permit optimization of control measures.

Although designed originally for analysis of CSOs, the model is equally useful for evaluating the effectiveness of detention facilities in reducing the frequency of runoff peaks for longer catchments.

STORM uses a simple runoff coefficient method for generation of hourly runoff depths from hourly rainfall inputs, and uses the buildup and washoff formulation for simulation of six pre-specified pollutants. However, the model can be manipulated to provide loads for arbitrary pollutants (Najarian et al. 1986). A microcomputer version has been demonstrated (Bontje et al. 1984), and a mainframe version is available. The HEC has also provided application guidelines (Abbott 1977).

7. SWMM

The original version of the Storm Water Management Model (SWMM) was developed for EPA as single-event model specifically for the analysis of combined sewer overflows (Metcalf and Eddy Inc. 1971). Through continuous maintenance and support, the model now is well suited to all types of storm water management from urban drainage to flood routing and floodplain analysis. Version 4 (Huber and Dickinson 1988; Roesner et al. 1988) performs both continuous and single-event simulation throughout the whole model; can simulate backwater, surcharging, pressure flow and looped connections (by solving the complete dynamic wave equations in its Extran Block); and has a variety of options for quality simulation, including traditional buildup and washoff formulations as well as rating curves and regression techniques. Subsurface flow routing (constant quality) may be performed in the Runoff Block in addition to surface quantity and quality routing, and treatment devices may be simulated in the Storage/Treatment Block using removal functions and sedimentation theory. A hydraulic design routine is included for sizing of pipes, and a variety of regulator devices may be simulated, including orifices (fixed and variable), weirs, pumps, and storage. A bibliography of SWMM usage is available (Huber et al. 1986) that contains many references to case studies.

SWMM is segmented into the Runoff, Transport, Extran, Storage/Treatment and Statistics blocks for rainfall-runoff, routing, and statistical computations. Water quality may be simulated in all blocks except Extran, and metric units are optional. Since the model is nonproprietary, portions have been adapted for various specific purposes and locales by individual consultants and other federal agencies, e.g., FHWA. Mainframe and microcomputer versions are available from EPA in Athens, Georgia.

8. TR55

The original Soil Conservation Service (SCS) methodology developed for general application (USDA 1971) was later adapted specifically to urban areas, and the latter procedure has come to be known as TR55 (USDA 1975). An updated user's manual is available (USDA 1986a), along with a microcomputer version (USDA, 1986b). (The 1975 version was not computerized by the SCS, although microcomputer versions are available from private vendors.)

Unit hydrographs are used to convert rainfall into runoff. If required, flow routing in channels must be performed separately by another model such as by the companion TR20 program (USDA 1983). SCS methods are widely used in the United States due to the wealth of soil information provided by the agency. Additional background on the method is provided by Viessman et al. (1989) and McCuen (1982). Information on application is usually available from local SCS offices as well.

9. HEC-1

The HEC-1 model developed by the Corps of Engineers Hydrologic Engineering Center (HEC 1985) is designed to simulate the surface runoff response of a river basin to precipitation by representing the basin as an interconnected system of hydrologic and hydraulic components. Each component models an aspect of the precipitation-runoff process within a portion of the basin, commonly referred to as a sub-basin. A component may represent a surface runoff entity, a stream channel, or a reservoir. The result of the modeling process is the computation of streamflow hydrographs at desired locations in the river basin. Multiplan-multiflood analysis allows the simulation of up to nine ratios of a design flood for up to five different plans (or characterizations) of a stream network in a single computer run. Dam-break simulation provides the capability to analyze the consequences of dam overtopping and structural failures. The depth-area option computes flood hydrographs preserving a user-supplied precipitation depth versus area relation throughout a stream network.

10. HEC-2

This program developed by the Corps of Engineers Hydrologic Engineering Center (HEC 1982), is intended for calculating water surface profiles for steady, gradually varied flow in natural or man-made channels. Both subcritical and supercritical flow profiles can be calculated. The effects of various obstructions such as bridges, culverts, weirs, and structures in the flood plain may be considered in the computations. The computational procedure is based on the solution of the one-dimensional energy equation with energy loss due to friction evaluated with Manning's equation. The computational procedure is generally known as the standard-step method (see Chapter 6). The program is also designed for application in flood plain management and flood insurance studies to evaluate floodway encroachments and to designate flood hazard zones. It is also capable of assessing the effects of channel improvements and levees on water surface profiles. Input and output units may be either U.S. customary units or metric.

Two supplementary programs are also available:

(a) A data edit program which checks the data cards for various input errors.
(b) A Fortran graphics program which produces HEC-2 cross section and profile plots in interactive or batch modes.

HEC-1 and HEC-2 are supported by the HEC and available from several private vendors.

IX. MODEL SELECTION

The brief abstracts in the previous section, information in Table 7.2, and the comparative reviews referenced previously may help in selection of a model, but the choice is often made on much more pragmatic grounds. For instance, an agency may specify that a certain model be used, or local support may be available. Probably the most important factor is familiarity of the potential user with techniques employed by the model. Inferior techniques applied by a knowledgeable engineer will often produce much more reliable results than a sophisticated model that the user does not understand and therefore treats as a "black box." Data availability is another important consideration. For instance, complex flow routing cannot be performed in a drainage system without extensive information on invert elevations, etc. Such data are not always available, which may lead the engineer to a simpler technique that is not so data intensive. The need for data should not be ignored, however. If the problem is sufficiently complex, there may be no alternative to the use of a sophisticated model and its attendant data collection requirements.

Finally, the number of modeling options is very large; the reviews provided here are representative of the best known operational models, but are not all-inclusive. The potential modeler should consult current journals for information on availability of new models.

X. COMPUTER AIDED DRAFTING AND DESIGN (CADD)

Computer Aided Drafting and Design (CADD) includes three general types of computer applications: drafting; design, analysis and simulation of hydrologic processes and hydraulic structures; and geographic information systems (GIS) or special mapping or feature analysis.

Computer Aided Drafting (CAD) has become synonymous with microcomputer workstations and high quality graphics or drawings transferable from the computer to mylar, and then to blueprints. It is possible to computerize standard design details, such as manholes, catch basins, headwalls and piping, and to store them in the computer for later recall and insertion on a detail sheet. Many manufacturers of drainage products have developed design drawings that can be modified on the user's computer to customize them for an individual project.

In hydrology and hydraulics, computer simulation has been significantly enhanced by the use of graphic displays to aid in data entry and editing, and in analysis and interpretation of model results. (It is actually possible, for instance, to follow graphically changes in the hydraulic gradient as the simulation progresses.) Moreover, these capabilities have been expanded and merged with surveying and drafting packages to provide even greater flexibility. For example, after digitizing topographic data into the computer, a detention basin can be graphically designed on the computer screen. The basin's volume then can be calculated from the drawings and interactively input into the hydaulic model (in the model's required format). Other programs are available that can combine standard input data (curve numbers, rainfall amounts and distribution) with data supplied for the individual design by the user and calculate times of concentration, travel times, inflow/outflow hydrographs, stage/storage/outflow curves, etc. The rapid improvements in both software and computer hardware mean that, in the future, the possibilities will be limited only by the imagination and skill of the user.

Geographic information systems (GIS) enable the user to incorporate a wide range of information about the physical system into a computer data base. This can include not only information about the ground surface, but details of the urban infrastructure (water/wastewater, streets, electric, gas). The potential exists to integrate all the elements described above into a complete mapping and hydrology/hydraulics analysis and design package that can:

(a) Provide watershed physical feature mapping.
(b) Compute hydrologic model input parameters from the GIS.
(c) Model the rainfall/runoff process to determine design flows.
(d) Provide the capability for on-screen design of the system, including conveyance structures and appurtenances.
(e) Optimize the final design.
(f) Map or draw the system as designed, including plan and profile drawings of all structural components.

It should be noted that the requirements for checking and verification of designs so developed will still be necessary (or perhaps even more important).

XI. SUMMARY

Some form of modeling will almost assuredly become part of routine analyses performed at some portion of the thousands upon thousands of CSO and stormwater discharge locations around the country. Several modeling options exist, but none of them are truly "deterministic" in the sense of fully characterizing the physical, chemical, and biological mechanisms that underlie conceptual buildup, erosion, transport, and degradation processes that occur in an urban drainage system. Even if fully deterministic models were available, it is doubtful that they could be routinely applied without calibration data. But this is essentially true of almost all methods. Because a method is simple, e.g., constant concentration, does not make it more correct. Rather, the assumption is made that there will be some error in prediction regardless of the method, and there may be no point in compiling many hypothetical input parameters for a more complex model lacking a guarantee of a better prediction. For example, a study in Denver showed that regression equations could predict about as well as DR3M-QUAL given the available quality information (Ellis and Lindner-Lunsford 1986). But physically-based (conceptual) models do have certain advantages, discussed below.

Physically-based models depend upon conceptual buildup and wash-off processes incorporated into the quality algorithms. Such models have withstood the test of time and have been applied in major urban runoff quality studies. However, the relative lack of fundamental data on buildup and washoff parameters has led to simpler methods more often being applied, starting with the assumption of a constant concentration and becoming more complex. For example, the derived distribution approach of the EPA Statistical Method provides very useful screening information with minimal data; but it requires more data than by just assuming a constant concentration. With the mass of NURP and other data, regression approaches are now more viable but still subject to the usual restrictions of regression analysis. Spreadsheets are ubiquitous on microcomputers and serve as a convenient mechanism to implement several of the simple approaches, especially those that rely upon sets of coefficients, unit loads and/or EMCs as a function of land use or other demographic information.

Minimal data requirements and ease of application are the principal advantages of simpler simulation methods (constant concentration, unit loads, statistical, regression). However, in spite of their more complex data requirements, conceptual models (DR3M-QUAL, HSPF, STORM, SWMM) have advantages in terms of simulation of routing effects and control options as well as the superior statistical properties of continuous time series. For example, the EPA Statistical Method assumes that stream flow is not correlated with the urban runoff flow. This may or may not be true in a given situation, but it is not necessary to require such an assumption when running a model such as HSPF or SWMM. The four conceptual models discussed in detail all have a means of simulating storage and treatment effects. Other than a constant removal, this is difficult to do with the simpler methods. The conceptual quality models generally have very much superior hydrologic and hydraulic simulation capabilities (not true for STORM except that it can also use real rainfall hyetographs as input). This alone usually leads to better prediction of loads (product of flow times concentration). It should also be borne in mind that even complex models such as SWMM and HSPF can be run with minimal quality (and quantity) data requirements, such as using only a constant concentration. Finally, some of the case studies imply that transferability of coefficients and parameters is easier with buildup and washoff than with rating curve and constant concentration methods.

If a more complex conceptual model is to be applied, which one should it be from among those described herein? SWMM is certainly the most widely used and probably the most versatile for urban areas. HSPF may be more appropriate in areas with more open space where groundwater contributions increase in importance, where rainfall-induced erosion occurs, or where quality interactions are important along the runoff pathway. The simplicity of STORM remains attractive, and various consultants have used their own version as a planning tool. The USGS DR3M-QUAL model has been successfully applied in several USGS studies but has not seen much use outside the agency.

It contains useful techniques for quality calibration. SWMM and HSPF get limited support from the EPA Center for Exposure Assessment Modeling (CEAM) at Athens, Georgia. STORM and DR3M-QUAL will remain useful, but it is unlikely that either of these two models will enjoy enhancements or support from their sponsoring agencies in the near future.

What is a reasonable approach to simulation of urban runoff quality? The main idea is to use the simplest approach that will address the project objectives at the time. This usually means to start simple with a screening tool such as constant concentration (usually implemented in a spreadsheet) or regression or statistical approach. If these methods indicate that more detailed study is necessary or if they are unable to address all the aspects of the problem, e.g., the effectiveness of control options, then one of the more complex models must be run. No method currently available (or likely to be available) can predict absolute (accurate) values of concentrations and loads without local calibration data, including complex buildup and washoff models. Thus, if a study objective is to provide input loads to a receiving water quality model, local site-specific data will probably be required. On the other hand, several methods and models might be able to compare the relative contributions from different source areas, say, or to determine the relative effectiveness of control options (if the controls can be characterized by simple removal fractions). When used for purposes such as these, the methods, including buildup and washoff models, can usually be initiated on the basis of NURP and/or the best currently available source of quality data.

When properly applied and their assumptions respected, models can be tremendously useful tools in analysis of urban runoff quality problems. Methods and models are evolving that use the large current data base of quality information. As increasing attention is paid to urban runoff problems in the future, the methods and models can only be expected to improve.

XII. REFERENCES

Aron, G. (1987). *Penn State runoff model for IBM-PC*. Pennsylvania State University, University Park, PA.

Abbott, J. (1977). "Guidelines for calibration and application of storm." *Training Document No. 8*, Hydrologic Engineering Center, Corps of Engineers, Davis, CA.

Adams, B.J. and Howard, C.D.D. (1985). "The pathology of design storms." *Pub. 85-03*, University of Toronto, Dept. of Civil Engineering, Toronto, Ontario.

Alley, W.M. (1986). "Summary of experience with the distributed routing rainfall-runoff model (DR3M)." *Urban drainage modeling: Proc. of International*

Symposium on Comparison of Urban Drainage Models with Real Catchment Data, Dubrovnik, Yugoslavia, C. Maksimovic and M. Radojkovic, eds., Pergamon Press, New York.

Alley, W.M. and Smith, P.E. (1982a). "Distributed routing rainfall-runoff model—Version II." *USGS Open File Report 82-344*. Gulf Coast Hydroscience Center, NSTL Station, MS.

Alley, W.M. and Smith, P.E. (1982b). "Multi-event urban runoff quality model." *USGS Open File Report 82-764*, Reston, VA.

Ambrose, R.B. et al. (1988). "Waste allocation simulation models." *Water Poll. Cntrl Fed*, 60(9).

Ambrose, R.B., Jr. and Barnwell, T.O. Jr. (1989). "Environmental software at the U.S. Environmental Protection Agency's Center for Exposure Assessment Modeling." *Environmental Software*, 4(2).

APWA. (1969). "Water pollution aspects of urban runoff." *Report 11030DNS01/ 69 (NTIS PB-215532)*, American Public Works Association. Federal Water Pollution Control Administration, Washington, D.C.

Arnell, V. (1982). "Rainfall data for the design of sewer pipe systems." *Report Series, A:8*, Chalmers University of Technology, Goteborg, Sweden.

Balmer, P., Malmqvist, P-A. and Sjoberg, A., eds. (1984). *Proc. of the Third International Conf. on urban storm drainage*. Chalmers University, Goteborg, Sweden.

Barnwell, T.O., Jr. (1984). "EPA's center for water quality modeling," *Proc. of the third intl conf. on urban storm drainage*, Vol. 2, Chalmers University, Goteborg, Sweden, Vol. 2, 463–466.

Bedient, P.B. and Huber, W.C. (1989). *Hydrology and floodplain analysis*. Addison-Wesley Publishers, Reading, MA.

Bensen, M.A. (1962). "Spurious correlation in hydraulics and hydrology." *Journal of the Hydraulics Division, Proc. ASCE*, 91(HY4), 57–71.

Bontje, J.B., Ballantyne, I.K. and Adams, B.J. (1984). "User interfacing techniques for interactive hydrologic models on microcomputers." *Proceedings of the conference on stormwater and water quality management modeling, Burlington, Ontario*, Report R128, Computational Hydraulics, McMaster University, Hamilton, Ontario, 75–89.

Brandstetter, A.B. (1977). "Assessment of mathematical models for storm and combined sewer management." *EPA-600/2-76-175a (NTIS PB-259597)*, Environmental Protection Agency, Cincinnati, OH.

Brown, R.G. (1984). "Relationship between quantity and quality of storm runoff and various watershed characteristics in Minnesota, USA." *Proceedings of the Third International Conference on Urban Storm Drainage*, Vol. 3, Chalmers University, Goteborg, Sweden, 791–800, June 1984.

Camp, Dresser and McKee. "Boggy Creek basin nonpoint source water quality study." (1987). Report to the South Florida Water Management District from Camp, Dresser and McKee, Maitland, FL.

Crawford, N.H. and Linsley, R.K. (1966). "Digital simulation in hydrology: Stanford watershed model IV." *Technical Report No. 39*, Civil Engineering Dept., Stanford University, Palo Alto, CA.

Dawdy, D.R., Lichty, R.W. and Bergmann, J.M. (1972). "A rainfall-runoff simulation model for estimation of flood peaks for small drainage basins." *USGS Professional Paper 506-B*, Washington, D.C.

DeBarry, Paul A. (1990). "Computer watersheds." *Civil Engineering*, 60(7),

Delleur, J.W. and Dendrou, S.A. (1980). "Modeling the runoff process in urban areas." *CRC Critical Reviews in Environmental Control*, 10(1), 1–64.

Di Toro, D.M. (1979). "Statistics of advective dispersive system response to runoff." *Proc. of natl. conf. on urban stormwater and combined sewer overflow impact on receiving water bodies*, Y.A. Yousef et al., eds., EPA-600/9-80-056 (NTIS PB81-155426), Environmental Protection Agency, Cincinnati, OH.

Di Toro, D.M. (1984). "Probability model of stream quality due to runoff." *Journal of Environmental Engineering*, 110(3), 607–628.

Di Toro, D.M. and Small, M.J. (1979). "Stormwater interception and storage." *Journal of the Environmental Enginering Division*, 105(EE1), 43–54.

Donigan, A.S., Jr. and Crawford, N.H. (1976). "Modeling nonpoint pollution from the land surface." *EPA-600/3-76-083*, Environmental Protection Agency, Athens, GA.

Donigian, A.S., Jr. and Davis, H.H., Jr. (1978). "User's manual for agricultural runoff management (ARM) model." *EPA-600/3-78-080*, Environmental Protection Agency, Athens, GA.

Donigian, A.S. and Huber, W.C. (1990). "Modeling of nonpoint source water quality in urban and non-urban areas." *Report to the EPA Office of Research and Development, Contract No. 68-03-3513*, U.S. Environmental Protection Agency, Athens, GA.

Donigian, A.S., Jr. et al. (1984). "Application guide for hydrologic simulation program fortran (HSPF)." *EPA-600/3-84-065*, Environmental Protection Agency, Athens, GA.

Driscoll, E.D. (1979). "Benefit analysis for combined sewer overflow control." *Seminar Publication, EPA-625/4-79-013*, Environmental Protection Agency, Cincinnati, OH.

Driscoll, E.D. (1981). "Combined sewer overflow analysis handbook for use in 201 facility planning." Two Volumes, *Final Report for Contract No. 68-01-6148*, Environmental Protection Agency, Facility Requirements Division, Policy and Guidance Branch, Washington, D.C.

Driscoll, E.D. (1986). "Lognormality of point and nonpoint source pollutant concentrations." *Proceedings of the stormwater and water quality model users group meeting*, EPA/600/0-86/023 (NTIS PB87-117438/AS), Environmental Protection Agency, Athens, GA.

Driscoll, E.D., Shelley, P.E., and Strecker, E.W. (1989). "Pollutant loadings and impacts from highway stormwater runoff." *Vol. I—Design Procedure (FHWA-*

RD-88-06), Office of Engineering and Highway Operations R&D, Federal Highway Administration, McLean, VA.

Driver, N.E. et al. (1985). "U.S. Geological Survey urban stormwater data base for 22 metropolitan areas throughout the United States." *U.S.G.S Open File Report 85-337*, Lakewood, CO.

Driver, N.E. and Tasker, G.D. (1988). "Techniques for estimation of storm-runoff loads, volumes, and selected constituent concentrations in urban watersheds in the United States." *U.S.G.S. Open File Report 88-191*, Denver, CO.

Ellis, S.R. and Lindner-Lunsford, J.B. (1986). "Comparison of conceptually based and regression rainfall-runoff models in the Denver metropolitan area, Colorado, USA," in *Urban drainage modelling; Proc. of intl. symp. on comparison of urban drainage models with Real Catchment Data*, C. Maksimovic and M. Radojkovic, eds., Pergamon Press, New York, NY, 263–273.

Feldman, A.D. (1981). "HEC models for water resources system simulation: theory and experience," *Advances in Hydroscience*, vol. 12, Academic Press, New York, NY, 297–423.

Goforth, G.F., Heaney, J.P. and Huber, W.C. (1983). "Comparison of basin performance of modeling techniques." *Journal of Environmental Engineering*, 109(5), 1082–1098.

Hall, M.J. (1984). *Urban hydrology*. Elsevier Applied Science Publishers, New York, NY.

Harremöes, P., ed. (1983). "Rainfall as the basis for urban runoff design and analysis." *Proceedings of the Seminar*, Pergamon Press, New York, NY.

Heaney, J.P. et al. (1977). "Nationwide evaluation of combined sewer overflows and urban stormwater discharges, vol. II: Cost assessment and impacts." *EPA-600/2-064b (NTIS PB-266005)*, Environmental Protection Agency, Cincinnati, OH.

Heaney, J.P. et al. (1975). "Urban stormwater management modeling and decision making. *EPA-670/2-75-022 (NTIS PB-242290)*, Environmental Protection Agency, Cincinnati, OH.

Hinson, M.O., Jr. and Basta, D.J. (1982). "Analyzing surface receiving water bodies." *Analyzing natural systems, analysis for regional residuals—environmental quality management*, D.J. Basta and B.T. Bower, eds., Johns Hopkins University Press, Baltimore, MD (also available from NTIS as PB83-223321), 245–338.

Howard, C.D.D. (1976). "Theory of storage and treatment plant overflows." *Journal of the Environmental Engineering Division, Proc. ASCE*, Vol. 102(EE4), 709–722.

Huber, W.C. (1980). "Urban wasteload generation by multiple regression analysis of nationwide urban runoff data." *Proc. of a workshop on verification of water quality models*, R.V. Thomann and T.O. Barnwell, eds., EPA-600/9-80-016 (NTIS PB80-186539), Environmental Protection Agency, Athens, GA, 167–175.

Huber, W.C. and Heaney, J.P. (1982). "Analyzing residuals generation and discharge from urban and nonurban land surfaces." *Analyzing natural systems, analysis for regional residuals — Environmental quality management*, D.J. Basta and B.T. Bower, eds., Johns Hopkins University Press (also available from NTIS as PB83-223321), 121–243.

Huber, W.C. et al. (1982). "Urban rainfall-runoff-quality data base." *EPA-600/ 2-81-238 (NTIS PB82-221094)*, Environmental Protection Agency, Cincinnati, OH.

Huber, W.C. (1985). "Deterministic modeling of urban runoff quality." *Urban runoff pollution*, Series G: Ecological Sciences, NATO ASI Series, H.C. Torno, J. Marsalek, and M. Desbordes, eds. Vol. 10, Springer-Verlag, New York, NY, 167–242.

Huber, W.C. (1986). "Modeling urban runoff quality: State of the art." *Proceedings of conference on urban runoff quality, impact and quality enhancement technology*, B. Urbonas and L.A. Roesner, eds., Engineering Foundation, ASCE, New York, NY.

Huber, W.C., Cunningham, B.A. and Cavender, K.A. (1986) "Use of continuous SWMM for selection of historic rainfall design events in Tallahassee." *Proc. of the stormwater and water quality model users group meeting*, EPA/600/9-86/023 (NTIS PB87-117438/AS), Environmental Protection Agency, Athens, GA, 295–321.

Huber, W.C., Heaney, J.P. and Cunningham, B.A. (1986). "Storm water management model (SWMM) bibliography." EPA/600/3-85/077 (NTIS PB86-136041/AS), Environmental Protection Agency, Athens, GA.

Huber, W.C. and Dickinson, R.E. (1988). "Storm water management model user's manual, version 4. *EPA/600/3-88/001a (NTIS PB88-236641/AS)*, Environmental Protection Agency, Athens, GA.

Hydrologic Engineering Center. (1977). "Storage, treatment, overflow, runoff model, STORM, user's manual." *Generalized Computer Program 723-S8-L7520*, Corps of Engineers, Davis, CA.

Hydroscience, Inc. (1979). "A statistical method for assessment of urban stormwater loads — impacts — controls." *EPA-440/3-79-023 (NTIS-PB-299185/9)*, Environmental Protection Agency, Washington, D.C.

Jackson, T.J. (1982). "Application and selection of hydrologic models." *Hydrologic modeling of small watersheds*, C.T. Haan, H.P. Johnson and D.L. Brakensiek, eds., Monograph No. 5, American Society of Agricultural Engineers, St. Joseph, MI, 473–504.

James, L.D. and Burges, S.J. (1982). "Selection, calibration, and testing of hydrologic models." *Hydrologic modeling of small watersheds*, C.T. Haan, H.P. Johnson and D.L. Brakensiek, eds., Monograph No. 5, American Society of Agricultural Engineers, St. Joseph, MI. 11, 435–472.

James, W. and Boregowda, S. (1985). "Continuous mass-balance of pollutant buildup processes." *Urban runoff pollution*, NATO ASI Series G: Ecological Sciences, H.C. Torno, J. Marsalek and M. Desbordes, eds., Vol. 10, Springer-Verlag, New York, NY, 243–271.

James W., and Robinson, M.A. (1982). "Continuous models essential for detention design." *Proc. of an engineering foundation conf. on stormwater detention facilities planning, design, operation and maintenance*. ASCE, New York, NY.

James, W. and Shtifter, Z. (1981). "Implications of storm dynamics on design storm inputs." *Proceedings stormwater and water quality management modeling and SWMM users group meeting, Niagara Falls, Ontario*, McMaster University, Hamilton, Ontario, 55–78.

Johanson, R.C. et al. (1984). "Hydrological simulation program—Fortran (HSPF): user's manual for release 8." *EPA-600/3-84-066*, U.S. Environmental Protection Agency, Athens, GA.

Keifer, C.J. and Chu, H.H. (1957). "Synthetic storm pattern for drainage design." *Journal of the Hydraulics Division, Proc. ASCE*, 83(HY4).

Kibler, D.F., ed. (1982). "Urban stormwater hydrology." *Water Resources Monograph 7*, American Geophysical Union, Washington, D.C.

Kibler, D.F. and Aron, G. (1980). "Observations on kinematic response in urban runoff models." *Water Resources Bulletin*, V1. 16(3), 444–452.

Kibler, D.F., Froelich, D.C. and Aron, G. (1981). "Analyzing urbanization impacts on Pennsylvania flood peaks." *Water Resources Bulletin*, 17(3), 270–274.

Kohlhaas, C.A., ed. (1982). *Compilation of water resources computer program abstracts*, U.S. Committee on Irrigation, Drainage and Flood Control, Denver, CO.

Lager, J.A. et al. (1977). "Urban stormwater management and technology: update and users' guide." *EPA-600/8-77-014 (NTIS PB-275654)*, Environmental Protection Agency, Cincinnati, OH.

Linsley, R.K. and Crawford, N.H. (1974). "Continuous simulation models in urban hydrology." *Geophysical Research Letters*, 1(1), 59–62.

Loganathan, V.G. and Delleur, J.W. (1984). "Effects of urbanization on frequencies of overflows and pollutant loadings from storm sewer overflows: A derived distribution approach." *Water Resources Research*, 20(7), 857–865.

Maalel, K. and Huber, W.C. (1984). "SWMM calibration using continuous and multiple event stimulation." *Proc. of the third intl. conference on urban storm drainage*, Chalmers University, Goteborg, Sweden, 2, 595–604.

Manning, M.J., Sullivan, R.H. and Kipp, T.M. (1977). "Nationwide evaluation of combined sewer overflows and urban stormwater discharges—Vol. III: characteristics of discharges." *EPA-600/2-77-064c (NTIS PB-272107)*, Environmental Protection Agency, Cincinnati, OH.

McCuen, R.H. (1982). *A guide to hydrologic analysis using SCS methods*. Prentice-Hall, Englewood Cliffs, NJ.

McElroy, A.D. et al. (1976). "Loading functions for assessment of water pollution from non-point sources." *EPA-600/2-76-151 (NTIS PB-253325)*, Environmental Protection Agency, Washington, D.C.

McPherson, M.B. (1974). "Innovation: A case study." *Technical Memo. No. 21, (NTIS PB-232166)*, ASCE Urban Water Resources Research Program, ASCE, New York, NY.

McPherson, M.B. (1978). "Urban runoff control planning." *EPA-600/9-78-035*, U.S. Environmental Protection Agency, Washington, D.C.

Metcalf and Eddy, Inc., University of Florida and Water Resources Engineers, Inc. (1971). "Storm water management model, volume I—Final report." *EPA Report 11024DOC07/71 (NTIS PB-203289)*, Environmental Protection Agency, Washington, D.C.

Miller, R.A., Mattraw, H.C., Jr. and Jennings, M.E. (1978). "Statistical modeling of urban storm water processes, Broward County, Florida." *Proc. of the intl. symp. on urban storm water management*, University of Kentucky, Lexington, KY, 269–273.

Murphy, C.B. Jr. et al. (1982). "The Use of Event Simulation Models for the Protection of Long-Term Continuous Impacts in Combined Sewer Overflow Abatement Planning." Reprint of paper presented at the ASCE National Conference on Environmental Engineering, Minneapolis, MN.

Najarian, T.O., Griffin, T.T. and Gunawardana, V.K. (1986). "Development impacts on water quality: A case study." *Journal of Water Resources Planning and Management*, 112(1), 20–35.

National Oceanic and Atmospheric Administration. (1987). *The National Coastal Pollutant Discharge Inventory, Urban Runoff Methods Documents*. Office of Oceanography and Marine Assessment, NOAA, Rockville, MD.

Noel, D.C. and Terstriep, M.L. (1982). "Q-ILLUDAS—A continuous urban runoff/washoff model." *Proc. intl. symp. on urban hydrology, hydraulics, & sediment control*, Lexington, KY, University of Kentucky, Lexington, KY.

Novotny, V. (1985). Discussion of "Probability model of stream quality due to runoff," by D.M. Di Toro. *Journal of Environmental Engineering*, 111(5), 736–737.

Orlob, G.T., ed. (1983). *Mathematical modeling of water quality: streams, lakes, and reservoirs*. John Wiley and Sons, New York, NY.

Overton, D.E. and Meadows, M.E. (1976). *Stormwater modeling*. Academic Press, New York, NY.

Pisano, W.C. and Quieroz, C.S. (1977). "Procedures for estimating dry weather pollutant deposition in sewerage systems." *EPA-600/2-77-120 (NTIS PB-270695)*, Environmental Protection Agency, Cincinnati, OH.

Pitt, R. (1979). "Determination of non-point pollution abatement through improved street cleaning practices." *EPA-600/2-79-161 (NTIS PB80-108988)*, Environmental Protection Agency, Cincinnati, OH.

Proctor and Redfern, Ltd. and J.G. MacLaren, Ltd. (1976a). "Storm water management model study—vol. I, final report." *Research Report No. 47*, Canada-Ontario Agreement on Great Lakes Water Quality, Environmental Protection Service, Environment Canada, Ottawa, Ontario.

Proctor and Redfern, Ltd. and J.G. MacLaren, Ltd. (1976b). "Storm water management model study—vol. II, technical background." *Research Report No. 48*, Canada-Ontario Agreement on Great Lakes Water Quality, Environmental Protection Service, Environment Canada, Ottawa, Ontario.

Reckhow, K.H., Clements, J.T. and Dodd, R.C. (1990). "Statistical evaluation of mechanistic water-quality models." *Journal of Environmental Engineering*, 116(EE2).

Renard, K.G., Rawls, W.J. and Fogel, M.M. (1982). "Currently Available Models," *Hydrologic Modeling of Small Watersheds*, C.T. Haan, H.P. Johnson and D.L. Brakensiek, eds., Monograph No. 5, American Society of Agricultural Engineers, St. Joseph, MI, 505–522.

Roesner, L.A. et al. (1974). "A model for evaluating runoff-quality in metropolitan master planning." ASCE Urban Water Resources Research Program, *Technical Memo. No. 23 (NTIS PB-234312)*, ASCE, New York, NY.

Roesner, L.A., Shubinski, R.P. and Aldrich, J.A. (1981). "Storm water management model user's manual version III: addendum I, EXTRAN." *EPA-600/2-84-109b (NTIS PB84-198431)*, Environmental Protection Agency, Cincinnati, OH.

Roesner, L.A. and Dendrou, S.A. (1985). Discussion of "Probability Model of Stream Quality Due to Runoff," by D.M. Di Toro. *Journal of Environmental Engineering*, 111(5), 738–740.

Roesner, L.A., Aldrich, J.A., and Dickinson, R.E. (1988). "Storm water management model user's manual, version 4: EXTRAN addendum." *EPA-600/3-88/001b (NTIS PB88-236658/AS)*, U.S. Environmental Protection Agency, Athens, GA.

Sartor, J.D. and Boyd, G.B. (1972). "Water pollution aspects of street surface contaminants." *EPA-R2-72-081 (NTIS PB-214408)*, U.S. Environmental Protection Agency, Washington, D.C.

Schilling, W. (1985). "Urban runoff quality management by real-time control." *Urban runoff pollution*, NATO ASI Series, Series G: Ecological Sciences, H.C. Torno, J. Marsalek and M. Desbordes, eds. Vol. 10, Springer-Verlag, New York, NY, 765–817.

Schueler, T.R. (1987). *Controlling urban runoff: A practical manual for planning and designing urban BMPs*. Metropolitan Information Center, Metropolitan Washington, Council of Governments, Washington, D.C.

Small, M.J. and Di Toro, D.M. (1979). "Stormwater treatment systems," *Journal of the Environmental Engineering Division, Proc. ASCE*, 105(EE3), 557–569.

Stall, J.B. and Terstriep, M.L. (1972). "Storm sewer design—An evaluation of the RRL method." *EPA-R2-72-068*, Environmental Protection Agency, Washington, D.C.

Surkan, A.J. (1974). "Simulation of storm velocity effects on flow from distributed channel networks." *Water Resources Research*, 10(6), 1149–1160.

Tasker, G.D. and Driver, N.E. (1988). "Nationwide regression models for predicting urban runoff water quality at unmonitored sites." *Water Resources Bulletin*, 24(5), 1091–1101.

Terstriep, M.L., Bender, G.M. and Noel, D.C. (1982). "Evaluation of the effectiveness of municipal street sweeping in the control of urban storm runoff pollution." *CR #300*, Illinois State Water Survey, Champaign, IL.

Terstriep, M.L. and Stall, J.B. (1974). "The Illinois urban drainage area simulator, ILLUDAS." *Bulletin 58*, Illinois State Water Survey, Urbana, IL.

Thomann, R.V. (1982). "Verification of water quality models." *Journal Environmental Engineering Div., Proc. ASCE*, 108(EE5), 923–940.

Thomann, R.V. and Mueller, J.A. (1987). *Principles of surface water quality modeling and control*. Harper and Row, New York, NY.

U.S. Army Corps of Engineers. (1987). "Water quality models used by the Corps of Engineers." *Information Exchange Bulletin, Vol. E-87-1*, Water Operations Technical Support, Waterways Experiment Station, Vicksburg, MS.

U.S. Environmental Protection Agency. (1973). *Methods for identifying and evaluating the nature and extent of non-point sources of pollutants*. EPA-430/9-73-014, Environmental Protection Agency, Washington, D.C.

U.S. Environmental Protection Agency. (1976a). "Land use—water quality relationship." *WPD-3-76-02*, Water Planning Division, Environmental Protection Agency, Washington, D.C.

U.S. Environmental Protection Agency. (1976b). *Areawide assessment procedures manual*, Three Volumes, EPA-600/9-76-014 (NTIS PB-27183/SET), U.S. Environmental Protection Agency, Cincinnati, OH.

U.S. Environmental Protection Agency. (1983a). *EPA Environmental Data Base and Model Directory*. Information Clearinghouse, (PM 211A), U.S. Environmental Protection Agency, Washington, D.C.

U.S. Environmental Protection Agency. (1983b). "Results of the nationwide urban runoff program, volume I, final report. *NTIS PB84-185552*, Environmental Protection Agency, Washington, D.C.

U.S. Department of Agriculture. (1971). *SCS national engineering handbook, section 4, hydrology*. Soil Conservation Service, U.S. Dept. of Agriculture, U.S. Government Printing Office, Washington, D.C.

U.S. Department of Agriculture. (1983). "Computer program for project formulation—hydrology." *Technical Release 20*, 2nd ed., Soil Conservation Service, U.S. Dept. of Agriculture, Available from NTIS, Springfield, VA.

U.S. Department of Agriculture. (1968a). "Urban hydrology for small watersheds." *Technical Release No. 55*, 2nd ed., Soil Conservation Service, U.S. Dept. of Agriculture, *NTIS PB87-101580*, Springfield, VA.

U.S. Department of Agriculture. (1986b). "TR-55, Urban hydrology for small watersheds, 2nd ed., microcomputer version 1.11, executable modules only." *NTIS PB87-101598*, Soil Conservation Service, U.S. Dept. of Agriculture, Springfield, VA.

Viessman, W., Jr. et al. (1989). *Introduction to hydrology*. 3rd edition, Harper and Row, New York, NY.

Walesh, S.G., Lau, D.H., and Liebman, M.D. (1979). "Statistically-based use of event models." *Proc. of the intl. symp. on urban storm runoff*, University of Kentucky, Lexington, KY, 75–81.

Walesh, S.G., and Snyder, D.F. (1979). "Reducing the cost of continuous hydrologic-hydraulic simulation." *Water Resources Bulletin*, 15(3), 644–659.

Walker, J.F., Pickard, S.A., and Sonzogni, W.C. (1989). "Spreadsheet watershed modeling for nonpoint-source pollution management in a Wisconsin area." *Water Resources Bulletin*, 25(1), 139–147.

Warwick, J.J., and Wilson, J.S. (1990). "Estimating uncertainty of stormwater runoff concentrations." *Journal of the Water Resources Planning and Management Division*, 116(2).

Water Pollution Control Federation. (1989). "Combined sewer overflow pollution abatement." *Manual of Practice FD-17*, WPCF, Alexandria, VA.

Watkins, L.H. (1962). "The design of urban sewer systems." *Road Research Technical Paper No. 55*, Dept. of Scientific and Industrial Research, Her Majesty's Stationery Office, London, England.

Whipple, W.J. et al. (1983). *Stormwater management in urbanizing areas.* Prentice-Hall, Englewood Cliffs, NJ.

Wischmeier, W.H., and Smith, D.D. (1958). "Rainfall energy and its relationship to soil loss." *Trans., American Geophysical Union*, 39(2), 285–291.

Wisner, P.E. (1982). "The IMPSWM procedures for urban drainage modelling and some applications." *Proc. stormwater and water quality management modeling and SWMM users group meeting*, University of Ottawa, Ottawa, Ontario, 225–248.

Wisner, P.E. et al. (1986). "Application of inlet control devices and dual drainage modelling for new subdivisions." *Proc. of stormwater and water quality model users group meeting*, EPA/600/9-86/023 (NTIS PB87-117438/AS), Environmental Protection Agency, Athens, GA, 275–294.

Woodward-Clyde Consultants. (1989). *Synoptic Analysis of Selected Rainfall Gages Throughout the United States*, Report to EPA, Woodward-Clyde Consultants, Oakland, CA.

Yen, B.C. (1986). "Hydraulics of sewers." *Advances in hydroscience*, B.C. Yen, ed., vol. 14, Academic Press, New York, NY, 1–122.

Zison, S.W. (1980). "Sediment-pollutant relationships in runoff from selected agricultural, suburban and urban watersheds." *EPA-600/3-80-022*, Environmental Protection Agency, Athens, GA.

Zukovs, G., Kollar, J., and Shanahan, M. (1986). "Development of the HAZPRED model." *Proc. of the stormwater and water quality model users group meeting*, EPA/600/9-86/023 (NTIS PB87-117438/AS), Environmental Protection Agency, Athens, GA. 128–146.

Chapter 8

DESIGN OF DRAINAGE CONVEYANCES

I. INTRODUCTION

A typical urban drainage system has two separate and distinct components, the major and the minor systems. Storm sewers, a part of the minor system, often have been the only **planned** portion of these drainage works. When their capacity is exceeded, the excess often flows overland, causing damage and losses. The purpose of the major drainage system is to accommodate the runoff that exceeds the capacity of the minor system.

The storm drainage system uses the energy available from the difference between its upstream and downstream elevations, or the energy added by pumps. The total available energy normally is used to maintain proper flow velocities, including sufficient velocities to insure a self-cleansing system, with minimum head loss. The design of storm drainage systems therefore involves the balancing of hydraulic losses, which must be kept within the limits of available energy, with the need for adequate energy to maintain self-cleansing velocities. The wider the variations in the rates of flow, the more difficult it is to meet both requirements (ASCE 1969).

This chapter focuses on the design of conveyance facilities common to the major and minor systems, including storm sewers, streets, open channels, culverts, and bridges. Details on their hydraulic design are presented in Chapter 6, and one should be completely familiar with that Chapter (and with Chapter 5, Hydrology and Introduction to Water Quality) before proceeding with any design. All of the facilities described closely interrelate, and major and minor system planning and design should be conducted concurrently.

The designer should treat the design effort as an iterative process involving the developer, regulators, planners and landscape architects, the general project civil engineer, and others (National Association of

225

Homebuilders 1991; Urban Land Institute 1976). All of these individuals have much to contribute, and optimum drainage solutions usually are those generated by various parties with different points of view (Spirn 1984; American Society of Civil Engineers 1985).

II. FREQUENCY OF DESIGN RUNOFF

Historically, storm drainage systems were designed solely for the rapid removal of stormwater, and hydrological considerations were largely restricted to sizing conduits (American Society of Civil Engineers 1985). The performance criterion was how effectively the land had been drained. It is common now, however, to require that these systems meet several regulatory criteria, including containment of flows and their pollutant burdens. The designer therefore is faced not only with the problem of assuring that flooding will be minimized, but also that the system will meet established limits on the discharge of pollutants.

The selection of a frequency of design runoff, commonly (and erroneously) called the "design storm," to satisfy the various regulatory criteria must be carefully done. As noted in Chapter 5, it can be easily shown (McPherson 1978; WPCF 1989) that, for a given rainfall event, the return frequencies of peak runoff flow, flow volume and the pollutants associated with that flow will probably be different from each other, as well as different from the return frequency of the rainfall itself.

The normal level of protection expected from the **minor system** ranges from once in two years to once in ten years (used in most cities in North America). Typical return frequencies were presented in Table 5.1. It must be recognized that site-specific conditions can justify deviations from Table 5.1, and "blind" acceptance of these values is not appropriate. As noted in Chapter 1, there has not been, over the years, any adequate rationale developed for the selection of any particular design return frequency for either the minor or the major system.

The return frequency of design runoff used for the **major system** typically ranges from 25- to 100-years, and should be selected based on an evaluation of the specific project under consideration and any local or regulatory requirements. Once the overall design return frequency has been set, the system should be examined for points where deviation may be justified or necessary. If this analysis reveals serious problems with the existing major system, then remedial measures should be considered (Denver Urban Drainage and Flood Control District 1984), such as over-designing the minor system to compensate for deficiencies that may exist in the major system. The marginal economic benefits to be obtained should be calculated, if possible, to provide information for the economic (political) decisions that inevitably follow.

III. GENERALIZED DESIGN PROCEDURES

These procedures apply generally for new drainage systems. For existing systems, one or more of the steps may not be necessary.

A. Preliminary Design

1. Define Project Goals and Objectives

Define the goals and objectives (water quality and quantity) that will guide the project. What does the project proponent want the design to accomplish? What is expected of the design engineer? What is the scope of the assignment? Reference: Chapter 4.

2. Define Pertinent Regulations and Criteria

Identify all applicable federal, state, and local regulations and criteria that will affect the design. For contemplated drainage structures, list criteria that will apply (major/minor system return frequencies, freeboard in channels, pollutant removal requirements, etc.). If local criteria are not available, obtain guidance from local regulators (note that national and/or state criteria may apply). Reference: Chapter 2.

3. Collect Basic Data

Collect background information (soil types, locations of wetlands, existing drainage structures, rainfall characteristics, historic flood information, water quality data, groundwater situation, survey/mapping/GIS data, etc.), including prior studies in the area. Reference: Chapter 3.

4. Determine Limits of Basin and Analyze Preliminary Data

Classify probable future development within the basin as it affects both hydrology and hydraulic design (see Chapter 5). Off-site areas that drain onto the site, not just the site itself, must be included. Classify streets as to storm water drainage carrying capacity. Determine design frequency for initial design. Obtain or develop rainfall data (intensity-duration-frequency curves, design storms, or time-series data, depending on hydrologic methods being used) and calculate associated runoff for both the minor and the major system.

Identify location of discharges (outfalls) for the project, along with their capacity and downstream constraints. Identify natural drainageways through the site.

When sub-dividing the total drainage area of interest, remember that at various inlets on a continuous grade, only a portion of street flow will be diverted to the storm sewer system (due to "carryover"), and at intersections of collector streets or arterial streets it may be necessary to remove 100% of the minor drainage from the road surface to preclude cross-street flow. The sub-basins should vary according to the actual storm sewer system layout being considered.

5. Obtain Site Development Plan and Formulate Conceptual Alternatives

Obtain initial site development plan from the project proponent or planner and formulate conceptual alternative drainage approaches. Based on "back of the envelope" hydrology and field inspection, identify potential "fatal flaws" in the development plan, which can include:

(a) Obliteration of existing major drainageways.
(b) Adverse effects of increased runoff (peak flow/volume).
(c) Site development plan that conflicts with the natural hydrologic characteristics of the site rather than enhances these characteristics.
(d) Lack of attention to whether the site drains to an outfall with assured capacity. Where will the water go once it is discharged off-site and what hazards are posed to downstream properties?
(e) Off-site flows that enter but do not safely pass through the site.
(f) Insufficient attention to local, state or federal permitting requirements.
(g) Failure to set aside sites for detention storage.

It may be desirable to modify the proposed layout to minimize the need for drainage and sanitary sewer facilities, and for earthwork. Conceptual alternatives will vary according to such factors as magnitude and type of detention; major/minor system alignments (flow paths); and channelization to convey full major discharge versus tight land use controls in floodplains.

6. Refine Conceptual Alternatives to 2–3 Preferred Strategies

Review the reasonable alternative layouts for both the major and minor systems, selecting those that appear most practical, economical, and consistent with project objectives. A number of iterations between the planner and drainage engineer may be required. The most promising concepts should be selected for analysis.

Existing drainage, whether clearly defined channels or more subtle "swales," should normally be set aside as major drainageways. Rolling, hilly terrain usually has natural drainage patterns that cannot be significantly modified. If existing major drainageways are to be blocked by land development, alternative drainage capacity must be provided. Transferring the problem from one drainage basin to another, however, should be strictly avoided (Denver Urban Drainage and Flood Control District 1984).

The safety, convenience and cost effectiveness of preserving natural drainage to serve as "outfalls" for street and storm sewer discharges are well recognized, and the designer is urged to identify and utilize the natural system to its fullest advantage (Sheaffer et al. 1982; Urban Land Institute 1975).

7. *Preliminary Design.*

Proceed with preliminary design of the major/minor systems for each of the remaining alternatives. The methods for determining design runoff are discussed in Chapter 5 (see also Section XI, this chapter). Design runoff rates may need to be adjusted to reflect detention requirements, in accordance with the procedures described in Chapter 11.

Preliminary design consists of the following kinds of activities; this list is not meant to be comprehensive and the reader is referred to Chapters 4, 8, 10, and 11 for further detail. These tasks should be conducted for each of the alternatives.

    a. Define alignments and grades for storm sewers and open channels. For pipes, identify factors that will influence alignment and grade, such as other utilities (crucial), railroads, tree preservation, embankments, buildings, etc. Other factors that influence channel grade include existing slope (of natural channels), erodibility, available right of way, and channel lining. Several preliminary layouts may be considered. Pipe alignments should probably coincide with collector or arterial streets, since it usually will be necessary to preclude cross street flow on these streets. Experiment with different approaches to optimize the number and size of storm water conveyance structures required.

    b. Define detention strategy and storage locations, how detention will influence the outfall system for each alternative (APWA 1981), and the role detention storage will have in water quality enhancement (Schueler 1987).

    c. Define the inception points of storm sewers—the locations where specified street carrying capacity is exceeded. Examine ways to reduce flows in headwater areas so as to be able to move the starting point of storm sewers downstream (to save money).

Preliminary street grades and cross-sections must be available to the designer. Beginning at the upper end of the basin in question, calculate the flow in the street until the allowable street carrying capacity matches the design runoff. The storm sewer system will begin at this location if there is no alternate method of removing runoff from the street surface (Malcom 1989). It is not necessary to remove all of the flow from the street surface at the beginning of the storm sewer system, or at any given location along the system, unless the intersection of streets dictates that no cross-street flow may occur. It is, however, necessary that the allowable street capacity plus the storm sewer capacity equal or exceed the design flow.

This process should account for the fact that some portion of the total runoff will not reach the streets, but will instead be conveyed through natural drainage which is unaltered by development, by manmade open channels, or simply by "inadvertent" grading of yards and open spaces.

The portion of flow removed from the street surface becomes the design flow for the storm sewer. For preliminary design purposes, a

Manning's "n" value, or other roughness coefficient, about 25% above that contemplated for final design should be used in calculation. Using this n value, the preliminary grades established earlier, and assuming the sewer is flowing full, a pipe size for the design flow may be determined from an applicable chart or formula, or by using one of the many commercially-available computer programs developed for this purpose. This method for sizing pipes should be adequate for preliminary design purposes, since the increased roughness coefficient offsets the effect of minor losses. Flow velocities will be sufficiently accurate for routing calculations. Cost estimates can then be based on these sizes and assumed depths of cut.

   d. Starting in the study area headwaters and proceeding to the downstream point of interest, calculate flows at ultimate development, using the major system design runoff, in the major drainageways, and determine the right-of-way required to convey the flows (with an allowance for freeboard).

Determine if the combined capacity of the street and storm sewer system is sufficient to maintain surface flows within acceptable limits during the major storm. The combined total of the allowable street carrying capacity for the major storm and the storm sewer capacity should equal the major design runoff. At any given point along the storm sewer system, the capacity of the sewer can usually be assumed to be the same for the major runoff as for the minor runoff, unless special considerations indicate otherwise.

In the uppermost area of a drainage basin, major drainageways may not consist of readily distinguishable channels—they may be homeowners yards, swales or streets or whatever "low ground" exists— and for the entire study area, the designer should determine the path that the major system design runoff event will take. This land should be set aside as a perpetual "flow easement." Preliminary design-level channel sizing can normally be done via hand calculations using normal depth techniques and assumptions regarding location of hydraulic "controls."

   e. Define all hydrologic "design points" for channel and storm sewer reaches, including drainageway confluences, street intersections, detention storage locations, junction structures, culverts, etc. When establishing the design points, inspect future condition site mapping to establish flow paths for all parcels on the site.
   f. Conduct detailed hydrologic computations to quantify design flows at all drainage structures of interest.
      1. Compute off-site and on-site flows.
      2. Calculate ultimate development flows and "historic" flows for floods ranging from the 2-year event to the major system design runoff. Under particular circumstances, it may be necessary to consider runoff events smaller than the 2-year (for water quality design purposes) or larger than the major system design runoff.

If it is impractical to accommodate the major system design flows using the combination of the street and storm sewer carrying capacity, the engineer has these options: (1) change the major system design basis (which may require an important policy decision), (2) increase the minor system capacity, (3) elevate or otherwise protect buildings, (4) begin a formal major drainageway, or (5) provide upstream detention to decrease downstream discharge. Each of these choices will have economic implications.

The urban basin size required to effectively mandate that a formal major drainageway be reserved varies from location to location. In Denver, Colorado, for example, policy states that a properly designed, engineered and maintained open channel (an "outfall") will be established for all urban basins larger than 130 acres.

3. Based on the analysis in f. 2, applicable local criteria, and interviews with regulators, establish whether or not stormwater detention is required. Size detention facilities in accordance with the procedures described in Chapters 5 and 11.

g. Given a storm drainage system alignment and estimates of runoff discharges for the specified minor system return frequency, locate and size inlets. Pay particular attention to intersections where carryover is not acceptable.

h. Given storm drainage system alignment, grades, and inflows (via inlets), the hydraulic grade line and energy grade line for the storm sewer should be computed. Adjust alignment and grade as required to comply with criteria. Assure that the hydraulic grade line remains a specified distance below the ground surface for the minor event.

i. Not all of the minor system will be below ground. The major drainageways and associated hydraulic structures (culverts, transitions, etc.), surface swales and detention facilities will also need to function properly during the minor storm. Accordingly, evaluate their behavior during the minor event.

j. Check the function of the overall minor system during the design event and assure that adequate outfall capacity is available.

k. Given designated major drainageways (with right-of-way set aside to accommodate the design runoff) and a functioning minor system, evaluate where the flood waters, in excess of those being conveyed in the storm sewer, will flow. Also evaluate how site topography must be adjusted to assure that these flows will be conveyed safely downgradient without threatening lives, safety, or property.

l. Evaluate behavior of detention facilities during the major system design flow condition. Assure that detention dams, which could pose a threat to human safety or property if they fail, are designed to handle flows larger than major system design flows (Sheaffer et al. 1982).

m. Special evaluation of culverts and bridges, in accordance with the procedures described later in this chapter, will be required. Assure that applicable criteria are not violated. Adjust the size and characteristics of the conveyances until the amount of backwater during the major system design runoff event meets target levels. Define the area affected by backwater flooding during the major event and indicate that this land should remain undeveloped for perpetuity (Jones and Jones 1987).

n. Define right-of-way requirements for alternatives.
o. Prepare preliminary-design level capital and operation/maintenance costs for the alternatives. If feasible and appropriate, use life cycle cost theory (see discussion later in this chapter) for economic evaluation (Urban Drainage and Flood Control District 1987). The design cost objective should be to minimize the total annual costs of the drainage facilities and flood-related damage.
q. Evaluate the alternatives with respect to important qualitative criteria such as preservation of open space, enhanced wildlife habitat, impact on wetlands and water quality benefits (if no formal attempt has been made to contrast the pollutant removal capabilities of various alternatives).
r. Prepare a preliminary design report that contrasts the alternatives quantitatively and qualitatively (in a form suitable for submission to the client and regulators).

8. Review Alternates and Formulate Preferred Alternative for Final Design

Meet with client, project planner, general project civil engineer and others to review preliminary design work and to select a preferred alternative. Unbiased evaluation of the good and bad points of each alternative is necessary if the most desirable system is to be identified. Quantitative and qualitative factors should be assessed. If any potential problems exist, such as effects on downstream entities, large conflicts with utilities, or difficulty in acquiring easements, they should be thoroughly reviewed and resolved.

Adjust the preferred alternatives (still at the preliminary-design level) to accommodate the requests of the client and his advisors. Submit preferred preliminary design to regulators. Adjust preliminary design as required to fulfil regulatory requirements and obtain approval of the preliminary design.

B. Overview of Final Design

The preceding steps constitute the preliminary design effort. Up to this point, the engineer has been compiling the information necessary to make an informed decision on which system to use for final design (Urban Drainage and Flood Control District 1984). The following steps will complete the process.

1. Review all Preliminary Work

Hydrologic assumptions, basin boundaries, sub-basin delineations, street classifications, assumed/calculated pollutant removal efficiencies, and any other preliminary design values that will be used subsequently in final design should be reviewed for accuracy and applicability to final design. Unresolved questions must be answered at this time, and the designer should step back and ask himself if his design constitutes sound engineering practice. Fundamental compatibility of site layout with drainage needs must be assured at this stage (Urban Land Institute 1975).

2. Obtain Final Street Grades, Geometry, Elevations, Etc.

Often it will be necessary to revise street construction details to facilitate drainage. This may include eliminating cross fall on streets, raising required ground elevation at buildings adjacent to streets to accommodate major drainage, or increasing street gradients to achieve sufficient capacity within the street. It is especially important to assure that first floor elevations of buildings are well above street crown elevations to prevent repeated flooding (Jones 1980).

3. Hydraulically Design the Open Channel and Storm Sewer Systems

The final hydraulic design of a system should be on the basis of procedures set forth later in this Chapter and in Chapter 6. A realistic "n" value for final design should be determined and applied, treating the conduits as either open channels or pipes flowing full, as appropriate. For open channel flow, the energy grade line should be used as the base for calculation. For conduits flowing full, the hydraulic grade line should also be calculated. The energy grade line governs for major drainageways.

For storm sewers, the design engineer must review the hydraulic grade line for various runoff conditions (initial design runoff and others that are larger) to insure that the hydraulic grade line is consistent with desired system performance.

4. Complete all Other Aspects of the Design Effort

The final design should address all other factors, including structural and geotechnical design, permitting, water quality, construction documents, coordination with other aspects of the project, etc.

IV. SYSTEM LAYOUT

Design begins with system layout—approximately defining the minor and major flow routes. System layout includes the selection of an outfall, delimiting drainage area boundaries, and identifying the locations of trunk and main sewers, or outfall channels that will feed the outfall. Initial layouts can usually be done from topographic maps.

The ease with which the designer will be able to locate and utilize an "outfall" to discharge into will vary widely depending principally upon the extent of pre-existing development and whether the overall basin drainage was properly planned to assure that adequate outfall capacity would be available as development occurs. An inadequate outfall may force the design towards detention, whereas an oversized outfall could result in reduced on-site detention requirements. The layout of the minor system should generally conform to the following standards, which will vary from locale to locale.

A. Location Requirements

1. Main Location

Storm sewers are normally located a short distance behind the curb, or in the roadway near the curb. On occasion, it may be necessary to locate them in right-of-way easements, but locating them on public property is preferred.

2. Alignment

Storm sewers should be straight between manholes where possible (ASCE 1969). Where curves are necessary to conform to street layout, the radius of curvature should not be less than 100 feet. Radius of curvature specified should coincide with standard curves available for the type of material being utilized. Short radius curves at manholes can be designed per Table 8.1. Short radius bends are not recommended on sewers 21 inches or less in diameter.

3. Crossings

Crossings with other underground utilities should be avoided whenever possible, but, if necessary, should be at an angle greater than 45 degrees. Utility crossings are a major design factor (and cost) when retrofitting in urban areas, and considerable effort is required to comprehensively define and locate utilities in the field.

B. Manholes (Cleanout Structures)

Manholes are normally located at the junctions of sewers and often at changes in grade or alignment. Manhole spacing is usually specified locally, but Table 8.2 provides general values. Short radius bends are

TABLE 8.1.   Typical Allowable Radius of Curvature for Short Radius Curves (Boulder County 1984).

Diameter of Pipe (1)	Minimum Radius of Curvature (2)
24" to 54"	28.50 ft.
57 to 72"	32.00 ft.
78" to 108"	38.00 ft.

Note: Short radius bends should rarely be necessary because pipe alignments usually follow street alignments. In the rare situation where a short radius bend is necessary, discharge energy and geotechnical conditions should be evaluated to determine need for thrust blocks.

TABLE 8.2.   Typical Manhole Spacing (Boulder County 1984).

Pipe Size (1)	Maximum Spacing (2)
15" or less	400 feet
18" to 36"	500 feet
42" or greater	600 feet

Note: With pipes larger than 24"–30", manholes usually may be spaced as far apart as possible, consistent with sewer cleaning limitations.

often used on 24" and larger pipes when flow must undergo a direction change at a junction. Reductions in head losses at manholes may be realized in this way. A manhole should always be located at the end of such short radius bends.

C. Grade and Cover

The minimum slope should be capable of producing a velocity of at least 2–3 feet per second when the sewer is flowing full. Consideration should be given to the capacity required, sedimentation problems, and other design parameters.

The sewer grade should assure that there is a minimum of 3 feet of cover over the crown of the pipe (depending on local practice). Uniform slope between manholes is desirable for ease of maintenance. The sewer system must be structurally reinforced if sufficient vertical clearance (cover) is not available. Railroad and highway crossings pose a particular concern. Structural loading calculations are recommended for all such installations. For further structural details, see Chapter 14.

V. HYDRAULIC DESIGN OF STORM SEWERS

A. Design of Closed Conduits

The following design procedure is applicable to closed conduits flowing with a free water surface. Although it is theoretically possible for a pipe flowing less than full to carry more than when it is full, it is impractical and imprudent to assume that this larger conveyance will be available.

1. Range of Applicability

The following design procedure is based on the assumption of a uniform hydraulic gradient within pipe reaches. Where conduits are sufficiently large, or a higher degree of accuracy is needed, surface water profiles can be calculated (see Chapter 6 for water surface profile computations). In local drainage design, however, the computation of water surface profiles will rarely be required as flows and conveyance capacities are small (Urban Drainage and Flood Control District 1984).

2. Design Procedures

The basic approach to design of closed conduits consists of calculating the energy grade line along the system profile. In most situations, one can assume that the energy grade line is parallel to the pipe grade, and any losses other than pipe friction may be accounted for by assuming point losses at each manhole (this assumption is not valid, however, whenever a hydraulic jump is likely to occur within a pipe).

Once the discharge has been determined, and a pipe size and slope assumed for a given section, the d/D and v/V_{full}, ratios can be determined from a graph of hydraulic elements and the energy grade line calculated as described in Chapter 6.

At each manhole the energy grade lines of all pipes should coincide, allowing reasonable losses at the junction.

3. Losses at Manholes

Methods for calculating losses in manholes were described in Chapter 6. Recent studies by Marsalek and the American Public Works Association (Marsalek 1986, 1987, 1988) have shown that properly designed bottom sections can reduce manhole losses significantly. It should be noted that it is common to find large storm drainage conduits converging at a junction at an acute angle. Such junctions should be carefully designed (see, for instance, the City of Los Angeles 1968) to avoid large head losses.

Clean-Out or Manhole Structures with Straight Flow   Flow goes straight through the structure, which may be square or circular in plan (Figure 8.1a). The pipe diameter is the same on both the inlet and outlet side, with several internal shapes (Figure 8.1b) possible. Shape Types 2 and 3 have a bench constructed in the bottom of the manhole to improve flow paths. In Type 2, the bench is at the centerline of the pipe, and in Type 3, the bench is at the top of the pipe.

Head losses can be calculated using Equation (6-33). Values of K_c vary with the ratio of structure width (W) to pipe diameter (d), and are given in Figure 8.1c. Head losses in Types 2 and 3, which are benched, are less than those in Type 1.

Clean-Out Structures at 90° Bends   Head losses coefficients for Equation (6-33) in structures at 90° bends (with and without benching) are shown in Figure 8.2.

Coefficients are for fully submerged flow, and open-channel flow coefficients are about ⅓ less.

Alignment of Pipe in Manholes   For straight-through flow, aligning the inverts of the pipes is generally advantageous as the manhole bot-

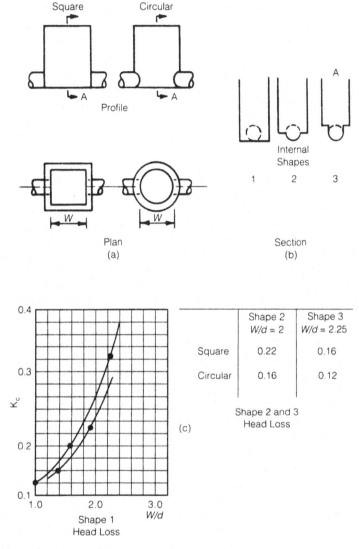

Figure 8.1—Manhole losses (straight-through flow).

tom then supports the bottom of the jet issuing from the upstream pipe.

When two laterals intersect a manhole, the alignment should be quite different. If lateral pipes are aligned opposite one another so the jets may impinge upon each other, losses are extremely high. If directly opposed laterals are necessary, the installation of a deflector as shown in Figure 8.3 will result in significantly reduced losses. Some examples of inefficient manhole shapes are provided in Figure 8.4.

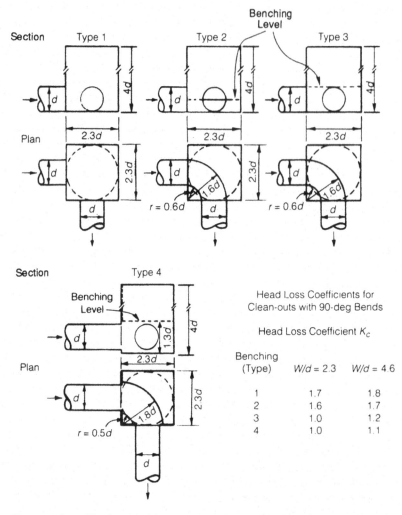

Figure 8.2—Clean-out structures.

Entrances   Rounding entrances or using pipe socket entrances to provide smooth transitions between pipe and manhole will reduce losses, though installation costs may be high in retrofitted systems.

B. Pressurized Storm Sewers

Storm sewers are often planned as pressure conduits for the design runoff, or to operate under pressure when runoff occurs that is greater than design flow. The fact that such storm sewers usually have manhole and storm inlet appurtenances that provide a direct hydraulic connec-

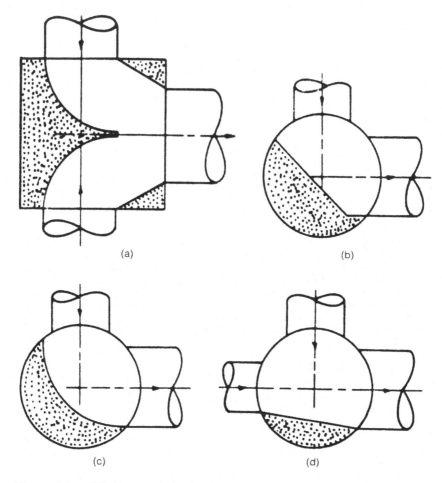

Figure 8.3—Efficient manhole shaping.

tion to the street surface means that special care must be taken in their analysis.

If the hydraulic grade line rises above the ground surface, storm inlets will not function, and storm water will be discharged from the storm sewer to the street surface via the inlets and manholes. Over-design of upstream storm inlets, which is encouraged as good practice, can lead to this situation (a frequent cause of "popping" manhole covers). Sometimes manholes with anchored and sealed covers are designed for a hydraulic gradient higher than street level with limited storm inlet connections.

The following design procedures are applicable where it is found advisable or necessary to have storm sewers flow full as pressure con-

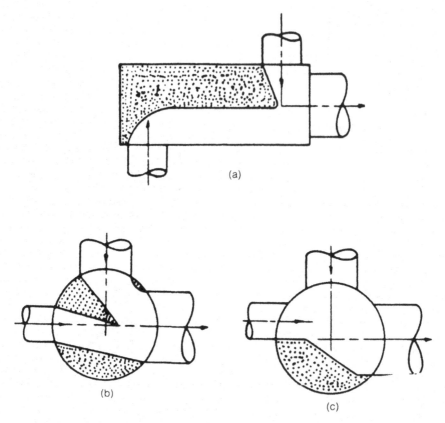

Figure 8.4—Inefficient manhole shaping.

duits. Checks must continually be made to verify if the conduit is in fact flowing full. Often the storm sewer system will alternate between pressurized and open channel flow. In such cases, it will be necessary to establish the type of flow and design accordingly.

1. Basic Design Procedures

The basic design procedure for a pressurized storm sewer consists of: (1) determining allowable pressures, (2) determining type of flow, (3) computing pipe junction and form losses, (4) computing the hydraulic grade line, and (5) integrating minor and major system behavior during the major system design runoff event.

The initial sizing of a pressurized system typically begins in the upstream portion of the basin, where the above-ground flow capacity is exceeded and underground pipes are needed. The design process proceeds in a downstream direction. After initial sizing, the system is checked for adequacy by calculating the hydraulic and energy grade lines, beginning at the downstream end of the system (i.e. the receiving water surface) (Clark et al. 1977). All this information is plotted on a profile that shows the pipe, the ground surface, utility crossings, etc. This process may require adjusting pipe sizes until a desired hydraulic grade line is achieved.

2. Allowable Pressures

Two major considerations limit the maximum allowable pressure in a sewer. First, the structural limitations must not be exceeded for a given pipe (this rarely is the controlling factor). When considering structural limitations, both the pipe and the joint must be analyzed. Second, the hydraulic grade line should be kept below ground level (or below the basement floor level, for buildings connected by drains to the sewer system) for the design runoff event, unless special precautions are taken to prevent water from escaping from inlets and manholes, or to handle it once it does escape. A further limitation is to assure a sufficient drop into inlets to allow them to function properly.

3. Discharge Point

The discharge point of the system usually establishes a control point. If the discharge is submerged, as when the receiving water level is above the crown of the sewer, the exit loss should be added to the hydraulic grade line and calculations for head loss in the system started from this point, as illustrated in Figure 8.6. If the hydraulic grade line is above the pipe crown at the next upstream manhole, full flow calculations may proceed. If the hydraulic gradient is below the pipe crown at the upstream manhole, then open channel flow calculations must be used at the manhole.

When the discharge is not submerged, a flow depth must be determined at some control section to allow calculations to proceed upstream. As shown in Figure 8.5, the hydraulic grade line is then projected to the upstream manhole or inlet. Full flow calculations may be used at the manhole if the hydraulic gradient is above the pipe crown.

Urbonas and Roesner (1991) provide a good discussion of how coastal (tidal) variability and lake level fluctuation affect the selection of starting water surface elevations. Regardless of the point of discharge, however, the designer must define the conditions likely to prevail under "design" flows.

4. Within the System

At each manhole, a procedure similar to that outlined for the discharge point must be repeated. The water depth in each manhole is calculated to verify that the water level is above the crown of all pipes.

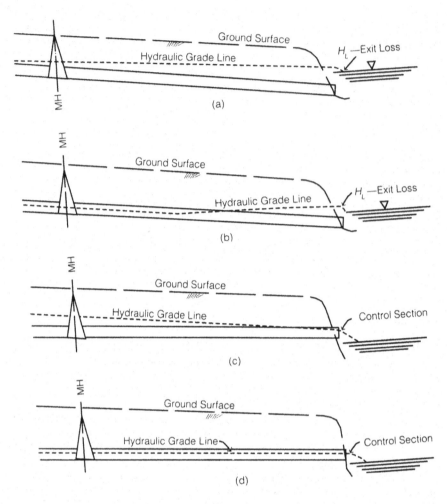

Figure 8.5—Determining type of flow.

Whenever the level is below the crown of a pipe full flow methods are not applicable.

5. Friction and Form Losses

Friction and form losses should be calculated in accordance with the methods outlined in Chapter 6.

VI. STORM SEWER INLETS

A. Categories, Definitions, and Applications

Inlets are used to intercept and convey surface runoff to closed conduit hydraulic systems. When used in streets, inlets are generally placed in the street curb and gutter. Gutter inlets can be classified as one of four types: curb; grate, or yard opening; curb and grate (combination); or special purpose inlets. Gutter inlets can be placed on a continuous grade, a partially continuous grade with a local depression, or in a sump. Several combinations of these inlet types are illustrated in Chapter 6, which includes a detailed discussion of inlet hydraulics.

The hydraulic capacity of a gutter inlet depends upon inlet geometry (the capacity of which can be affected by expected debris loadings and maintenance), and the characteristics of gutter flow (which can be a function of roadway configuration).

Inlets placed on a continuous grade are usually designed to allow runoff bypass. For a given design flow, gutter inlet capacity can be increased by using a gutter/inlet opening depression, steeper street cross slope, and/or a flatter longitudinal street slope. Inlet capacity can be increased at locations with a local sump condition where a certain amount of gutter ponding is allowed before runoff spills and is conveyed downstream. Length, width, depth and slope of a local depression have significant effects on the capacity of an inlet. Finally, inlets can be installed in a sump, where ponding is allowed to a regulated depth. The effect of a local depression on inlet capacity is graphically illustrated in Figure 8.6 (Johns Hopkins University 1956). Inlet capacity increases for both grate and curb opening inlets with local depressions on flatter longitudinal grades.

Reduction factors are generally used to reduce theoretical hydraulic inlet capacities to account for potential debris blockage and future street resurfacing. Reduction factors are usually obtained from local criteria manuals. The efficiency of an inlet is defined as the ratio of intercepted runoff divided by the total runoff in the upstream roadway cross-section. During inlet design, overflow paths should be identified at proposed inlet locations, especially where a sump condition will occur. This precaution will lessen the chance for incurring damages when runoff exceeds the inlet capacity or if debris blockage should occur.

B. Curb Opening Inlets

Curb opening inlets are most effective, and have several advantages over grate- and combination-type inlets. They do not encroach into the travelled way of either motorized vehicles or bicycles. Where there is pedestrian access, curb opening inlets are safer than other types of inlets. They are also more effective than other types of inlets where debris flows are expected. Opening heights should not exceed 6", to reduce risks to children.

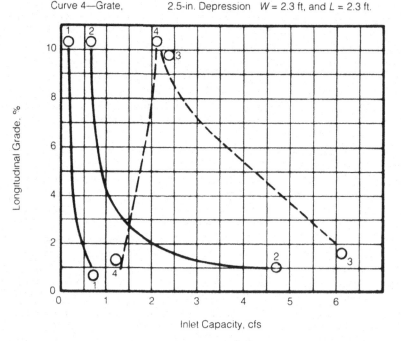

Inlet Capacity
at 95% Capture of Gutter Flow

Manning's n = 0.013		Cross Slope = 0.041 7 ft/ft:
Curve 1—Curb Opening,	No Depression, and	L = 10 ft;
Curve 2—Curb Opening,	2.5-in. Depression, and	L = 10 ft;
Curve 3—Grate,	No Depression, W = 2.3 ft, and L = 2.3 ft; and
Curve 4—Grate,	2.5-in. Depression W = 2.3 ft, and L = 2.3 ft.

Figure 8.6—Comparison of inlets with and without local depressions.

A disadvantage of curb opening inlets is their sensitivity to changes in street grade, both along continuous grades and at local depressions. Curb inlets, like other types of inlets, lose capacity both with steeper longitudinal street slopes and flatter street cross slopes. Generally, curb opening inlets on continuous grades without a local depression are very inefficient. Their capacity and efficiency can be estimated using methods presented in Chapter 6. Additional information can be found in various publications (Izzard 1977; Bauler and Woo 1964; U. S. Department of Transportation 1972).

C. Grate Inlets

Grated inlets are less sensitive to street grade changes than curb opening inlets. Disadvantages of grated inlets include their interference with motorized vehicles and bicycles and their tendency for debris blockage, with resultant loss of capacity. Some cities have discontinued their use for these reasons.

The capacity of non-depressed grated inlets on continuous street grades can be estimated from published equations and/or charts. If the grate is long enough, the flow intercepted by the grate will consist of the runoff flowing within the width of the grate and possibly a portion of the runoff flow from the street along the length of the grate. An efficient grate has a bar arrangement that does not cause splashing and resultant flow bypass, and that provides a sufficient area of opening to accommodate the design flow.

A safety factor should be applied to theoretical grate inlet capacities, especially if the inlet does not have a curb opening (combination inlet).

D. Curb and Grate (Combination) Inlets

Curb and grate inlets offer the advantages of both curb inlets and gutter inlets. Combination inlets are frequently used in a sump location and provide a greater capacity than the curb or grate type inlet alone.

E. Special Purpose Inlets

1. Bicycle-Safe Grates

Studies by the FHWA (U. S. Department of Transportation 1978) have developed bicycle-safe grate configurations. These studies investigated the hydraulic, structural and debris handling characteristics of seven different inlets, each of which was placed at grade with the gutter pavement. The study results compare hydraulic characteristics of these types of inlets and provide equations for estimating hydraulic capacities and efficiencies.

2. Pipe Drop Inlets

Pipe drop inlets consist of a grate flush with the adjacent ground surface, which drains into a vertical section of concrete or corrugated metal pipe. This type of inlet is not designed for use within roadways. Since a pipe drop inlet has a round surface area, it can intercept flow from any direction, and is most effective for flows that are deepest at the grate center. Generally, pipe drop inlets are economical. They should not be used where vandals may remove the grate, leaving a hazardous open pipe.

3. Slotted Drain Inlets

Slotted drain inlets consist of a corrugated steel pipe cut along the longitudinal axis and reinforced with a grate of solid spacer bars. Slotted drains are often used to intercept overland sheet flow from wide, flat areas, but can also be used in a typical curb and gutter. Design nomographs and criteria are available.

F. Inlet Design Criteria and Practices

Storm sewer inlet criteria are most often available from the entity responsible for the formal review and approval of design submittals. Common governmental sources of inlet criteria include the federal government (Federal Highway Administration), the state (Department of Highways), the county, and the township/city. Intergovernmental organizations, such as drainage and flood control districts, often develop broad criteria for those entities within the district's jurisdiction. Finally, criteria are often available from professional and technical society publications (American Society of Civil Engineers), manufacturer catalogs, and organizations with technical programs related to a particular material (American Iron and Steel Institute, American Concrete Pipe Association).

Location and Spacing   The placement of curb inlets is generally a trial-and-error procedure since both inlet size and spacing can be varied. The goal in the placement of inlets is the most hydraulically effective system at the most economical cost. The following general rules apply to inlet location design:

(a) Intersections—As discussed earlier, inlets are normally required at intersections to intercept 100% of runoff. This is necessary to prevent street cross flow, which could cause a traffic hazard. Inlets should generally be placed on tangent curb sections and near corners.
(b) Superelevation Transitions—Inlets are generally required where the street cross slope begins to superelevate. These inlets reduce the traffic hazard of street cross flow to the opposite side of the street.
(c) Side Drainage Entrances—Inlets should be placed downstream from points where side drainage enters streets and overloads gutter capacity. Side drainage often results from parking lots. A better practice is to intercept side drainage before it enters the street.

As discussed previously in this section, an inlet is required at the uppermost point in a street where gutter capacity criteria are violated. The inlet location is established by a trial-and-error procedure by which the drainage area is adjusted until a point is located where the gutter capacity is exceeded. The proposed inlet is then sized to intercept a portion of the total flow. Although the references listed previously contain tables summarizing interception capability, 70–80% is a reasonable rule of thumb for flow interception (Urban Drainage and Flood Control District 1984). Succeeding inlets are designed similarly by locating the point where the sum of bypassing flow and flow from the intervening street (and side drainage, if applicable) exceed the gutter capacity. Inlets should not be placed in driveways, or directly in front of store fronts or private residences.

Finally, the designer must assure that there is an adequate number of inlets to fully utilize the available capacity of the storm sewer.

G. Inlet Selection

When selecting inlets, hydraulic considerations are sometimes sacrificed relative to potential for clogging, nuisance to traffic, convenience, safety and cost. The following factors should be considered before choosing an inlet type:

(a) Likelihood of Clogging—If clogging due to debris is not expected, a grate or combination type inlet will provide more capacity than a curb opening inlet. Otherwise, a curb opening inlet may be favorable. A local depression at either inlet location will increase the inlet capacity.

(b) Traffic Considerations—If traffic is expected close to the curb and the street slope is steep, use a deflector inlet. If the street slope is relatively flat and a potential exists for debris clogging the deflector slots, use a gutter or combination having a grate with longitudinal bars only (these may be hazardous to bicycle traffic).

(c) Safety—This includes traffic and pedestrian safety (affected by the spread of flows around the inlet), as well as safety for bicyclists (Note that attempts to increase inlet efficiency by "dropping" the gutter up-slope from the inlet can be dangerous to cyclists).

VII. INFILTRATION AND EXFILTRATION

Infiltration and exfiltration should always be considered during storm sewer design. There are two types: (1) that which is deliberately provided, and (2) that which is unanticipated and which may lead to grave difficulties. See also relevant portions of Chapter 12 for a further discussion.

A. Deliberate Infiltration/Exfiltration

Deliberate infiltration into a storm sewer may sometimes be useful in de-watering areas having persistent or intermittent high groundwater table elevations. It typically is accomplished by installing storm sewers with open or unsealed joints, bedded in granular materials, sometimes even using perforated pipe. Pipe materials selected should be compatible with corrosive and erosive potentials of the groundwater. Total opening area into pipes should be small so that the pipe will primarily serve to convey stormwater runoff and only secondarily, when runoff is negligible, provide an outlet for infiltrating groundwaters. To avoid undesirable groundwater recharge during runoff events, openings for infiltration into a pipe should be limited, consistent with the need to draw down the water table.

When infiltration is desired, the pipe normally should be bedded in granular materials wrapped in a permeable geotextile to provide a granular shell thickness of at least three inches around the pipe. Alternatively, graded granular materials might be provided as bedding, with greater thickness, to filter out soil fines and prevent their conveyance into the pipe. Gradation and cohesiveness of the natural in-place materials will indicate the extent to which their piping may be a potential problem and the extent to which anti-piping measures are appropriate.

Deliberate infiltration should not be used in soils that are expansive (montmorillonite clays), collapsible (naturally cemented loess), or dispersive (some sensitive clays) because of resultant potentially damaging soil movements. Similarly, deliberate infiltration either should not be used or should be used only as a calculated risk where tree or shrub roots might penetrate pipe openings and cause potential blockage.

A special case of deliberate infiltration is where facility perimeter drains or underdrains may be used to remove groundwaters to prevent their adverse effects upon expansive or collapsible soils. Such drains are often bedded in gravel or crushed stone, which requires accumulation of free water in the bedding before outflow can occur to the drain. This accumulation should particularly be avoided in the indicated soils. This involves bedding the drain pipes in lean but relatively impermeable concrete to the depth of openings in pipes, above the pipe invert, so that water cannot accumulate beneath the drain. Where such piping is used for perimeter drains, to help exclude water from beneath facilities, it may be desirable to line the facility side of the drain trench with an impermeable membrane, usually extended beneath the concrete bedding, before backfilling with gravel or crushed stone to within two feet of the ground surface; remaining backfill then should be an expansive clay or other relatively impermeable material, with ground surface grading that will assure rapid surface drainage.

Deliberate exfiltration is a technique that has proven useful for disposal of stormwater runoff from urban impervious surfaces, or to reduce runoff pollutant loadings. Where used, natural soils should be highly permeable, and the maximum potential level of the groundwater table should be no higher than about 2–4 feet (varies substantially, depending on the literature consulted) below the invert of the storm sewer pipe. The potential adverse effects of the stormwater on groundwater quality are an important consideration (Livingston et al. 1988; Urbonas and Roesner eds. 1986). The soils should be moderately (in non-seismic areas) to well (in seismic areas) consolidated open graded sands or gravels, ground surface slopes should not exceed about 5% (with pipe invert slopes typically no greater than 0.5–1.0%), and there should be no expansive, collapsible or dispersive soils present (University of Wisconsin 1990).

Deliberate exfiltration provides underground disposal of stormwater runoff in situations where its accumulation on the ground surface would be undesirable, frequently in locations where there is no defined surface drainageway available for drainage outfall. To accomplish this, perforated pipe is used, or the storm sewer is installed with open (cracked and unsealed) joints. A grit chamber (sediment trap), easily accessible for solids removal, should be installed downstream from stormwater inlets to prevent accumulation of settleable solids in the storage/exfiltration pipe.

The permeability of soils into which effluent will be exfiltrated should be determined prior to design. Since some fine street washings inevitably will pass the grit chamber, and may tend to reduce natural permeability over the long-term, the design infiltration rate used for determining the size of necessary storage/exfiltration piping should be less than the soil's natural infiltration rate. Location of a grit chamber downstream from storm inlets, rather than usage of combined inlets/catch basins, will increase the "catch" of suspended solids by deposition at locations free from the turbulence usually present in an inlet box.

The balance between in-pipe storage capacity and accumulation of runoff on the ground surface (ponding) is usually a balance between cost and acceptable inconvenience or risk, a determination normally made by the client, project owner, or regulatory authority.

Precautions noted above for deliberate infiltration are equally applicable to exfiltration. Deliberate exfiltration should not be considered unless stormwater runoff quality is consistent with long-term maintenance of groundwater quality (an NPDES permit may be necessary).

B. Inadvertent Infiltration/Exfiltration

Inadvertent or unanticipated infiltration occurs when storm drains lose their structural integrity, when tree or shrub roots penetrate a storm drain and cause blockage, or when natural trench materials are piped into storm drains creating cavities that may collapse with subsidence of the overburden and thereby damage the pipe. Where stormwater runoff must be treated to improve its quality prior to its discharge (especially in some combined sewer systems), infiltration can increase the amounts of water which must be processed to unacceptable levels. Deliberate (often surreptitious) introduction of process or other wastewaters into a separate storm sewer system also may be viewed as unanticipated infiltration, but its control is an administrative problem.

Inadvertent infiltration usually can be prevented or minimized. Prevention entails use of:

(a) Non-brittle pipe materials to the extent possible.
(b) Sealed pipe joints that can absorb some movement without leakage.
(c) Proper bedding of pipes to minimize their potential movements.
(d) Backfilling of trenches with compacted soils materials that will limit percolation of surface waters into trenches.
(e) Avoidance of stones or other materials, in trench backfills, that might result in point loads on pipes.
(f) Careful supervision of construction to assure watertightness of completed storm sewer systems, possibly with post-construction system pressure testing.
(g) Attention to these factors throughout the entire drainage system, from extreme headwater appurtenances to points of storm sewer system discharge. Enforced ordinances may be desirable to prevent unauthorized discharges into storm sewers.

Inadvertent exfiltration occurs when runoff waters leak from a storm sewer and cause undesirable soil movements or degradation of groundwater quality. Particular care should be exercised to avoid exfiltration (leakage) into expansive clay soils as the water may induce damaging volume increases (heave or swell), into collapsible soils such as cemented loess deposits, into dispersive soils subject to piping and resultant collapse, or into karst (limestone) formations subject to the formation of solution channels. Exfiltration from a pressure sewer (where the hydraulic gradient may be some distance above the sewer) may be unusually damaging if leakage under head erodes trench materials and causes storm sewer blowout. Precautions necessary to prevent unwanted exfiltration are the same as those listed above for infiltration.

VIII. STREET AND INTERSECTION DESIGN

The following provides an introduction to street and intersection design. More detailed guidance can be found in most large city drainage criteria manuals (several of which are listed in the references to this chapter), and in references published by AASHTO, the Federal Highway Administration, and many state highway departments.

Design criteria for the collection and disposal of runoff water on public streets is based on a reasonable frequency of traffic interference. Depending on the character of the street, certain traffic lanes may be fully inundated with a frequency not exceeding the minor system design return period. Good drainage design should provide direct traffic benefits, lower street maintenance costs, and protect street paving and the subgrade from unnecessary deterioration.

A. CLASSIFICATION OF STREETS

Street drainage practices are dependent upon the type of street use and construction (Urban Drainage and Flood Control District 1984).

(a) Local—A local street is a minor traffic carrier within a neighborhood characterized by one or two moving lanes and parking along curbs, with no through traffic moving from one neighborhood to another. Traffic control may be by use of stop or yield signs.

(b) Collector—A collector street collects and distributes traffic between arterial and local streets. There may be two to four moving traffic lanes and parking may be allowed adjacent to curbs. Traffic on collectors has right-of-way over traffic from adjacent local streets.

(c) Arterial—An arterial street permits rapid and relatively unimpeded traffic movement. There may be four to six lanes of traffic, and parking adjacent to curbs may be prohibited. The arterial traffic normally has the right-of-way over collector streets. Construction of an arterial will often include a median strip with traffic channelization and signals at numerous intersections.

(d) Freeway—Freeways permit rapid and unimpeded movement of traffic through and around a city. Access is normally controlled by inter-changes at major arterial streets. There may be eight or more traffic lanes, frequently separated by a median strip. Parking normally is not permitted on the freeway right-of-way.

B. Effect of Stormwater Runoff on Street Traffic Capacity

Storm runoff that influences the traffic-carrying capacity of a street can be classified as follows (see Figure 8.7) (Urban Drainage and Flood Control District 1984):

(a) Sheet flow across the pavement once rain strikes the pavement and flows to the edge of the street.
(b) Runoff flowing adjacent to the curb.
(c) Stormwater ponded at low points.
(d) Flow across the traffic lane from external sources or cross street flow, (as distinguished from rain falling on the pavement surface).
(e) Splashing on pedestrians.

C. Storm Drainage Design Criteria for Urban Streets

1. General Guidance

The gutter grade refers to the grade of the gutter parallel to the direction of flow. The maximum allowable grade for a gutter is not governed by drainage, but the allowable capacity of gutters on exces-sively steep slopes is limited. The minimum allowable gutter grade, to facilitate proper drainage should be 0.4%.

The maximum allowable crown slope is not affected by drainage requirements, but the minimum crown slope of the street should be 2% to facilitate drainage from the pavement.

The standard gutter, together with a curb, should be at least 6 inches deep and 2 feet wide, with the deepest portion adjacent to the curb. Other gutter configurations may be used as conditions require. Some cities use higher curbs to allow for maintaining runoff capacity after repaving, though there are no recognized design criteria for maintaining such capacity. The conservative designer may choose to allow for the fact that each repaving reduces capacity.

Driveway entrances should be recessed into curbs. The driveway should slope up to an elevation equal to the top of the curb so runoff within the street cannot flow onto adjacent property through the drive-way entrance.

Inverted crown or dished streets should not be used, since during times of flooding they cannot function as a vehicular traffic carrier. Other disadvantages of dished streets include:

(a) Concentration of snowmelt runoff at center of street where it can then refreeze, causing a significant traffic hazard.
(b) Lawn irrigation and other low flows continually affect traffic.

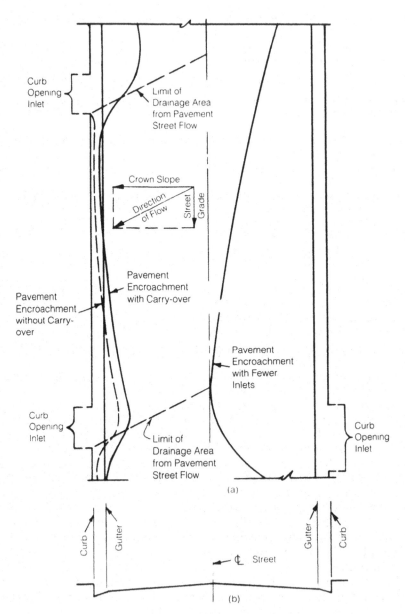

Figure 8.7—Diagram of gutter and pavement flow patterns.

(c) Any pavement cracking allowing leakage to the subbase affects the traffic-carrying portion of the street first.
(d) Difficulty in designing stormwater inlets.
(g) Lack of damage-resistant pavement designs and materials.
(h) Safety hazard during flooding.

2. Cross Fall

The term cross fall refers to the difference in elevation between the gutter flow lines on opposite sides of a street. Under most conditions streets are designed with zero cross fall. In hilly areas, and particularly at intersections in hilly areas, it may be necessary to construct the curbs at different elevations, resulting in cross fall.

Figure 8.8 (Urban Drainage and Flood Control District 1984) illustrates the loss of capacity of the higher gutter in a street with cross fall. When calculating allowable flow for the higher gutter, the actual flow prism configuration must be used. The capacity of the lower gutter may or may not be decreased, depending upon the street design. As with the upper gutter, the actual flow prism must be used in calculating allowable capacity.

When calculating the volumes of flow in each gutter, note that the upper gutter may fill quickly, by virtue of its location on the side of the street which will be receiving drainage from adjacent areas. Flow will then proceed across the crown of the street and into the opposite gutter. On local streets this is acceptable. However, on major streets, the interference to traffic movement due to the water flowing across the traffic lanes usually is unacceptable.

To prevent low flows from flowing across the traffic lanes, adequate capacity must be maintained in the higher gutter. To preserve this capacity the crown should be maintained within the one-quarter points of the street as shown by Section B-B of Figure 8.9.

On local streets, where cross fall is necessary due to the existing topography, inlets may be placed in the lower curb, and the street crown removed to allow flow from the upper curb to reach the inlet in the lower curb at specified locations.

3. Street Capacity for Minor Storms

Determination of street carrying capacity for the minor storm is based on two considerations; (1) pavement encroachment for computed theoretical flow conditions, and (2) an empirical reduction of the theoretical allowable rate of flow to account for practical field conditions.

Pavement Encroachment   The pavement encroachment for the minor storm is often limited as shown in Table 8.3, although there is considerable variability from community to community, and local standards should be consulted. The storm sewer system should begin where the maximum encroachment is reached.

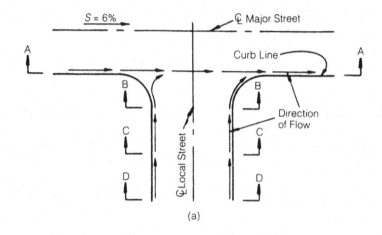

(a)

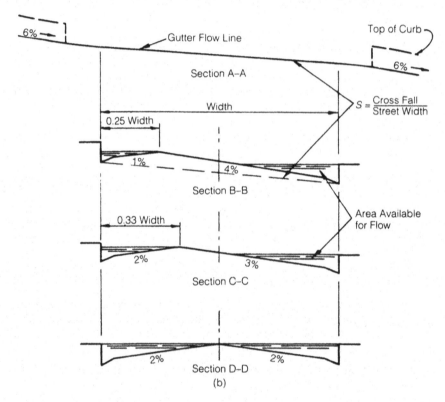

(b)

Figure 8.8—Typical intersection construction at junction of local and major street.

TABLE 8.3. Typical Allowable Pavement Encroachment for Minor Storm Runoff (Urban Drainage and Flood Control District 1984).

Local	No curb overtopping.* Flow may spread to crown of street.
Collector	No curb overtopping.* Flow spread must leave at least one lane free of water.
Arterial	No curb overtopping.* Flow spread must leave at least one lane free of water in each direction.
Freeway	No encroachment is allowed on any traffic lanes.

*Where no curb exists, encroachment onto adjacent property should not be permitted.

Calculating Theoretical Capacity   When the allowable pavement encroachment has been determined, the theoretical gutter carrying capacity for a particular encroachment can be computed using the integrated form of the Manning equation, described in Chapter 6. To simplify computations, graphs for particular street shapes may be plotted, or computer programs can be used.

Allowable Gutter Flow   The actual flow rate allowable per gutter can be calculated by multiplying the theoretical capacity by the corresponding factor obtained from Figure 8.9. The designer can then develop discharge curves for standard streets.

4. Street Capacity for the Major System Design Runoff

Determination of the allowable flow for the major system design runoff is based on allowable depth and inundated area, and the reduced allowable flow due to velocity considerations. In sump areas, overflow outlets (to parking or other graded areas) should be provided to prevent water in sumps, particularly when the sump is clogged, from entering adjoining buildings.

Allowable Depth and Inundated Area   The allowable depth and inundated area for the major storm are normally limited per criteria such as in Table 8.4. The theoretical street carrying capacity then can be calculated using the Manning equation.

Ponding   Ponding refers to areas where runoff is restricted to the street surface by sump inlets, street intersections, low points, intersections with drainage channels, or for other reasons. Limitations on pavement encroachment by ponding can be based upon recommended flood depths in Tables 8.3 and 8.4.

5. Cross-Street Flow

There are two general categories of cross-street flow. The first is runoff that has been flowing in a gutter and then flows across the street to

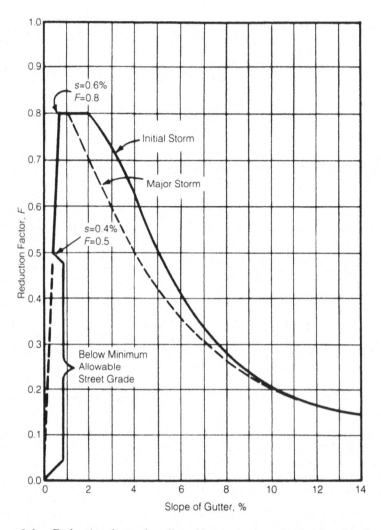

Figure 8.9—Reduction factor for allowable gutter capacity local and collector streets.

the opposite gutter or to an inlet. The second is flow from some external source, such as a drainageway, which will flow across the crown of a street when the conduit capacity beneath the street is exceeded. Cross-street flow depth is often limited. Table 8.5 gives an example of such limits, and local standards should also be consulted (note that, in Table 8.5 no cross street flow is permitted for the minor design runoff).

TABLE 8.4.   Typical Allowable Street Inundation Criteria for Major Storm Runoff (See Also Local Criteria) (Urban Drainage and Flood Control District 1984).

Street Classification (1)	Allowable Depth and Inundated Areas (2)
Local and collector	Residential dwellings, public, commercial, and industrial buildings shall not be inundated at the ground line, unless buildings are flood-proofed. The depth of water over the gutter flow line shall not exceed an amount specified by local regulation, often 12 inches.
Arterial and freeway	Residential dwellings, public, commercial, and industrial buildings shall not be inundated at the ground line, unless buildings are flood-proofed. Depth of water at the street crown shall not exceed 6 inches to allow operation of emergency vehicles. The depth of water over the gutter flow line shall not exceed a locally-prescribed amount.

TABLE 8.5.   Typical Allowable Cross Street Flow Depths (Urban Drainage and Flood Control District 1984).

Street Classification (1)	Minor Design Runoff (2)	Major Design Runoff (3)
Local	6-inch depth at gutter or in cross pans	18 inches of depth above gutter flow line
Collector	Where cross pans are allowed, depth of flow shall not exceed 6 inches	18 inches of depth above gutter flow line
Arterial	None	6 inches or less over crown
Freeway	None	6 inches or less over crown

6. Intersections

The various criteria presented herein for street inundation, ponding and cross street flow can be used in combination for intersection design in a procedure that ultimately determines the number, type, and size of inlets required.

When local streets intersect major streets, the grade of the major street should not be interrupted, if possible. Figure 8.10 illustrates the typical street cross-sections necessary for such an intersection. The figure assumes that the major street grade is 6%, the maximum allowable crown slope is 4%, the minimum allowable crown slope is 1%, and the crown must be maintained within the one-quarter points of the street.

Storm Sewer System   When a storm sewer will be located in an intersection, inlets should be placed and sized so that encroachment on the intersection is equivalent to that allowed on the street for the design runoff. Figure 8.11 illustrates typical inlet locations for various categories of intersections.

Drop Inlet Culverts   When a storm sewer system is not required, and sufficient grade is available, a drop inlet culvert may be used to transport runoff beneath a street. Encroachment on the intersection should be limited to that allowed for the street.

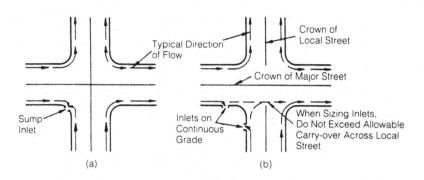

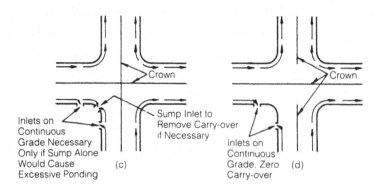

Figure 8.10—Typical street intersection drainage to storm sewer system.

Cross Drains Conventional cross drains (cross pans) may be used to transport runoff across local streets when a storm sewer system is not required. The cross pan size and slope should be sufficient to restrict encroachment to that allowed on the street. If absolutely necessary, pans may be used on collector streets.

Covered cross pans may be used where frequent low flows are anticipated, although they are difficult to maintain, and therefore should be avoided where possible. The covered pans should carry at least the low flows, and must allow larger (design runoff) flows to pass over the pan without exceeding depth of flow allowed on the street. No form of cross pan should be constructed across an arterial street.

7. Special Considerations

No specific limitations are set for sheet flow. Designers should be aware of its existence and effects and take precautions to limit its occurrence.

Where there is heavy pedestrian traffic, depth and area limitations may need modification. As an example, streets adjacent to schools, while considered local from a vehicle traffic standpoint, are arterials from a pedestrian standpoint, and should be designed accordingly. Designing for the pedestrian is at least as important as designing for vehicular traffic.

Where commercial buildings are constructed to property lines, the reduced clearance between buildings and heavy traffic must be considered. Splash from vehicles striking gutter flow may damage store fronts and make walking on sidewalks impossible. Ponded water and gutter flow exceeding 2 feet in width are difficult to negotiate by pedestrians (Urban Drainage and Flood Control District 1984).

As emphasized repeatedly in this manual, the criteria listed above apply principally to newly-developed areas. In established areas, it may be appropriate to accept the system as it exists, depending on the degree of hazard present, or to make some modifications (but not necessarily in accordance with the criteria).

D. Storm Drainage Design Criteria for Rural Streets

Rural areas (or areas with low-density development) have streets that may use roadside ditches for drainage purposes, as opposed to curbs and gutters. Most requirements set forth for typical urban streets apply equally to rural streets.

Determination of rural street carrying capacity for the minor and major storms is based upon the following considerations:

(a) Pavement encroachment allowed.
(b) Maximum allowable velocity to prevent scour.

Site specific standards apply for these factors. As a guide, the designer can use Tables 8.3, 8.4, and 8.5, and adjust the criteria to reflect rural conditions. Note, however, that future curb/gutter installation may reduce roadside runoff detention significantly, completely changing drainage responses. It is prudent to anticipate future requirements.

Drainage ditches adjacent to rural roads are common. As reach-by-reach evaluation of probable channel stability is not economically feasible, the guidance in Table 8.6 should be helpful. Note that velocities shown in Table 8.6 are maximum velocities. Design velocities should be no more than 75% of these values.

Design velocities for all linings should not fall below 2 feet per second for the minor storm runoff to minimize sediment deposition problems, unless lower velocities are required for water quality enhancement— in which case provision should be made to control erosion/sedimentation. The allowable capacity for the drainage ditch should be calculated using Manning's formula with an appropriate n value. If the natural channel slope would cause excessive velocity, drop structures, checks, riprap, or other suitable channel protection should be employed. Design depths are ideally limited to 1.5 feet, and preferably less than 1.0 foot.

IX. MAJOR DRAINAGEWAYS (OPEN CHANNELS)

Major drainage is the cornerstone of an urban storm runoff system, and will function whether or not it has been planned and designed. If a conveyance route for the major flood is not provided, flood waters will move downgradient on their own accord—often through yards,

TABLE 8.6(a).   Typical Permissible Velocities for Roadside Drainage Ditches with Erodible Linings (Site Specific Analysis Recommended) (Chow 1959).

Soil Type or Lining (earth, no vegetation) (1)	Permissible Velocity (fps) (2)
Fine sand (noncolloidal)	2.5
Sandy loam (noncolloidal)	2.5
Silt loam (noncolloidal)	3.0
Ordinary firm loam	3.5
Fine gravel	5.0
Stiff clay (very noncolloidal)	5.0
Graded, loam to cobbles (noncolloidal)	5.0
Graded, silt to cobbles (noncolloidal)	5.5
Alluvial silts (noncolloidal)	3.5
Alluvial silts (colloidal)	5.0
Coarse gravel (noncolloidal)	6.0
Cobbles and shingles	5.5
Shales and hard pans	6.0

TABLE 8.6(b). Typical Permissible Velocities for Roadside Drainage Ditches Lined with Uniform Stand of Grass Covers (Well-Maintained Grass) (Chow 1959).

Cover (1)	Slope Range (percent) (2)	Permissible Erosion Resistant Soils (3)	Velocity (fps) Easily Eroded Soils (4)
Bermuda grass Crested wheat grass Buffalo grass Kentucky bluegrass Smooth brome Grass mixture	0–5 5–10 over 10 0–5 5–10	6.0 5.0 4.0 4.0 3.0	5.0 4.0 3.0 3.0 2.5
Lespedeza sericea Weeping lovegrass Yellow bluestem Alfalfa Crabgrass Common lespedeza Sudangrass	0–5	3.0	2.0

homes and businesses. Thus, major drainage systems must be given high priority when considering drainage improvements. The major system may include features such as natural and artificial channels, long underground conduit outfalls, streets, property line drainage easements, and other water-carrying routes.

In small urbanized basins (typically less than 20–50 acres, depending upon topography, imperviousness, rainfall and other factors), it may be feasible to contain the major flow within storm sewers and/or streets. At some point in the system, however, major flows eventually exceed the combined capacity of storm sewers and streets, and a major drainageway is necessary.

The major system and minor drainage systems should be planned concurrently. A good major system can reduce or eliminate the cost of an underground storm sewer system, whereas an ill-conceived major system can make a storm sewer system very costly. The 2- or 10-year runoff can flow in the major system, but only a small portion of the major system design runoff will flow in the minor system.

The planner and engineer have great opportunities when working on a major drainageway to provide a better urban environment for all citizens. Benefits, in addition to flood control that often can be achieved, include creation of wetlands and other kinds of wildlife habitat, parks, trails and other recreation areas, and groundwater recharge zones. Properly designed, installed, and maintained channels can also reduce pollutant loads and mitigate sediment problems. The following discussion of open channels is, of course, not limited to the major system, and many of the concepts apply to the minor system as well.

Open channels have significant advantages with regard to cost, capacity, multiple use for recreational, aesthetic, and other purposes, and potential for providing transient storage. Disadvantages include right-of-way needs and maintenance costs. The ideal channel is an old (geologically) natural one. The benefits of such a channel are that:

(a) Velocities are usually low, due to relatively flat longitudinal slopes, resulting in longer concentration times and lower downstream peak flows.
(b) Channel storage tends to decrease peak flows.
(c) Maintenance needs are usually low because the channel is somewhat stabilized.
(d) The channel provides social benefits (greenbelt, recreational benefits).

The more an artificial channel can be made to resemble a natural channel, generally the better it will be. It must be recognized however, that few natural channels would respond favorably to the hydrologic impacts of urbanization without man-made improvements, with channel stability leading the list of concerns.

In many areas about to be urbanized, the runoff has been so minimal that natural channels do not exist. However, thalwegs nearly always exist that provide an excellent basis for location and construction of channels. Good land planning should reflect these thalwegs and natural channels to reduce development costs and minimize drainage problems. In some cases, the wise use of natural water routes in the development of a major drainage system will obviate the need for an underground storm sewer system.

Other considerations with respect to open channel evaluation and selection include:

(a) Preservation/destruction of wetlands.
(b) Improvements in water quality.
(c) Trails/public recreation.
(d) Impacts on aquatic life.
(e) Wildlife habitat improvement or removal.
(f) Protection of views or open space.
(g) Historic/cultural preservation.

A. Choice of Channel

The choices of channel available to the designer are almost infinite. However, from a practical standpoint, the basic choice to be made initially is whether or not the channel is to be lined—structurally to accommodate higher velocities, or grassed to accommodate intermediate velocities. A variation of grass-lined channels is to simply add stabilization measures to natural channels, if increases in peak discharge are moderate. Concurrent with this evaluation is consideration of channel slope modification (cutting or drop construction) to reduce velocities. The choice must be based on a variety of factors which include:

(a) Regulatory—Federal, state and local regulations (see also Chapter 2).
(b) Hydraulic—Slope of thalweg, right-of-way, capacity needed, basin sediment yield, topography, ability to drain adjacent lands.
(b) Structural—Costs, availability of material, areas for wasting excess excavated materials.
(c) Environmental—Neighborhood character and aesthetic requirements, need for new green areas, street and traffic patterns, municipal or county policies, wildlife, water quality.
(d) Sociological—Neighborhood social patterns and child population, pedestrian traffic, recreational needs.

Prior to choosing the channel type the designer should be sure to consult with experts in related fields to assure that the channel chosen will create the greatest overall benefits. Whenever practical, the channel should have slow flow characteristics, be wide and shallow, and be natural in its appearance, and functioning (Urban Drainage and Flood Control District 1984; Livingston et al. 1988; Urbonas and Roesner 1986).

For open channel design, the general approach is to prepare profiles of routes that appear satisfactory and make rough cost estimates of each, using approximations of the character and location of channels or conduits. Include costs of bridges, drops, special structures, utility relocations, land acquisition, and other such factors. The advantages and disadvantages of potential routes are then examined with an environmental design team, which may include an urban planner, an attorney knowledgeable in drainage law, a landscape architect, and an urban sociologist.

The reader is referred to Chapters 9, 13, and 14 for detailed discussion on special structures commonly found in channels (riprap, energy dissipators, drop structures, etc.), channel linings, scour, structural design, and construction.

B. Hydraulic Analysis

The following sorts of information will be needed in the design of open channels, (discussed in detail in Chapter 6) (University of Colorado 1989):

1. Channel Geometry

(a) The cross-section.
  • Perpendicular to the direction of flow.
  • Encoded as station and elevation pairs.
  • Related to each other by distance between sections.

- Forms a three-dimensional digital model of floodplain and channel.
(b) Guidelines for cross-section spacing.
   - Cross-sections should be located where there is an appreciable change in cross-sectional area, roughness, or gradient.
   - Detailed cross-sections should be located above, below, and within bridges to define transitions.
   - Cross-sections are needed at all control sections.
   - Cross-sections are required immediately above and below a confluence on a main stream and above the confluence on the tributary.
   - In general, more cross-sections are needed to define energy losses in urban areas as opposed to rural, where steeper slopes are encountered, and on smaller streams.
   - Based on computed results, additional cross sections may be required to provide accurate results. The reach is too long if the energy slope decreases by more than 50% or increases by more than 100%, and flow distribution should be reasonable from section to section.
(c) Sources of topographic data.
   - Aerial photography/photogrammetry.
   - Field surveys.

2. Hydraulic Roughness Values and Loss Coefficients

(a) Sources of roughness values.
   - U.S. Geological Survey Water Supply Papers 1849 and 1898-B (See references for other studies).
   - Tabulation of n-values in textbooks (French 1985; Chow 1959).
   - Photographs of channels and floodplains with calibrated n-values are useful to compare with field conditions.
   - Formulas for computing roughness—these require additional data and may not account for all roughness factors.
   - Can be calculated using HEC-2 if measured water surface profile elevations are available.
(b) Use of "n" values.
   - May reflect more than just friction loss, including the effect of bed forms, vegetation and debris, and urbanization in the floodplain.
   - Complex roughness conditions—can assign "n" values to streets, lawns and sidewalks and deduct the space occupied by buildings, or use composite values and manipulate the "n" values to account for blockage (for example, if 50% blockage by buildings then n-values increase by factor of 2).
   - Design "n" should reflect minimal maintenance.
(c) Contraction and expansion losses.
   - Referred to as "minor losses" and are applied to the change in velocity head.
   - 0.1 to 0.3 for contractions.
   - 0.3 to 0.5 for expansions.
   - Are not significant if velocity changes are small.
   - Are significant at encroachments where large velocity head changes occur such as at bridges.

3. Stage/Discharge Data

(a) Use of control sections for determining stage/discharge relationships.

(b) Controls due to critical depth at changes in thalweg profile from mild to steep, at weirs and other hydraulic drops, and at severe contractions in thalweg section.

(c) Controls due to uniform flow—artificial prismatic channels (or natural channels that have a constant cross-section) with a uniform grade.

4. Data Sensitivity

(a) Cross-section data.
  - Scale of mapping—larger scale mapping (USGS quads) can differ substantially from higher resolution mapping.
  - Spacing also has a direct impact on computed results.

(b) Roughness data.
  - Probably the most important parameter since conveyance is inversely proportional to the "n" value.
  - "n" values should be selected to be conservative for the objectives of the project—upwardly conservative values should be used for flood studies, and downwardly conservative for scour and erosion analysis.
  - Consistent with **observed** past maintenance performance.

(c) Stage/discharge data.
  - Error in the estimated value for initial water surface elevation rapidly decreases as the calculation progresses.
  - Errors in discharge estimation can be significant. If uncertainty exists then a sensitivity analysis should be conducted, using confidence bands for probability distribution.
  - Relatively large error in discharge may have only moderate effect on stage forecasts. Do not split hairs.

C. Concrete-Lined Channels

Where the project requires a lined channel because of hydraulic, topographic, or right-of-way needs, concrete lining is usually chosen, although soil cement and roller-compacted concrete are gaining acceptance (see also Chapters 9 and 13). Whether the flow will be supercritical or subcritical, the lining must be designed to withstand the various forces and actions which tend to overtop the bank, deteriorate the lining, and erode unlined areas. Figure 8.11 provides a cross-section for a typical concrete channel. While concrete channels have high hydraulic capacities, they are usually the most expensive to maintain over time, because of settlement, cracking, and weed growth in joints.

1. Supercritical Flow

Supercritical flow in an open channel in an urbanized area creates certain hazards which the designer must take into consideration (Williams 1990; Urban Drainage and Flood Control District 1984), and is

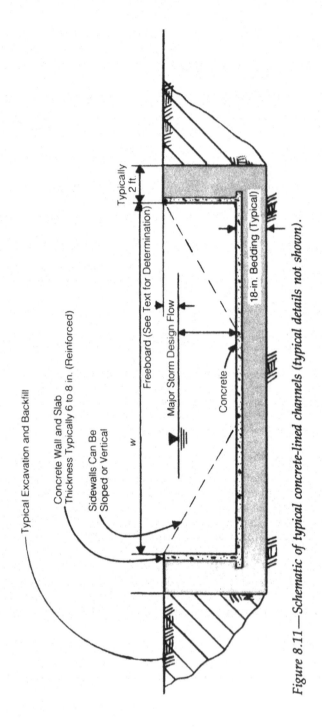

Figure 8.11—Schematic of typical concrete-lined channels (typical details not shown).

best avoided altogether. From a practical standpoint it is generally not advisable to have any curvature in such a channel. Careful attention must be taken to insure against excessive oscillatory waves which may extend down the entire length of the channel from only minor obstructions upstream. Imperfections at joints may cause their rapid deterioration, which in turn may cause a complete failure of the channel.

All channels carrying supercritical flow should be lined with concrete, which is continuously reinforced both longitudinally and laterally (San Diego County 1985). There should be no diminution of wetted area cross-section at bridges or culverts. Freeboard should be adequate to provide a suitable safety margin (a calculation procedure is provided in the next section). Bridges or other structures crossing the channel must be anchored to withstand the full dynamic load that might be imposed on the structure in the event of major trash plugging.

The concrete lining must be protected from hydrostatic uplift forces, which are often created by a high water table or momentary inflow behind the lining from localized flooding. Generally, a perforated, free-draining underdrain pipe will be required under the lining. With supercritical flows, minor downstream obstructions do not create backwater effects. Backwater computation methods are applicable for computing the water surface profile or the energy gradient in channels having a supercritical flow, however, the computations must proceed in a downstream direction. The designer must insure against the possibility of unanticipated hydraulic jumps forming in the channel. Flow at Froude numbers near 1 is unstable and should be avoided.

Roughness coefficients for lined channels are particularly important when dealing with supercritical flow. Rough channels may aggravate oscillatory wave tendencies. Once a particular roughness coefficient is chosen, construction inspection must be carried out in a manner to insure that the particular roughness is obtained. Because of field construction limitations, the designer should not use a Manning's "n" lower than 0.013 for a well-trowelled concrete finish. Other finishes should have proportionately larger "n" values assigned to them. Problems have arisen in the past (Williams 1990) when high bed loads have increased the effective "n" value to the point that flow shifts from supercritical to subcritical, leading to over-bank flooding. The designer may wish to adopt considerably larger "n" values as a safety margin to account for sediment and debris. For example, some individuals advocate "n" values of 0.04 to 0.05 for concrete channels specifically for this reason.

2. Subcritical Flow

Where subcritical flow is anticipated, wide, naturally-vegetated channels are normally preferable, if available thalweg slopes are steep enough, as channels of this nature are safer, provide open space and recreational opportunities, and may be more harmonious with natural geomorphologic processes (NAHB 1991; Urban Land Institute 1975). Concrete-lined channels may be needed in unusual circumstances (such as where right-of-way limitations exist).

Where the available slope is so small as to require that a smooth lining be used, the designer may find that desired roughness coefficients can be obtained by lining only the channel bottom. Such a practice should be confined to sites with stable soils.

The following recommended criteria for the design and construction of concrete-lined channels are typical recommendations that should be evaluated for applicability based on site-specific conditions.

Hydraulics

(a) Freeboard—Adequate channel freeboard above the designed water surface should be provided. One suggested method of calculating minimum freeboard is (Chow 1959):

$$H_{FB} - \sqrt{Cd} \qquad (8\text{-}1)$$

where

H_{FB} = freeboard (feet).
C = a coefficient varying from 1.5 (for channel capacities of 20 cfs) to 2.5 (for channel capacities of 3000 cfs or more).
d = depth (feet).

Minimum freeboard is generally 1 foot, and additional freeboard must be provided to accommodate superelevation, standing waves, and/or other water surface disturbances. Freeboard usually should not be obtained by the construction of levees.

(b) Superelevation—Superelevation of the water surface must be determined at horizontal curves and design of the channel section adjusted accordingly.

An approximation of the superelevation can be obtained from the following equation:

$$h = \frac{V^2 Tw}{gr_c} \qquad (8\text{-}2)$$

where

V = velocity (feet per second).
r_c = centerline radius of curvature (feet).
Tw = top width of channel (feet).

Curved, concrete-lined channel. Note the overflow on the right side of the channel into the dry pond below.

    (c) Velocities—Flow velocities should not exceed 7 fps or result in a Froude number greater than 0.8 during the 100-year flood for non-reinforced linings, and usually should not exceed 18 fps for reinforced linings (Urban Drainage and Flood Control District 1984).

Discussion regarding concrete materials, concrete lining sections, joints, finish, curing, and steel reinforcement is provided in Chapter 13. Earthwork, bedding, underdrains, and other structural and geotechnical design considerations are presented in Chapter 14.

D. Grass-Lined Artificial Channels

Grass-lined channels are desirable in many respects (Urban Land Institute 1975; Williams 1990). Channel storage, low velocities, and ecological (water quality enhancement) and aesthetic benefits create significant advantages over other types, although they generally require more land area. Their design must give full consideration to sediment deposition and scour, as well as hydraulics. Figures 8.12, 8.13, and 8.14 provide typical grass-lined channel cross-sections. Figure 8.15 provides a cross-section of a roadside ditch, which is a variation of a grass-lined channel.

1. Channel Stability/Protection

Allowable velocities, compatible with various types of soils and ground cover, typically published in handbooks, should be recognized as **max-**

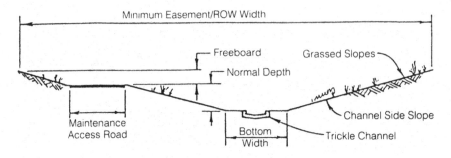

Figure 8.12—Typical grass-lined channel section, Type A.

imum velocities. Such information sources usually are silent with regard to the occurrence frequency and flow velocity durations related to the published data. These considerations remain essentially unquantified. The engineer is advised to be conservative when selecting design flow velocities, unless there are adequate **observed** erosion data available on nearby installations.

In non-cohesive soils, especially in arid areas lacking significant stabilizing vegetation, side slopes of channels typically will undermine during significant discharges and collapse (slough, slide) into the channel, and then be moved downstream. Bank stabilization, often to depths below scour depths, is important in such situations. The steeper the thalweg gradient, the more important such stabilization becomes.

Channel thalweg gradients must often be controlled to limit potential flow velocities (USBR 1974; USDA 1962). This can be done by various means, such as check dams, with stilling basins or other energy-absorbing structures provided immediately downstream. Such facilities should be designed so that the maximum foreseeable discharges will not overtop them, nor result in their undermining and loss. Where "head-cutting" (serious, progressive upstream bottom degradation) may occur, it is necessary to provide facility protection or extend gradient controls downstream to a location below which degradation will not occur.

Generally, channels in non-cohesive soils in urban areas should have sufficient width to assure that their allowable velocities are not exceeded under maximum foreseeable discharges. This may result in unusual costs, unsightliness, or other undesirable effects, but is the price paid for continuing channel stability (U.S. Soil Conservation Service 1954, 1977).

Potential sediment loads always should be considered when planning any kind of ponded storage, or channel with a mild invert slope. Up-

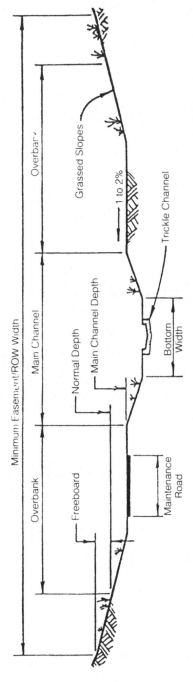

Figure 8.13— Typical grass-lined channel section, Type B.

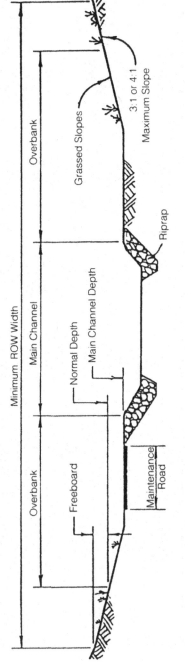

Figure 8.14—Typical grass-lined channel section, Type C.

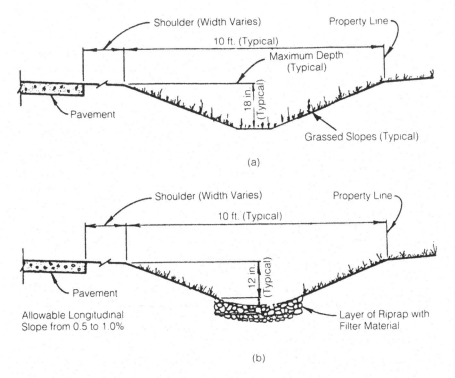

Figure 8.15—Typical roadside ditch sections.

stream erosion may have particularly adverse consequences where relatively small ponds are developed on-line with large upstream contributory areas. Development of ponds as off-line facilities, with inlets operated to fill the ponds when suspended solids loads in the waterway are minimal, is a wise approach when feasible.

There is a foreseeable relationship between stage-frequency characteristics and potential sediment loads for any given type of upstream soils (Urban Land Institute 1978). Where significant deposition occurs naturally, following mean-annual or smaller discharges, caution is advised. There is no practical way to overcome Stokes' Law. Bed loads will deposit when velocity diminishes. One of the most frequent deposition areas is upstream of culvert or bridge entrances, where significant headwater depth (with its associated velocity reduction) may be necessary to develop design discharge capacities.

2. Preliminary Design Criteria

The maximum velocity for the major storm design runoff should be low enough to keep scour problems within reasonable limits. Until a satisfactory grass cover is established, design flows will cause serious channel cutting and bank cutting at bends. Bends, transitions and the like merit careful evaluation, and often the designer will find it prudent to install erosion protection at key locations. More discussion of how to determine whether or not protection is required at channel discontinuities is provided in Chapter 9. Where the natural topography is steeper than desirable, drops should be used to keep velocities within desired limits. Drops are discussed in Chapter 9.

Channels will function better with longer radii of curvature. In general, the centerline radius of curvature should be at least twice the design flow top width, and not less than 100 feet.

Bridge deck bottoms and exposed pipe channel crossings often control the freeboard along urban channels. Where this is not the case, the allowance for freeboard should depend somewhat upon the conditions adjacent to the channel. For instance, localized overflow in certain areas may be desirable because of ponding benefits. In general, a minimum freeboard of 1 foot should be allowed (see discussion of freeboard on next page).

The grass species chosen must be sturdy, inundation and drought resistant, easy to establish, and able to spread after establishment. A thick root structure is necessary to control weed growth and erosion. The Soil Conservation Service and local landscape architects can provide assistance in selecting grass mixtures suitable for actual site conditions.

3. Channel Cross-Sections

Any channel shape suitable to the location and the environmental conditions may be used, so long as channel stability, public safety, and maintenance are not impaired. Often the channel configuration can be chosen to provide open space, recreational opportunities and wildlife habitat (Grove 1990). For example, in an effort to maximize open space, some communities have established 2–5 year capacity trapezoidal channels, and then formally designated adjacent overflow areas as regulatory floodplains (Urban Drainage and Flood Control District 1989, 1990). Varying channel characteristics from reach to reach is especially appealing to the public. Limitations within which design should normally fall for the major storm design flow include:

Side slopes   The flatter the side slope, the better. Under special conditions, such as where development exists and right-of-way is a problem, the slopes may be as steep as 3:1 (5:1 is generally the safe limit for mowing equipment). Slope stability must of course be maintained, particularly during periods of rapid drawdown.

Bottom width   The bottom width should be designed to accommodate the hydraulic capacity of the cross-section, recognizing the limitations on velocity and depth. Width must be adequate to allow necessary maintenance.

Trickle channels   Trickle channels or underdrain pipes are sometimes used on urban grass channels to convey low flows (1–5% of design flow). In some circumstances, their use is being expanded to facilitate

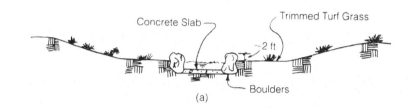

(a)

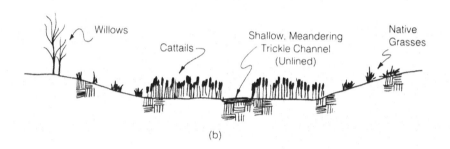

(b)

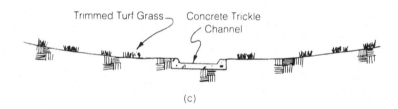

(c)

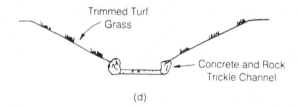

(d)

Figure 8.16—Channel designs with trickle channels.

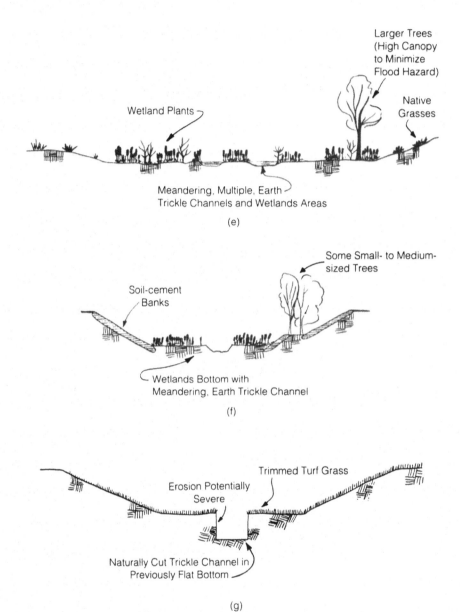

Figure 8.16—*Continued*

Denver, Colorado—Grass-lined open channel with a rock-lined "trickle channel." Note the grouted-rock drop structures.

maintenance and to avoid creation of urban "wetlands." Under the current federal definition, nearly any location supporting water-loving vegetation can be classified as a "regulatory" wetland, subject to federal and state regulatory constraints (see also Chapter 2). In other situations, trickle channels are discouraged, so as to foster wetlands in channel bottoms (for water quality enhancement, groundwater recharge, wildlife habitat, etc.).

Concrete trickle channels are common because they facilitate maintenance and control base flow erosion. Other types are acceptable if properly designed. Trickle channels may not be practical on major streams and rivers or in large channels through sandy soils. Figure 8.16 provides typical trickle channel details, and some examples of how trickle channels can be incorporated into channel design.

4. Freeboard

Except where localized overflow (via side channel spillways or weirs) is desirable for additional ponding benefits or other reasons, freeboard should be provided. Freeboard requirements are normally specified locally; however, the following equation is generally applicable, at least for planning purposes.

$$H_{FB} = 0.5 + \frac{V^2}{2g} \qquad\qquad (8\text{-}3a)$$

or, in metric units,

$$H_{FB} = 0.152 + \frac{V^2}{2g} \qquad\qquad (8\text{-}3b)$$

The minimum freeboard should normally be at least 1 foot above the maximum design water surface elevation, although more is usually desirable to allow for a margin of error due to computational inaccuracies. An approximation of the superelevation can be obtained as indicated previously for concrete-lined channels.

5. Curvature

The centerline radius of curvature should be at least twice the top width of the channel, but not less than 100 feet.

6. Right-of-way Width

The minimum right-of-way width should include freeboard and adequate access for maintenance (including heavy equipment). Upstream from structures such as bridges or culverts, where floodwaters can leave the formal channel, additional right-of-way may be set aside to limit land development (Jones and Jones 1987).

7. Roughness Coefficients

The hydraulic roughness of man-made grass-lined channels depends on the length of cutting, if any, the type of grass, and the depth of flow, as well as the state of maintenance. Poorly maintained channels with large accumulations of sediment and debris have very high roughness coefficients. Typical roughness coefficients are given in Table 8.7. The 0.7 to 1.5 foot depth in Table 8.7 is generally suitable for computing the wetted channel portion for the minor storm runoff, while the greater than 3 foot depth is suitable for the major runoff computations. A depth of flow of 2.0 feet or more will usually lay the grass down to form a relatively smooth bottom surface (U.S. Department of Agriculture 1962).

Care must be exercised in operation and maintenance during periods following completion of construction, and before the grass stand has matured. While an "n" factor of 0.07 might be chosen for lower flows, before the grass is up, the effective "n" may be as low as 0.025. Runoff during this period would have higher velocities and erosion damage might result.

TABLE 8.7.   Manning Roughness Coefficients, n, For Typical Grasses
For Well-Maintained Straight Channels Without
Shrubbery or Trees (Chow 1959).

Grass Type (1)	Depth of Flow	
	0.7–1.5 ft. (2)	> 3.0 ft. (3)
Bermuda grass, buffalo grass, Kentucky bluegrass		
a. Mowed to 2 inches	0.035	0.030
b. Length 4–6 inches	0.040	0.030
Good stand any grass		
a. Length of 12 inches	0.070	0.035
b. Length of 24 inches	0.100	0.035
Fair stand any grass		
a. Length of 12 inches	0.060	0.035
b. Length of 24 inches	0.070	0.035

8. Erosion Control

Practice has shown that it is uneconomical to design a grassed channel completely protected from erosion. It is preferable to provide a reasonable erosion-free design that includes allowances for additional erosion control measures and corrective steps after the first year of operation. However, the use of erosion control cutoff walls at regular intervals in a grassed channel is desirable (U.S. Department of Agriculture 1962). They will safeguard a channel from serious erosion in case of a large runoff prior to the grass developing a good root system and are also useful in containing the trickle channel.

Erosion control cutoff walls (see also Chapter 9) are often constructed of reinforced concrete, approximately 8 inches thick and from 1.5–3 feet deep, extending across the entire bottom of the channel. They can be shaped to fit a slightly sloped bottom to help direct water to the trickle channel or to an inlet. Depending upon scour, it may be necessary to anchor the wall into bedrock. Sloping rock structures are also common.

Grass will not grow under bridges, and the erosion tendency may be large. A cutoff wall at the downstream edge of a bridge is good practice, or the designer might choose to soil-cement the entire bottom width under the bridge deck.

At bends in the channel, special erosion control measures may be needed. However, once a good growth of grass is established and if the design velocities, depths, and curvatures are adhered to, erosion at bends will normally not be a problem.

To achieve the appropriate channel slope, the designer may find it necessary to use frequent drops. Erosion tends to occur at the edges and immediately upstream and downstream of drops, even though they may be only 6–18 inches high. Proper use of riprap at these drops is necessary. Remember that infrequent flows may override the drops, with resultant loss of the drop structure, unless deep cutoff walls and extensive toe erosion protection are provided.

9. Water Surface Profile

A water surface profile should be computed for all channels. Open channel flow in urban drainage is usually nonuniform because of bridge openings, curves, and structures (French 1985; Rouse 1950). Computation of the water surface profile should therefore use standard backwater methods, taking into consideration all losses due to changes in velocity, drops, bridge openings and other obstructions (see Chapter 6 for a discussion of these computations). Computations begin at a known point and extend in an upstream direction for subcritical flow. For this reason, hydraulic design of the channel should proceed in an upstream direction. The energy gradient should be shown on all preliminary drawings to help insure against errors. The water surface profiles and energy gradients may be shown on the final drawings (local practice will usually dictate the level of detail required for water surface profiles).

E. Natural and Composite Channels

If a natural channel is to be used for carrying storm runoff from an urbanized area, it may be assumed initially that the changed runoff regime will result in erosion (Urban Land Institute 1975). Careful hydraulic analyses must be made of natural channels to evaluate these tendencies. In many cases some modification of the channel will be required to create a more stabilized condition for the channel. Natural channels that include structural improvements are technically "composite" channels, although the distinction is subtle.

The investigations necessary to insure that a natural channel will be adequate are different for every waterway; however, the designer will generally find it necessary to prepare cross-sections of the channel for the major design runoff, to investigate the bed and bank material as to the particle size classification and to generally study the stability of the channel under future conditions of flow. Note that supercritical flow usually does not occur in natural channels (except in mountainous or other steeply sloping areas) and frequent checks should be made during the course of the backwater computations to insure that supercritical flow conditions are not present. Figure 8.17 provides a flowchart for the design procedures for high-gradient natural channels.

With many natural waterways it is necessary to construct drops and/ or erosion cutoff check structures at regular intervals to decrease the thalweg slope and to control erosion. However, these channels should be left in as near a natural condition as possible, and extensive modifications should not be undertaken unless they are deemed necessary

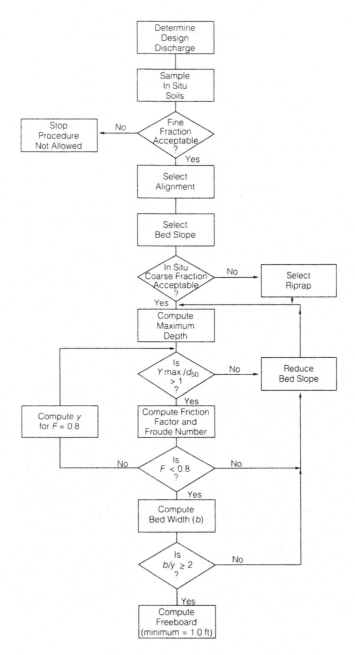

Figure 8.17—Design procedure schematic for high gradient natural channels.

to avoid excessive erosion with subsequent deposition downstream (Note also that modification of the channel within the normal high water line may require a U.S. Army Corps of Engineers Section 404 permit).

The guidelines for design criteria and evaluation techniques for natural channels include:

(a) Channel and overbank areas should normally have adequate capacity for the major system design runoff (Urban Drainage and Flood Control District 1984). Freeboard is usually desirable.

(b) Natural channel segments with a Froude number greater than 0.80 for the major system design peak flows should be protected from erosion (Urban Drainage and Flood Control District 1984).

(c) Water surface profiles should be computed so that the floodplain can be zoned and protected (Federal Flood Insurance Administration 1981).

(d) Filling of the flood fringe reduces valuable channel storage capacity and tends to increase downstream runoff peaks. Filling of the flood fringe is usually subject to restrictions imposed by floodplain regulations (Federal Flood Insurance Administration 1981).

(e) Roughness factors (n) which are representative of **unmaintained** channel conditions should be used for the analysis of water surface profiles (Jones and Jones 1987).

Denver, Colorado—Small "pilot" channel that accommodates less than the 1-year runoff event with higher flows conveyed in the overbank area. Effective if the objective is maximizing the amount of open space to be set aside for the major drainageway.

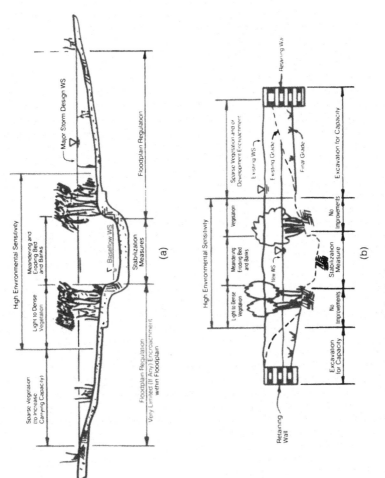

Figure 8.18—Typical natural channels.

(f) Roughness factors (n) representative of maintained channel conditions should be used to determine **velocity** limitations.
(g) Plan and profile drawings of the floodplain should be prepared. Appropriate allowances for future bridges or culverts, which can raise the water surface profile and cause the floodplain to be extended, should be included in the analysis (Sheaffer et al. 1982).

The usual rules of freeboard, depth, curvature, and other guidelines applicable to artificial channels do not necessarily apply to natural channels. All structures constructed along the channel should be a minimum of 1 foot above the major system design water surface, or higher if the natural channel is in a mountainous area (subject to supercritical flow). Relative to maximizing open space, significant advantages may accrue if the designer incorporates into his planning the overtopping of the channel and localized flooding of adjacent areas, providing they are laid out and developed for the purpose of being inundated during the major storm runoff.

One variation of the natural channel is to leave the main channel area undisturbed (i.e., that area containing the base flow plus the immediate vegetation area) and to improve the overbank conveyance capabilities by excavating the floodplain area. This "naturalized" channel conveys the base flow and increases the capacity of the total channel to convey the major discharge (ASCE 1990). Figures 8.18 and 8.19 provide examples of slightly improved natural channels.

F. Other Channels

Other channels/materials discussed in Chapters 9 and 13, although not covered explicitly here, include:

(a) Rock/riprap channels.
(b) Soil cement/roller compacted concrete.
(c) Synthetic fabrics, such as woven fabric forms filled with concrete or grout, or discrete blocks on continuous backing.
(d) Special vegetation linings.
(e) Channel "rundown" (to convey minor flows from parking lots into outfalls). See Figure 8.20 for detail.

After the designer is thoroughly familiar with the applicability and limitations of the particular channel liner under consideration, and lacking specific local guidance for the subject material, he can adjust general criteria presented herein for concrete, grass, and natural channels to suit the particular application. For example, soil cement, roller-compacted concrete, synthetic fabrics and others would have general design parameters (freeboard, side slopes, maintenance requirements) **similar** to those for concrete channels.

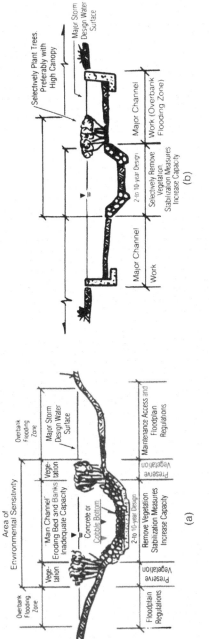

Figure 8.19—Typical natural composite channels.

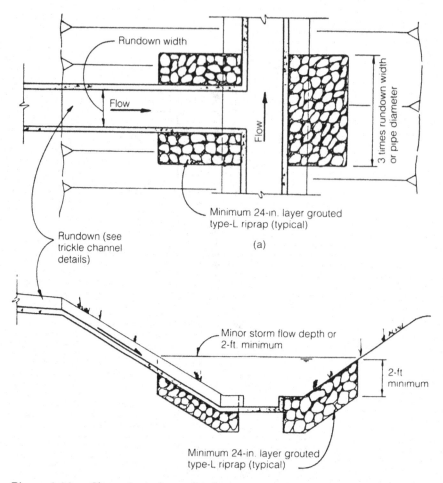

Figure 8.20—Channel rundown details.

X. CULVERTS AND BRIDGES

A. Culverts

Much of the following text on culverts has been adapted from Federal Highway Administration publications (U.S. Department of Commerce 1965a and b; U.S. Department of Transportation 1972 and 1985). The reader should consult these references for more detail. The type of culvert used depends on such factors as roadway profiles, channel characteristics, flood damage evaluations, construction and maintenance costs, estimates of service life and public safety considerations.

Culverts are constructed from a variety of materials and are available in many different shapes and configurations (see Figure 8.21). The shape selection is based on construction costs, limitations on upstream water surface elevation, roadway embankment height and hydraulic performance. The selection of a culvert material depends upon its required structural strength, hydraulic roughness, durability, and corrosion and abrasion resistance. The three most common culvert materials are concrete (reinforced and non-reinforced), corrugated aluminum and corrugated steel. Culverts may also be lined with other materials to inhibit corrosion and abrasion or to reduce hydraulic resistance.

Many different inlet configurations are used on culvert barrels. They can either be prefabricated or constructed in place, and common inlet configurations include projecting culvert barrels, cast-in-place concrete headwalls, precast or prefabricated end sections, and culvert ends mitered to conform to the fill slope (see Figure 8.22). Structural stability, aesthetics, erosion control, and other factors influence the selection of various inlet configurations.

The hydraulic capacity of a culvert may be improved by appropriate inlet selection. Since the natural channel is usually wider than the culvert barrel, the inlet is a flow contraction and may be the primary flow control. The provision of a more gradual flow transition will reduce energy loss and create a more hydraulically efficient inlet condition. A flowchart of the desirable culvert design procedure is provided in Figure 8.23.

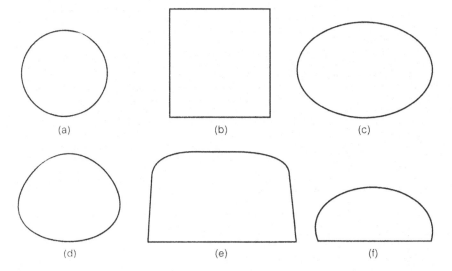

Figure 8.21—Commonly used culvert slopes.

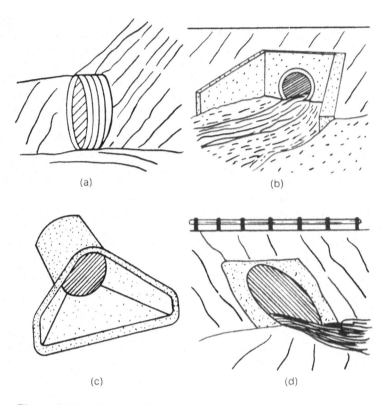

(a)

(b)

(c)

(d)

Figure 8.22—Four standard inlet types.

1. Culvert Hydraulics

Culverts can flow under inlet control or outlet control (see also Chapter 6). Under inlet control, the cross-sectional area of the barrel, the inlet configuration or geometry, and the amount of headwater are the factors affecting capacity. Outlet control involves the additional consideration of the tailwater in the outlet channel and the slope, roughness, and length of barrel. Under inlet control conditions, the slope of the culvert is steep enough so that the culvert does not flow full and tailwater does not affect the flow.

Nomographs have been used extensively in culvert design. There are a number of easy to use computer programs available now which replace these nomographs.

Inlet Control Condition Inlet control for culverts may occur in two ways (see Figure 8.24).

(a) Unsubmerged—the headwater is not sufficient to submerge the top of the culvert and the culvert invert slope is steep (can sustain supercritical

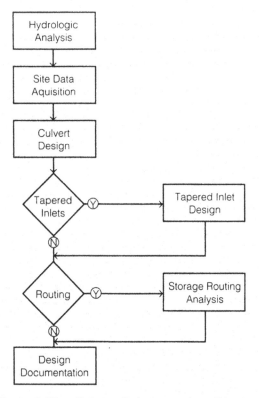

Figure 8.23—Culvert design procedure flowchart.

flow). The culvert inlet effectively acts like a weir (Condition A, Figure 8.24.

(b) Submerged—the headwater submerges the top of the culvert but the pipe does not flow full. The culvert inlet acts like an orifice (Condition B, Figure 8.24).

In the submerged inlet condition, the equation governing the culvert capacity is the orifice flow equation:

$$Q = C_d A \sqrt{2gh} \qquad\qquad (8\text{-}4)$$

where

Q = flow.
C_d = orifice coefficient.
A = area.
g = gravitational constant.
h = head on culvert measured from pipe/barrel centerline.

The orifice coefficient, C_d, varies with head on the culvert as well as the culvert type and entrance geometry. Inlet control rating curves for several culvert materials, shapes, and inlet configurations are given in Figures 8.25 and 8.26. These nomographs were developed empirically by the pipe manufacturers, Bureau of Public Roads, and the Federal Highway Administration. The nomographs (or computer programs that replace them) are recommended for use, rather than Equation 8-4, due to the uncertainty in estimating the orifice coefficient, and because they are applicable to a wide range of depths. Equation 8-4 is applicable only when the ratio of water depth to culvert height (diameter) is ≥ 2.

Outlet Control Condition   Outlet control will govern if the headwater is deep enough, the culvert slope sufficiently flat, and the culvert sufficiently long. There are three types of outlet control flow conditions:

(a) The headwater submerges the culvert top, and the culvert outlet is submerged under the tailwater. The culvert will flow full (Condition A, Figure 8.24).
(b) The headwater submerges the top of the culvert and the culvert is unsubmerged by the tailwater (Condition B or C, Figure 8.24).
(c) The headwater is insufficient to submerge the top of the culvert. The culvert slope is subcritical and the tailwater depth is lower than the pipe critical depth (Condition D, Figure 8.24).

The factors affecting the capacity of a culvert in outlet control include the inlet geometry and associated losses, the culvert material with friction losses, and the tailwater condition.

The capacity of the culvert under outlet control is calculated using the principle of the conservation of energy (Bernoulli's Equation). An energy balance is determined between the headwater at the culvert inlet and at the culvert outlet, which includes inlet losses, friction losses, and the velocity head (see Figure 8.27). The equation is then expressed as:

$$H = h_e + h_f + h_v \qquad (8\text{-}5)$$

where

H = total energy head (feet).
h_e = entrance head losses (feet).
h_f = friction losses (feet).
h_v = velocity head (feet) = $V^2/2g$.

For entrance losses the governing equation is:

$$h_e = K_e \left(\frac{V^2}{2g} \right) \qquad (8\text{-}6)$$

where K_e = is the entrance loss coefficient.

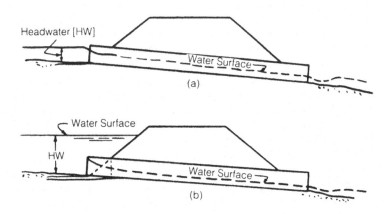

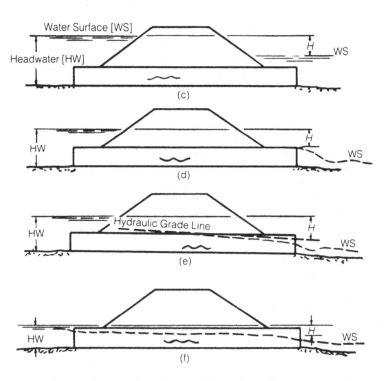

Figure 8.24—Inlet and outlet conditions for culverts.

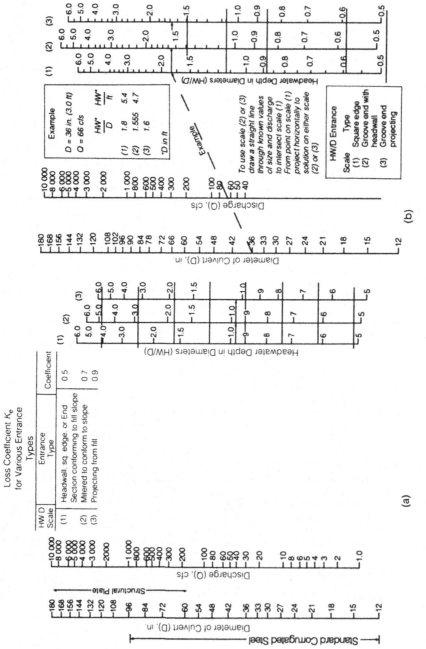

Figure 8.25— Inlet control nomograph circular pipe.

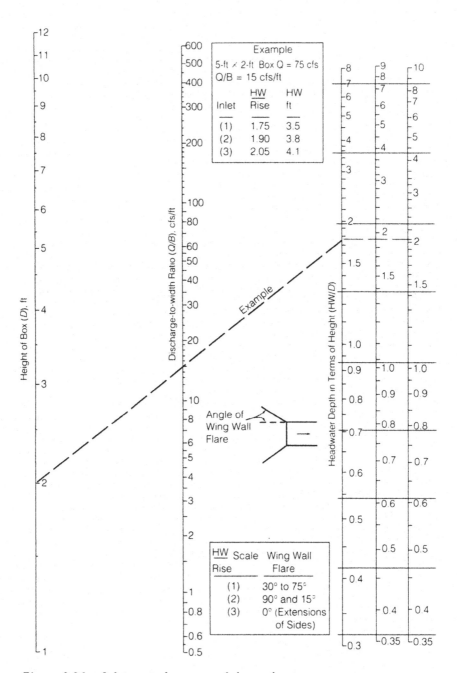

Figure 8.26—Inlet control nomograph box culverts.

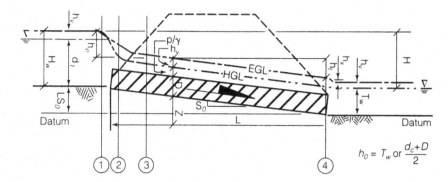

Figure 8.27—Hydraulics of a culvert under outlet condition.

Typical inlet loss coefficients recommended for use are given in Table 8.8.

Friction loss is the energy required to overcome the roughness of the culvert and is expressed as follows:

$$h_f = \frac{29n^2L}{R^{1.33}} \left(\frac{V^2}{2g}\right) \tag{8-7a}$$

and, in SI units,

$$h_f = \frac{19.5n^2L}{R^{1.33}} \left(\frac{V^2}{2g}\right) \tag{8-7b}$$

where

n = Manning's coefficient.
L = length of culvert.
R = hydraulic radius.

Combining Equations 8-5, 8-6, and 8-7 and simplifying gives:

$$H = \frac{K_e + 1 + 29n^2L}{R^{1.33}} \left(\frac{V^2}{2g}\right) \tag{8-8a}$$

and, in SI units,

$$H = \frac{K_e + 1 + 19.5n^2L}{R^{1.33}} \left(\frac{V^2}{2g}\right) \tag{8-8b}$$

Equation 8-8 can be used to calculate the culvert capacity directly when the culvert is flowing under outlet conditions A or B as shown on Figure 8.24. For conditions C or D, the HGL at the outlet is approximated by averaging the critical depth and the culvert diameter, which is used if the value is greater than the tailwater depth (Tw) to compute headwater depth (Hw).

A series of outlet control nomographs for various culvert materials and shapes have also been developed. The nomographs are presented in Figures 8.28, 8.29, and 8.30 (computer programs are also available that greatly simplify the calculations). When rating a culvert, either the outlet control nomographs or Equation 8-8 can be used to calculate the headwater requirements.

When using the outlet nomographs for corrugated steel pipe the data must be adjusted to account for the variation in the n-value between the nomographs and the culvert being evaluated. The adjustment is made by calculating an equivalent length according to the following equation:

$$L' = L \left(\frac{n'}{n}\right)^2 \qquad (8-9)$$

where

L' = equivalent length.
L = actual length.
n = Manning's n.
n' = actual n-value of culvert.

2. Design Procedure and Example

Computer programs (proprietary programs, available from several vendors) suitable for use on a microcomputer are readily available, and are easy to use. If a computer is not available, the following detailed procedure for selection of culvert size is adopted directly from Hydraulic Engineering Circular No. 5: Hydraulic Charts for the Selection of Highway Culverts (U.S. Department of Commerce 1965a).

Design Example

Step 1: List design data. (see suggested tabulation form, Table 8.9).

(a) Design discharge Q, in cfs, with average return period (i.e., Q_{25} or Q_{50}, etc.).
(b) Approximate length L of culvert, in feet.
(c) Slope of culvert. (If grade is given in percent, convert to slope in feet per foot).
(d) Allowable headwater depth, in feet, which is the vertical distance from the culvert invert (flow line) at the entrance to the water surface elevation permissible in the headwater pool or approach channel upstream from the culvert.
(e) Mean and maximum flood velocities in natural stream.

TABLE 8.8.   Hydraulic Data for Culvert: Culvert Entrance Losses
(UDFCD 1984).

Type of Entrance	Entrance Coefficient, K_e
Pipe	
Headwall	
Grooved edge	0.20
Rounded edge (0.15D radius)	0.15
Rounded edge (0.25D radius)	0.10
Square edge (cut concrete and CMP)	0.40
Headwall & 45° Wingwall	
Grooved edge	0.20
Square edge	0.35
Headwall with Parallel Wingwalls Spaced 1.25D apart	
Grooved edge	0.30
Square edge	0.40
Beveled edge	0.25
Projecting Entrance	
Grooved edge (RCP)	0.25
Square edge (RCP)	0.50
Sharp edge, thin wall (CMP)	0.90
Sloping Entrance	
Mitered to conform to slope	0.70
Flared-end Section	0.50
Box, Reinforced Concrete	
Headwall Parallel to Embankment (no wingwalls)	
Square edge on 3 edges	0.50
Rounded on 3 edges to radius of $\frac{1}{12}$ barrel dimension	0.20
Wingwalls at 30° to 75° to barrel	
Square edged at crown	0.40
Crown edge rounded to radius of $\frac{1}{12}$ barrel dimension	0.20
Wingwalls at 10° to 30° to barrel	
Square edged at crown	0.50
Wingwalls parallel (extension of sides)	
Squared edged at crown	0.70

NOTE: The entrance loss coefficients are used to evaluate the culvert or sewer capacity operating under outlet control.

(f) Type of culvert for first trial selection, including barrel material, barrel cross-sectional shape and entrance type.

Step 2:   Determine the first trial size culvert
Since the procedure given is one of trial and error, the initial trial size can be determined in several ways:

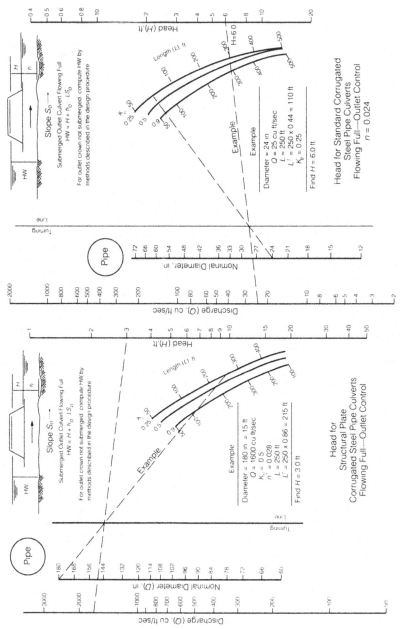

Figure 8.28—Outlet control nomograph, circular CSP.

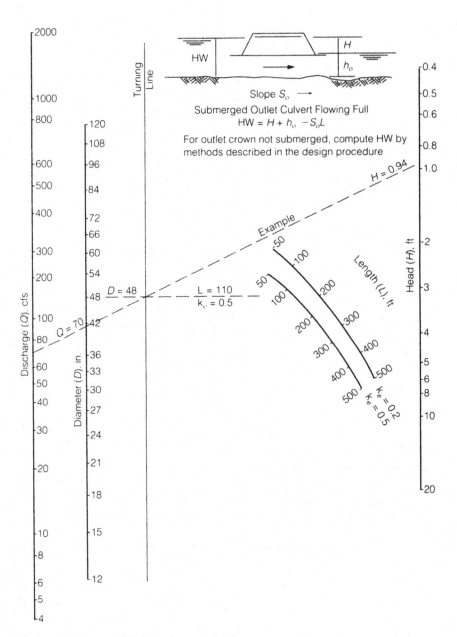

Figure 8.29—Outlet control nomograph, circular RCP.

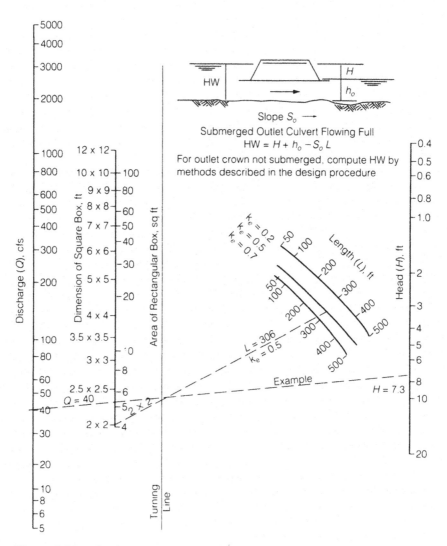

Figure 8.30—Outlet control nomograph, box culverts.

TABLE 8.9.   Culvert Design Form (UDFCD 1984)

PROJECT: _____   DESIGNER: _____

                        DATE: _____

HYDROLOGIC AND CHANNEL INFORMATION   SKETCH   STATION: _____

$Q_1 = $ _____ $TW_1 = $ _____
$Q_2 = $ _____ $TW_2 = $ _____

$\left(\begin{array}{l}Q_1 = \text{DESIGN DISCHARGE, SAY } Q_{25} \text{ OR } Q_{100} \\ Q_2 = \text{CHECK DISCHARGE, SAY } Q_{50}\end{array}\right)$

MEAN STREAM VELOCITY = _____
MAX. STREAM VELOCITY = _____

EL. _____   EL. _____ $S_o = $ _____ $L = $ _____   EL. _____   AHW= _____   TW= _____

Culvert Description (Entrance type)	Q	Size	Inlet Cont.		Headwater Computation									Controlling HW	Outlet Velocity	Cost	Comments
			$\dfrac{HW}{D}$	HW	Outlet Control ($HW = H + h_o - LS_o$)												
					K_c	H	d_c	$\dfrac{d_c + D}{2}$	TW	h_o	LS_o	HW					

Summary & Recommendations:

(a) By arbitrary selection.
(b) By using an approximating equation such as $Q/10 = A$ from which the trial culvert dimensions are determined.
(c) By using inlet control nomographs (Figures 8.26 or 8.27) for the culvert type selected. If this method is used an HW/D must be assumed, say HW/D = 1.5, and using the given Q a trial size is determined.

If any trial size is too large because of limited embankment height or availability of pipe, multiple culverts may be used by dividing the discharge equally among the number of barrels used. Raising the embankment height or the use of pipe arch and box culverts with width greater than height should be considered. Final selection should be based on an economic analysis.

Step 3: Find headwater depth for trial size culvert

(a) Assuming INLET CONTROL.
   1. Using the trial size from step 2, find the headwater depth, HW, by use of the appropriate inlet control nomograph. Tailwater (TW) conditions are to be neglected in this determination. HW in this case is found by multiplying HW/D obtained from the nomographs by the height of culvert D.
   2. If HW is greater or less than allowable, try another trial size until HW is acceptable for inlet control before computing HW for outlet control.
(b) Assuming OUTLET CONTROL.
   1. Approximate the depth of tailwater TW, in feet, above the invert at the outlet for the design flood condition in the outlet channel.
   2. For tailwater TW elevation equal to or greater than the top of the culvert at the outlet, set h_o equal to TW and find HW by the equation given in Table 8.10.
   3. For tailwater TW elevation less than the top of the culvert at the outlet, find headwater HW by the equation given in Table 8.10, except that:

$$h_o = \frac{d_c + D}{2}$$

or TW, whichever is the greater, where:

d_c = critical depth in feet
D = height of culvert opening, in feet

(c) Compare the headwaters found in Step 3a and Step 3b (Inlet Control and Outlet Control). The higher headwater governs and indicates the flow control existing under the given conditions for the trial size selected.
(d) If outlet control governs and the HW is higher than is acceptable, select a larger trial size and find HW as instructed under Step 3b. (Inlet control need not be checked, since the smaller size was satisfactory for this control as determined under Step 3a.)

TABLE 8.10. Culvert Rating Form (City of Norman 1989)

PROJECT: _____   LOCATION: _____   STATION: _____

CULVERT DATA

TYPE: _____
INLET: _____   n: _____
K_c: _____ Q_{FULL}: _____
            V_{FULL}: _____

OUTLET CONTROL EQUIPMENT

(1) $H_W = H + h_o - LS_o$

(2) For $T_w < D, h_o = \dfrac{d_c + D}{2}$ or T_w (whichever is greater)

For $T_w = D, h_o = T_w$

(3) For Box Culvert: $d_c = 0.315(Q/B)^{2/3} \leq D$

LOW POINT
ELEV. 1151.9

ELEV. 1135.5

ELEV. 1140.0

S_o 0.030
L 150
S_oL 4.5

Q	Inlet Control		H	T_w	d_c	Outlet Control			Cont.	Control	Elev.
						$T_w < D$	$T_w > D$				
	$\dfrac{H_w}{D}$	H_w				$\dfrac{d_c + D}{2} = h_o$	h_o	H_W	H_W		
1	2	3	4	5	6	7	8	9	10	11	12
70	1.0	4	1.9	1.5	2.5	3.3		0.7	4	Inlet	1144.0
115	1.5	6	5.5	2.0	3.0	3.5		4.5	6	Inlet	1146.0
145	2.0	8	8.9	2.5	3.4	3.7		8.1	8.1	Outlet	1148.8
170[1]	2.5	10	12.5	3.0	3.7	3.9		11.9	11.9	Outlet	1151.9
195[2]	3.0	12	16.0	3.5	4.0	4.0		15.5	15.5	Outlet	1155.5

Outlet Velocity, $V = Q/A = 170$ cfs/12.6 ft² $= 13.5$ fps

Step 4: Try a culvert of another type or shape and determine size and HW by the above procedure.

Step 5: Compute outlet velocities for size and types to be considered in selection and determine need for channel protection.

  (a) If outlet control governs in Step 3c above, outlet velocity equals Q/A_o, where A_o is the cross-sectional area of flow in the culvert barrel of the outlet. If d_c or TW is less than the height of the culvert barrel, use A_o corresponding to d_c or TW depth, whichever gives the greater area of flow. A_o should not exceed the total cross-sectional area A of the culvert barrel.
  (b) If inlet control governs in Step 3c, outlet velocity can be assumed to equal mean velocity in open channel flow in the barrel as computed by Manning's equation for the rate of flow, barrel size, roughness and slope of culvert selected.
  Note: Charts and tables are helpful in computing outlet velocities (see references noted in the first paragraph of this section).

Step 6: Record final selection of culvert with size, type, required headwater, outlet velocity, and economic justification.

Step 7: (Not contained in reference). Assess flood hazard posed by recommended culvert to upstream and downstream properties, particularly if a change in the status quo, such as increased flood levels, is going to occur. Have client and review authorities make an informed decision as to appropriate culvert size.

Design Example—Rating an Existing Culvert A sample calculation for rating an existing culvert is presented in Table 8.10. The required data are as follows:

  (a) Culvert size, length, and type (48" CMP, L = 150, n = .024).
  (b) Inlet, outlet elevation, and slope (1140.0, 1135.5, S_o = 0.030).
  (c) Inlet treatment (flared end section).
  (d) Low point elevation of embankment (elevation = 1151.9).
  (e) Tailwater rating curve (see Table 8.10, Column 5).

From the above data, the entrance loss coefficient, K_e, (Table 8.8) and the n-value are determined. The full flow Q and the velocity are calculated from these values for comparison. The rating then proceeds in the following sequence:

Step 1: Headwater values are selected and entered in Column 3. The headwater to pipe diameter ratio (HW/D) is calculated and entered in Column 2. If the culvert is not circular, the height of the culvert is used.

Step 2: For the HW/D ratios, the inlet rating is read from the various figures based upon culvert type (see Figure 8.26 for the example) and entered into Column 1. This completes the inlet condition rating portion.

Step 3: For outlet condition, the Q values in Column 1 are used to determine the head values (H) in Column 4 from the appropriate outlet rating curves (see Tables 8.9 and 8.10 for examples).

Step 4: The tailwater depths (TW) are entered into Column 5 for the corresponding Q values in Column 1 according to the tailwater rating curve. If a tailwater rating curve is not available, compute the normal depth (subcritical or critical only) of a trapezoidal channel approximating the existing drainageway. If the tailwater depth (TW) is less than the diameter of the culvert (d), Columns 6 and 7 are to be calculated (go to Step 5). If TW $>$ d, the tailwater values in Column 5 are entered into Column 8 for the h_o values and proceed to Step 6.

Step 5: The critical depth (d_c) for the corresponding Q values in Column 1 are read from appropriate figures or computed and entered into Column 6. The average of the critical depth and the culvert diameter is calculated and entered into Column 7 as the h_o values.

Step 6: The headwater values (HW) are calculated according to the equation:

$$H_w = H + h_o - LS_o \qquad\qquad (8\text{-}10)$$

where H is from Column 4 and h_o is the value from Column 8 (for Tw $>$ D) or the larger value between Column 5 and Column 7 (for Tw $<$ D). The values are entered into Column 9.

Step 7: The final step is to compare the headwater requirements (Column 9 and 3) and to record the higher of the two values in Column 10. The type of control is recorded in Column 11, depending upon which case gives the higher headwater requirements. The headwater elevation is calculated by adding the controlling H_w (Column 10) to the upstream invert elevation. A culvert rating curve can then be plotted from the values in Column 12 and 1.

Step 8: Compute outlet velocity for designing downstream protection. For full flow conditions, $V = Q/A$; for partially full conditions, see Figure 6.13 for hydraulic properties of pipe. Design channel protection as described in Chapter 9.

To size a culvert crossing, the same form can be used, with some variation in the basic data. First, a design Q value is selected and the maximum allowable headwater is determined. An inlet type (i.e., headwall) is selected and the invert elevations and culvert slope are estimated based upon site constraints. A culvert type is then selected and first rated for inlet control then outlet control. If the controlling headwater exceeds the maximum allowable headwater, the input data are modified and the procedure repeated until the desired results are achieved.

3. Design Considerations

Culvert Sizing   Factors to be considered in the sizing of the culvert include, but are not limited to: (1) minimum design frequency, (2) minimum culvert size, (3) allowable cross street flow in the street for which the culvert is being considered, (4) public safety, (5) maintenance, (6) allowable frequency of inundation of upstream and downstream properties, and (7) potential for debris accumulation.

The minimum design frequency for culverts is often specified by such entities as city or county governments, state highway departments, or the Federal Highway Administration. Frequency often is stated in terms of a maximum headwater depth for a given return frequency flood. It is essential for the engineer to recognize that the design frequency must be selected for each culvert on a case-by-case basis; a 10-year culvert may be acceptable under certain conditions, whereas a culvert with much greater than 100-year capacity may be required in others. Recognition of the nature of development both upstream and downstream is of foremost importance. For instance, a nuclear power facility or hospital immediately upstream from a location requiring culverts would argue for a conservative design. By contrast, if the engineer has assurance that upstream land will stay undeveloped for decades and if periodic inundation of the subject land will not otherwise be harmful (or if the developer has obtained the rights for backwater storage), then a culvert with only nominal capacity could suffice. In any event, the engineer should obtain review authority and client approval of his tentative selection of design return frequency.

Localities often state the minimum culvert size (commonly in the 12–24 inch range). Debris potential of the tributary basin is an important consideration when determining the minimum acceptable size for a culvert. A viable alternative to increased culvert capacity may be to lengthen the roadway area subject to overflow.

When the flow in a channel exceeds the capacity of the culvert and overtops the cross street, the flow across the street must not exceed stipulated local inundation criteria. If the cross street flow exceeds the limits for the minimum design frequency or the minimum culvert size, then the culvert must be increased in size until all criteria are met.

The debris potential of the tributary basin, along with whether or not the culvert will have a debris/safety rack, are critical considerations when determining minimum size of the pipe. Localities often stipulate that, for example, the engineer should assume 25% debris blockage when computing the required size of the culvert. Under certain conditions, greater than 25% debris blockage may be justified. Guidance on debris/safety rack installation follows in this section.

Velocity Limitations   When designing culverts, both the minimum and maximum velocities should be considered. A minimum velocity of flow is required to assure self cleaning; this minimum velocity is normally taken to be between 2–3 feet per second at the outlet.

The maximum velocity in a culvert is controlled by two factors, the channel protection provided at the outlet and the maximum allowable headwater. If the outlet velocities are less than a stipulated value (based on site specific channel erosion factors) then only minimal protection is required due to the eddy currents generated by flow transitions. As outlet velocity is increased however, additional protection is required such as more extensive riprap or an energy dissipator structure. Discussion of outlet protection is provided in Chapter 9 of the Manual.

Headwater Depth   Maximum headwater depths are normally stipulated by local governments, generally in terms of a particular factor times the culvert diameter or culvert rise dimension for shapes other than round. Considerations governing headwater selection are similar to those described above for design return frequency. Essentially, the engineer must carefully consider the consequences to upstream properties for a wide variety of floods for different headwater depths. Of particular importance is the historic precedent with respect to maximum headwater depths associated with the tributary area in question. Engineers are cautioned against down-sizing the conveyance capacity of a culvert for a given headwater depth in an attempt to save money. Such a change would increase flood risks to established upstream properties that have developed "reliance" (in a legal sense) on the existing conveyance capacity. Decreasing culvert capacity under these conditions leaves the engineer and his client potentially liable for damages in the event that a flood occurs that damages upstream properties due to the use of a replacement culvert, but would not have damaged the upstream properties had the original facility been in place. Finally, the design should consider the implications of the fact that a Federal District Court (Rooney et al. 1986) has defined an embankment with a culvert as a "dam" during times of flooding, with attendant hazard considerations.

Inlet and Outlet Protection   Generally, the engineer should assume that inlet and outlet protection is required. Culvert inlets normally include a headwall with wingwalls or a flared end section. The outlet also typically includes a headwall with wingwalls or an end section in addition to riprap protection.

Structural Design   All culverts should be designed to withstand minimum loading requirements in accordance with design procedures summarized by the American Concrete Pipe Association in its publication "Standard Specifications for Highway Bridges" (Hendrickson 1957) and with pipe manufacturers' recommendations. Minimum depth of cover of culverts typically ranges from 1–3 feet.

4. Debris/Safety Racks

The engineer is obligated to consider the advantages and disadvantages of debris racks on a case-by-case basis. Guidance on the proper design of trash racks for detention pond outlet structures is found in "Stormwater Detention Outlet Control Structures" (ASCE 1985). Highlights from this guidance include the following:

(a) Trash racks should be hinged at the top to permit lifting and cleaning, and should slope at 3:1 to 5:1 (horizontal to vertical) to permit debris to "ride up" as the water level rises.
(b) The trash rack should be located a considerable distance out from the entrance to the culvert or pipe to assure that entrance velocities will be low enough that a person will be able to lift himself up.
(c) Bar spacing should be such that a child will not be able to pass between the bars (typically maximum 6 inches clear spacing).
(d) The net open surface area of the trash racks should be at least four times the cross-sectional area of the pipe.

Culverts normally run short distances beneath roads, railroads, etc. In cases where drainage ditches transition to drainage pipes that run for hundreds or even thousands of feet (common in the Midwest), a

Denver, Colorado—Typical debris rack at a culvert. Such racks must be designed with low entrance velocities, and in a way that the debris floats upward as the water level rises. Access to the channel bottom must be provided to remove debris between flood events.

safety rack at the pipe entrance is recommended because of the consequences of debris blockage within the pipe and for public safety.

B. Bridges

The designer of urban stormwater management systems will often find it necessary to characterize the hydraulic aspects of bridge crossings of major drainageways. Procedures described within this chapter with respect to culvert evaluation will often suffice. However, if a more rigorous evaluation is merited, special bridge evaluation techniques may be necessary. For a detailed discussion of this subject, the reader is referred to "Hydraulics of Bridge Waterways" (U.S. Department of Transportation 1970). The following discussion and figures regarding bridges are based upon edited excerpts from this publication. Other valuable bridge design references include the AISI Handbook of Steel Drainage and Highway Construction Products (1983), Neil et al. (1973), and "Hydrology for Transportation Engineers" (U.S. Department of Transportation 1980).

The bridge designer, even in comparatively small urbanized basins, will frequently find it economically advantageous to extend approach embankments into the floodplain to reduce construction costs, recognizing that, in so doing, the embankments will constrict the flow of the stream during flood stages. The greater the free span opening, the less the amount of backwater during the design flood, and vice versa. This has been considered acceptable practice for decades, so long as it is done within reason (Jones and Jones 1987).

When designing a new bridge or replacing an existing one, the designer is urged to conduct an economic evaluation of various bridge alternatives that accounts not only for the comparative construction costs of the bridges, but also includes corresponding flood damages to upstream properties posed by the alternatives. Such an evaluation permits the developer, municipality, or regulatory officials to make an informed decision as to:

(a) How much of a capital expenditure should be made on the bridge.
(b) Expected average annual flood damages for properties affected by the bridge.
(c) Whether or not it would be cost effective to purchase properties subject to backwater flooding upstream of the bridge to save on bridge construction costs (Federal Flood Insurance Administration 1981).

1. Types of Flow Through Bridges

Figure 8.31 provides plan and profile drawings for a typical bridge crossing of a major drainageway. Note the flow is divided into three regions, as labelled Q_a, Q_b, and Q_c. Q_b is that portion of the flow which is confined to the main channel and which passes through the bridge opening in essentially unhindered fashion. Important sections are labelled 1, 2, 3, and 4 (proceeding from upstream to downstream).

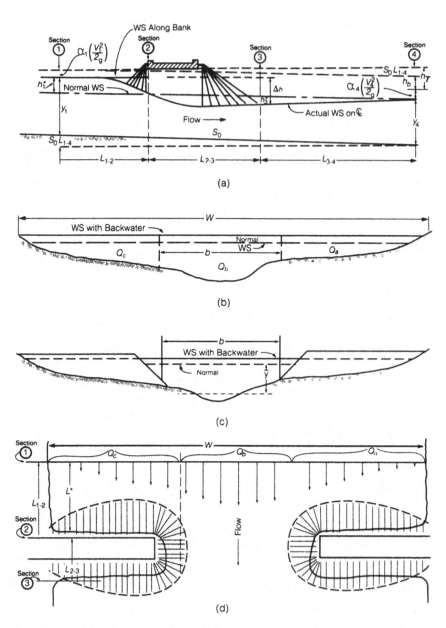

Figure 8.31—Normal crossings: Spillthrough abutments.

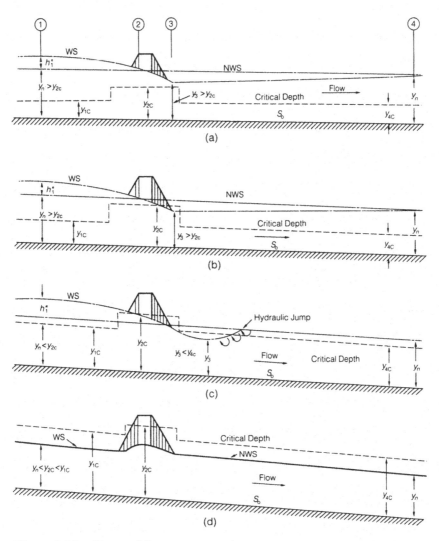

Figure 8.32—Types of flow encountered.

The manner in which flow is contracted in passing through a bridge constriction is illustrated in Figure 8.32. A very marked change is evidenced near the abutments since the momentum of the flow from both sides (or floodplains) must force the advancing central portion of the stream over to gain entry to the constriction. Upon leaving the constriction, the flood gradually expands until normal conditions in the stream are again reestablished.

Constriction of the flow causes a loss of energy, the greater portion occurring in the re-expansion downstream. This loss of energy is reflected in a rise in the water surface upstream of the bridge. The magnitude of this rise, or backwater, is one of the most important bridge design criteria (other important criteria include downstream channel impacts, potential scour of the bridge abutments, piers, caissons, etc., and the potential for roadway overtopping during extreme floods).

Figure 8.32 indicates the four types of flow that may be encountered at bridges. These are labelled types I, IIA, IIB and III. The long dashed lines shown on each profile represent the normal water surface, or the stage the design flow would assume prior to placing a constriction in the channel. The solid lines represent the configuration of the water surface, on centerline of channel in each case, after the bridge is in place. The short dashed lines represent critical depth, or critical stage in the main channel (Y_{1c} and Y_{4c}) and critical depth within the constriction, Y_{2c}, for the design discharge in each case.

Type I Flow Referring to Figure 8.32, it can be observed that normal water surface is everywhere above critical depth. This has been labelled type I or subcritical flow, the type usually encountered in practice. The backwater expression for type I flow is obtained by applying the conservation of energy principle between sections 1 and 4.

Type IIA Flow For type IIA flow, the normal water surface in the unconstricted channel again remains above critical depth throughout, but the water surface passes through critical depth in the constriction. Once critical depth is penetrated, the water surface upstream from the constriction, and thus the backwater, becomes independent of conditions downstream (even though the water surface returns to normal stage at section 4).

Type IIB Flow The water surface for type IIB flow starts out above both normal water surface and critical depth upstream, passes through critical depth in the constriction, next dips below critical depth downstream from the constriction and then returns to normal. The return to normal depth can be rather abrupt (hydraulic jump) since the normal water surface downstream is above critical depth. A backwater expression applicable to both types IIA and IIB flow has been developed by equating the total energy between section 1 and the point at which the water surface passes through critical stage in the constriction.

Type III Flow In type III flow, the normal water surface is everywhere below critical depth and the flow throughout is supercritical. This is an unusual case normally found only in mountainous regions. Theoretically, backwater should not occur for this flow type, since the flow throughout is supercritical. It is likely that an undulation of the water surface will occur in the vicinity of the constriction, however.

2. Hydraulic Evaluation

A practical expression for backwater caused by bridges has been formulated by applying the principle of conservation of energy between the point of maximum backwater upstream from the bridge, section 1, and a point downstream from the bridge at which normal stage has

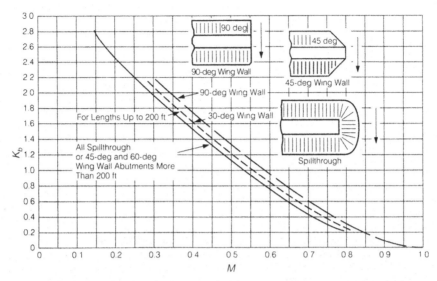

Figure 8.33—Backwater coefficient base curves (subcritical flow).

been reestablished, section 4 (see Figure 8.33). The expression is reasonably valid if the channel in the vicinity of the bridge is essentially straight, the cross-sectional area of the stream is fairly uniform, the gradient of the bottom is approximately constant between sections 1 and 4, the flow is free to contract and expand, there is no appreciable scour of the bed in the constriction and the flow is in the subcritical range.

The expression for computation of backwater upstream from a bridge constricting the flow follows:

$$h_1^* = K^* a_2 \left(\frac{V_{n2}^2}{2g} \right) + a_1 \left(\frac{A_{n2}^2}{A_4} \right) - \frac{A_{n2}^2}{A_1} \left(\frac{V_{n2}^2}{2g} \right) \tag{8-11}$$

where

h_1^* = total backwater (feet).
K^* = total backwater coefficient. K^* is comprised of K_b (the principal loss due to the fact that a bridge opening confines flow) and minor losses that account for piers (K_p), skew (K_s) and eccentricity (K_e).
a_1 & a_2 = as defined in U.S. Department of Transportation (1970).
A_{n2} = gross water area in constriction measured below normal stage (square feet).
V_{n2} = average velocity in constriction or Q/A_{n2} (feet/second)
A_4 = water area at section 4 where normal stage is reestablished (square feet).

A_1 = total water area at section 1, including that produced by the backwater (square feet).

To compute backwater, it is necessary to first obtain the approximate value of $h_1{}^*$ by using the first part of Equation (8-11):

$$h_1{}^* = K^* a_2 \left(\frac{V_{n2}^2}{2g} \right) \qquad (8\text{-}12)$$

The value of A_1 in the second part of Equation 8-11, which depends on $h_1{}^*$, can next be determined and the second term of the expression then evaluated:

$$A_1 \left[\frac{A_{n2}^2}{A_4} - \frac{A_{n2}^2}{A_1} \right] \frac{V_{n2}^2}{2g} \qquad (8\text{-}13)$$

This part of the expression represents the difference in kinetic energy between sections 4 and 1, expressed in terms of the velocity head, $V_{n2}^2/2g$.

The symbol K_b is the backwater coefficient for a bridge in which only the bridge opening ratio, M, is considered. M is defined as the degree of stream constriction involved, expressed as the ratio of the flow that can pass unimpeded through the bridge constriction to the total flow of the river. Referring to Figure 8.33,

$$M = \frac{Q_b}{Q_a + Q_b + Q_c} = \frac{Q_b}{Q_{total}} \qquad (8\text{-}14)$$

K_b may be thought of as a base coefficient and the curves on Figure 8.35 are called base curves. The value of the overall backwater coefficient, K^*, is likewise dependent on the value of M but also affected by:

(a) Number, size, shape, and orientation of piers in the constriction
(b) Eccentricity of asymmetric position of bridge with respect to the channel cross-section
(c) Skew (if bridge crosses stream at other than 90° angle).

K^* consists of a base curve coefficient, K_b, to which is added incremental coefficients to account for the effect of piers, eccentricity and skew. The value of K^* is nevertheless primarily dependent on the degree of constriction of flow at a bridge.

Since the designer will commonly use a computer program like HEC-2 to calculate the water surface profile through bridges, it is recommended that bridge behavior be carefully checked for reasonableness using hand techniques. Plotting the water surface and energy grade line profiles is always instructive.

A good understanding of the watershed enables a sound decision to be made as to debris blockage. Debris blockage may be accounted for by increasing the cross-sectional area of the bridge opening.

XI. APPLICATION OF THE RATIONAL METHOD IN DESIGN

(Note: Check to be sure this is consistent w/ design example)

The Rational Formula, or one of its many variants, has been used for many years for sizing storm drainage systems. For small basins, it continues to be a reasonable method, **provided that it is used properly and that results and design concepts are assessed for reasonableness** (for a detailed discussion on the application/limitations of the method, see Chapter 5).

After the minor system is preliminarily designed and checked for its interaction with the major system, reviews are made of alternates, hydrological assumptions are verified, new computations are made, and final data obtained on street grades and elevations. The engineer then should proceed with final hydraulic design of the system. To establish street carrying capacities and storm sewer requirements at a preliminary design level, a procedure similar to that shown in Figure 8.34 (adapted from Urban Drainage and Flood Control District 1984), is suggested. The procedure is for the average situation; variations may be necessary to fit actual field conditions. It should be noted that procedures like this are particularly suitable for computer spreadsheets.

(a) Column 1—Determine design point location and list. This design point should correspond to the sub-basin illustrated on the preliminary layout map.

(b) Column 2—List basins contributing runoff to this point that have not previously been analyzed.

(c) Column 3—Enter length of flow path between previous design point and design point under consideration.

(d) Column 4—Determine the inlet time for the particular design point. For the first design point of a system, the inlet time will be equal to the time of concentration (t_c). Remember that t_c is the wave travel time, and includes both overland flow time and travel time in a discrete channel. For subsequent design points, inlet time should also be tabulated to determine if it may be of greater magnitude than the accumulated time of concentration from upstream basins. If the inlet time exceeds the time of concentration from the upstream basin, and the area tributary to the inlet is of sufficient magnitude, the inlet time should be substituted for time of concentration and used for this and subsequent basins. In other words, at each design point in the system, the engineer should ascertain whether the total drainage area with a composite t_c or the given individual upstream basin (with a different t_c) produces the higher discharge.

(e) Column 5—Enter the appropriate flow time (wave travel time) between the previous design point and the design point under consideration. The flow time of the street should be used if a significant portion of the flow from the above basin is carried in the street.

(f) Column 6—Pipe flow time should generally be used unless there is significant carry-over from above basins in the street.

Location of Design Point	Basins	Length ft.	Inlet Time min.	Flow Time		Time of Concentration min.	Coefficient "C"	Intensity "I" in./hr.	Acre "A" acre	Direct Runoff cts	Other Runoff cts	Summation Runoff cts	Street		Pipe			Street		Pipe		Remarks
				Street min.	Pipe min.								Slope %	Allowable Capacity cts	Slope %	Size in.	Capacity cts	Design cts	Velocity fps	Design cts	Velocity fps	
1	2	3	4	5	6	7	8	9	10	11	12	13	14	15	16	17	18	19	20	21	22	23

STORM DRAINAGE SYSTEM PRELIMINARY DESIGN DATA

Figure 8.34—Typical form for storm drainage system preliminary design data.

(g) Column 7—The time of concentration is the summation of the previous design point time of concentration and the intervening flow time.

(h) Column 8—Rational Method Runoff Coefficient, "C", for post-development conditions for the basins listed in Column 2, should be determined and listed. The "C" value should be weighted if the basins contain areas with different "C" values.

(i) Column 9—The intensity to be applied to the basins under consideration is obtained from the intensity-duration-frequency curve developed for the specific area under consideration based upon depth-duration-frequency information. The intensity is determined from the time of concentration and the return frequency for this particular design point.

(j) Column 10—The area in acres of the basins listed in Column 2 is tabulated here. Subtract ponding areas which do not contribute to direct runoff such as rooftop and parking lot ponding areas.

(k) Column 11—Direct runoff from the tributary basins listed in Column 2 is calculated and tabulated here by multiplying Columns 8, 9, and 10 together.

(l) Column 12—Runoff from other sources, such as controlled releases from rooftops, parking lots, base flows from groundwater, and any other source are listed here.

(m) Column 13—The total of runoff from the previous design point summation plus the incremental runoff listed in Column 11 and 12 is listed here.

(n) Column 14—The proposed street slope is listed in this column.

(o) Column 15—The allowable capacity for the street is listed in this column. Allowable capacities should be calculated in accordance with procedures set forth in the "Streets" section of this chapter.

(p) Column 16—List the proposed pipe grade.

(q) Column 17—List the required pipe size to convey the quantity of flow necessary in the pipe (round up to the next commercially available pipe size).

(r) Column 18—List the capacity of the pipe flowing full (with the slope expressed in Column 16).

(s) Column 19—Tabulate the quantity of flow to be carried in the street.

(t) Column 20—List the actual velocity of flow for the volume of runoff to be carried in the street.

(u) Column 21—List the quantity of flow determined to be carried in the pipe.

(v) Column 22—Tabulate the actual velocity of flow in the pipe for the design Q.

(w) Column 23—Include any remarks or comments that may affect or explain the design. The allowable quantity of carry-over across street intersections, if any, should often be listed for the minor design storm. When routing the major storm through the system, required elevations for adjacent construction can often be listed in this column.

Inlet design is not specifically accounted for with Figure 8.34, although it can be by designating each inlet a "Design Point" in Column 1 of Figure 8.34. Inlets are designed using local values for tc (flows to individual inlets are from small areas) whereas the procedure in Figure 8.34 applies to an entire storm sewer/street conveyance system, including the total tributary drainage area.

Figure 8.34 results in a preliminary design only. Final design involves computing the hydraulic grade line for the proposed storm sewer system for the design flow, with pipe size adjustments as necessary until the desired hydraulic grade line is obtained.

XII. ECONOMIC CONSIDERATIONS

It is axiomatic in the public works profession that: (1) all works should be built for permanency, and (2) nothing should be built which cannot be maintained. Surveys (Urban Drainage and Flood Control District 1989) have shown that a weighted mean recommended project life was 58 years. Others have recommended service lives for urban storm drainage projects of between 50 and 100 years (see Chapter 14 for further discussion of design life).

The designer of stormwater management facilities focuses on two kinds of economic evaluation:

(a) Cost-effectiveness of alternatives during conceptual and preliminary design.
(b) Preparation of detailed cost estimates for the preferred alternative.

A. ECONOMIC COMPARISON OF ALTERNATIVES

The stormwater master plan for any development typically presents various alternatives from which the preferred alternative is selected. One criterion in the selection process (normally the most important) is the cost of each alternative. Costs for local drainage systems consist of capital, operation and maintenance and replacement expenses, as well as the cost of money, reflected in inflation and interest rates.

The economic evaluation of alternatives must consider frequency of flooding. If the "least cost" alternative subjects property owners to frequent flooding, increases the historic likelihood of flooding, or otherwise fails to accomplish the principal goal of the design effort—protection of lives and property—the alternative should be rejected. The level of protection that each alternative offers is an integral part of the cost of that alternative. When sufficient data are available, the local regulatory review process provides latitude with respect to design frequency selection, and when the engineer's work scope and budget are sufficiently broad, the engineer can calculate the total annual cost of all alternatives being considered.

1. Procedures for Economic Evaluation

One cannot assume that the alternative with the lowest capital cost is the most economically advantageous because flood damage reduction, pollution abatement, and other measurable benefits, along with operation and maintenance and replacement costs, must also be considered. For this reason, the engineer should quantify the present worth or annualized costs of the alternatives studied on the basis of capital, operation and maintenance, and replacement costs. Furthermore, cost studies should consider inflation and interest rates, material service life, and project design life. Finally, on larger projects, the engineer should consider the average annual flood losses (expressed in dollars) for the various alternatives, and adjust costs accordingly. On local projects, however, such flood damages are not normally calculated.

Cost comparisons for alternatives are often conveniently presented in matrices (or on "spreadsheets" if a microcomputer is available). These matrices compare alternative costs with respect to such factors as:

(a) Land requirements, including easements.
(b) Pipe installation requirements and pumping facilities.
(c) Open channel requirements and size of culverts/bridges.
(d) Excavation quantity.
(e) Detention Facility costs, and costs for water quality enhancement measures.
(f) Amount of de-watering required.
(g) Number of replacements of selected pipe materials in a hypothetical planning period (50–100 years).
(h) Annual operation and maintenance expenses.
(i) Special environmental or other permitting costs.
(j) Engineering and other costs for design and construction supervision.
(k) Contingencies.
(l) "Costs of construction," such as traffic delays or service disruption.

Specifically, estimates are prepared for capital, O&M and replacement costs, as follows:

2. Capital Costs

Capital costs typically include:

a. Land.
b. Easements (and related legal/administrative work).
c. Materials acquisition.
d. Site preparation and erosion control.
e. Facility Construction and site restoration.
f. Mapping and surveying.
g. Engineering/design fees.
h. Inspection.
i. Permit acquisition and compensatory measures.

It is not difficult to generate reliable local data for each of these elements, unless the project has unusual features. City and county governments and regional drainage authorities normally keep comprehensive bid price breakdowns covering items c, d, and e. City/county budget and finance offices have historical data on land and easement costs. Finally, local engineering firms can provide information on items f through i. The engineer should use all of these sources to generate price ranges for each capital cost, and then consult with the client to establish "most likely" cost estimates.

3. Operation and Maintenance Costs

Operation and maintenance costs should reflect both scheduled and unscheduled maintenance. For instance, the engineer knows that a grass-lined open channel will require mowing and debris removal a certain number of times per year (scheduled), but it is more difficult to quantify the replacement of riprap removed by exuberant children (unscheduled). Thought should be given to how to best estimate annual fees for the maintenance activities described, based on local considerations. Reliable O&M data are normally available from local governments.

4. Replacement Costs

Replacement costs for culverts, open channel linings, inlets, storm sewers and other elements of urban stormwater management systems can be reliably estimated in most cases. It is prudent to assume a project life of between 50 and 100 years for the various alternatives studied so that replacement factors are appropriately accounted for, recognizing that it is difficult to justify frequent (more than once every 40–50 years) tearing up of streets and embankments to replace deteriorated pipe.

5. Material Service Life

Material service life can be defined as that point in time when one of several things occur, including:

(a) Annual operation and maintenance (O&M) costs reach the amortized annual capital replacement cost. Money would be spent more efficiently by replacing the conveyance rather than continuing to maintain it at increasing levels.
(b) For storm sewers, pipe joint displacement causes debris blockage that may be sufficient to reduce capacity by as much as 20% (or more).
(c) Using a rating system of 0–5 with 0 being the original condition and 5 being a completely deteriorated invert, the service life may be defined as the point at which the condition reaches level 4 (replacement is warranted due to risks posed by excessive rust, small holes or deep pitting in the pipe invert, unusual deformation in plastic piping, or similar problems).

6. Life Cycle Cost Analysis

The different alternatives possible in any urban storm drainage design can have significantly different costs, which will include a number of widely-varying components in addition to the initial cost, including durability or service life, maintenance costs, rehabilitation or replacement costs, inflation, and changes in interest rates. For this reason, it may be advisable to use a technique like Life Cycle Cost (LCC) analysis (James and Lee 1971) when performing economic evaluations of design alternatives, or to at least indicate to the project proponent that costs can be calculated in this way.

The various parameters involved in performing LCC analyses include:

(a) Economic study period, (np).
(b) Conveyance to be evaluated and associated material service life (n).
(c) Initial cost (IC).
(d) Interest rate (i).
(e) Inflation rate (I).
(f) Replacement cost (RE) at end of material service life (n).
(g) Number of material replacements (m) necessary within the economic study.
(h) Residual value, if any, of material at end of economic study period.
(i) Routine annual maintenance costs, if data are available.

The formula for the total life cycle, exclusive of maintenance is:

$$LCC = IC + RE\left[\frac{1+I}{1+i}\right]^{n} + \ldots + RE\left[\frac{1+I}{1+i}\right]^{mn}$$
$$- \left[\frac{(m+1)n - np}{n}\right]RE\left[\frac{1+I}{1+i}\right]^{np} \tag{8-15}$$

IC is the initial construction cost. The second term is the present value of any future replacement costs that will be necessary within the economic study period. The last term is the present worth of any residual value.

The term $(1+I)/(1+i)$, which will generally be a value less than 1.0, is the inflation/interest factor. The rate of interest and rate of inflation are interrelated because of federal monetary and fiscal policies. Life cycle cost analyses are typically performed for interest/inflation differentials from 0–5% to test the sensitivity of the rates on the analysis.

7. Non-quantifiable Factors

One word of caution with respect to the economic comparison among alternatives follows. Factors that cannot readily be quantified in terms of dollars often influence the decision-making process. For instance, although an alternative might not be the most cost effective, it may preserve a historic landmark, valuable wetlands, or unusual wildlife habitat. A dollar value cannot necessarily be assigned to such factors. Nevertheless, local decision-makers and citizens may deem these factors to be of overriding importance, in which case the engineer would adopt other than the least cost alternative. To avoid selection strictly on the basis of cost, the engineer should involve a wide array of interest groups with the analysis and make certain that the client's needs, limitations, desires and regulatory constraints are understood. Local policies and social desires normally do more to influence design (and cost) than any other factors.

B. Engineer's Estimate of Construction Cost

As the engineer moves through the design process, the level of refinement (and accuracy) of the cost estimate steadily improves. For instance, at the conceptual design level, cost estimates can be "rule of thumb" numbers that are based largely on the experience and judgment of the designer, and a 30% contingency may be appropriate. At the preliminary design level, the engineer will have sufficiently refined the design, perused representative bids on other projects, and obtained prices from suppliers and contractors to project estimates that are accurate to within ±20% of the final design estimate. For the final design estimate, the engineer thoroughly quantifies all construction aspects of the project in question including excavation quantities, special structures, length of pipe, probable areas requiring de-watering prior to construction, site mobilization expenses, etc.

It is important for the engineer to provide the client with the best estimate of project costs at each phase of the investigation so that the client can make necessary adjustments. Limitations on the accuracy of construction costs need to be impressed upon the client as reconnaissance level estimates simply may not reflect final design estimates. Cost information can be obtained from catalogs, interviews with suppliers and contractors, bid documents, and regular review of trade journal periodicals.

Caution must be exercised when extrapolating costs from one locale to another. Hydrologic and geotechnical factors change considerably over even short distances, and the designer must assure that cost estimates apply to the specific site in question.

XIII. REFERENCES

"Alternative report for planning of First Creek, Irondale Gulch and DFA 0055 outfall systems." (1988). Denver Urban Drainage and Flood Control District, Denver, CO.

American Iron and Steel Institute. (1983). *Handbook of steel drainage and highway construction products*. AISI, Washington, D.C.

APWA. (1981). "Urban storm water management." *APWA Special Report #49*, American Public Works Association, Chicago, IL.

American Society of Civil Engineers. (1969). "Design and construction of sanitary and storm sewers." *Manuals and Reports on Engineering Practice No. 37, WPCF MOP. No. 9*, ASCE, New York.

American Society of Civil Engineers. (1983). "Existing sewer evaluation & rehabilitation." *Manuals and Reports on Engineering Practice No. 62*, ASCE, New York, NY.

American Society of Civil Engineers. (1985). "Stormwater detention outlet control structures." Report of the Task Committee on the Design of Outlet Control Structures of the Committee on Hydraulic Structures of the Hydraulics Division, ASCE, New York, NY.

American Society of Civil Engineers. (1990). "Urban stormwater quality management." *Continuing Education Course Notes* (Jonathan Jones and Larry Roesner, co-instructors), New York, NY.

Bauler, W.J. and Woo, D.C. (1964). "Hydraulic design of depressed curb opening inlets." *Hydraulic Research Board Record #58*, Federal Highway Administration, Washington, D.C.

"Beebe draw and Bar Lake tributary outfall systems planning study (alternative evaluation study)." (1990). Urban Drainage and Flood Control District, Denver, CO.

Brighton, J.P. et al. (1989). "Economic targeting of nonpoint pollution abatement for fish habitat protection." *Water Resources Research*, 25(12).

Boulder County. (1984). *Drainage criteria manual.* Boulder County, CO.

Chow, V.T. (1959). *Open channel hydraulics.* McGraw-Hill Book Company, Inc., New York.

City of Los Angeles. (1968). "Hydraulic analysis of junction." *Engineering design manual on storm drain design*, Department of Public Works, Storm Drain Design Division, Los Angeles, CA.

City of Norman. (1989). *Storm drainage criteria manual.* Norman, OK.

City of Stillwater. (1979). *Drainage criteria manual.* Public Works Department, Stillwater, OK.

City of Tulsa. (1991). *Storm drainage criteria manual.* Department of Storm Water Management, Tulsa, OK.

Clark, J.W., Viessman, W. and Hammer, M.J. (1977). *Water supply and pollution control.* Harper and Row, New York, NY.

FEMA. (1983). "Questions and answers on the national flood insurance program. *FIA-2*, Federal Emergency Management Agency, Washington, D.C.

Federal Flood Insurance Administration. (1981). *Evaluation of the economic, social and environmental effect of floodplain regulations*, FFIA, Washington, D.C.

Florida Department of Environmental Regulations. (1988). *The Florida Development Manual—A guide to sound land and water management, volumes 1 and 2.* Tallahassee, FL.

French, R.H. (1985). *Open-channel hydraulics.* McGraw-Hill, New York.

Grove, N. (1990). "Greenways—paths to the future," *National Geographic Magazine.* 177(6).

Heed, B.H. (1966). "Design, construction and cost of rock check dams." *U.S. Forest Research Paper (RM-20)*, Rocky Mountain Forest and Range Experiment Station, Fort Collins, CO.

Hendrickson, J.G., Jr. (1957). *Hydraulics of culverts*. American Concrete Pipe Association, Chicago, IL.

Izzard, C.F. (1977). "Simplified method for design of curb opening inlets." *Transportation research record 631: Geometrics, water treatment, utility practices, safety of pertinencies and outdoor advertisement*, Federal Highway Administration, Washington, D.C.

James, L.D. and Lee, R.E. (1971). *Economics of water resources planning*. McGraw-Hill Book Co., New York.

Johns Hopkins University. (1956). *The design of stormwater inlets*. Johns Hopkins University Press, Baltimore, MD.

Jones, D.E. (1967). "Urban hydrology—A redirection." *Civil Engineering*, 37(8).

Jones, D.E. (1980). "Locally important economic considerations in flooding and flood plain management." *Improved hydrologic forecasting—Why and how, proceedings of an Engineering Foundation Conference*, ASCE, New York, NY.

Jones, J.E. and Jones, D.E., Jr. (1987). "Floodway delineation and management." *Journal of water resources planning and management*, 113(2).

King County. (1990). *Surface water design manual*. Department of Public Works, Seattle, WA.

Los Angeles County. (1971). *Hydrology manual*. Los Angeles County Flood Control District, Los Angeles, CA.

Malcom, H.R. (1989). *Elements of stormwater design*. North Carolina State University, Raleigh, NC.

Marsalek, J. (1985). "Head losses at selected sewer manholes." *Special Report No. 52*, American Public Works Association, Chicago, IL.

Marsalek, J. (1986). "Hydraulically efficient junction manholes." *APWA Reporter*, 53(12).

Marsalek, J. (1987). "Improving flow in junction manholes." *Civil Engineering*, 57(1).

Marsalek, J. and Greck, B. (1988). "Head losses at manholes with a 90° bend." *Canadian Journal of Civil Engineering*, 15(5).

McCrory, J.A. and Jones, D.E., Jr. (1976). "Dealing with variable hazard flood." *Journal of Water Resources Planning and Management Division*, 102(WE2).

National Association of Home Builders. (1991). *Manual for developing on difficult sites*. NAHB, Washington, D.C.

Neil, C.R., ed. (1973). *Guide to bridge hydraulics*. University of Toronto Press, Toronto.

Ontario Ministry of the Environment. (1987). *Urban drainage design guidelines*. Toronto.

Portland Cement Association. (1964). *Handbook of concrete pipe hydraulics*. PCA, Chicago, IL.

"Rooney, William P., et al., v. Union Pacific Railroad Company." (1986). (*No. C85-0535*) U.S.D.C. Wyoming.

Rouse, H., ed. (1950). *Engineering hydraulics*. John Wiley and Sons, New York.

St. Johns River Water Management District. (1989). *Applicant's handbook—management and storage of surface waters*. Palatka, FL.

San Diego County. (1985). *Design and procedure manual*. San Diego County Flood Control District, San Diego, CA.

San Diego County. (1990). "Design criteria—standards for inundation of roads and time of concentration." San Diego County, San Diego, CA.

Schueler, T.R. (1987). *Controlling urban runoff: A practical manual for planning and designing urban BMPs*. Metropolitan Washington Council of Governments, Washington D.C.

Sheaffer, J.R. et al. (1982). *Urban storm drainage management*. Marcel Dekker, Inc., New York.

South Florida Water Management District. (1990). "Management and storage of surface waters." *Permit Information Manual Volume IV*, West Palm Beach, FL.

Spirn, A.W. (1984). *The granite garden: Urban nature and human design*. Basic Books, Inc., New York.

"Storm sewer pipe material technical criteria." (1987). Denver Urban Drainage and Flood Control District, Denver, CO.

University of Colorado. (1989). Course notes from the University of Colorado at Denver/Urban Drainage and Flood Control District Short Course on hydraulic model, HEC-2, Denver, CO.

University of Wisconsin. (1990). "Urban stormwater management." *Professional Development Short Course Notes*, Madison, WI.

Urban Drainage and Flood Control District, rev. ed. (1984). *Urban storm drainage criteria manual*. Denver Regional Council of Governments, Denver, CO.

Urban Land Institute. (1975). *Residential storm water management, objectives, principles and design considerations*. Published jointly with the American Society of Civil Engineers and National Association of Home Builders, New York.

Urban Land Institute. (1978). *Residential erosion and sediment control: Objectives, principles and design considerations*. Published jointly with the American Society of Civil Engineers and The National Association of Home Builders, New York.

Urbonas, B. and Roesner, L., eds. (1986). "Urban runoff quality—impact and quality enhancement technology." *Proceedings of an Engineering Foundation Conference, New England College, Henniker, New Hampshire*, ASCE, New York.

Urbonas, B., and Roesner, L. (1991). "Hydrologic design for urban drainage and flood control." *The McGraw-Hill handbook of hydrology*, David Maidment, ed., McGraw-Hill, New York.

U.S. Bureau of Reclamation. (1974). *Design of small canal structures*. USBR Water Resources Technical Publication, Washington, D.C.

U.S. Department of Agriculture. (1962). *National engineering handbook*. Soil Conservation Service (SCS), Washington, D.C.

U.S. Department of Agriculture. (1973). *National engineering handbook, section 3, sedimentation*. Soil Conservation Service, Washington, D.C.

U.S. Department of Commerce. (1965a). "Hydraulic charts for the selection of highway culverts." *Hydraulic Engineering Circular No. 5 (HEC-5)*, Bureau of Public Roads, U.S. Government Printing Office, Washington, D.C.

U.S. Department of Commerce. (1965b). "Capacity Charts for the Hydraulic Design of Highway Culverts." *Hydraulic Engineering Circular No. 10 (HEC-10)*, Bureau of Public Roads, U.S. Government Printing Office, Washington, D.C.

U.S. Department of the Interior. (1977). *Design of small dams*. Bureau of Reclamation, Government Printing Office, Washington, D.C.

U.S. Department of Transportation. (1970). *Hydraulics of bridge waterways*. Federal Highway Administration, Hydraulic Design Series No. 1, Washington, D.C.

U.S. Department of Transportation. (1972). "Hydraulic design of improved inlets for culverts." *Hydraulic Engineering Circular (HEC-13)*, Federal Highway Administration, U.S. Government Printing Office, Washington, D.C.

U.S. Department of Transportation. (1978). "Hydraulic and safety characteristics of selected grate inlets on continuous grades." *Bicycle Safe Grate Inlets Study*, Vol. 2, FHWA-RD-78-4, Federal Highway Administration, Washington, D.C.

U.S. Department of Transportation. (1979). *"Design of urban highway drainage— The state of the art."* FHWA-TS-79-225, Federal Highway Administration, Washington, D.C.

U.S. Department of Transportation. (1980). *Hydrology for transportation engineers*. Federal Highway Administration, Washington, D.C.

U.S. Department of Transportation. (1985). "Hydraulic design of highway culverts." *Hydraulic Design Series No. 5 (HDS-5)*, Federal Highway Administration, National Technical Information Service, Springfield, VA.

U.S. Soil Conservation Service. (1954). "Handbook of channel design for soil and water conservation." *Publication No. SCS-TP-61*, U.S. Government Printing Office, Washington, D.C.

U.S. Soil Conservation Service. (1977). "Design of open channels." *Technical Release No. 25*, U.S.C.S., Washington, D.C.

U.S. Soil Conservation Service, (1983). "Computer program for project formulation—hydrology." *SCS Technical Release 20 (TR-20)*, Washington, D.C.

Water Pollution Control Federation. "Combined sewer overflow pollution abatement." (1989). WPCF, *Manual of Practice FD-17*, Alexandria, VA.

Whipple, W. et al. (1983). *Stormwater management in urbanizing areas*. Prentice-Hall, Inc., Englewood Cliffs, NJ.

Williams, P.B. (1990). "Rethinking flood-control channel design." *Civil Engineering*, 60(1).

CHAPTER 9

SPECIAL STRUCTURES AND APPURTENANCES

I. INTRODUCTION

Stormwater conveyance systems require a variety of structures and appurtenances to control, divert and redirect flows, and to control velocities to minimize erosion and scour. This chapter provides some general principles, and presents basic guidelines for design of these structures. References are included for design procedures beyond the scope of this manual.

There are many hydraulic situations that fall outside the range of design parameters for which the structures in this chapter are intended. In such cases, a hydraulic model study may be necessary. Model studies will give the designer a much higher degree of certainty that his structure will perform as intended, particularly when conventional designs have been extended beyond their intended limits. Typical structures that are modeled include spillways, outlet-works, energy dissipators, stilling basins, drop structures, canal structures, river channels, fish ladders, and boat chutes. Details on hydraulic modeling can be found in Chapter 7 and in several references (U.S. Department of the Interior 1980; Davis and Sorensen 1969).

II. EROSION AND SCOUR

Erosion and local scour can result in channel degradation, in undermining and structural failure, or in loss of channel bed materials and damage to channel linings. Excessive suspended sediment in streams may result in undesirable environmental impacts, aesthetic problems, and burdensome maintenance costs.

Most unlined natural or man-made channels are affected by tractive forces of flowing waters and are subject to erosion. Channels are subject to intense local erosion or scour at obstructions, sudden changes in channel cross-section, drops, regions of changes in channel bed materials, and other similar conditions. The design of channels, conduits and any other structure that results in changes in flow regime should consider the following factors and provide measures for channel and outlet protection.

The main factors that provide favorable conditions for erosion and scour in a channel are high flow velocities, particularly at shallow depths, and soft and/or fine bed materials. Velocities are higher in steep channels, at changes in channel configuration, in smooth channels, and at higher discharges. Soil type largely determines the erosion potential of bed materials.

A. Determination of Scour Potential

1. Maximum Permissible Flow Velocities—Unlined Channel

Table 9.1 shows the maximum permissible velocities (those which do not cause scour) and the corresponding unit tractive force values published by Fortier and Scobey (1926) and converted by the U.S. Bureau of Reclamation (1952). This table was developed for straight and mature channels with relatively mild slopes and flow depths of less than 3 feet.

Permissible flow velocities were estimated for various cohesive and non-cohesive soils as shown in Figure 9.1. The correction factors for variations in flow depths that must be applied to values in Figure 9.1 are presented in Figure 9.2. Both figures are based on Hydrotechnical Construction (1936) and Chow (1959).

If newly designed channel or hydraulic structures, or additional discharges diverted to a drainage system, cause velocities that exceed the maximum permissible, design modifications or scour control structures are required.

2. Retardance and Permissible Velocities—Grassed Channels

Manning's roughness coefficient ("n") is also called the retardance coefficient. Various types of grasses have different retardance coefficients because of their density and length (see Table 9.2). There is a relationship between the retardance coefficient and the product of the mean flow velocity (V) and the hydraulic radius (R) (Chow 1964). This relationship is characteristic of vegetation and is practically independent of channel slope and shape. The "n" vs. VR relationship therefore is very useful in the design of vegetated channels which considers the various types of grasses that can be used in a particular climatic zone (see Figure 9.3).

The permissible velocity is that which will not result in significant erosion in a grassed channel for the design runoff event. Permissible velocities for different vegetative covers, channel slopes and soils conditions are shown in Table 9.3.

TABLE 9.1.   Maximum Permisible Velocities Recommended by Fortier
and Scobey and the Corresponding Unit-tractive-force
Values Converted by the U.S. Bureau of Reclamation*
(For straight channels of small slope, after aging).

Material	n	Clear Water		Water Transporting Colloidal Silts	
		V, fps	τ_0, lb/ft^2	V, fps	τ_0, lb/ft^2
Fine sand, colloidal	0.020	1.50	0.027	2.50	0.075
Sand loam, noncolloidal	0.020	1.75	0.037	2.50	0.075
Silt loam, noncolloidal	0.020	2.00	0.048	3.00	0.11
Alluvial silts, noncolloidal	0.020	2.00	0.048	3.50	0.15
Ordinary firm loam	0.020	2.50	0.075	3.50	0.15
Volcanic ash	0.020	2.50	0.075	3.50	0.15
Stiff clay, very colloidal	0.025	3.75	0.26	5.00	0.46
Alluvial silts, colloidal	0.025	3.75	0.26	5.00	0.46
Shales and hardpans	0.025	6.00	0.67	6.00	0.67
Fine gravel	0.020	2.50	0.075	5.00	0.32
Graded loam to cobbles when noncolloidal	0.030	3.75	0.38	5.00	0.66
Graded silts to cobbles when colloidal	0.030	4.00	0.43	5.50	0.80
Coarse gravel, noncolloidal	0.025	4.00	0.30	6.00	0.67
Cobbles and shingles	0.035	5.00	0.91	5.50	1.10

*The Fortier and Scobey values were recommended for use in 1926 by the Special
Committee on Irrigation Research of the American Society of Civil Engineers.

3. Tractive Force

The tractive force is defined as a shear force or drag force. This force
is applied on the submerged portion of the channel bed and side slopes.
This force acts in the direction of flow. The unit tractive force τ, or the
average value of the tractive force per unit wetted area, can be expressed
(Chow 1959) as:

$$\tau = wys \qquad (9\text{-}1)$$

where:

w = unit weight of water
y = flow depth
S = slope

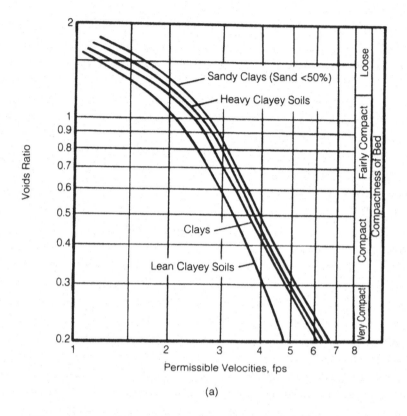

(a)

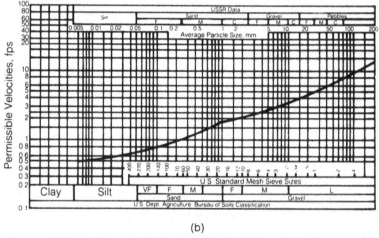

(b)

Figure 9.1—Curves showing U.S. and U.S.S.R. data on permissible velocities for noncohesive soils; and U.S.S.R. data on permissible velocities for cohesive soils.

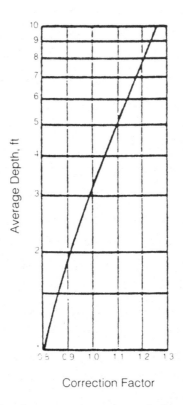

Correction Factor

Figure 9.2—Curves showing corrections of permissible velocity for depth for both cohesive and noncohesive materials.

The unit tractive force in channels varies between the channel bottom and the side slopes. The maximum tractive force on the channel bottom is approximately wyS, and on the side slope is approximately $0.76\ wyS$. Figure 9.4 shows the maximum unit tractive forces on the channel bottom and side slopes (b = channel width). The maximum permissible unit tractive forces in non-cohesive materials were developed by the U.S. Bureau of Reclamation and are shown in Figure 9.5. The permissible unit tractive forces for channels in cohesive materials are presented in Figure 9.6 (Hydrotechnical Construction 1936). For further discussion of tractive force see French (1985).

B. Channel Side Slopes

From the standpoint of channel stability the minimum allowable side slopes of channels depend mainly on the soil type (As discussed in Chapter 8, many factors influence channel side slope selection including maintenance, safety, recreational use of the channel, land availability, etc.). The more cohesive the soil, the steeper the channel slope, and

TABLE 9.2.   Permissible Velocities for Channels Lined with Grass

Cover	Slope Range, %	Permissible Velocity, fps	
		Erosion-Resistant Soils	Easily Eroded Soils
Bermuda grass	0–5	8	6
	5–10	7	5
	>10	6	4
Buffalo grass, Kentucky bluegrass, smooth brome, blue grama	0–5	7	5
	5–10	6	4
	>10	5	3
Grass mixture	0–5	5	4
	5–10	4	3
	Do not use on slopes steeper than 10%		
Lespedeza sericea, weeping love grass, ischaemum (yellow bluestem), kudzu, alfalfa, crabgrass	0–5	3.5	2.5
	Do not use on slopes steeper than 5%, except for side slopes in a combination channel		
Annuals–used on mild slopes or as temporary protection until permanent covers are established, common lespedeza, Sudan grass	0–5	3.5	2.5
	Use on slopes steeper than 5% is not recommended		

Remarks.   The values apply to average, uniform stands of each type of cover. Use velocities exceeding 5 fps only where good covers and proper maintenance can be obtained.
(U.S. Soil Conservation Service 1954.)

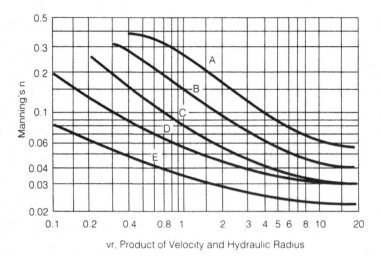

Figure 9.3—Relation between Manning's roughness coefficient and the product of velocity and hydraulic radius.

TABLE 9.3.   Guide to Selection of Vegetal Retardance.

Grass Density (1)	Average Length of Grass (inches) (2)	Degree of Retardance (3)
Good	30	A Very high
	11–24	B High
	6–10	C Moderate
	2–6	D Low
	2	E Very low
Fair	30	B High
	11–24	C Moderate
	6–10	D Low
	2–6	D Low
	2	E Very low

(U.S.S.C.S. SCS-TP-61 1954)

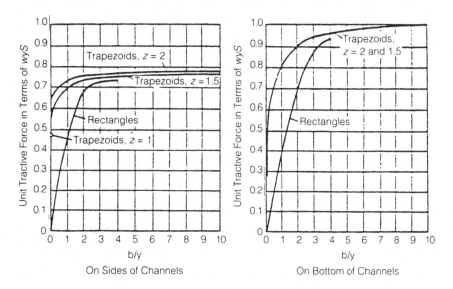

Figure 9.4—Maximum unit tractive forces in terms of wyS.

vice versa. Table 9.4 illustrates this concept. However, when the channel is designed in erodible materials the channel side slopes should be analyzed with the maximum permissible velocity and the principle of tractive force in mind.

C. Local Scour

Local scour occurs in non-uniform flow regions where pressure forces, lift forces, and shear forces fluctuate. For example, local scour around

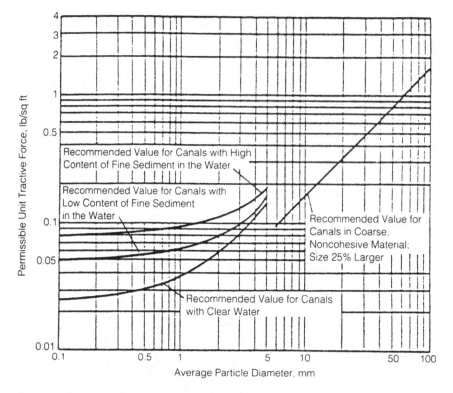

Figure 9.5—Recommended permissible unit tractive forces for canals in noncohesive material.

bridge piers is caused by the vortex resulting from water piling up on the upstream edge and subsequent acceleration of flow around the nose of the pier.

Local scour is a function of a combination of several of the following factors:

(a) Slope of the channel.
(b) Characteristics of bed materials.
(c) Characteristics of the flood hydrograph.
(d) Characteristics of man-made hydraulic structures.
(e) Direction of the flow in relation to its depth.
(f) Characteristics of transported materials.
(g) Accumulation of ice and drift.

The numerous scouring factors in combination with various channel conditions and structures such as embankments, bridge piers and walls, have resulted in the development of numerous empirical relationships, based on laboratory observations, each of which applies to specific conditions. Even though the most critical factors are the flow velocity

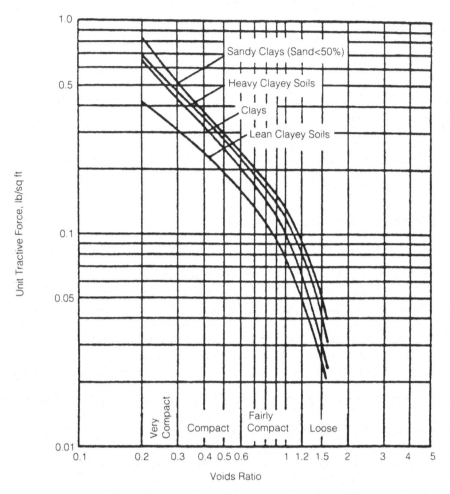

Figure 9.6—Permissible unit tractive forces for canals in cohesive material as converted from U.S.S.R. data on permissible velocities.

and mean size and specific gravity of the sediment mixture, a comprehensive analysis should be performed for each case study using all the relevant equations and applying judgment. Most of these equations are available in ASCE (1975) and Simons and Senturk (1977). Equations and procedures for estimating scour at pipes and culvert outlets have also been published by the U.S. Federal Highway Administration (1983), and are very useful for assessing the potential scour for drainage systems including culverts and channels. These procedures are used for estimating the scour geometry based on coefficients dependent upon desired parameter such as length, width, depth, or volume of scour.

TABLE 9.4. Suitable Side Slopes for Channels (Based strictly on channel stability considerations).

Material (1)	Side Slope (horizontal to vertical) (2)
Rock	Nearly Vertical
Muck and peat soils	1/2:1
Stiff clay or earth with concrete lining	1/2:1 to 1:1
Earth with stone lining, or earth for large channel	1:1
Firm clay or earth for small ditches	1-1/2:1
Loose sandy earth	2:1
Sandy loam or porous clay	3:1

(Chow 1959).

D. Structural Measures for Channel Protection

Channel protection must be provided to suit the local physical and scour characteristics. Channel vegetation is perhaps the most simple erosion and scour control measure. Where the flow velocities exceed the velocities at which the vegetation is effective, other erosion protection measures or structures such as energy dissipators, drops, and drop shafts should be considered.

III. EROSION PROTECTION MEASURES FOR CHANNELS

A. Definitions, Categories, and Applications

Erosion protection is required for channel linings in reaches where the maximum permissible flow velocities or critical tractive forces are exceeded under the design discharge conditions.

1. Classification of Erosion Protection Measures

There are a variety of natural and man-made materials available to the engineer for erosion protection. According to "Bank and Shore Protection in California Highway Practice" (State of California 1970), bank protection systems generally can be classified as armor protection, retard protection, retaining walls, groins, and baffles. Many of these systems also can be used for bed protection. Listed below are four of the more common classifications, including examples of bank protection systems for each.

(a) Armor and Lining Protection—Armor protection is a protective surface placed to resist erosive forces. Popular armor protection includes riprap, grouted riprap, gabions, concrete, concrete rubble, sacked concrete,

shot-crete, asphalt, stone masonry, soil-cement, precast concrete revetments, and other precast materials. Lining protection can also consist
of man-made and natural materials formed into mats. Several of the
materials used in mat construction include concrete, plastic, geotextiles,
woody plants and wood excelsior, jute, coconut, straw, and other plant
fibers.

(b) Retard Protection—Retards are designed to reduce velocities and promote siltation near the toe of a bank to decrease bank erosion. Retard
devices can be permeable or impermeable and include steel or concrete
tetrahedrons, timber or steel jacks, fences, rock filler fences, piling, pile
bents, and woody plants such as willows.

(c) Retaining Walls—Retaining walls are near-vertical structures supporting embankments or vertical side slopes in urban areas where the right-
of-way is limited. Timber, concrete, masonry, and steel are all used to
construct retaining walls. Retaining walls can be constructed on footings, as piling, as crib walls, and in combinations.

2. Flexible and Rigid Erosion Protection

The U.S. Department of Transportation (1988) has noted that the
primary difference between rigid and flexible channel linings from an
erosion control standpoint is their response to changing channel shape.
Flexible linings are able to conform to change in channel shape while
rigid linings can not. The result is that flexible linings can sustain some
change in channel shape while maintaining the overall integrity of the
channel lining. Rigid linings tend to fail when a portion of the lining
is damaged. Damage to a lining is often from secondary forces such as
frost heave or slumping. Rigid linings can be disrupted by these forces
whereas flexible linings, if properly designed, will retain their erosion-
control capabilities.

Flexible linings also have several other advantages compared to rigid
linings. They are generally less expensive, permit infiltration/exfiltra-
tion, and have a natural appearance. Hydraulically, flow conditions in
channels with flexible lining generally conform to those found in natural
channels, and thus provide better habitat opportunities for local flora
and fauna. In some cases, flexible linings may provide only temporary
protection against erosion while allowing vegetation to be established.
The vegetation will then provide permanent erosion control in the
channel. The presence of vegetation in a channel can also provide a
buffering effect for runoff contaminants.

Flexible linings have the disadvantage of being limited in the magnitude of erosive force they can sustain without damage to either the
channel or the lining. Because of this limitation, the channel geometry
(both in cross-section and profile) required for channel stability may
not fit within the acquired right-of-way. A rigid channel can provide a
much higher capacity and in some cases may be the only alternative."

3. Temporary and Permanent Erosion Protection

Erosion protection measures can also be classified as temporary or permanent. Temporary linings provide erosion protection until vegetation is established or for temporary projects usually associated with construction. In most cases the temporary protection will deteriorate over the period of one growing season which means that successful revegetation is essential. Temporary erosion protection measures include straw mats, curled wood mats, jute, paper, or synthetic nets, synthetic mats, and fiberglass roving (U.S. Department of Transportation 1988).

4. Vegetative Erosion Protection

Vegetative protection is suitable where uniform flow exists and shear stresses are moderate. Vegetative channel linings are not suited to sustained flow conditions or long periods of submergence. Vegetative channels with sustained low flow and intermittent high flows are often designed with a composite lining, including a riprap or concrete low-flow section (U.S. Department of Transportation 1988).

When vegetation establishes itself in a ditch line or drainage channel, then that channel is capable of handling drainage flow velocities far in excess of that handled by the soil lining alone. It is common practice to size drainage channels for vegetative linings where the two-year or ten-year rainfall event produces a design velocity of 5 fps or less and a Froude number less than 0.8. In some cases, they are analyzed first on their ability to carry a two-year storm as a soil-lined channel without vegetation, assuming that the grass will establish itself within the first two-year period after installation. After the two-year period, the ditch is assumed to be lined with a good vegetative cover and is analyzed for a ten-year or larger storm to determine its ability to handle erosive forces and the design flows under those conditions.

5. Flow Duration

Flow duration is a significant aspect of channel protection design. Design hydrographs should, therefore, be studied prior to design, and discharges and corresponding durations should be considered before designing a channel's erosion protection. For instance, it may not be cost-effective to riprap above a certain stage in a grass-lined channel that will only experience velocities in excess of those permissible for very short periods of time. Minor damage and the risk that goes with it may be acceptable to the owner in such cases.

6. Permissible Shear Stress

The permissible shear stress, p, indicates the force required to initiate movement of the protective lining. Prior to movement of the lining, the underlying soil is relatively protected. Therefore, permissible shear stress is not significantly affected by the erodibility of the underlying soil. However, if the lining is eroded and moved, the bed material is exposed to the erosive force of the flow. The consequence of lining failure on highly erodible soils is great, since the erosion rate after failure is high compared to soils of low erodibility (U.S. Department of Transportation 1988).

Values of permissible shear stress for linings are based on research conducted at laboratory facilities and in the field. The values presented here are judged to be conservative and appropriate for design use. Table 9.5 gives permissible shear stress values for manufactured, vegetative, and riprap lining types.

B. Channel Bank and Bed Protection

1. Riprap

The most common and relatively inexpensive lining in most areas is riprap. Riprap is a facing or protective layer of stones randomly placed to prevent erosion, scour, or sloughing of a structure or embankment. Riprap used for slope protection includes dumped stone, hand-placed stone, wire-enclosed stone (gabion), grouted riprap, concrete riprap in bags and concrete-slab riprap. Dumped and hand-placed riprap and

TABLE 9.5. Permissible Shear Stresses for Lining Materials (U.S.D.O.T. 1967)

Lining Category (1)	Lining Type (2)	Permissible Unit Shear Stress	
		(lb/ft²) (3)	(Kg/m²) (4)
Temporary	Woven paper net	0.15	0.73
	Jute net	0.4	2.20
	Fiberglass roving:		
	Single	0.60	2.93
	Double	0.85	4.15
	Straw with net	1.45	7.08
	Curled wood net	1.55	7.57
	Synthetic mat	2.00	9.76
Vegetative	Class A	3.70	18.06
(Grass, with degree of	Class B	2.10	10.25
retardance)	Class C	1.00	4.88
	Class D	0.60	2.93
	Class E	0.35	1.71
Gravel Riprap	1 inch	0.33	1.63
	2 inch	0.66	3.25
Rock Riprap	6 inch	2.00	9.76
	12 inch	4.00	19.53

gabions are the most common for drainage and stormwater management projects because of their relatively low initial costs and ease of replacement. While the strength of a riprap wall is somewhat limited, such a channel lining has several advantages over more impermeable linings. Riprap's porous nature protects a channel from uplift or flotation concerns, since the groundwater under pressure is able to enter the channel. In addition, riprap has the ability to conform to slight irregularities created by erosion and scour, and thus minimize additional damage.

Negative aspects of riprap linings include the difficulty of maintenance, the collection of weeds and debris, and the susceptibility to theft and vandalism in some urban areas. The economics of riprap are highly dependent upon the distance the rock must be hauled from its source to the project site.

(a) Hydraulic Considerations. Stream flow velocities, discharge, flow depth and channel bed and side slopes are the main hydraulic parameters considered in riprap design. Since riprap is subject to flow velocities and tractive forces in the channel, the size of the riprap can be estimated using tractive force criteria or design flow velocity criteria. However, in practice, the design of riprap in channels uses empirical relationships developed by various researchers and adopted by public agencies. The empirical approach, based on observations that have resulted in a great variety of charts, suggests relationships of the mean diameter or riprap to hydraulic parameters. Several references (U.S. Department of Transportation 1988; Dickinson 1968; ASCE 1975) provide comparisons of the various methods.

(b) Design Criteria. Generally, riprap may be applied to channel slopes of less than 10%. Riprap can be used on steeper slopes, but the cost may be prohibitive. Riprap selection must consider the design runoff, stone size and location, riprap gradation, thickness of riprap lining, side slopes, filter requirements and quality of stone. The design criteria are discussed below:

1. Design Runoff—The riprap is designed for the design discharge in the channel. However, it is recommended (National Academy of Science 1970) that a minimum design return interval of 25 years be used for riprap design.

2. Stone Size and Location—As indicated above there are a variety of procedures for riprap design based on empirical relationships. The latest methodology (National Academy of Science 1970) used by various states, and also recommended by the U.S. Environmental Protection Agency (1976), provides a means for estimating a design stone size (median stone diameter-d_{50}), such that the stone is stable under the design flow conditions. The procedures developed for this methodology are based on the tractive force method. Figure 9.7 shows the P/R (wetted perimeter/hydraulic radius) versus b/d (channel bottom width/flow depth) relationship (NAS 1970). The estimated P/R value is then used in Figure 9.8 to estimate the median stone diameter for a trapezoidal channel. Figure 9.9 is used for estimating the median stone diameter (d_{50}) for triangular channels.

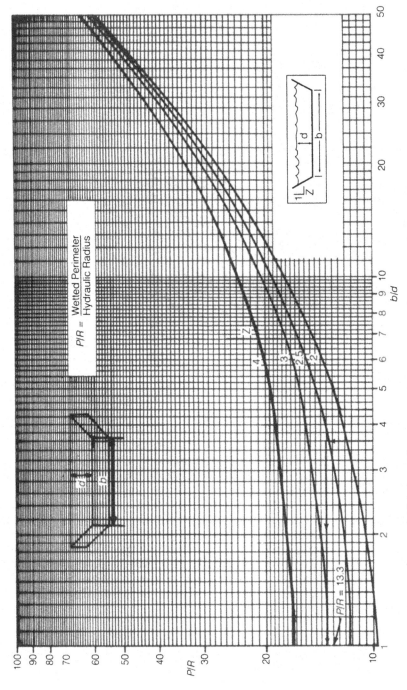

Figure 9.7—P/R for trapezoidal channels.

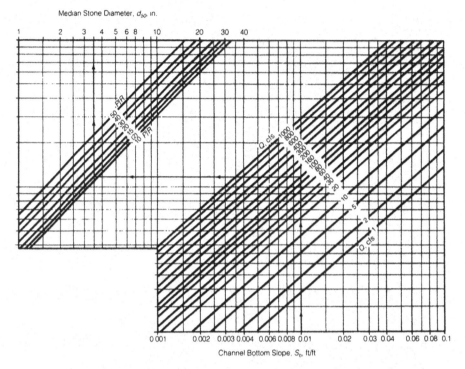

Figure 9.8—Median riprap diameter for straight trapezoidal channels.

Since the erosive forces of flowing water are greater in bends than in straight channels, larger riprap must be used at the bend. Figure 9.10 can be used for increasing the d_{50} for the bend condition. It is recommended that the riprap size used in a bend extend upstream from the point of curvature and downstream from the point of tangency a distance equal to five times the channel bottom width and extend across the bottom and up both sides of the channel. As an absolute minimum, the riprap should extend to the level of the design runoff. Freeboard is recommended, however.

In channels where no riprap is required on the bottom and where the bottom consists of unconsolidated materials, the toe of the bank riprap should extend at least eight times the maximum stone size below the channel bottom. A schematic drawing for streambank riprap is shown in Figure 9.11.

3. Riprap Gradation—The riprap should be well graded and 50% of the mixture by weight should be larger than the d_{50}. The diameter of the largest stone should be 1.5 times the d_{50}. The minimum size of stone is that which is just stable under the

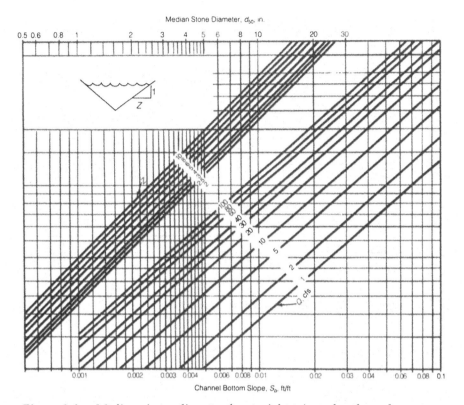

Median Stone Diameter, d_{50}, in.

Channel Bottom Slope, S_b, ft/ft

Figure 9.9—Median riprap diameter for straight triangular channels.

design flow condition. The U.S. Department of Transportation (1967) provides riprap gradation that can be used as a guide for selection of the minimum size stone.

4. Thickness of Riprap Lining—Various parameters such as discharge, size of channel, size and gradation of riprap and construction techniques should be considered when estimating the thickness of riprap lining. The following minimum criteria should be met:

    (a) A thickness of at least three times the d_{50} if a filter layer is not used. (A filter is recommended in nearly all cases, however.)

    (b) A thickness of at least two times the d_{50} if a filter layer is used.

The minimum riprap thickness to be used is 1.0 feet.

5. Side Slopes—It is recommended that the maximum design side slopes do not exceed 2:1. Figure 9.12 shows the maxi-

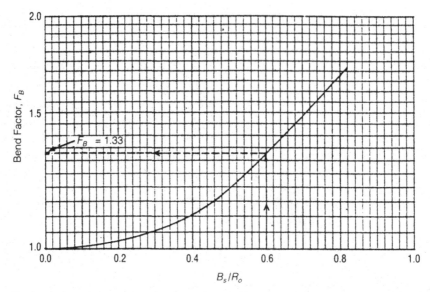

Figure 9.10—*Riprap size correction factor for flow in channel bends.*

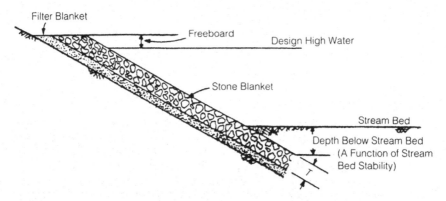

Figure 9.11—*Stream bank riprap.*

mum side slopes with respect to riprap size and stone geometry. Crushed rock is more stable than rounded stone on
steeper slopes. Likewise, larger stones are more stable on
steeper slopes.

6. Filter—Fine soil materials underneath the riprap are subject
to erosion and piping. On steep slopes, highly erodible soils,
loose sand or at high flow velocities, a filter should be used
or riprap thickness increased beyond the minimum referenced above.

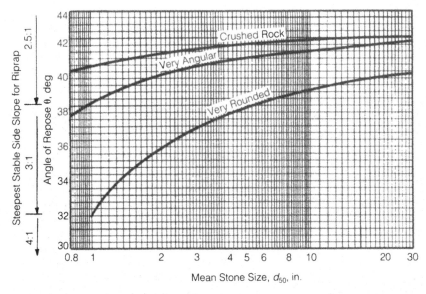

Figure 9.12—Maximum riprap side slope with respect to riprap size.

The filter consists of either graded granular soil material that blends with the local soil or synthetic filter fabric. Criteria for detail of filters are included in a National Academy of Science (1970) publication.

Many riprap failures are caused by a lack of a filter or by use of an improperly designed filter.

7. Quality—Stone for riprap should consist of field or quarry stone of approximately rectangular shape. The stone must be hard and angular and able to withstand damage due to weathering. The specific gravity of the stone must be at least 2.5. Further discussion of riprap quality is provided in Chapter 13.

8. Installation. The proper installation of riprap and filter is critical and its importance is often overlooked. During installation, careful monitoring of riprap and filter size, gradation and thickness and assuring that the specifications are met is important. Further discussions of riprap installation is provided in Chapter 16.

2. Gabions

Gabions consist of multi-celled rectangular wire mesh boxes filled with stones. Gabions are provided folded flat and bundled together for ease of shipment and handling. Each gabion is placed in position and wired to adjoining gabions, filled with rock at one third increments of height and wire cross-braced for strength and stability. After filling, the top is folded and wired shut. They have been used extensively in Europe on drainage projects of all sizes, and their use is gaining acceptance in the United States. They can be used at locations where the only rock economically available is too small for rip-rap, or where steeper side slopes are required. Another advantage is the versatility that results from the regular geometric shapes of wire-enclosed rock. The rectangular blocks and mats can be fashioned into almost any shape that can be formed with concrete.

The durability of wire-enclosed rock is generally limited by the service life of the galvanized binding wire which, under normal conditions, is considered to be about 15 years. Water carrying silt, sand, or gravel can reduce the service life of the wire, and water that rolls, or otherwise moves cobbles and large stones, breaks the wire with a hammer-and-anvil action and considerably shortens the life of the wire. The wire has been found to be susceptible to corrosion by various chemical agents and is particularly affected by high sulfate soils. If corrosive agents are known to be in the water or soil, a plastic coated wire should be specified.

The following criteria should be met when gabions are designed:

(a) The design runoff should be at least the same as for riprap, if not higher, to offset the problems with gabions noted above.

(b) The design flow velocity should not exceed the criteria indicated in Table 9.6. Table 9.6 values were taken from the New Jersey State Soil Conservation Committee (1987), and are considered minimum thicknesses permissible for the stated velocities.

(c) The gabions should not be exposed to abrasion from larger debris and/or bedload transported by the flow. The abrasion can damage the wire mesh and lead to failure.

(d) Some believe that plastic coated gabions can be used to discourage corrosion at the wire mesh, although the longevity of the plastic is uncertain.

(e) All gabions should be underlain by filter fabric or a gravel filter. Limitations on the applicability of filter fabric are provided in Table 9.7. If bottom slopes are steeper or velocities greater than those shown in Table 9.7, the gabions should be underlain with a properly designed gravel filter.

TABLE 9.6.   Gabion Thickness and Maximum Velocity.

Gabion Thickness (ft.) (1)	Maximum Velocity (ft./sec.) (2)
1/2	6
3/4	11
1	14

TABLE 9.7.   Maximum Allowable Channel Bottom Slopes Using Geo-textile Fabrics (New Jersey State Soil Conservation Committee 1987)

Soil Texture (1)	Maximum Permissible Velocity (fps) (2)	Allowable Invert Slope (ft/ft) (3)
Sandy loam	2.5	0.029
Silt loam	3.0	0.041
Sandy clay loam	3.5	0.056
Clay loam	4.0	0.074
Clay, fine gravel	5.0	0.115
Cobbles	5.5	0.139

(f) Rock used to fill the gabion should be angular, block-shaped rock. Minimum rock sizes are generally 3" to 4" or ⅓ the basket depth. The maximum stone size should not exceed ⅔ the basket depth or 12", whichever is smaller.

Gabions are not maintenance-free and must be periodically inspected to determine whether the wire is sound. If breaks are found while they are still relatively small, they may be patched by weaving new strands of wire into the wire cage. Wire-enclosed rock installations need to be inspected at least once a year under ideal circumstances and may require inspection every three months in vandalism prone areas (in conjunction with a regular maintenance program). Mattresses on sloping surfaces must be securely anchored to the surface of the soil.

Where aggregate is readily available at reasonable cost, gabions can be a very economical solution to channel lining. A major factor is the cost of labor, since much of the placement of the stone, and the work in general, is very labor-intensive.

Gabions are very useful for small localized drainage problems where immediate lining solutions are needed, and which must be accomplished by local and inexperienced workmen. By installing a grout over the finished mattress installation, the effect of a paved ditch can also be gained, with possible savings in initial construction costs.

3. Man-made Protection Materials

In addition to riprap and gabions, several man-made erosion protection products are available. They may have advantages over riprap and gabions in specific situations, although the performance of the products is not always well documented. The designer is, therefore, advised to proceed with caution when specifying these products. It is suggested that trial installations or other projects using these materials be evaluated.

The advantages of these materials may include consistent quality and dimensions, ease of installation, ease of underwater installation, desirable aesthetics, ease of maintenance, ease of quantity estimation, and low cost. Disadvantages may include difficulty in estimating product life, high maintenance and/or replacement costs, ease of removal by vandals, and comparatively high cost.

Maintenance of man-made products is often easier than that of riprap because they present an even surface to drive or walk over while mowing or conducting other maintenance activities. They should also be considered in locations where rodent infestation could be a problem in riprap voids.

A number of currently available products are described below. Case-by-case evaluation of each, coupled with performance evaluation and regular maintenance, is essential.

(a) Concrete Filled Fabric Mats—The mats are constructed of heavy-duty fabric filled with concrete. The flexible double-walled fabric framework is pressure injected with fine grain concrete. The fabric forms are available in a number of sizes, with or without filter points, and in thicknesses ranging from 3–12 inches. The mats can be installed with low headroom, underwater, and with relative speed. Several of the mats are available with reinforcing cords which provide tensile strength.

(b) Fabric Soil Stabilization Mats—The mats are a three-dimensional product made from heavy fabric monofilaments fused at their intersections. The bulky open construction allows for backfilling with soil, gravel, or other appropriate materials. Vegetation is then planted within the mat which acts as permanent turf reinforcement. The mats are generally available in thicknesses up to 0.75".

(c) Precast Concrete Revetments—The revetments are available in various shapes and sizes from a number of suppliers. Many are designed to allow vegetation to grow up in void areas. Several have interlocking features or are secured with mats or cords which reduce vandalism and potential removal by high velocity flows.

(d) Fabric Grids—Fabric grids are manufactured as a three-dimensional, semi-rigid matrix. The honeycomb-type designs are made in sizes up to 4 inches deep, with hexagon shapes having side dimensions of up to 8 inches. After the grids are unrolled on a surface, they are filled with native soils, gravel, cobble, or mixtures of soil and rock. In submerged applications, gravel or cobble is preferred, while the use of native soils that can support vegetation is an option above normal water surface elevations. Rocks can be mixed with the soil for increased erosion protection. The fabric grid adds strength and erosion resistance to the material placed in it.

The grids are anchored to the slope to protect against movement. According to one manufacturer, the grids are inexpensive, easy to install, permeable, lightweight, and rot-proof.

4. Common Channel Protection Measures Other Than Riprap and Gabions

(a) Bagged Concrete—In this technique, bags filled three-quarters full with concrete are laid in close contact, with staggered joints and tied ends turned in. The consistency of the concrete must be as stiff as satisfactory discharge from the mixer and the bagging process permit. Bagged concrete may be used when all the following conditions are met:
  1. The design runoff, riprap size and location, and filter criteria are met.
  2. The weight of the filled bags is at least equal to the weight of the maximum stone size required for rock riprap.
  3. Settlement or lateral movement of foundation soils is not anticipated.
  4. Ice conditions are not severe.
  5. Slopes should be 2:1 or flatter. However slightly steeper slopes may be permitted under special circumstances.

(b) Soil-Cement—Soil cement is a concrete product formed by the mixing of on-site soils with portland cement in a "pug-mill" erected at the project site. The resulting mixture is a low-slump soil cement concrete with a low compressive strength, frequently between 600 and 900 pounds per square inch. A typical mixture contains about 6–12% cement and 8–12% moisture by weight. Soil-cement requires a less costly aggregate than concrete because more fines are acceptable. Soil-cement is normally placed and compacted in 6–8-inch lifts in a stair-stepped fashion (or on a single level on gentle slopes). Its primary advantages are low cost, durability, and low permeability. Soil cement may be used along with wire mattresses and other industrial fabrics to provide a stable ditch lining. Factors which tend to erode soil cement linings include rapidly changing water depths as well as freezing and thawing conditions.

Soil cement has been employed in the construction of levees, channel bank protection, and drop structures, and merits consideration as a substitute for riprap protection in areas where rock is not economically available, or where flow velocities would be too high for riprap linings.

(c) Soil-Bioengineering—Soil-bioengineering is a bank protection method which involves the use of live woody plant material such as willows. The plant materials are used in various systems and configurations to meet specific project requirements. The native plant material is designed to root and grow into structures including live crib walls, live stakes, joint planting, live soft gabions, brush-layering, brush-mattresses, and live fascines. The plant material can be used by itself or in conjunction with other bank protection materials. These systems provide the most "natural" form of bank protection and may offer low initial costs, but do require specialized maintenance. The level of maintenance is dependent on the type of system, the local climate, and desired appearance.

Reference materials that are useful in preparing site-specific designs are limited. A number of publications are available from the U.S. Army Corps of Engineers, U.S. Fish and Wildlife Service, U.S. Forest Service, and the U.S. Environmental Protection Agency. District Soil Conservation Service offices are another source. Private publications from practicing professional hydrologists and wetland biologists are a good source of applied information. Specialists trained in bioengineering have recently begun working in the field of urban storm drainage, and their projects and publications will also be of interest.

(d) Retaining Walls. Various types of traditional retaining walls are used in special situations where the higher cost is warranted. Types of walls include crib walls, sheet pile walls, timber pile walls, concrete walls, and boulder walls. Special precautions such as filter layers are advised to guard against loss of native soils through the walls.

(e) Concrete. Concrete is by far the most widely used channel lining where grass or soil are not sufficient, especially when high Froude numbers and/or velocities indicate the need for a very reliable structure. Although more expensive than other alternatives, concrete has distinct advantages such as low maintenance, ease of installation, excellent durability, and resistance to high velocities. Architectural concrete finishes are available to improve appearance.

Most conditions allow for poured in-place concrete, although precast panels have been successfully used in channel lining applications. Concrete can be used to line all channel shapes from rectangular to trapezoidal and "V" ditch sections. A typical section of structural concrete lining may be a trapezoidal section using 6" thicknesses of concrete with welded wire fabric reinforcement laid over stone or some other porous material. Expansion joints must be provided at routine intervals as well as control joints for crack control. Weep holes are routinely installed at 6–10 foot intervals allowing the release of water which has accumulated around the outside of the concrete surface. This reduces the effects of hydrostatic uplift as well as the effects of freezing and thawing which can cause deterioration of the lining. Low strength concrete (around 2500 psi) is often used for these purposes. The finish is a rough broom finish. While side slopes as steep as 1:1 are possible, they are normally 2:1 or flatter.

A concrete channel lining provides for a reduced resistance to flow (a Manning's "n" smaller than most other lining materials), which can allow the channel to carry significantly more flow and allow for flatter slopes. An important engineering consideration is the matter of safety, since such channels are often specified in urban or urban transition areas that allow direct access to the drainage ways by children and others. The characteristic high velocities and turbulence created by drainage flow in these channels creates a dangerous situation which must be taken into account by using fences or other barricade-type devices.

(f) Roller compacted concrete (RCC) is similar to soil cement except that a mixture of sand and coarse aggregate replaces the on-site soils. RCC is placed in the same manner as soil cement and has become popular for use in massive concrete structures such as dams. The use of RCC eliminates or greatly reduces the need for concrete forming.

(g) Bituminous Mixtures. Bituminous materials offer an economical and easily installed method of lining channels. Unfortunately, there are many problems with the use of this material that cannot be overlooked by the engineer. Oils in the lining material tend to be leached as the material degenerates, causing environmental problems. Bituminous materials, being flexible pavement, are generally kept "alive" by use. This quality makes bituminous materials excellent for road surfacing, but when they are used in a ditch without such repetitive load conditions, bituminous materials have a tendency to degenerate. In areas subject to freezing and thawing, the bituminous surface should be kept sealed to prevent the degenerative action of water and ice. Vegetation may also root through cracks causing crumbling of the surface. Bituminous paving of channels for the above reasons can only be expected to produce a serviceable channel lining for a limited life span. Where channel aesthetics are not a factor and if a limited life span is desired, bituminous linings should be considered. This solution then represents an easily installed low cost lining of a temporary nature.

(h) Corrugated Steel Sheet Channel Liner. Corrugated steel sheets may offer an economical means of lining open channels or flumes. Corrugated steel sheets can be fabricated in many sizes and shapes with a wide range of protective coatings designed to resist soil and water conditions. Fittings include tees, wyes, laterals and elbows fabricated from the same material. The open curved sheets are made up on standard lengths of 25–1/2" which are lapped, resulting in usable lengths of even 2' multiples. The sections are connected at the lap by field bolting through pre-punched holes. Flanges are provided at the longitudinal edges for anchorage.

C. Riprap Protection at Outlets

If the flow velocity at a conduit outlet exceeds the maximum permissible velocity for the local soil or channel lining, channel protection is required. This protection usually consists of an erosion resistant reach, such as riprap, between the outlet and the stable downstream channel to provide a stable reach at the outlet in which the exit velocity is reduced to a velocity allowable in the downstream channel. The design of such protection is normally based on a 25-year design runoff event.

If protection is needed at the outlet, a horizontal (zero slope) apron must be provided.

1. Apron Dimensions

(a) The length of an apron (L_a) is determined using the following empirical relationships that were developed for the U. S. Environmental Protection Agency (1976):

$$L_a = \frac{1.8Q}{D_0^{3/2}} + 7D_0, \text{ for } TW < \frac{D_0}{2} \qquad (9\text{-}2a)$$

and

$$L_a = \frac{3Q}{D_0^{3/2}} + 7D_0, \text{ for } TW > \frac{D_0}{2} \tag{9-3a}$$

where:

D_o = maximum inside culvert width (ft)
Q = pipe discharge (cfs)
TW = tailwater depth (ft)

In metric units, these equations become:

$$L_a = \frac{3.26Q}{D_0^{3/2}} + 7D_0 \tag{9-2b}$$

and

$$L_a = \frac{5.44Q}{D_0^{3/2}} + 7D_0 \tag{9-3b}$$

respectively.

(b) Where there is no well defined channel downstream of the apron, the width, W, of the outlet and of the apron (as shown in Figure 9.13) should be as follows:

$$W = 3D_0 + 0.4L_a, \text{ for } TW \geq \frac{D_0}{2} \tag{9-4}$$

and

$$W = 3D_0 + L_a, \text{ for } TW < \frac{D_0}{2} \tag{9-5}$$

The width of the apron at the culvert outlet should be at least 3 times the culvert width.

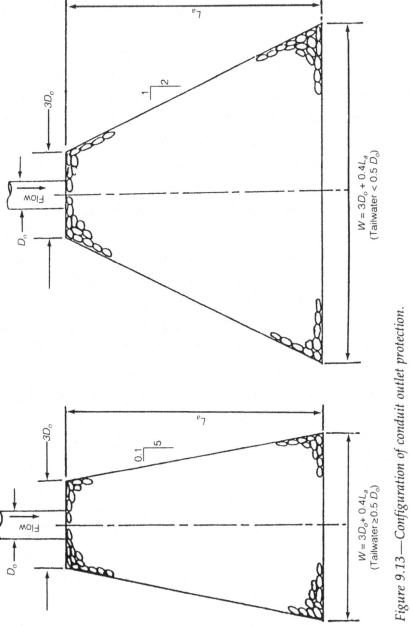

Figure 9.13—Configuration of conduit outlet protection.

(c) Where there is a well-defined channel downstream of the apron, the bottom width of the apron should be at least equal to the bottom width of the channel and the lining should extend at least one foot above the tailwater elevation and at least two-thirds of the vertical conduit dimension above the invert.
(d) The side slopes should be 2:1 or flatter.
(e) The bottom grade should be level.
(f) There should be overfall at the end of the apron or culvert.

2. Apron Materials

(a) The median stone diameter, d_{50} is determined from the following equation:

$$d_{50} = \frac{0.02\ (Q)^{4/3}}{TW\ (D_0)}$$ (9-6a)

and, in metric units:

$$d_{50} = \frac{0.066\ (Q)^{4/3}}{TW\ (D_0)}$$ (9-6b)

(b) Existing scour holes may be used where flat aprons are impractical. Figure (9.14) shows a general design of a scour hole. The stone diameter is determined using the following equations:

$$d_{50} = \frac{0.0125\ (Q)^{4/3}}{TW\ (D_0)},\ \text{for}\ Y = \frac{D_0}{2}$$ (9-7a)

and, in metric units:

$$d_{50} = \frac{0.041\ (Q)^{4/3}}{TW\ (D_0)}$$ (9-7b)

Also,

$$d_{50} = \frac{0.0082\ (Q)^{4/3}}{TW\ (D_0)},\ \text{for}\ Y = D_0$$ (9-8a)

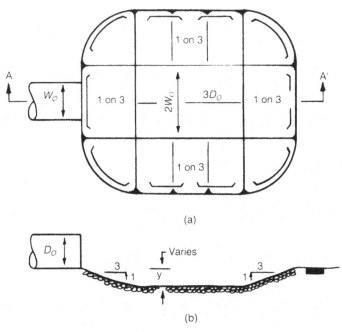

(a)

(b)

Figure 9.14—Preformed scour hole.

and, in metric units:

$$d_{50} = \frac{0.027\ (Q)^{4/3}}{TW\ (D_0)} \qquad (9\text{-}8b)$$

where Y = depth of scour hole below culvert invert

   (c) The other riprap or gabion requirements are as indicated in the previous
       sections for channel lining.
   (d) Aprons constructed of man-made materials are often a viable alterna-
       tive. Refer to the above discussion of man-made materials for design
       consideration.

IV. CHECK DAMS

A. Categories, Definitions, and Applications

   Check dams are commonly used for energy dissipation and/or sediment control in streams. They are commonly constructed using gravel

or rock, gravel overlaying a sand core, timber, or gabions. Many check dams are constructed of materials available on-site or at a nearby quarry. One of the most important design considerations for check dams is the hazard to downstream properties posed by dam failure, and the designer cannot dismiss the problem as insignificant simply because "the dam is low" or "the volume of water stored is small."

Check dams constructed of earth fill are shown in Figure 9.15. These check dams are constructed of rock and soil materials which could be clean excavation spoil. The core of some of these structures contains finer size particles enclosed by coarse gravel, cobbles, and small boulders. Depending on the design, the check dam may be porous and act entirely as a filter, contain an outlet structure, or operate as a filter and overflow structure in combination. Both weirs and drop inlets may be used. Where large storage is anticipated in the impoundment created

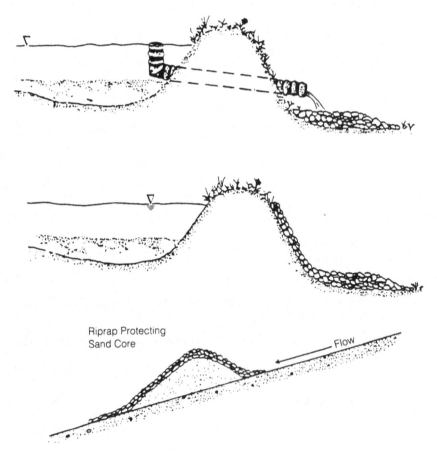

Figure 9.15—Alternative earthfill check dam designs.

by earth fill check dams, procedures delineated in U.S. Bureau of Reclamation (1977) and Chapter 11 should be followed.

Check dams can change the flow regime from supercritical to subcritical when placed at appropriate intervals along the channel. The
intervals are determined by dividing the steep channel into reaches in
which backwater conditions are created by the dams.

B. Porous Check Dams

Porous check dams are primarily used for sediment trapping, however, they can also be used for energy dissipation in steeply sloped
channels. Check dams designed for sediment control can be used temporarily during construction, or as a permanent measure in watersheds
that are subject to significant erosion and channel degradation. Check
dams can be used as filters and impoundment structures that trap the
sediments, or as a means of promoting aggradation in the channel by
dissipating the flow energy and settling sediments.

Rock check dams may have large void spaces, therefore, the use of
rocks in combination with smaller stones such as gravel or with filter
cloth anchored to the check dam surface increases sediment trap efficiency. Gabions can also be used in check dams as shown in Figure
9.16. Gabions may be used strictly as porous dams, or as dam/weir
combinations.

C. Impervious Check Dams

Impervious check dams are primarily used for energy dissipation and
water level control upstream of the dam. Erosion and scour can be
controlled in channels with steep slopes, and the degradation process
replaced by aggradation. As shown in Figure 9.17, a series of check
dams can be used to change the flow regime from supercritical to
subcritical while providing backwater ponding and quiescent settling
conditions.

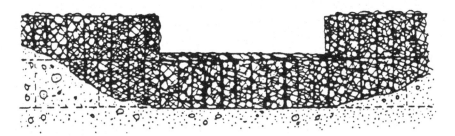

Figure 9.16—Gabion check dam.

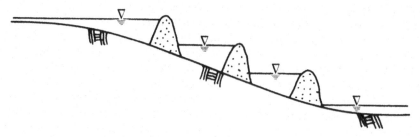

Figure 9.17—A series of check dams.

D. Check Dam Design Considerations

Porous check dams are designed to maximize sediment trap efficiency. The efficiency depends on the relationship between the size distribution of the sediment inflow and the voids in the dam. A sediment check dam is usually designed to provide quiescent flow and settling conditions by creating backwater detention ponds. A porous check dam design methodology developed by Hirschi (undated) is based on backwater surface profiles and Stokes' Law. This methodology considers the following factors:

(a) Inflow sediment size distribution.
(b) Design flow.
(c) Check dam porosity.
(d) Channel slope.
(e) Manning's "n."
(f) Channel configurations.

The design analysis may include the following steps:

(a) Determination of the backwater curve.
(b) Designation of the quiescent settling zone.
(c) Calculation of the lateral distance available for settling.
(d) Estimation of flow velocity in the stream.
(e) Estimation of travel time.
(f) Estimation of flow depths.
(g) Estimation of trap efficiency.

Impervious check dam designs generally are based on the same considerations, except for the steps related to sediment trap efficiency, and require the following additional design elements.

(a) Minimum tailwater at the dam.
(b) Overflow weir level that provides minimum tailwater and results in the desired backwater surface profile.
(c) Flow velocity that is below maximum permissible velocity in the channel.

Maintenance at check dams normally includes dredging and off-site disposal. The design should estimate the frequency of dredging required before assuming that check dams are feasible.

V. ENERGY DISSIPATORS

A. Definitions, Applications, and Categories

Energy dissipators are required in the immediate vicinity of hydraulic structures where high impact loads, erosive forces and severe scour are expected. Said another way, they are usually required where the flow regime changes from supercritical to subcritical or where the flow is supercritical and the tractive force or flow velocities are higher than the maximum allowable values. The basic hydraulic parameter that identifies the flow regime, and that is used in connection with energy dissipators in general, and with hydraulic jump dissipators in particular, is the Froude number. A Froude number equal to one indicates a critical flow condition, while numbers greater or less than one indicate supercritical or subcritical flow, respectively. The Froude number (F) is estimated using the following equation (Rouse 1949):

$$F = \frac{v}{\sqrt{gy}} \qquad (9\text{-}9)$$

Where:

v = flow velocity
g = acceleration due to gravity
y = hydraulic mean depth (cross-section area/top width)

The Froude number is a product of the flow velocity and the wave celerity. In rectangular channels and some large rivers, the equation may be rewritten in the following form:

$$F^2 = \frac{1}{B}\left(\frac{Q^2}{gy^3}\right) \qquad (9\text{-}10)$$

Where:

Q = discharge
B = width of channel

Energy dissipation structures act as transitions which reduce high flow velocities that may exist under a range of flows. The energy dis-

Denver, Colorado—Innovative energy dissipator in an open channel.

sipators localize hydraulic jumps and act as stilling basins. The use of energy dissipators is very common downstream of hydraulic structures where common channel protection cannot be used alone because of potential damage. If riprap or other protection is used for energy dissipation, it should be confined in a basin and secured in place with grout or mesh.

Energy dissipators range from simple horizontal concrete aprons and hydraulic jump basins to wave suppressors and flip buckets for large streams and structures as described in Peterka (1958) and McLaughlin Water Engineers (1986). Because of their special hydraulic characteristics, drop structures are discussed later in this chapter. The primary difference between energy dissipators in general and drop structures is that the energy dissipators are used to reduce high velocities at critical locations by hydraulic jumps, while drop structures are vertical structures used for controlling velocities in channel reaches by reducing channel slopes. Because of various appurtenances such as sills, baffles, weirs, and because of variations in the geometry of the stilling basins, a wide variety of energy dissipators is used, and a number of them are discussed below.

B. Riprap Basins for Small Culvert Outlets

The most common energy dissipators used for stormwater management are riprap basins (Figure 9.18). Their advantages include simplicity, low cost, and wide application.

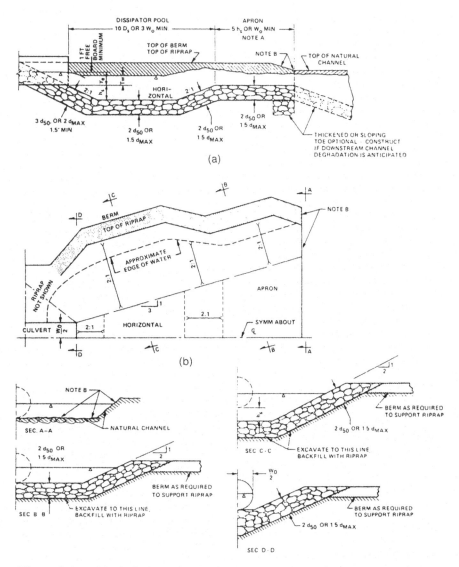

Figure 9.18—Typical riprap basin.

The riprap placed in the basin must be inspected and repaired, if necessary, after major storms. The median stone diameter can be estimated based on the exit velocity of the pipe or culvert as shown in McLaughlin Water Engineers (1986) and the AASHTO Drainage Handbook (1987). The length of the basin is estimated based on the width or diameter of the conduit. The depth of the basin is based on the median stone diameter.

C. Stilling Basins

If a hydraulic jump is used for energy dissipation, it should be confined to a heavily-armored channel reach, the bottom of which is protected by a solid surface such as concrete to resist scouring. Since the cost of concrete structures is relatively high, the length of the hydraulic jump is usually controlled by accessories that not only stabilize the jump action and increase the factor of safety, but reduce the cost of the structure.

1. Design Considerations

There are several considerations that should be included in designing hydraulic jumps and stilling basins (Chow 1959; U.S. D.O.T., 1983):

(a) Jump Position—There are three positions or alternative patterns that allow a hydraulic jump to form downstream of the transition in the channel. These positions are controlled by tailwater.
(b) Tailwater Conditions—Tailwater fluctuations due to changes in discharge complicate the design procedure. They should be taken into account by classification of tailwater conditions using tailwater and hydraulic jump rating curves (see also Chapter 6).
(c) Jump Types—U.S. Bureau of Reclamation (1955) and Bradley and Peterka (1957) identify various types of hydraulic jumps that may occur. These are summarized in Figure 9.19. Oscillating jumps in a Froude number range of 2.5 to 4.5 are best avoided, or specially designed wave suppressors may be used to reduce wave impact.

The greater the Froude number the higher the effect of tailwater on the jump. Therefore for a Froude number as low as 8 the tailwater depth should be greater than the sequent depth downstream of the jump so that the jump will stay on the apron. When the Froude number is greater than 10, the common stilling basin dissipator may not be as cost-effective as a special bucket type dissipator (see Peterka 1958).

2. Control of Jumps

Jumps can be controlled by several types of appurtenances such as sills, chute blocks, and baffle piers. The purpose of a sill located at the end of a stilling basin is to induce jump formation and to control its position under most probable operating conditions. Sharp crested or broad crested weirs can be used to stabilize and control the jump.

Chute blocks are used at the entrance to the stilling basin. Their function is to furrow the incoming jet and lift a portion of it from the floor, producing a shorter length of jump than would occur without them.

Baffle piers are blocks placed in intermediate positions across the basin floor for dissipating energy mostly by direct impact action. They are useful for small structures with low flow velocities. High flow velocities may result in cavitation action on the piers and basin floor downstream.

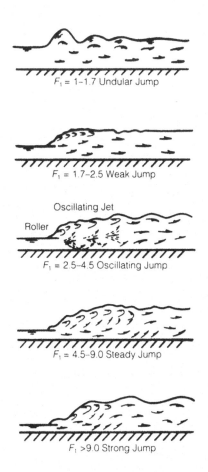

$F_1 = 1\text{--}1.7$ Undular Jump

$F_1 = 1.7\text{--}2.5$ Weak Jump

Oscillating Jet

Roller

$F_1 = 2.5\text{--}4.5$ Oscillating Jump

$F_1 = 4.5\text{--}9.0$ Steady Jump

$F_1 > 9.0$ Strong Jump

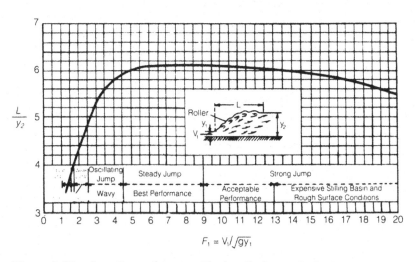

Figure 9.19—Lengths and types of hydraulic jumps in horizontal channels.

363

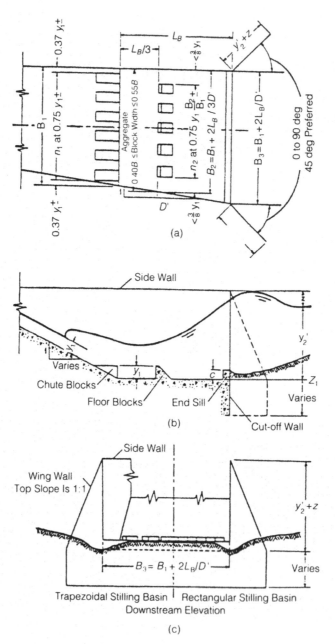

Figure 9.20—Proportions of the SAF basin.

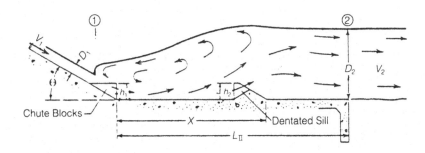

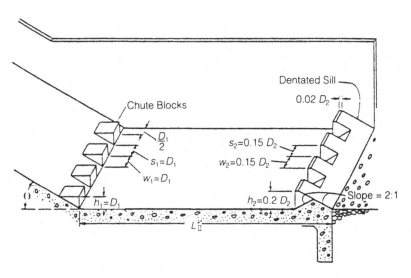

Figure 9.21—Proportions of the USBR basin II.

3. Stilling Basin Categories

The following three major categories of basins are used for a range of hydraulic conditions. Design details can be found in the AASHTO Drainage Handbook (1987), Chow (1959), and U.S. D.O.T. (1983).

(a) The SAF ("Saint Anthony Falls") Stilling Basin (Chow 1959)—This basin, shown in Figure 9.20, is recommended for use on small structures such as spillways and outlet works where the Froude number varies between 1.7 and 17. The appurtenances used for this dissipator can reduce the length of the basin by approximately 80%. This design has great potential in stormwater management because of its applicability to small structures.

Stilling Basin III developed by the U.S. Bureau of Reclamation (USBR) is similar to the SAF basin, but it has a higher factor of safety.

(b) The USBR Stilling Basin II—This basin, shown in Figure 9.21, is recommended for use with jumps with Froude numbers greater than 4.5 at large spillways and canals. This basin may reduce the length of the jump by a third. This basin is used for high-dam and earth-dam spillways. Appurtenances used in this basin include chute blocks at the upstream end of the basin and a dentated sill at the downstream end. No baffle piers are used in this basin because of the cavitation potential.

(c) The USBR Stilling Basin IV—This basin (Figure 9.22) is used where jumps are imperfect or where oscillating waves occur with Froude numbers between 2.5–4.5. This design reduces excessive waves by eliminating the wave at its source through deflection of directional jets using chute blocks. When a horizontal stilling basin is constructed without appurtenances, the length of the basin is made equal to the length of the jump.

D. Simple Energy-Dissipating Headwalls

Another simple type of energy dissipator that can be used at culvert outlets are energy dissipating headwalls. Three typical such headwalls are shown in Figures 9.23 through 9.25.

E. Design Criteria and Practices

Most of the design criteria for stilling basin dissipators are included in the previous paragraphs. Table 9.8 provides a summary of selected parameters, and may be used for preliminary identification of alternative energy dissipators that may used. Because of the great variety and combination of types of energy dissipators and appurtenances, the designer should review available references in sufficient detail to arrive at a design that is suited to his specific field conditions.

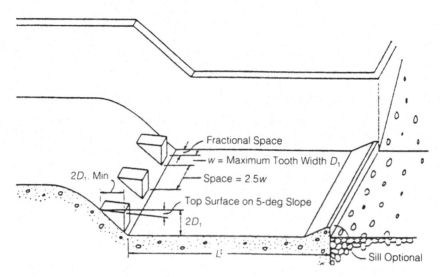

Figure 9.22—Proportions of the USBR basin IV.

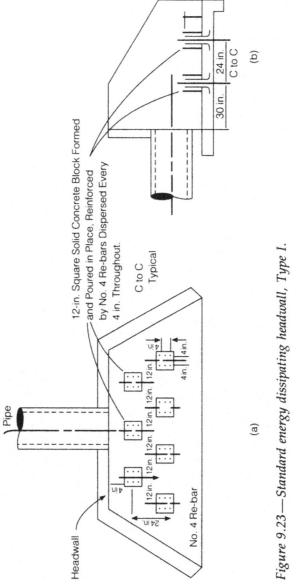

Figure 9.23 — Standard energy dissipating headwall, Type 1.

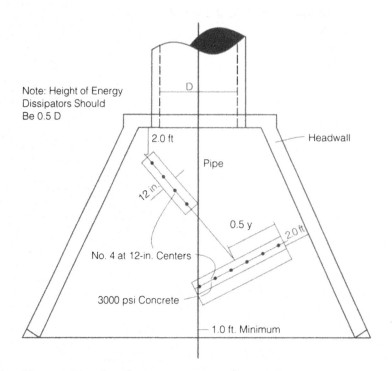

Note: Height of Energy
Dissipators Should
Be 0.5 D

D

Headwall

2.0 ft

Pipe

12 in.

0.5 y

2.0 ft

No. 4 at 12-in. Centers

3000 psi Concrete

1.0 ft. Minimum

Figure 9.24—Standard energy dissipating headwall, Type II.

Drop/grade control structures sculptured to fit the urban park setting.

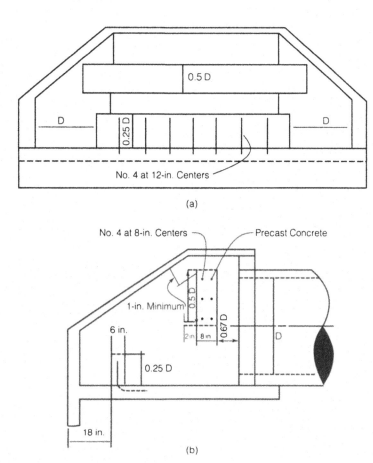

Figure 9.25—Standard energy dissipating headwall, Type III.

VI. DROP STRUCTURES

A. Definitions, Categories, and Applications

Vertical drop structures are controlled transitions for energy dissipation in steep channels where riprap or other energy dissipation structures are not as cost effective. Drop structures used for drainage and storm water management can be categorized primarily as either open channel transitions (drop spillways) or transitions between open channel and closed conduits (drop shafts).

Drop structures are usually constructed of concrete because of the forces involved, however, it is possible to utilize riprap stilling basins or gabions where physical, economic and other conditions permit.

TABLE 9.8.   Dissipator Criteria (U.S. DOT 1983)

Dissipator Type	Froude Number Fr	Allowable Debris			Tailwater TW	Special Consideration
		Silt Sand	Boulders	Floating		
Free Hydraulic Jump	>1	H	H	H	Required	
CSU Rigid Boundary	<3	M	L	M	—	$4 < S_o < 25$
Tumbling Flow	>1	M	L	L	—	Check Outlet
Increased Resistance	—	M	L	L	—	Control HW
USBR Type II	4 to 14	M	L	M	Required	
USBR Type III	4.5 to 17	M	L	M	Required	
USBR Type IV	2.5 to 4.5	M	L	M	Required	
SAF	1.7 to 17	M	L	M	Required	
Contra Costa	<3	H	M	M	$<0.5D$	
Hook	1.8 to 3	H	M	M	—	
USBR Type VI	—	M	L	L	Desirable	$Q < 400$ cfs, $V < 50$ fps
Forest Service	—	M	L	L	Desirable	$D < 36$ inch
Drop Structure	<1	H	L	M	Required	Drop < 15 ft.
Manifold	—	M	N	N	Desirable	Note:
Corps Stilling Well	—	M	L	N	Desirable	N = none
						L = low
Riprap	<3	H	H	H	—	M = moderate
						H = heavy

B. Open Channel Drops (Drop Spillways)

Drop structures in open channels change the channel slope from steep to mild by combining a series of gentle slopes and vertical drops. Flow velocities are reduced to non-erosive velocities, while the kinetic energy or flow velocity gained by the water as it drops over the crest of each spillway is dissipated by an apron or stilling basin.

Open channel drop structures generally require aerated nappes and subcritical flow conditions at both the upstream and downstream section of the drop. The stilling basin can vary from a simple concrete apron to baffle blocks or sills as described previously.

Figure 9.26 shows the flow geometry and important variables at a vertical (straight) drop structure. The flow geometry at such drops can be described by the drop number (D) which is defined (Chow 1959) as:

$$D = \frac{q^2}{gh^3} \tag{9-11}$$

Where:

q = discharge per unit width of crest overfall, cfs/ft
g = acceleration of gravity, ft/sec^2
h = height of the drop, ft

The drop functions are:

$$\frac{L_d}{h} = 4.30D^{0.27} \tag{9-12}$$

$$\frac{Y_p}{h} = 1.00D^{0.22} \tag{9-13}$$

$$\frac{Y_1}{h} = 0.54D^{0.425} \tag{9-14}$$

$$\frac{Y_2}{h} = 1.66D^{0.27} \tag{9-15}$$

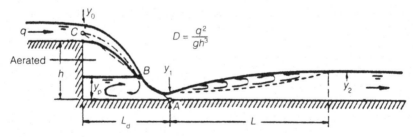

Figure 9.26—Flow geometry of a straight drop spillway.

Where:

L_d = drop length
Y_p = pool depth under the nappe
Y_1 = the depth of the toe of nappe
Y_2 = tailwater depth sequent to Y_1

For a given drop height, h, and discharge, q, the sequent depth Y_2 and the drop length L_d can be estimated by Equations 9-2 and 9-5, respectively. The length of the jump can be estimated by techniques discussed in Section V. If the tailwater is less than Y_2, the hydraulic jump will recede downstream. Conversely, if the tailwater is greater than Y_2, the jump will be submerged. If tailwater is equal to Y_2 no supercritical flow exists on the apron and the distance L_d is minimum.

When the tailwater depth is less than Y_2 it is necessary (according to U.S. Department of Transportation 1983) to provide either 1) an apron at the bed level and a sill or baffles, or 2) an apron below the downstream bed level and an end sill.

The choice of design type and dimensions depends on the unit discharge (q), drop height (h), and tailwater depth. The design should take into consideration the geometry of the undisturbed flow. If the spillway (overflow crest) length is less than the width of the approach channel, the approach channel must be designed properly to reduce the effect of the end contractions to avoid scour.

The two most common vertical open channel drops are the straight drop structure and the box inlet drop structure.

1. Straight Drop Structure

Figure 9.27 shows a layout of typical straight drop structure and hydraulic design criteria developed by the U.S. Soil Conservation Service. McLaughlin Water Engineers (1983) provides specific criteria and reviews design considerations related to hydraulic, geotechnical, and structural design of drop structures.

2. Box Inlet Drop Structure

The box inlet drop structure is a rectangular box open at the top and downstream end as shown in Figure 9.28 (U.S. D.O.T. 1983). Water is directed to the crest of the box inlet by earth dikes and a headwall. Flow enters over the upstream end and two sides. The long crest of the box inlet permits large flows to pass at relatively low heads. The width of the structure should not be greater than the downstream channel. Box inlet drop structures are applicable to drops from 2–12 feet.

Design data and criteria for these structures based on U.S. Soil Conservation Service and St. Anthony Falls Hydraulic Laboratory are available (U.S. Department of Transportation 1983; Blaisdell and Donnelly

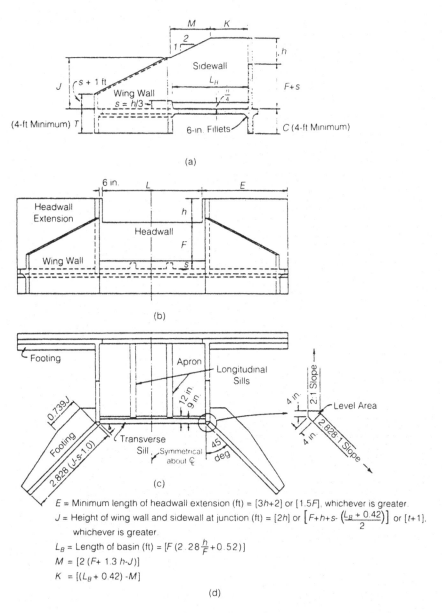

Figure 9.27—Typical drop spillway and some hydraulic design criteria.

1956). The parameters to consider for the hydraulic design of the drop are:

(a) Section (length) of the crease of the box inlet
(b) Opening the headwalls

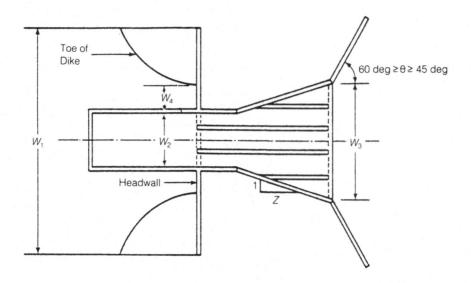

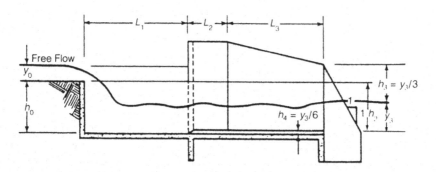

Figure 9.28—Box inlet drop structure.

    (c) Discharge, discharge coefficients and flow regime changes
    (d) Box inlet length and depth
    (e) Minimum length and width of stilling basin

C. Drop Shaft Structures

    Storm sewer drop shafts are commonly used to dispose of storm runoff in deep large pipes or tunnels. They consist of an inlet structure, the drop shaft, vertical bend, horizontal conduit with or without a de-aeration chamber, outlet structure and auxiliary structures. Schematics

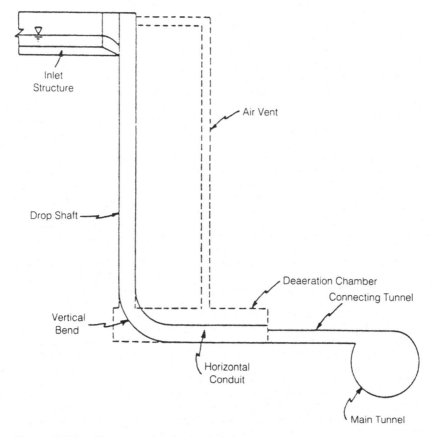

Figure 9.29—Components of a typical drop structure.

of most of these components are shown in Figure 9.29 (Jain and Kennedy 1983). The drop shaft structure design can vary tremendously.

According to Bushey (1952), there are three methods of dropping flows down a deep vertical shaft (Figure 9.30).

1. Completely Flooded Drop Shafts

This type of structure has an inlet tank and a vertical shaft and has the hydraulic grade line above the top of the shaft. Under this condition, pressures throughout the structure are hydrostatic and the water does not fall freely. Changes in pressure in the conduit or tunnel connected to the drop are transmitted through the column of water in the shaft, and result in an immediate corresponding change in the water depth at the inlet tank. With this design the water surface in the tank remains level, there is no drawdown in the shaft, and air cannot be entrained. The discharge is easily computed by Bernoullis' equation and conventional methods of pipe flow analysis.

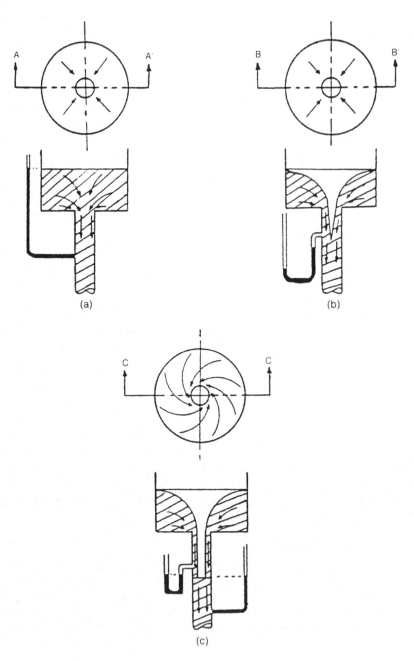

Figure 9.30—Three types of flow drops in shafts.

2. Subatmospheric Drop Shafts

This type of drop shaft is characterized by a radial flow toward the shaft. The water surface is drawn down as the velocity of the flow increases in the approach toward the center of the tank. Large quantities of air are entrained as the water drops into the shaft. The water appears to be mixed with air and there are also voids in the freely-falling water. The pressure at the shaft inlet is subatmospheric. This negative pressure results in large discharge under low head in the tank. However, since these negative pressures are the source of air entrainment and cavitation, this method of entry is considered undesirable. The subatomspheric drop shaft results in unsteady flow with large slugs of air breathed in, which result in considerable noise. To achieve radial flow, it is important that the entrance to the tank have radial fins or guide vanes. A bell-mouth or a morning-glory entrance to the shaft can improve the hydraulic performance of the system.

The U.S. Department of Transportation (1983) has noted the following undesirable characteristics of radial entrances, and suggests that radial flow be avoided.

(a) Negative pressure near the inlet of the shaft.
(b) Intake of large quantities of air.
(c) Need for radial guide vanes to provide steady motion, depth and pressure.

If radial fins or guide vanes are not used, the radial flow may change to spiral or vortex flow.

3. Spiral (Vortex) Drop Shafts

This type of drop shaft is characterized by spiraling water that flows around the shaft walls with an air void that is connected to the atmosphere and extends into the shaft down to the hydraulic grade line. The pressure in this region is essentially atmospheric. Air is drawn into the shaft constantly, however, it is drawn in at a lower rate than with radial flow at the same discharge.

The undesirable conditions created by the radial entrance are avoided if the spiral or vortex approach is used. The following conditions usually prevail at a vortex drop shaft:

(a) Pressures are always atmospheric above full shaft.
(b) Smaller quantities of air are entrained.
(c) The vortex is very stable and does not vary with time.

Since air entrainment is responsible for cavitation and impact on the integrity of the structure, the spiral or vortex drop shaft is recommended for use. Blaisdell and Donnelly (1956) provide a review of a variety of vortex drop shafts, modeling techniques, and design considerations. Figure 9.31 shows a plan and profile of one of the many types of vortex inlets.

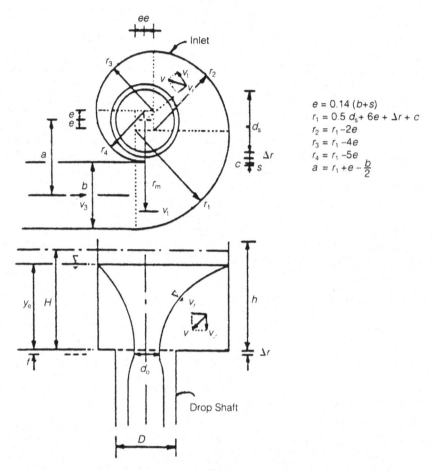

$$e = 0.14\,(b+s)$$
$$r_1 = 0.5\,d_s + 6e + \Delta r + c$$
$$r_2 = r_1 - 2e$$
$$r_3 = r_1 - 4e$$
$$r_4 = r_1 - 5e$$
$$a = r_1 + e - \frac{b}{2}$$

Figure 9.31—Standard scroll inlet for a vortex flow drop structure.

4. Design Considerations

The design of drop shafts focuses on the dimensions of the various components of the drop, and on maintaining the structural integrity of the system. Design charts are available from Bushey (1952), Jain and Kennedy (1983), and Anderson and Dahlin (1968). Structural integrity issues include the following:

(a) Elimination of potential cavitation that may result from air entrainment and subatmospheric pressure in the drop shafts. This can be done by using atmospheric pressure vortex drop shafts, keeping drop shaft heads low, and aerating the flow.
(b) Energy dissipation, which can be provided by:
   (1) Wall friction.

(2) Annular hydraulic jumps produced by creating constrictions in the drop shaft.
(3) Drop shaft baffles.
(4) Plunge pools at the bottom of the shaft.
(5) Impact cups located at the bottom of the shaft.
(6) Wear-resistant linings such as iron-fiber reinforced concrete or stones.

Even though extensive literature is available on the subject, the designer should apply judgment relative to the applicability of available experimental data to specific design. Because of site conditions and design considerations not covered by previous experiments, physical hydraulic modeling may be required.

VII. SIPHONS

A. Definitions, Categories, and Applications

Any conduit that drops under an obstruction such as railroad tracks, depressed roadways or utilities, and regains elevation at the downstream side of the obstruction is referred to as a depressed sewer, or inverted siphon.

Because of the inverted bottom, the siphon stands full of storm water even when there is no flow. Some drainage districts discourage the use of siphons on the basis that the siphon requires more frequent maintenance including removal of debris that may clog the conduit. Nevertheless, siphons have advantages in particular settings, usually in urban areas where other solutions such as flow re-routing may result in disruptions to traffic and higher costs.

Siphons are normally single- or multi-barrel and consist of an entrance section, drop, depressed reach, rise, and outlet structure. Siphons require hydraulic head to operate properly and the adequacy of available head should be assessed early in the design process. Siphons can be simple or sophisticated, and the related design effort can be nominal or complex. The following examples apply to larger, sophisticated siphons with multiple appurtenances—some of which may not always be necessary.

B. Single-Barrel Siphons

Single barrel siphons can be used for conveying stormwater flows where there are periods of no flow during which maintenance can be provided.

Even though some agencies limit the slope of the rising leg of the siphon to 15%, steeper slopes and even vertical drops and risers are acceptable, if maintenance chambers with debris collection sumps at the bottom are provided at the drop and riser of the siphon, as shown in Figure 9.32.

Sloping legs of siphons (Figure 9.33) have been designed without maintenance chambers, however, the chambers provide maintenance

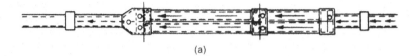
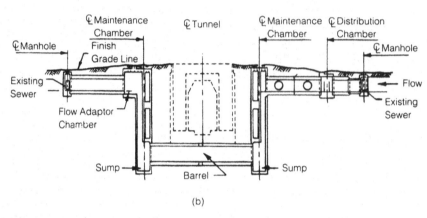

Figure 9.32—Profile and plan of double barrel siphon vertical legs.

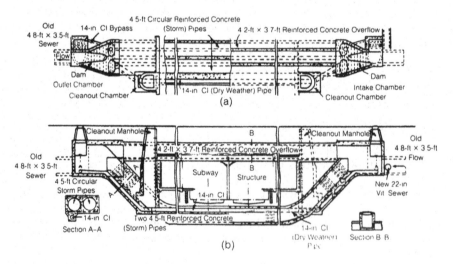

Figure 9.33—Profile and plan of double barrel siphon sloping legs.

flexibility with direct access to service the siphon. The steeper the legs of the siphon, the more difficult it is to clean the siphon from shallow manholes located near the ground surface, and deep maintenance chambers reaching to barrel inverts may be required.

Where a vertical drop and riser are provided, they should serve as maintenance chambers and include access down to the barrels and sumps. Sumps located at the bottom of the maintenance chambers trap the debris that accumulates in the siphon.

C. Multi-Barrel Siphons

In channels or sewers that convey a continuous flow, where one barrel does not have sufficient capacity and the flow has to be divided, or where redundancy is required by local agencies, the multi-barrel siphon is applicable. Plan and profiles of such a siphon are shown in Figure 9.33.

Where redundancy is required for maintenance purposes, one additional equal capacity barrel is sufficient. To fulfil its functions, the multi-barrel siphon requires special equipment and structures, including gates that close the barrel to be maintained while the other barrel is open.

Special structures may also include a flow distribution chamber and a flow adaptor chamber. These chambers are used to contract and expand the flows. The distribution chamber serves to direct the flow from one sewer to the two barrels of the siphon alternatively used, while the flow adaptor chamber serves to direct the flow from the two barrels of the siphon to one conduit.

D. Design Criteria and Practices

One of the critical criteria for the design of siphons is the maintenance of self-cleansing velocities under widely varying flow conditions (ASCE 1969). Siphons used for conveying storm water are usually designed for a flow velocity of 3 fps for a 5-year return interval design flow. Siphons with water containing abrasive suspended materials should be designed for a flow velocity less than 10 fps.

The head losses through each of the siphon components must be estimated for the purpose of plotting the hydraulic grade line. Upstream surcharging should be avoided, and therefore one of the main design objectives should be to minimize the head losses through the siphon. The friction losses can be estimated by using the combined Darcy-Weisbach and Manning's Equation. The equation is useful in the following form:

$$h_f = \frac{29.1 n^2 \, LV^2}{r^{4/3} \, 2g} \qquad (9\text{-}16a)$$

and, in metric units:

$$h_f = \frac{19.5n^2\,LV^2}{r^{4/3}\,2g}$$
(9-16b)

where:

h_f = head loss (ft)
n = Manning's friction factor
L = length of conduit (ft)
r = hydraulic radius (ft)
V = velocity (fps)
g = acceleration of gravity (ft/sec^2)

Minor losses such as bend, contraction and expansion, and entrance and exit losses can be estimated using tables and charts appearing in several publications, such as Brater and King (1976) and City of Baltimore (1972). These losses are also discussed in Chapter 6. It should be noted that head losses in siphons can be significant, particularly in flat coastal areas, where the low terrain does not allow for surcharge and the available project corridor is narrow.

The size of the barrel or conduit can be determined initially based on the minimum required flow velocity. However, the barrel can be sized accurately only after the hydraulic losses are estimated. If the head loss under the design flow condition is excessive, an increase in size of the conduit should be considered.

VIII. SIDE-OVERFLOW WEIRS

A. Definitions and Applications

Side-overflow weirs facilitate overflow and diversion of stormwater by directing the discharges away from the original channel. Such structures are commonly used in combined sewers where the weirs permit the combined flows to build up to a design level, at which time overflow is initiated and directed from the side weir into relief sewers or natural water courses. Since the design overflow level is located above the level of the sanitary flow, the dry-weather sanitary flow is discharged continuously toward sewage treatment facilities. Flow diversions occur only during storms (Figure 9.34). Side overflow weirs are also used to direct channel discharges above predetermined levels into off-line stormwater detention facilities.

B. Design Considerations

The design of side weirs is based on empirical equations which quantify the relationship between the discharge over the weir and geometric parameters at the weir, including the length of the weir and head (Hager 1987). Figure 9.35 (Metcalf and Eddy 1972) shows three head or water-surface profile conditions that can prevail at a side weir. These conditions were defined by DeMarchi in 1934 (Collinge 1957):

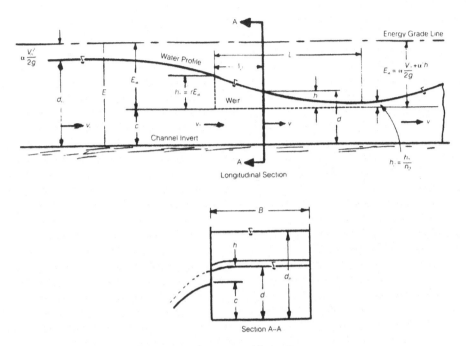

Figure 9.34—Typical cross sections at a side weir.

(a) Condition 1: The channel bed slopes steeply, producing supercritical flow. Under this condition the weir has no effect upstream and along the weir there is a gradual reduction in depth. Downstream of the weir the flow depth in the original channel increases, tending asymptotically to the normal depth corresponding to the remaining discharge.

(b) Condition 2: The channel bed slopes mildly. Under this condition subcritical flow prevails and the weir impact is noticed upstream of the weir only. The water surface profile downstream of the weir corresponds to the normal depth of the remaining discharge. Along the weir there is a gradual increase in depth and upstream of the weir the flow depth tends asymptotically to the normal depth for the initial discharge.

(c) Condition 3: The channel bed slopes mildly, but the weir crest is below critical depth corresponding to the initial flow, and the flow at the weir is supercritical. Recent studies (Frazer 1957) indicate that conditions 1 and 3 may result in development of a hydraulic jump at the weir.

The most common condition that a designer will encounter is Condition 3, where the weir elevation is below the critical depth. When only a relatively small amount of the flow is diverted, a rising water surface profile occurs. According to Metcalf and Eddy Inc. (1972), the falling profile results if the ratio of the height of the weir (C) to the channel specific energy (E_w) referenced to the top of the weir is less than 0.6.

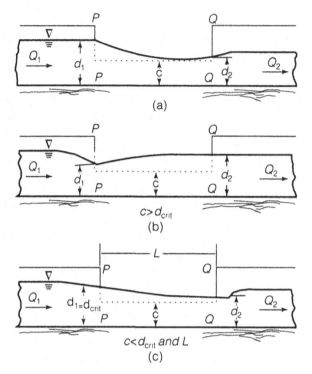

Figure 9.35—Possible types of water surface profiles at a side weir.

C. Design Practices

1. Falling Water Surface

The equations and procedures for computing weir length for the falling water surface profile were developed by Ackers (Chow 1959). These equations combine Bernoulli's theorem with a weir discharge formula. Metcalf and Eddy Inc. (1972) suggest using:

$$L = 2.03B\left(5.28 - 2.63\,\frac{c}{E_w}\right) \tag{9-17}$$

Where:

L = length of weir
B = channel width
c = height of weir
E_w = channel specific energy

and,

$$E_w = \alpha\,\frac{v^2}{2g} + \alpha'\,(d_n - c) \tag{9-18}$$

where:

α = velocity coefficient
V = normal velocity in the approach channel, fps
g = acceleration due to gravity, ft/sec^2
α' = pressure-head correction
d_n = normal depth of flow in approach channel, ft.
c = height of weir above channel bottom, ft.

α and α' of 1.2 and 1.0, respectively, can be used in the approach channel, while at the lower end of the weir values of 1.4 and 0.95 can be used for α and α', respectively.

2. Rising Water Surface

The analysis for estimating the weir length for rising water surface is based on the theoretical equations developed by DeMarchi (Collinge 1957):

$$L = \frac{B}{C}\left[\phi\left(\frac{d_2}{E}\right) - \phi\left(\frac{d_1}{E}\right)\right] \qquad (9\text{-}19)$$

where:

L = length of weir, ft.
B = channel width, ft.
C = constant (0.35 for a free nappe).
$\phi(d/_E)$ = varied flow function taken from Fig. 3 in Collinge (1957).
d_1, d_2 = depths in channel, ft.
E = specific energy, ft.

This equation is recommended for use only in case of a rising water surface. Metcalf and Eddy Inc. (1972) indicates that this equation works best when the Froude number is between ± 0.3–0.92.

IX. FLOW SPLITTERS, JUNCTIONS, FLAP GATES, AND MANHOLES

A. Flow Splitters

A flow splitter is a special structure designed to divide a single flow and divert the parts into two or more downstream channels. A flow splitter can serve two functions:

(a) Reduction in water surface elevation—By dividing the flow from a large pipe into multiple conduits, the height of flow (measured from the flow line to the water surface (or for pipes flowing full, the inside diameter) can be reduced. This may be necessary to route flows under immovable obstructions.
(b) Dividing flows wherever necessary. Examples of this include division of existing large special-design conduits, such as arches or horseshoes, into less expensive multiple-pipe continuations, and division of flow

between low- and high-flow conduits at the intake of an inverted siphon.

Two major considerations exist for the design of flow-splitting devices:

(a) Head Loss—Hydraulic disturbances at the point of flow division result
in unavoidable head losses. These losses, however, may be reduced
by the inclusion of proper flow deflectors in the design of the structure.
Deflectors minimize flow separation by providing a gradual transition
for the flow, rather than by forcing abrupt changes in flow direction.
(See also the discussion in Chapter 8.)

(b) Debris—In all transitions from large to smaller pipes, debris accumulation is a potential problem. Tree limbs and other debris that flow
freely in the larger pipe may not fit in the smaller pipe(s) and may
restrict flow. In addition, flow splitters cause major flow disturbances
resulting in regions of decreased velocity. This reduction causes material suspended in the stormwater flow to settle in the splitter box.
Although the deflector design should minimize velocity reduction as
much as possible, total elimination of the problem is unlikely. Therefore, positive maintenance access must be provided. Because flow split-
ting devices are maintenance-intensive, their use should be judiciously
controlled by the engineer.

B. Junctions

A junction is a region of converging or intersecting flow occurring
in either closed-conduit or open-channel systems. Since one or more
of the incident flows must undergo a change in velocity, energy losses
are experienced at junctions. Junction losses are also discussed in Chapters 6 and 8. For further information on closed-conduit junctions in
storm drainage systems, the engineer is referred to the Los Angeles
Manual on the Hydraulic Analysis of Junctions (City of Los Angeles
1968), Street & Highway Drainage, Volume 2, Design Charts (University
of California 1965) and the Denver Urban Storm Drainage Criteria Manual, (Urban Drainage and Flood Control District 1984).

C. Flap Gates

Flap gates, usually made of cast iron or cast steel, are used to permit
flow in only one direction. A small differential pressure on the back of
the gate will open it, allowing discharge in the desired direction. When
water on the front side of the gate rises above that on the back side,
the gate closes to prevent backflow. The seat or ring of the flap gate is
attached to a headwall or directly to the pipe that forms the opening
through which flow passes. Larger flap gates are often mounted to a
cast iron thimble, cast in the headwall or other structure through which
flow is to pass. Flap gates are available for round, square, and rectangular openings and in various designs.

Flap gates can act as a skimmer and cause brush and trash to collect
between the flap and the seat at low flow. Rubber flap gates (Brombach
1990) and "duck bill" (Freeman et al. 1990), which are less susceptible

to this sort of clogging, are also being used. Periodic inspection and cleaning should be scheduled when the water flowing through the gate carries floating material. If the gate is to be kept clear of debris, it should be mounted 12–18 inches (30–46 centimeters) above the apron in front of the gate. This allows room at the bottom for floating material to work its way through the gate. When this clearance is not available, more frequent inspection and cleaning must be provided.

For those drainage structures that have a flap gate mounted on a pipe projecting into a stream, the gate must be protected from damage by floating logs or ice during high flows. In these instances, protection must be provided on the upstream side of the gate.

D. Manholes

The primary function of a manhole is to provide convenient access for inspection, maintenance, and repair of storm drainage systems. Secondary functions include provision for multiple pipe intersections, ventilation, and pressure relief. Manhole design is covered in Chapters 6 and 8.

1. Size

On small lines, a minimum inside diameter of 4' (see Figure 9.34), tapering to a cast-iron frame, that provides a clear opening usually specified as 24", has been widely adopted. Immediately under the frame, the diameter should be at least 24", generally enlarging to 30" within 2' of the surface, but some authorities allow a diameter of 24" for a distance of 24" below the frame. Occasionally, the working space of a manhole is a rectangular vault, but the opening and access usually remain circular.

It is common practice to use eccentric cones, especially in precast manholes, to provide a vertical side for the steps and to avoid interference with street curbs. Most often the orientation places the steps on the wall over the bench, but some designs place the steps on the wall opposite the outlet pipe.

Instead of extending the 4-foot diameter section up to a cone near the surface, some engineers prefer a design that maintains the 4-foot section to a suitable working height, tapering then to 3' as shown in Figure 9.36. The cast iron frame in this case has a broad base to rest on the 36" diameter shaft. Still another design uses a removable flat reinforced concrete slab instead of a cone, as shown in Figure 9.34(d). This is applicable whether the working space is circular or rectangular. The slab must be suitably reinforced to withstand traffic loads.

2. Frame and Cover

The manhole frame and cover are normally made of cast iron. The cover is designed to provide a good fit between cover and frame, and with adequate strength to support superimposed loads. The closure should be relatively tight, and the cover should be designed to prevent

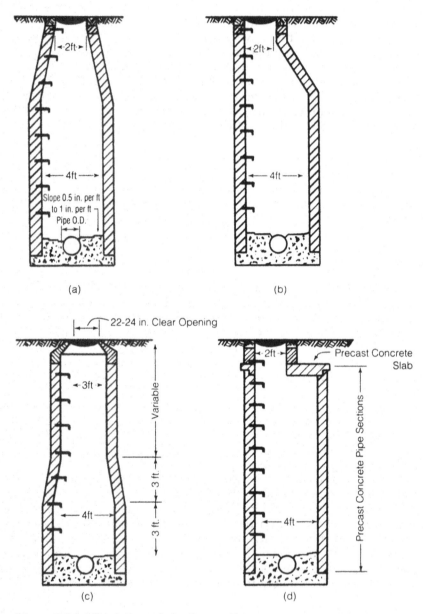

Figure 9.36—Typical manholes for small sewers.

unauthorized entry. It is a good practice to identify manholes with covers with the words "Storm Sewer" or equivalent cast into the top surface to differentiate them from those on sanitary sewers, communication conduits, or other underground accessways.

3. Steps

It is common practice in many areas to abandon the use of manhole steps in favor of having maintenance personnel supply their own ladders. Reasons for this include danger from rust-damaged steps, ease of access to children, and debris snagging. Steps coated with neoprene or epoxy or steps fabricated from a rust-resistant metal such as stainless steel or aluminum are preferable to steps made from reinforcing steel.

4. Manholes on Pipes Flowing Full

It often becomes necessary to design a storm drainage conduit to flow full. This poses special problems to the designer because of the need to minimize head loss (see Chapters 6 and 8 for a more complete discussion). If the hydraulic grade line is designed to rise above the ground surface at a manhole site, special consideration must be given to the design of the frame and cover. The cover must be secured so that it remains in placing during peak flooding periods, avoiding a manhole "blow-out." This may be accomplished with a bolted manhole cover or by use of a cover with a locking mechanism. Leakage from around the edges of such manhole covers is usually minimal and not of concern to the designer.

5. Deep Manholes

Deep manholes must be carefully designed to withstand soil pressure loads. If the manhole is to extend very far below the water table, it must also be designed to withstand the associated hydrostatic pressure (see Chapter 14) or excessive seepage may occur. Access must be provided, and very deep manholes must be supplied with either steps or built-in ladders, because long ladders would be cumbersome and dangerous in deep manholes.

X. STORMWATER PUMPING

A. Introduction

Stormwater pumping is generally used only when gravity drainage is not feasible, although there may be cases where it is more economical to pump than to construct gravity conveyance systems. Storage in storm drains, wet wells, or ponds can reduce the pumping capacity required to handle peak runoff rates. The following is adapted from the "Manual for Highway Stormwater Pumping Stations" (USDOT 1982), which, along with other publications, should be consulted for the final design of stormwater pump stations.

B. Planning and Site Considerations

Several important considerations affect planning and site selection for pump stations. The access necessary for safe operation, maintenance, and emergency functions must be available at all times. Hydraulic conditions will have primary importance in site selection, but site appearance and sound attenuation should also be assessed. Foundation investigations are essential (although they rarely reveal conditions unstable enough to merit rejection of a site), and the effects of soil erosion and contaminated runoff on the pump station influent should be evaluated, as well as the rate of discharge and water quality of the effluent. Enough space must be provided in the area outside the station to accommodate parking as well as movements of large machinery.

A dependable energy source is essential. The primary source of electrical power for most stormwater pump stations is a public utility. Underground service is preferred for safety and aesthetic reasons, and overhead lines into the station should be avoided, as they present potential safety hazards during large equipment operation.

C. Design Features

Some features to evaluate when designing a pump station depend on the size of the facility, while others relate to site conditions, economic considerations, and operating needs. Table 9.9 lists these features and rates the parameters of each in terms of high, medium, and low conditions. A stormwater pump station selection matrix is presented in the USDOT manual (1982).

D. Storage

Stormwater storage can reduce the required peak capacity of a pump station. Selection of a peak pumping capacity is a trial and error process that considers the inflow hydrograph, available storage, and possible pump discharge rates. An approximate method developed by Baumgardner (1981) yields results that can be evaluated during selection of final design conditions using computer program developed by the FHWA.

The approximate method includes adjusting the inflow hydrograph (developed using appropriate procedures presented in Chapter 5 of the manual) to an equivalent triangular hydrograph, as shown in Figure 9.37. An estimate of the storage required to reduce the peak pumping rate to a desired value can be found by assigning a peak pumping rate and plotting it as a horizontal line, as shown in Figure 9.38 (U.S. D.O.T. 1982).

The area of the triangular hydrograph above the peak pumping rate represents an estimate of the storage volume required. This assumes that storage below the last pump-on elevation will not affect the design. The effect of this storage on final design can be considered using the FHWA computer program, or a mass curve routing procedure as presented in Section F. The FHWA program is currently available only for microcomputer applications. The required storage can be estimated by the equation:

$$\frac{V_s}{V_t} = \left(\frac{\Delta Q}{Q_p}\right)^2$$ (9-20)

where:

V_s = Required storage volume, ft^3
V_t = Volume of triangular inflow hydrograph, ft^3
ΔQ = Peak flow reduction, cfs
Q_p = Peak flow of triangular inflow hydrograph, cfs

A graphical presentation of the relationship in Equation 9-20 is shown in Figure 9.39. By selecting a peak reduction ratio ($\Delta Q/Q_p$), the storage ratio (V_s/V_t) can be obtained directly. When the inflow hydrograph volume (V_t) is known, the storage required is estimated as the product of the storage ratio and the inflow hydrograph volume.

E. Pumps

Pumps are commonly used in storm drainage systems, though the designer is cautioned that they should be used only when gravity flow is not possible.

1. Pump Types

Stormwater pumping typically uses large pumps that operate at relatively low heads. The main types of pumps include:

(a) Vertical (propeller and mixed flow)
(b) Submersible (vertical and horizontal)
(c) Centrifugal (horizontal non-clog)
(d) Screw
(e) Volute or Angleflow (vertical)

All of these pumps generally are driven by electric motors, but diesel or gas engines are sometimes used for vertical, screw, or angleflow pumps.

Vertical Pumps   Vertical pumps, the type used most frequently in stormwater pump stations, are available in many capacities and discharge heads. Single-stage propeller pumps are used for low heads; mixed-flow pumps for higher heads. Two-stage propeller pumps are available that will approximately double the head capacity of a single-stage pump. A typical vertical stormwater pump is shown in Figure 9.40.

Submersible Pumps   Until recently, submersible pumps, except at small stations, have been mainly used as accessory pumps to handle low (dry weather) flows, groundwater infiltration, and cleanup and pumpout of

TABLE 9.9.   Pump Station Design Features and Considerations (U.S. DOT 1982).

Feature or Consideration	High Condition	Medium Condition	Low Condition
1. Station Design Capacity	Maximum exceeding 300 ft³/sec	Maximum between 100 and 300 ft³/sec	Maximum less than 100 ft³/sec
2. Station Design Head	Over 35' TDH	Between 15' and 35' TDH	Less than 15' TDH
3. Storage Upstream of Pumps	For velocity reduction, settlement of solids, minimizing equipment	Used if available	Not required or available
4. Quality of Pumped Water	Turbid and sand-laden inflow	Moderate contamination	Minimal contamination
5. Inflow Rate	Rapid increase	Normal hydrograph	Slow increase
6. Discharge Conditions	Long rising outfall from each pump	Short free outfall from each pump	Limitation of discharge rate
7. Sump Dewatering	Sump pump required	Vacuum truck preferred	No provision
8. Electric Power Reliability	Completely dependable; dual service	Very dependable; single service	Undependable; frequent outages
9. Natural Gas/LPG Desired as Fuel	Completely dependable; dual service	Very dependable; good storage	Not readily available; supply unreliable
10. Station Siting	Good access from frontage road or similar	Good access from highway	Poor access, alongside highway
11. Soil Conditions	Rock	Hard, steep unshored cuts	Clay or soft soil
12. Foundation Conditions	Acceptable bearing strata	Piling required for bearing	Extensive dewatering with piling required because of uplift
13. Above-Ground Structure	Large acceptable	Moderate preferred	Smallest possible desired
14. Structure Visibility	Large structure acceptable	Modest structure desired	Minimum only acceptable

15.	Initial Cost	High capital cost acceptable	Moderate cost acceptable	Lowest cost mandatory
16.	Maintenance Capability	Excellent, with complex machinery	Reasonably good	Mediocre
17.	Operating Cost	High cost acceptable	Moderate budget desired	Lowest budget desired
18.	Equipment Handling Devices (Built-in)	Elaborate type considered essential	Single type acceptable	Minimum or none required
19.	Equipment Handling Devices (Mobile)	Use preferred for all requirements	Used to supplement built-in	Not required due to elaborate built-in
20.	Trash Handling Devices	Elaborate built-ins preferred	Simple built-in adequate	Vacuum trucks preferred
21.	Pre-Screen Inflow Debris	—	—	—
22.	Hazardous Spill Vulnerability	—	—	—
23.	Epoxy Coatings and Lining of Pumps	—	—	—
24.	Grease Lubrication for Pumps	—	—	—
25.	Steelwork Galvanizing	—	—	—
26.	Manifold to Pressure Discharge	—	—	—
27.	Sediment and Hydrocarbon Removal from Discharge	—	—	—
28.	Emergency Generator	—	—	—
29.	Supervisory Control (Telemetering)	—	—	—

(USDOT 1982).

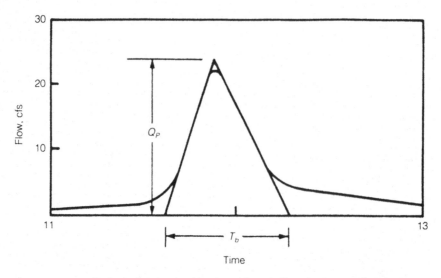

Figure 9.37—Triangular approximation of an inflow hydrograph.

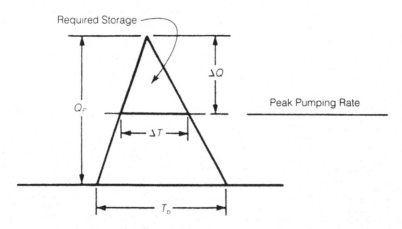

Figure 9.38—Estimation of required storage based on a selected peak pumping rate.

the pump pit after a storm. This has been changing as larger submersible pumps have become available, and large stormwater pump stations can now be designed that rely solely on submersible pumps.

Centrifugal Pumps   A special type of centrifugal pump is used in dry-pit stations. Designed to handle suspended solids and debris which pass the screens, these pumps are the end suction, non-clog type. Enclosed impellers and a hand-hole on the casing permit inspection and removal of foreign material. Small centrifugal pumps combined with upstream storage are well suited to small catchments.

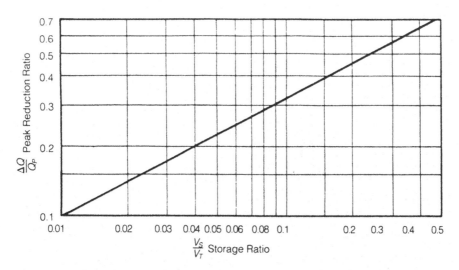

Figure 9.39—Relationship between peak reduction ratio and storage ratio.

Screw Pumps   Screw pumps are appropriate when pressurized transfer is not required. They are typically used in agricultural areas and offer a simple and reliable method of stormwater pumping when site conditions are suitable.

Volute or Angleflow Pumps   The volute or angleflow pump, also called a dry-pit angleflow pump or a single-suction centrifugal mixed-flow pump, is usually mounted in a vertical position, with the motor above the pump room operating floor and the pump as much as 25' below. Vertically mounted volute or angleflow pumps have been a standard in sewage pump station installations for many years. However, because of the complex accessories needed with these pumps, they are not recommended for stormwater pumping unless the level of operating head makes it necessary.

2. Pump Selection

A pump characteristic known as specific speed should be considered during pump selection. Pump impellers for high heads have low specific speeds, while impellers for low heads usually have high specific speeds. Figure 9.41 (U.S. D.O.T. 1982) shows the relationship of different types of impellers to the useful range of specific speeds.

Pumps of the axial-flow type have impellers shaped like ship propellers, and as the name implies, the liquid is discharged axially through the impeller. Pumps of the mixed-flow type have impellers with vanes integral with a conical hub. The pumping head is developed partly by centrifugal force and partly by a lifting action of the vanes. The application of the mixed-flow pump is similar to that of the vertical axial-flow pump. Pumps of the centrifugal type have impellers that develop head entirely by centrifugal force.

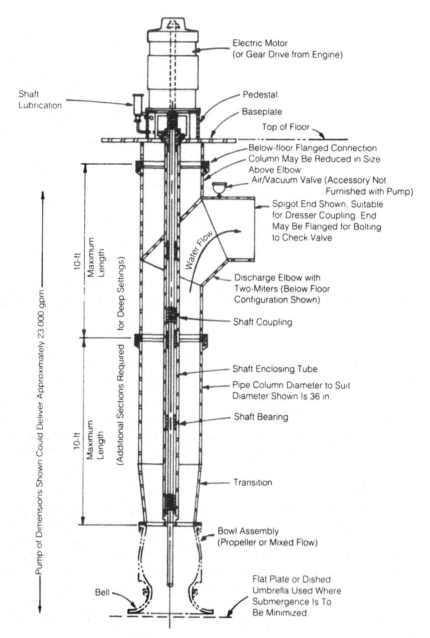

Figure 9.40—Typical vertical stormwater pump.

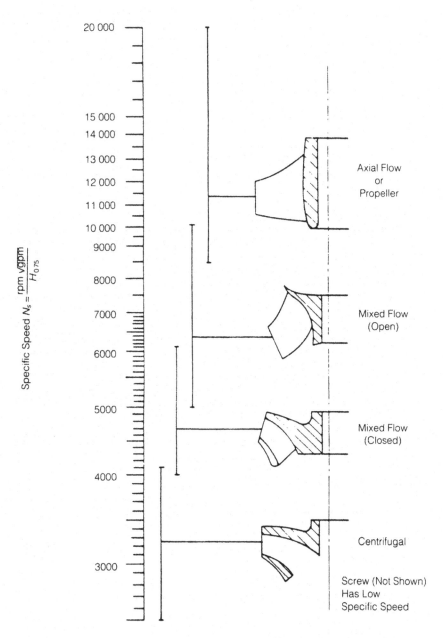

Figure 9.41—Specific pump speed versus impeller types.

The performance curve developed by the manufacturer should be obtained before selecting a particular pump. Procedures on pump selection are presented in a Federal Highway Administration manual (U.S. Department of Transportation 1982). The advantage of using standard equipment, however, often outweighs the need to obtain a perfectly tailored design.

F Mass Curve Routing

A wide range of storage and pumping rate combinations should be evaluated when considering pumping cycles for frequent small volume storms and the potential for flooding from less frequent, large-volume storms. Usually, computer programs are best suited to this type of trial and error evaluation. A mass curve routing procedure developed by Baumgardner (1982) follows.

(a) Develop an inflow hydrograph, a stage-storage curve, and a stage-discharge curve for the range of pumping facilities being considered. If the pumping rate is constrained by downstream or environmental considerations, the required storage must be determined by various trials of the mass curve routing procedure.

(b) Establish the point at which the cumulative flow curve has reached the storage volume associated with the first pump-on elevation.

(c) Draw the pump discharge line from the intersection point located in Step 2, upward to the right at a slope equal to the discharge rate of the pump. The vertical distance between the mass inflow curve and the pump discharge curve represents the amount of stormwater which must be held in storage.

(d) To determine the maximum storage volume required, a line is drawn parallel to the pump discharge curve, and tangent to the mass inflow curve. The vertical distance between this tangent and the pump discharge line represents the maximum storage volume required.

(e) The storage pipe and wet well should be designed to handle sediment, while the pumping system should be designed to safely carry sediment which is flushed from the wet well.

XI. REFERENCES

American Association of State Highway and Transportation Officials. (AASHTO), (1987). *Drainage handbook*. Washington, D.C.

American Concrete Pipe Association. (1981). *Concrete pipe handbook*. ACPA, Arlington, VA.

American Society of Civil Engineers. (1969). "Design and construction of sanitary and storm sewers." *Manuals and reports on engineering practice No. 37*, ASCE, New York.

American Society of Civil Engineers. (1975). "Sedimentation engineering." *Manuals and reports on engineering practice No. 54*, ASCE, New York.

Anderson, A.G., and Dahlin, W.Q. (1968). *Model studies—Lawrence Avenue sewer system, City of Chicago*. University of Minnesota, Minneapolis.

Armco Steel Corporation. (1972). *Armco Water Control Gates*, Middletown, OH.

"Bank and shore protection in California highway practice." (1970). State of California, Department of Transportation, Sacramento, CA.

Barfield, B.J., Warner, R.C., and Maan, C.T. (1983). *Applied hydrology and sedimentology for disturbed areas*. Oklahoma Technical Press.

Bauer Engineering, Inc. (1972). *Report on Drop Shaft Investigation for the Crosstown Expressway (I-494) and Flood and Pollution Control Plan*.

Baumgardner, R.H. (1981). "Estimating required storage to reduce peak pumping rates at stormwater pumping stations." *Federal Highway Administration Workshop Paper*, Office of Engineering, Washington, D.C.

Baumgardner, R.H. (1982). "Hydraulic design of stormwater pumping stations considering storage." *Federal Highway Administration Workshop Paper*, Office of Engineering, Washington, D.C.

Blaisdell, F.W. and Donnelly, C.A. (1956). "The box inlet drop spillway and its outlet." *Trans., American Society of Civil Engineers*, ASCE, New York, 121, 955–986.

Bradley, J.J. and Peterka, A.J. (1957). Proceedings, American Society of Civil Engineers, *Journal of the Hydraulics Division*, 83 (HYS), ASCE, New York, NY.

Brater, E.F. and King, H.W. (1976). *Handbook of hydraulics*. McGraw-Hill Book Co., New York.

Brombach, H. (1990). "Equipment and instrumentation for CSO control." *Urban stormwater quality enhancement—source control, retrofitting, and combined sewer technology, Proc. of An Engineering Foundation Conference held in Davos, Switzerland*, ASCE, New York.

Lauskey, L.M. (1952). *Flow in vertical shafts*. Carnegie Institute of Technology, Pittsburgh, PA.

Chow, V.T. (1959). *Open-channel hydraulics*. McGraw-Hill Book, Co., New York.

Chow, V.T. (1964). *Handbook of applied hydrology*. McGraw-Hill Book Co., New York.

City of Baltimore. (1972). *Manual of design procedure and criteria*. Baltimore, MD.

City of Fort Worth. (1967). "Storm drainage criteria and design manual." *Drainage master plan for the public works department*, Knowlton-Ratliff-English Consulting Engineers, Fort Worth, TX.

City of Los Angeles. (1965). "Design charts for catch basin openings as determined by experimental hydraulic model studies." *Bureau of Engineering, Office Standard No. 108*, Storm Drainage Division, Los Angeles, CA.

Collinge, V.D. (1957). "The discharge capacity of side weirs." *Proceedings, Institute of Civil Engineers*, London, England, 6 (2).

Davis, C.V. and Sorensen, K.E. (1969). *Handbook of applied hydraulics*. McGraw-Hill Book Co., New York.

Dickinson, W.T. (1968). "Design of rock riprap for streambank stabilization." *Canadian Agricultural Engineering*, 10 (2).

Engineering News. (1916). 76 (10), 442–443.

Fletcher, B.P. and Grace, J.S., Jr. (1972). *Practical guidance for estimating and controlling erosion at culvert outlets*, U.S. Army Corps of Engineers, Vicksburg, Mississippi.

Fortier, J. and Scobey, F.C. (1926). "Permissible canal velocities." *Trans., American Society of Civil Engineers*, ASCE, 89, 940–956.

Frazer, W. (1957). "The behavior of side weirs in prismatic rectangular channels." *Proc., Institute of Civil Engineers*, London, England, 6 (2).

Freeman, P.A., Forndran, A.B. and Field, R. (1990). "Development and evaluation of a rubber "duck bill" tide gate." *EPA/600/S2-89-020, (NTIS PB 89-188 379/AS)*, U.S. Environmental Protection Agency, Washington, D.C.

French, R.H. (1985). *Open-Channel Hydraulics*, McGraw-Hill, New York.

Hager, W.H. (1987). "Lateral outflow over side weirs." *Journal of Hydraulic Engineering*, 113 (4).

The Institute of Transportation and Traffic Engineering. (1965). "Street and highway drainage." *Design charts, volume 2*, University of California, Berkeley, CA.

Jain, S.C., and Kennedy, J.F. (1983). "Vortex-flow drop structures for the Milwaukee Metropolitan Sewerage District inline storage system." *Report No. 264*, Iowa Institute of Hydraulic Research, Ames, IA.

Johns Hopkins University. (1956). *The design of stormwater inlets*. Department of Sanitary Engineering and Water Resources, Baltimore, MD.

Johnson, F.L. and Chang, F.F.M. (1984). "Drainage of highway pavements." *Hydraulic Engineering Circular No. 12, FHWA-TS-84-202*, Federal Highway Administration, Washington, D.C.

"The maximum permissible mean velocity in open channels." (1936). *Gidroteknicheskoie Stroitel'stvo, (Hydrotechnical Construction)*, Moscow, USSR, 5, May, 5–7.

McLaughlin Water Engineers, Ltd. (1989). *Evaluation of and Design Recommendations for Drop Structures in the Denver Metropolitan Area*, Urban Drainage and Flood Control District.

Metcalf & Eddy, Inc. (1972). *Wastewater engineering, collection, treatment, disposal*. McGraw-Hill Book Company, New York, NY.

National Academy of Science Highway Research Board. (1970). "Tentative design procedure for riprap-lined channels." *National Cooperative Highway Research Program Report No. 108*, Washington, D.C.

Neenah Foundry Co. (1987). *Inlet Grate Capacities*.

New Jersey State Soil Conservation Committee. (1987). *Standards for soil erosion and sediment control in New Jersey*. NJEPA, Trenton, NJ.

Pemberton, E.L. and Lara, J.M. (1982). "Guide for Computing Degradation and Local Scour," *Technical Guide 10-B2*, U.S. Bureau of Reclamation, Denver, CO.

Peterka, A.J. (1958). "Hydraulic design of stilling basins and energy dissipators." *Engineering Monograph No. 25*, U.S. Department of Interior, Bureau of Reclamation, Washington, D.C.

Rouse, H. (1949). *Engineering hydraulics*. John Wiley & Sons, Inc., New York.

Sangster, W.M. et al. (1958). "Pressure changes at storm drainage junctions." *University of Missouri Engineering Series Bulletin Number 41*, Columbia, MO.

Simons, D.B. and Senturk, F. (1977). *Sediment transport technology*. Water Resource Publications, Ft. Collins, CO.

Taylor, E.H. (1944). "Flow characteristics at rectangular open-channel junctions." *Trans. of the American Society of Civil Engineers*, 109, 893.

Urban Drainage and Flood Control District. (1984). *Urban storm drainage criteria manual*, rev. ed., Denver Regional Council of Governments, Denver, CO.

U.S. Department of Agriculture. (1954). "Handbook of channel design for soil and water conservation." *SCS-TP-61*, Stillwater Outdoor Hydraulic Laboratory, U.S. Soil Conservation Service.

U.S. Department of Agriculture. (1962). *National engineering handbook, section 11, drop stillways*. Soil Conservation Service.

U.S. Department of Commerce. "Hydraulic characteristics of slotted drain inlets." *Report RD79-106, Vol. 4*, Federal Highway Administration, Washington, D.C.

U.S. Department of the Interior. (1952). "Canals and related structures." *Design and Construction Manual*. Design Supplement No. 3, Vol. X, Bureau of Reclamation, April 17, 1952.

U.S. Department of the Interior. (1955). "Research studies on stilling basins, energy dissipators, and associated appurtenances." *Hydraulic Laboratory Report No. Hyd-39*, Bureau of Reclamation.

U.S. Department of Interior. (1977). *Design of small dams*. Bureau of Reclamation, U.S. Government Printing Office, Washington, D.C.

U.S. Department of the Interior. (1980). *Hydraulic laboratory techniques*. Water and Power Resources Service, Denver, CO.

U.S. Department of Transportation. (1989). "Design of riprap revetments." *Hydraulic Engineering Circular No. 11*, Washington, D.C.

U.S. Department of Transportation. (1982). "Manual for highway stormwater pumping stations, Volumes 1 and 2." *FHWA-IP-82-17*, Federal Highway Administration, Washington, D.C.

U.S. Department of Transportation. (1983). "Hydraulic design of energy dissipators for culverts and channels." *Hydraulic Engineering Circular No. 14*, Federal Highway Administration, Washington, D.C.

U.S. Department of Transportation. (1975). "Design of stable channels with flexible linings." *Hydraulic Engineering Circular No. 15,* Federal Highway Administration, Washington, D.C.

U.S. Environmental Protection Agency. (1976). "Erosion and sediment control, surface mining in the Eastern U.S." *EPA-625/3-76-006,* Washington, D.C.

Water Pollution Control Federation. (1981). *Design of wastewater and stormwater pumping stations.* Alexandria, VA.

CHAPTER 10

COMBINED SEWER SYSTEMS

I. GENERAL DESCRIPTION OF COMBINED SEWERAGE SYSTEMS

A. Introduction

A combined sewer consists of a single conduit that collects and transports domestic sewage and industrial wastewater, along with stormwater runoff. The dry weather flow, which consists of domestic sewage and industrial wastewater as well as groundwater infiltration, is intercepted and conveyed to a sewage treatment plant for processing. During wet weather (events such as a rainstorm or snowmelt), the combined stormwater runoff and sewage usually exceeds the capacity of the interceptor and the treatment plant and overflows through an outfall into receiving waters with minimum, or no, treatment. Because of their large volumes, these combined sewage overflows (CSOs) are a major source of pollution during and following wet weather events. A typical layout of an urban combined sewer system is shown on Figure 10.1 (Kibler 1982). In a combined sewer system, the maximum stormwater runoff typically exceeds dry-weather flow by 50–200 or more times. Therefore, combined sewers are designed to convey stormwater runoff, that is, as storm sewers.

While this chapter provides an introduction to combined sewer systems, the reader is encouraged to consult the excellent recent Manual of Practice entitled "Combined Sewer Overflow Pollution Abatement" (WPCF 1989) for more detailed information.

B. Flow Regulation Devices

1. Purpose

Combined sewers are typically designed to carry 2–3 or more times dry-weather flows (though in some cases they cannot even carry all of the dry-weather flows). Wet-weather flows are typically much greater

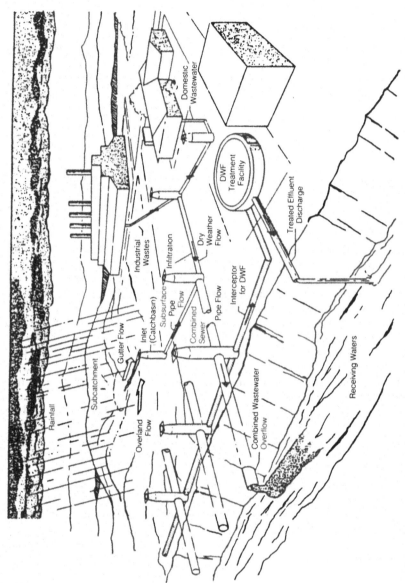

Figure 10.1—Typical urban combined sewer system.

than that, and regulators are used to control flows in the system. The combined sewer flow regulator is designed to:

(a) Divert dry-weather flows from tributary combined sewers to the interceptor for conveyance to the treatment plant.
(b) Direct excess flows to CSOs (or to CSO storage/treatment devices).
(c) Minimize the extent and frequency of combined sewage discharges into receiving waters by allowing overflows at selected locations, and by using the potential storage of the tributary sewer system.
(d) Prevent surcharging of the intercepting sewer.
(e) Prevent entry of debris into the intercepting sewers.

The regulators are of two types: static and dynamic. The static regulator, as the name implies, has no moving parts and must be adjusted manually. Flow through these devices increases with increases in upstream head. A dynamic regulator, on the other hand, functions semi-automatically or automatically, responding to the water level in the combined sewer to limit the flow diverted to the interceptor. The Water Pollution Control Federation (1989) Manual of Practice describes a variety of regulators.

2. Static Regulators

Static flow regulators are of many types, several of which are described herein. It should be noted that most static regulators operate efficiently over a limited range of flows, and care should be exercised in their selection.

Fixed Orifice   This regulator consists of a small dam constructed across the combined sewer, with a vertical orifice just upstream of the dam. The purpose of the dam is to divert the dry weather flow through the vertical orifice to the interceptor (Figure 10.2). During wet weather, the flow exceeding the capacity of the orifice flows over the dam and through the outfall to the receiving waters, or to CSO storage/treatment. The orifice is either circular or rectangular. The size of the opening is designed to intercept peak dry weather flow; however, the flow diverted to the interceptor during wet weather is higher. This type of regulator, generally used for flows less than 2 cfs, is not recommended because of its tendency to plug up with debris, and because the orifice discharge varies widely with head.

Leaping Weir   This is an opening located at the invert of the combined sewer. The opening is provided with an adjustable plate having a raised lip on the upstream side as shown on Figure 10.3. The opening of the leaping weir is designed to divert dry-weather flows from the combined sewer to the interceptor, or to CSO storage/treatment. During wet weather, the excess flow leaps over the weir for discharge through the outfall. To improve the interception of dry weather flow, the downstream end of the orifice is tipped up. The leaping weir is typically used for flows up to 4 cfs.

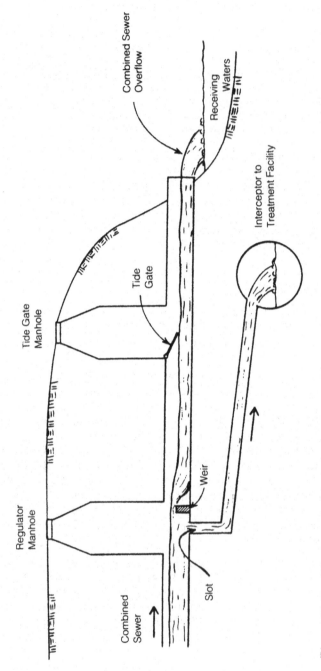

Figure 10.2 — Fixed orifice flow regulator.

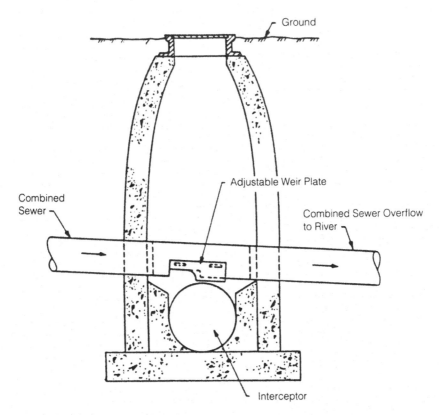

Figure 10.3—Leaping weir flow regulator.

Manually Operated Gate This regulator has an orifice with a gate as shown on Figure 10.4. The orifice opening can be varied by manual adjustment of the gate. The regulator consists of two chambers: diversion chamber and orifice chamber. The diversion chamber has a low dam placed across the combined sewer to divert the flow to the orifice chamber. The orifice chamber has one or more orifices with shear or slide gates, either circular or square in shape (a square shape is less subject to clogging than a circular shape). The setting of the gate is based on the maximum allowable flow that can be diverted to the interceptor during a storm.

Side-Spill Weir This consists of a weir constructed parallel to, or at a slight angle to, the axis of the combined sewer. The height of the weir crest is such that the peak dry-weather flow continues in the sewer to the interceptor. During wet weather conditions, the combined sewage overflows the weir to the outfall, or to CSO storage/treatment. These weirs cannot generally control flow accurately, but they are in common use because of their simplicity and low cost.

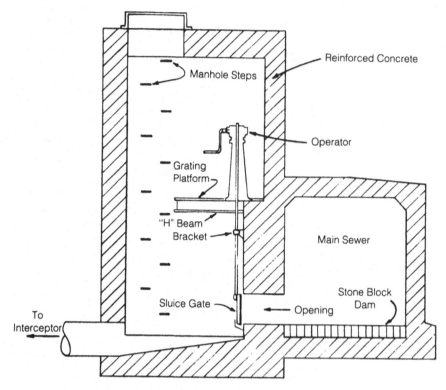

Figure 10.4—Manually-operated gate regulator.

Vortex Regulator   The vortex regulator is becoming widely used in combined sewer flow regulation. Its chief advantages are simplicity and the fact that it will produce a relatively constant discharge over a range of heads. See Water Pollution Control Federation (1989) for a detailed discussion.

3. Dynamic Regulators

There are two major types of dynamic regulators, semi-automatic and automatic.

Semi-automatic Dynamic Regulators   Semi-automatic regulators do not require an external source of energy for operation. The flow conditions in the combined sewer, regulator, and/or interceptor actuate mechanisms to restrict the flow being intercepted. These types of regulators include:

    (a) Float-Operated Gates: Automatic gates operated by floats that rise or fall as the water elevation in the sewer increases or decreases. When the interceptor is filled to its capacity, the gate could close entirely cutting off the flow to the interceptor. In common application, the regulator restricts the flow from the combined sewer to the interceptor to a predetermined peak dry weather flow. The regulator, as shown on Figure 10.5, includes a diversion chamber, a regulating chamber, and a tide gate chamber, if required.

The diversion chamber has a low (preferably 6″) weir perpendicular to the flow channel of the combined sewer for diverting dry-weather flow to the regulating chamber. Sometimes, the invert of the diversion chamber is depressed to prevent the overflow of peak dry-weather flow to the outfall. The regulating chamber houses the float, the regulating gate, and the connecting linkage between the float and the gate. The gate is installed on an opening in the common wall between the diversion and regulating chambers. The float is located in a well connected to the combined sewer or interceptor. For restricting the flow to the interceptor, the actuating mechanism operated by the float in the wet well connected to the combined sewer adjusts the gate opening to maintain the predetermined flow rate. In other types of arrangements

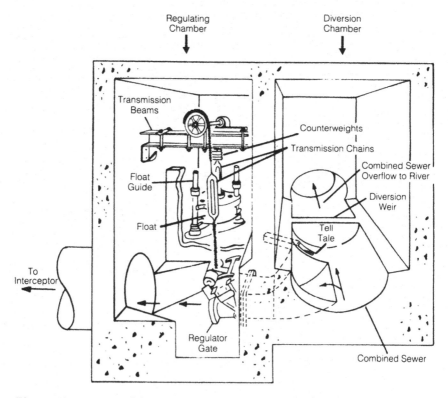

Figure 10.5—Typical float-operated mechanical regulator.

where the CSOs are to be restricted, the float well is connected to the interceptor. The overflows do not occur until the interceptor is fully loaded to a predetermined level.

(b) Tipping-Gate Regulator: This regulator consists of a plate which is pivoted off-center as shown on Figure 10.6. The motion of the plate is controlled by the difference between upstream and downstream water levels. The regulator structure is similar to the one required for manually-operated gates. The tipping plate is mounted on the vertical orifice through which the flow is to be controlled. The regulator can be adjusted in the field by adjusting the gate stop disk which controls the gate opening. This regulator is generally used for flows greater than 4 cfs.

Automatic Dynamic Regulators   Automatic dynamic regulators operate control gates using an external source of energy such as electricity or hydraulic power.

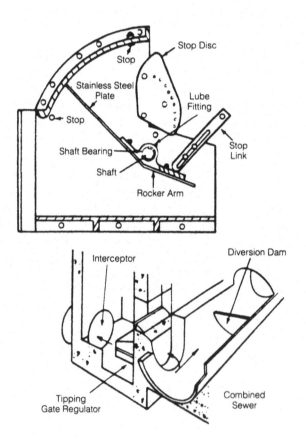

Figure 10.6—Tipping gate regulator.

(a) Motor-Operated Gates: This regulator consists of a low (typically 6″) weir at the invert across the combined sewer which diverts the combined sewage to a vertical opening such as an orifice. The size of the opening is based upon the maximum flow to be diverted. Either a tainter gate, sluice gate, or shear gate is mounted on the orifice to control the flow. The gate is operated by an electric motor activated by a sensing probe. The probe records the sewage level in the control section and transmits the signal to start the motor when a predetermined level is reached in the control section.

The regulator has two to four chambers: diversion chamber, regulator chamber containing the gate and the sensor, a motor chamber, and a tide gate chamber where required. A typical motor-operated gate flow regulator is shown on Figure 10.7.

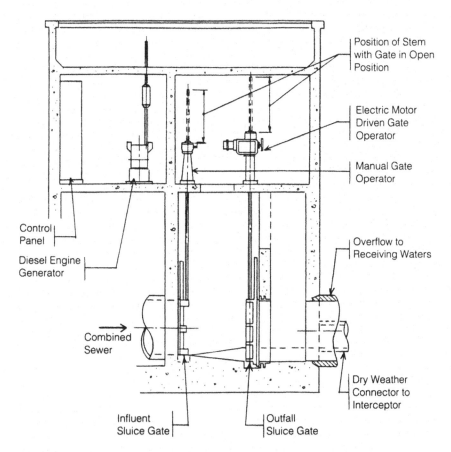

Figure 10.7—Typical motor-operated gate flow regulator.

Motor-operated regulators are used on larger combined sewer systems. They permit positive control of flow diverted to the interceptors as well as to outfalls. Remote control allows the storage within the sewer system to be used to minimize overflows as well as to prevent overloading of the treatment plant. This type of regulator is an integral part of many CSO abatement plans.

(b) Cylinder-Operated Gate Regulators: Similar to the motor-operated gate regulator, this gate is operated by hydraulic cylinder (which uses either oil, air or water). As the unit is self-enclosed, it can operate even when submerged.

C. Overflows

The purpose of an overflow structure is to convey combined sewage that exceeds the capacity of the intercepting sewer, CSO storage/treatment facilities and the wastewater treatment plant to receiving waters. Combined sewer outlets are usually located above the normal water surface of the receiving waters. However, in some cases, system hydraulic limitations require outlets to be submerged to a considerable depth.

1. Gravity Outfalls

Gravity outfalls are usually above the receiving water surface and are normally provided with a concrete or brick head wall. Riprap is typically provided to reduce erosion in the vicinity of the head wall. In some cases, the head walls are constructed on steel sheet piling.

Where receiving waters are subject to fluctuating water levels which result in submergence of the outlet, backwater (tide) gates are provided. These gates help to minimize the entry of receiving waters into the intercepting sewers and storage/treatment facilities during dry weather. During wet weather, the higher hydraulic grade in the sewerage system allows excess combined sewage to overflow into the receiving waters.

2. Submerged Outfalls

These outlets are partially or totally below the normal water surface of the receiving waters. Tide gates are installed to protect the intercepting sewers against flooding by receiving waters. In some cases, pumping stations are installed at the outfalls to provide positive drainage of combined sewers.

D. Interceptors

Intercepting sewers are designed to divert and convey dry-weather flow from combined sewers to the treatment plant. Regulators are used to divert and control flow into interceptors as discussed earlier.

The intercepting sewer typically is designed to convey the peak dry-weather flow, groundwater infiltration, and inflow generated within the service area at ultimate development. As the peak dry-weather flow is typically 2–4 times the average dry weather flow for the design year, the intercepting sewers are designed for some larger multiple of the average dry-weather flow to intercept a portion of storm runoff. Interceptors are, in some cases, designed for up to 10 times the average dry-weather flow to assure capture of the "first flush." The first flush of stormwater contains solids which have settled in sewers during the antecedent dry periods in addition to the solids contained in the runoff itself. The interception of the first flush can be important for minimizing pollution of the receiving waters from CSOs. The first flush phenomenon can be as variable as CSOs themselves, and may be difficult to observe at the outlet of very large catchments.

II. COMBINED SEWER OVERFLOW POLLUTION LOADS

The impact of pollution from CSOs can be significant, since they are generally located in some of our most heavily populated urban centers. The quality of combined sewage varies dramatically based on a number of natural and man-made influences. Among the natural influences are rainfall characteristics and inter-event times, ground topography, and vegetation. Man-made influences include land use (residential, commercial, industrial), street sweeping practices, air quality, and type of street surface. A summary of observed pollutant concentrations in CSOs is presented in Table 10.1.

CSOs occur as a result of rainfall events. While the runoff dilutes the sanitary flow, it can have a considerable pollutant load of its own (see Chapter 5 for a discussion of runoff quality). In addition, storm flows cause a re-suspension of materials settled in sewers, which contributes to the so-called "first flush" phenomenon mentioned in the preceding section. For an excellent discussion of CSO pollution, see also the Water Pollution Control Federation Manual of Practice (1989).

In early studies of CSO pollution loads, one of the problems was that samples only were taken for the first few minutes of a storm. Consequently, the measured loads of suspended solids and BOD were very high. As more data over entire storm events became available, it was evident that the concentration of pollutants often decreased during the runoff event. When such conditions exist, it is important to relate concentrations to their corresponding runoff discharges when evaluating the impact on receiving waters. Since the late 1960s, advancements have been made in the state-of-the-art of predicting the quality of combined sewage, primarily in the field of mathematical modeling (see also Chapter 7).

Three levels of refinement may be employed when estimating the pollution load from a combined sewer area. These levels represent the transition from a relatively simple average yearly loading to a quite detailed representation of stormwater pollution during storm events.

TABLE 10.1. Observed Pollutant Concentrations in Combined Sewer Overflows.

	Average Pollutant Concentration, mg/L									
	TSS	VSS	BOD	COD	Kjeldahl Nitrogen	Total Nitrogen	PO_4-P	OPO_4-P	Lead	Fecal Coliforms[a]
Des Moines, Iowa	413	117	64	—	—	4.3	1.86	1.31	—	—
Milwaukee, Wisconsin	321	109	59	264	4.9	6.3	1.23	0.86	—	—
New York City, New York										
Newtown Creek	306	182	222	481	—	16.6	4.5[b]	—	0.60	—
Spring Creek	347	—	111	358	—	43	17[b]	—	—	—
Poissy, France[c]	751	387	279	1005	—	—	—	—	—	—
Racine, Wisconsin	551	154	158	—	2.6	—	2.78	0.92	—	201
Rochester, New York	273	—	75	—	—	—	—	0.88	0.14	1140
Average (not weighted)	370	140	115	367	3.8	9.1	1.95	1.00	0.37	670
Range	273–551	109–182	59–222	264–481	2.6–4.9	4.3–16.6	1.23–2.78	0.86–1.31	0.14–0.60	201–1140

[a] 1000 organisms/1000 mL.
[b] Total P (not included in average).
[c] Not included in average because of high strength of municipal sewage when compared to the United States.
(Lager et al 1977).

The three levels of refinement have been defined (EPA 1976) as listed below:

(a) Level 1—Average Annual Storm Load: This level of refinement estimates the total stormwater pollution load from an urban area on an average annual basis. It is generally most useful in assessing potential long-term water quality problems such as sediment deposition, nutrient loading, or chronic toxic effects. It also may be useful in obtaining an order of magnitude comparison between annual stormwater pollution loads and point source pollutant loads.

(b) Level 2—Storm Event Loads: This level of refinement estimates the actual distribution of storm event pollution loads throughout the year and indicates the variability of the total stormwater pollution load generated by each storm event. This level of detail may be required for an assessment of transient water quality problems such as dissolved oxygen or bacterial concentrations.

(c) Level 3—Load Variation Within Storm Events: This level of refinement describes stormwater pollution loads as a function of time within each storm event. This level represents the effect of storm patterns and intensities on stormwater pollution and will indicate whether there is a first flush of pollutants.

The selection of an appropriate level of refinement for defining stormwater pollution loads should be based on the level of detail required to assess receiving water impacts. The impacts of the various types of pollutants discharged to receiving waters have characteristic time and space scales (EPA 1976). In general, long-term water quality problems, such as sediment deposition or nutrient loading, can adequately be evaluated using average annual storm loads (level 1), whereas more reactive pollutants (e.g., coliform bacteria or oxygen-consuming materials) usually require storm event load determinations (level 2). A level 3 determination of the load variation within storm events may be required to properly design stormwater pollution control structures.

Wu and Ahlert (1978a and b) have proposed categorizing the numerous methods for estimating stormwater pollutant loads as:

a) Zero-order methods—An estimate of an area's stormwater pollutant loads based exclusively on data reported in the literature is termed "zero-order," which indicates the relative flexibility of estimation from reported data. The method is the most crude, can be inaccurate, and is generally only useful for determining level 1 average annual loads. The best source is data from the Nationwide Urban Runoff Program (NURP) of the U.S. EPA (EPA, 1983). This data is available from the USGS and the Illinois State Water Survey. Other sources of stormwater data include Huber et al. (1980), Manning et al. (1977), Sartor and Boyd (1972), Bradford (1979), Polls and Lanyon (1980), and Collins and Ridgway (1980).

b) Direct or rational methods—If storm runoff and pollutant concentration are independent, a direct or rational calculation of the stormwater loading rate is the product of a mean concentration and a mean discharge.

The mean discharge is calculated, and the mean concentration is estimated using one of several possible approaches, which include:
(1) published data for a watershed with similar characteristics, i.e., zero-
order, (2) observed field data from the watershed of concern, and (3) a
regression equation, such as those proposed by Driver and Tasker (1988).
The direct calculation of stormwater pollutant loads can be applied to
estimate both level 1, average annual loads, and level 2, storm event
loads. Documentation regarding use of the direct calculation method is
presented by Driscoll et al. (1979), Young et al. (1978), and the U.S.
Environmental Protection Agency (1976).
c) Statistical methods: Statistical methods employ techniques such as
regression, correlation, and frequency analysis to predict stormwater
pollutant loads. Site-specific factors normally considered in statistical
methods include precipitation, watershed characteristics, land use patterns, and population densities. The best description can be found in
the report on the NURP published by the U.S. Environmental Protection
Agency (EPA 1983). Statistical methods can also be useful to examine
field data from a site for the purpose of identifying parameters which
affect nonpoint pollution loads (Griffin et al. 1978).
d) Descriptive or modeling methods: Descriptive methods for estimating
stormwater pollution loads deal with deterministic rainfall/runoff mechanisms that affect the quality of a stormwater discharged from a watershed. These mechanisms can usually be represented by mathematical
expressions which can be incorporated into programs for simulating
stormwater quantity and quality on a digital computer.

III. METHODOLOGY FOR EVALUATING COMBINED SEWER OVERFLOWS

A. General

CSOs are evaluated generally on the basis of impacts to receiving
water quality. Because of the complex and variable hydrologic, hydraulic, and water quality relationships associated with CSOs, mathematical simulation models are extensively used to facilitate these evaluations. As discussed in Chapter 7, these models simulate the physical
characteristics of the combined sewerage system and, if properly calibrated, can be used to predict CSO volumes and pollutant loadings to
a receiving water for a single storm event or over a long period. Receiving water models are used to predict pollutant levels in rivers,
estuaries or bays resulting from CSO discharges. They can interface
directly with models of runoff processes and simulate the effects of
advection, dispersion and individual constituent changes such as decay
or growth.

To assure accurate and representative models, input data must be
complete and correct. Data base development includes the collection of
all available information on the sewerage system, its tributary area and
the receiving water(s). This information is validated and expanded, as
required, through field investigations and surveys. All data is then

assembled and properly formatted for input into the various models. Computer data base management or spreadsheet programs are very useful (some are commercially available) for this purpose. In addition, schematic representations of the physical facilities within the system and, in some cases, the receiving waters are developed to graphically depict the models to facilitate their use.

Once developed, the land-based and receiving water models are calibrated and validated (see Chapter 7 for a description of this process). The land-based models can then be used with reasonable confidence to test system capacity under future dry-weather flow conditions, and to determine design flow rates for sizing CSO abatement facilities. Receiving water models can be used with somewhat less confidence to predict impacts of CSOs upon receiving water quality over a long-term period or for a specific storm event. The models together can also be used to test alternative CSO mitigation techniques (relative impacts).

B. Data Collection

Data collection is a formidable but very necessary and important task. Accurate and up-to-date information on the sewerage system and receiving waters is vital to assure proper model representation. Data collection procedures were described in Chapter 3, and the Water Pollution Control Federation Manual of Practice on Combined Sewer Overflow Pollution Abatement (WPCF 1989) provides excellent guidance.

C. Field Investigations

1. Inspections of Physical Facilities

Field inspections serve to verify and expand on information obtained under the data collection task. Typically, this includes obtaining or verifying dimensional data and determining the physical condition and operating deficiencies of regulators, tide gates, pumping stations, overflow pipes and other major system facilities. Internal television inspection of major interceptors is suggested to determine structural condition, location of major sources of infiltration and the extent of solids (grit) deposition. This latter item is vital since significant grit deposition reduces interceptor carrying capacity. This must be accounted for when calibrating the land-based computer models to assure accurate system representation.

2. Flow Monitoring and Sampling

Data on sewer system flow rates and wastewater quality under both dry weather and wet weather conditions is required for model calibration and to obtain a thorough understanding of system operation and response to wet-weather events. The selection of flow monitoring and sampling sites is dependent upon several factors, including the number of drainage basins within the system, land use within each basin, un-metered system boundary points (i.e., locations where wastewater en-ters the study area systems from adjacent, upstream systems and/or locations where wastewater leaves the study area system) and the extent of in-place flow and quality measurement facilities at boundary points, pumping stations, or wastewater treatment plants.

For systems having a large number of drainage basins and CSOs, it is not economically feasible nor is it necessary to monitor and sample each basin. Typically, a manageable number (3–5) are selected that represent a cross-section of land use throughout the entire system. Under dry-weather conditions, wastewater flow from each selected basin is continuously measured for at least a one-week period to determine average, maximum and minimum flow rates and to estimate infiltration. Samples are also taken over at least a 24-hour period and typically analyzed for conventional parameters and, if warranted, other param-eters that may be of concern to a particular receiving water (metals, PCBs, nutrients, etc.). This information, coupled with flow estimates, can be used to estimate dry-weather flow and quality in other, non-measured drainage basins. During this monitoring and sampling pe-riod, continuous flow gaging on major interceptors is warranted ᵗo verify total system dry weather flow estimates, and to achieve a system-wide flow balance. All dry-weather flow and quality data serve as base-flow input data to the land-based models.

A combined sewer system wet weather monitoring and sampling program is also conducted to determine the system's hydraulic response to a specific storm event, and the rate and quality of pollutant washoff from specific drainage basins. During several storm events (usually 3–5), continuous flow gauging is conducted at regulators or on overflow pipes to measure the rate and determine volumes of CSOs generated from the selected study area drainage basins. Samples are also taken at each CSO gauging location at specific time intervals following the start of the overflow. Sampling intervals are more frequent during the early stages of the overflow to determine whether there are water quality impacts due to "first flush" effects. In most cases, samples are analyzed for the same parameters measured for the dry-weather sam-pling program.

To assure that CSO monitoring and sampling results are typical and representative, criteria for selecting wet-weather events for conducting this field program must be established. Typically, the representativeness of each storm is based on antecedent dry period and the length and intensity of precipitation. Events occurring after a dry period of at least three days, that last a minimum of four hours and have intensities that produce measurable runoff are usually considered representative. A rain gauge located within the study area (if available) is used to develop storm hyetographs.

Information obtained from the wet-weather monitoring and sampling program is used to calibrate the land-based models.

3. Receiving Water Sampling

Receiving water sampling typically coincides with land-based monitoring and sampling under dry-weather conditions and during wet-weather events. For rivers and streams, flow monitoring may also be required. However, in many cases, stream flow data from existing gauges installed and maintained by the USGS or others can be used (and flow data adjusted as required) if located a reasonable distance from the study area.

Receiving water quality data is often available from a state authority, local planning agency or local university. If sampling is required, several aspects should be considered. Sampling locations are generally dependent on the type of receiving water (river, bay, or estuary), location of CSO discharges, model calibration needs and site-specific concerns and issues such as shellfish bed or beach closure problems. For rivers, it is important to measure quality at boundary points and at points upstream and downstream of the major CSO outlet(s). In bays or estuaries, sampling locations are more dependent on modeling needs. Some bay models do not generate results as precise as required near shorelines where, in many instances, CSO pollution is greatest. Consequently, more intensive near-shore sampling is sometimes warranted to assist in evaluating CSO impacts.

Pollutants of concern and, in turn, laboratory analytical requirements, are specific to the study area and receiving water issues and concerns. Common parameters include bacteria, solids, dissolved oxygen, BOD, pH, and selected metals.

D. Assessment of Existing and Future Land-Based Conditions

1. Population and Dry-weather Flows

Information on dry-weather flows in a wastewater collection system is essential to estimations of the quantity and quality of CSOs. Although the quantity of dry-weather flow is relatively minor in comparison to the large volume of stormwater required to produce an overflow, its high contaminant level has a significant impact on the quality of CSOs.

Dry-weather flow data is used in several ways during CSO evaluations. It is used with wet-weather data for calibration of the system computer models. Estimated average dry-weather flows are used to approximate available in-system storage capacity and simulate the mixing of dry-weather flow and stormwater runoff, providing estimates of overflow quality and quantity at each CSO outlet. Monitoring and sampling at the CSO outlets, discussed in the preceding section, provides verification of the mixed flows derived from the model.

An analysis of the existing and estimated future population and water consumption of the study area is necessary to estimate the area's dry-weather wastewater flow rates. Per capita water consumption estimates, obtained from metered water consumption data, multiplied by estimated population, results in an estimate of total average daily water consumption on a drainage basin or study area basis. This information is used to estimate average daily domestic wastewater flows, and when added to estimates of commercial, industrial, and institutional flows and infiltration, results in the estimated total average dry-weather flow. Dry-weather flow estimates should be confirmed with field measurements, since calculated dry-weather flows are based on several assumptions and estimates associated with population, water consumption, and infiltration.

2. Land Use and Zoning

Land use is directly related to the quantity of runoff expected to flow to receiving waters during wet-weather events. The type and quantity of pollutants contained in runoff varies considerably from one land use category to another, and a thorough knowledge of existing and potential future land uses is an important element of a CSO evaluation.

Depending on population densities, residential areas may contribute high coliform concentrations to CSOs due to the relatively high volume of domestic sanitary sewage discharged into the combined sewer system. Industrial areas may pose additional problems from toxic constituents, including the discharge of heavy metals, volatile organics, polychlorinated biphenyls (PCBs), and other toxic substances. Industries may also add to the total volume of combined sewer overflow if they contribute a major portion of the dry weather flow.

Parks and other unpaved open areas serve to reduce the amount of runoff that drains to either a combined sewer or storm drain because much of the rainwater that falls on grass-covered areas is able to infiltrate into the ground. However, open areas or parks that are not properly graded, and/or do not have sufficient grass cover, can increase the grit and suspended solids content of the runoff as a result of soil erosion. This can lead to sediment deposits within conduits and a high solids content in CSOs and their receiving water.

Areas designated for institutional land use can increase the pollutant content of CSOs in several ways. Institutional areas containing schools, churches, hospitals, or municipal buildings have a potential for discharging large slugs of sewage and associated pollutants to the collection system. If these discharges occur during a rainfall event, the sanitary portion of the total flow in the combined sewer can increase, increasing the pollutant concentrations of CSOs where they occur. Hospitals have the additional potential of producing waste streams containing high concentrations of pathogenic organisms that may become a health problem when discharged in the vicinity of bathing or shellfish areas.

Runoff from highways, major roadways and railroads must also be addressed. Oil, grease, and other automotive pollutants are usually present on road surfaces in sufficient quantities to contaminate stormwater runoff that eventually discharges to receiving waters. Hydrocarbons from engine exhaust systems plus sanding and salting operations during the winter season can also add contaminants to roadway runoff.

E. Definition of Receiving Water Issues and Goals

The techniques used to model receiving waters must be designed to address the local issues and federally approved water quality standards. "Issues" refer to the questions and concerns from the public and regulatory agencies regarding CSO pollution.

CSO abatement goals should be established to meet designated uses of receiving waters for all but an acceptable number of relatively infrequent events. Receiving waters can be designated for uses such as public water supply, recreation, protection and propagation of aquatic life, agriculture, and navigation. CSOs should not discharge directly into a public water supply, but generally can discharge to waters with other uses after suitable controls are in place to minimize the frequency, volume, and pollutant concentration of the discharge. The acceptable degree of control is usually established by a local, regional, or state agency with jurisdiction to protect water quality.

Wet weather water quality conditions are best evaluated in a probabilistic context. To do this, modeling must be done over some reasonable period of record. The use of specific historical records is obviously ideal, but synthetic rainfall/runoff sequence generation may be required if real data is unavailable or sparse. The purpose of such analyses is to quantify the relationship between runoff characteristics (intensity, duration, time of year), and the risk (probability) of receiving water quality problems. These analyses provide the framework for practical decision making and risk analysis (see also Chapter 8).

IV. COMBINED SEWER OVERFLOW MITIGATION TECHNIQUES

A. General

Techniques for reducing CSO pollution can be grouped into three broad categories; 1) source controls to reduce the quantity of pollutants entering the system, 2) collection system controls designed to increase the system's effectiveness in conveying and/or storing excess flow, and 3) off-line storage and treatment to remove pollutants from the overflow. The technique selected for any given situation may include components from each of these categories. Although not considered a separate improvement category, system maintenance is an essential part of any pollution control program.

This section discusses the identification of applicable control techniques, maintenance implications, and describes the different types of facilities and how to evaluate their effectiveness. For an extensive discussion of these issues, see the Water Pollution Control federation Manual of Practice (1989).

B. Identification of Applicable CSO Mitigation Techniques

Selection of CSO mitigation techniques is a very complex process. Not only are there numerous possibilities within each of the three major classifications noted above, but regulatory agency guidance, water use classification, and funding availability all must be considered. Further, some beneficial uses of receiving waters can be more economically addressed by treatment plant improvements.

1. Source Controls

Most source controls are nonstructural in nature, and are sometimes referred to as Best Management Practices (BMPs). Their common denominator is reduction of pollutant load which can reach the combined sewers and/or regulators. Control of illicit connections, street sweeping, catchbasin cleaning, waste oil collection, sewer cleaning/flushing, and control of runoff from oil storage facilities, car wash establishments, body shops, and the like, are examples. Another type of source control involves land use planning, which can impose limits on the flow volume and/or rate which can be conveyed to the combined sewer.

2. Collection System Controls

Generally, collection system controls are intended to ensure that the sewers are operating at peak efficiency. Elimination of illicit connections, adequate maintenance, and a complete understanding of how the system functions are nonstructural actions which should be the first step in a CSO control program. There then follows a set of possible improvements including regulator consolidation (conveying flow from several small structures to one large regulator) and control, storage, and/or treatment, both on- and off-line, and sewer separation (though from a pollution control standpoint, this generally is not an effective solution).

3. Off-line Storage and Treatment

These types of CSO abatement facilities are structurally intensive and costly, and are used as necessary following implementation of source controls and system optimization measures. Off-line storage can be provided using either underground or above-ground tanks. Because the runoff to be accommodated can vary greatly, it is most efficient to subdivide the tanks and to provide gravity flow into the unit. Treatment can vary from simple screening to complete treatment at a wastewater treatment plant.

C. Combined Sewer System Maintenance

In addition to normal maintenance activities involving sewers, regulators, manholes, pumping stations, metering facilities, etc., combined sewer systems have special maintenance needs involving regulators and tide gates. If CSO mitigation facilities have been installed, there is a substantially greater maintenance effort required. In particular, regulators can be difficult to maintain because of the harsh environment in which they must operate.

1. Records

Experience indicates that sewer maintenance records (or even records of regulator locations) are generally poor or nonexistent. Part of the problem can be traced to the age of the system, because most combined sewers are 40 or more years old. In the interim, field modifications and other changes may have resulted in a system that often bears little resemblance to that shown on original record drawings—if the drawings still exist.

Before the system can be maintained, it must be defined. Therefore, the first step in a maintenance program must be the development of a record system that not only describes what currently exists but also is easily updated to reflect new information that becomes available, and/or changes made by field personnel. As noted earlier, various computer programs are available which greatly aid in this process.

2. Regulators

By their very nature, regulators are prone to failure. The incoming combined sewer is larger than the dry-weather outlet and, as a result, the regulator is subject to clogging. All regulators should be inspected on a periodic basis. Those that are a particular problem, as demonstrated by maintenance and/or complaint records, must be inspected more often.

D. Nonstructural Techniques

Nonstructural CSO mitigation techniques focus on the operation of the existing combined system or regulatory control to minimize overflow quantity and pollution load. Nonstructural techniques can be an important element of a CSO mitigation plan, though significant reductions in CSO pollution for highly urbanized areas may not be achievable solely through nonstructural control.

1. Land Use

Historically, land use controls focused at mitigating CSOs have typically not been used, since most combined sewer areas are already developed. However, rationally conceived long-range land use programs can reduce or prevent the aggravation of CSO problems. The development or redevelopment of vacant land within a combined sewer area can provide a mechanism to regulate the stormwater runoff to a combined sewer system. Increasing open space and greenway areas (if possible) and/or the provision of detention facilities can both reduce the rate and quantity of stormwater runoff and decrease the nonpoint pollution load contribution to CSOs.

2. Interceptor Sediment Removal

Combined sewers, and the interceptors to which they are tributary, experience large ranges of flow velocity. During periods of low velocity (two ft/sec or less), sedimentation occurs. Additionally, irregularities in sewers such as cracks, breaks, misalignments, or root intrusions can reduce velocities and enhance solids deposition. While sediment removal by cleaning will increase the efficiency of the interceptor and somewhat reduce CSOs, the reduction will be short-term unless the factors causing low flow velocities and sedimentation are permanently corrected.

3. Regulator Modifications

Regulator modifications can be used to use the capacity of the interceptor system more efficiently and thereby mitigate combined sewer overflows. The proper maintenance and operation of regulators is one of the most important management practices that can be used for the reduction of CSOs. Regulators should be operated to assure that all dry-weather flow is intercepted and no dry-weather overflows occur.

4. Interceptor Surcharging

Interceptors are designed to convey all dry-weather flow from tributary combined sewers, and combined sewage and runoff in excess of the dry-weather flow during wet weather. The surcharging of an interceptor allows for flows in excess of design limits to be delivered to downstream outfalls, pumping stations, or the treatment plant. Interceptor surcharging can cause sewer backups and flooding if the hydraulic grade line is higher than the ground surface in low areas that are tributary to the system.

5. Inflow Diversion

Often downspouts from roof gutters are connected directly to combined sewer systems. Where adequate pervious area and drainage exists, downspout disconnection can reduce the stormwater inflow to the sewer system and reduce CSO volumes and pollution. Sump pumps for foundation and other clear water drains can also be directed to pervious areas.

E. Structural Alternatives

The growth of urban and combined sewer areas has often been underestimated in the design of combined sewers. Many times new areas have been annexed and served by main sewers and interceptors that had been designed for outdated and smaller system boundaries. Additionally, with development, the proportion of impervious area has increased with corresponding increase in stormwater runoff. As a consequence, most combined sewer systems do not have adequate conveyance and storage capacity, and the application of system maintenance and nonstructural techniques often is not adequate to allow attainment of applicable water quality standards. Generally, structural measures are required to address the inadequacies of large combined sewer systems and to solve major CSO problems.

Because of the great variations in combined sewer system characteristics, engineering studies are required to select appropriate CSO mitigation solutions. For large systems, water quality, hydraulic, and hydrologic modeling is often necessary as discussed earlier (see also WPCF 1989).

1. Storage

Storage involves the containment of combined sewage that normally would discharge to receiving waters (see also Chapter 11 for a broader general discussion of storage). When excess flow capacity is again available in the interceptor system, the stored combined sewage is discharged into the interceptor and conveyed to a treatment facility. Storage facilities have been used extensively for CSO mitigation, particularly in Europe (Lager et al. 1977; Brombach 1990). Three types of storage are:

Detention/Retention Detention facilities are used to regulate or prevent stormwater runoff from entering a combined sewer system. By controlling peak stormwater flows, CSOs can be decreased or eliminated. As relatively unpolluted stormwater runoff is captured and stored, detention facilities can be sited in most urban settings, and are extensively used to attenuate peak runoff flows in separated sewer areas.

Combined sewer system detention facility design considerations include anticipated precipitation (including consideration of storms of long duration with severe antecedent conditions), runoff rates, sewer system capacities, treatment plant capacities and CSO control standards. Additionally, safety, operation, maintenance, aesthetics, and mosquito control must be addressed. Care must be exercised to insure against the sewage contamination of the detention facility as the requirements for off-line combined sewage storage are not the same as those for detention. Detention facilities are often multi-purpose.

Ponding is the most frequently used detention practice. Typically, the release of water from a pond to the combined sewer system is controlled by a small diameter pipe, restrictor, or orifice, and the release of water is by gravity. When runoff from the pond's tributary area exceeds the release rate of the pond, water is impounded. Orifice sizes should be adequate to prevent blockages. The design release rate is a function of the depth of water and the orifice or restrictor size, and the required storage capacity is a function of the size and topography of the tributary area, land use, sewer system capacities and the level of CSO control required. Detention facilities can also be de-watered by pumping.

Sedimentation occurs in detention ponds, though in urban combined sewer areas, the sediment load may not significantly impact pond storage capacities.

Inline Storage   Inline storage uses the volume of a combined sewer and interceptor system that is not being used to transport combined sewage to accommodate the storage of additional stormwater runoff. Often the areal distribution of runoff results in CSOs, even though excess sewer capacity is present in portions of the system. A collector system which nominally has excess wet-weather system volume could most likely use inline storage.

Off-line Storage   Off-line storage involves the capture and storage of CSOs in tanks (usually large underground concrete tanks or tunnels), with the stored volume released (or pumped) into the interceptor system when interceptor capacity is available. This type of CSO storage/ treatment is intended to mitigate or eliminate CSOs generated by storms up to a specific intensity and/or duration, or to capture the first flush.

Most large combined sewer systems require off-line storage to eliminate or significantly reduce CSO pollution. Off-line storage facilities have demonstrated their effectiveness in controlling storm and CSOs. Many regional plans include storage or combinations of storage alternatives as an integral part of the overall control process.

Off-line storage systems can also provide for the following:

(a) The attenuation of combined sewage peak flows.
(b) More efficient use of existing treatment capacity.
(c) Improved treatment plant average effluent quality.
(d) The elimination or reduction of sewer backups.

Because off-line facilities store combined sewage, consideration must be given to preventing septicity or handling the stored wastewater in a septic state. Additionally, the deposition of solids in the storage facility requires designers to provide adequate means for solids removal.

Off-line storage can be used to reduce the required size of a treatment facility. Wastewater treatment plants are typically designed with excess capacity for diurnal, seasonal, wet weather or other characteristic flow variations. Plants with tributary combined systems have been required by the U.S. EPA National Pollution Discharge Elimination System (NPDES) permits to have 2.5 times average dry-weather treatment capacity. There is an optimum combination of storage and treatment for the most cost-effective system. As treatment capital and operational costs increase, it becomes more favorable to increase the storage capacity.

The off-line storage of combined sewage will result in the sedimentation of solids unless the storage facility has provisions for mixing. Accumulated solids can be washed out of the storage facility by inducing scouring velocities or agitation during the evacuation of the captured wastewater or by flushing with a potable or effluent water supply.

Storage facilities can be designed to hold the settled solids with periodic cleaning by dredging or other means, though prolonged storage can result in the settled solids becoming anaerobic with resulting odor problems. It is preferable to provide facilities for flushing out solids after each runoff event. When the solids are held in storage, some primary treatment (sedimentation) may be provided. The reduction of suspended solids could warrant the by-passing of primary unit operation during treatment.

Long detention times will also significantly reduce the BOD concentration (though if allowed to go anaerobic, BOD could increase), as well as reduce the dissolved oxygen concentration of the captured combined sewage and increase ammonia concentrations. Typically, the combined sewage stream from the off-line storage facility is blended with dry-weather influent prior to preliminary treatment to mitigate the possible impacts on the treatment unit operations of a relatively weak strength influent. If sedimentation is provided by an off-line facility, the discharge flows can be directed to bypass preliminary and primary treatment, and be directed to the secondary treatment unit operations, or if the flow quality permits, the discharge can be directed to receiving waters.

While some off-line storage facilities designs will allow some natural aeration, average combined sewage BOD concentrations will still exceed the initial dissolved oxygen concentrations. As a consequence, captured combined sewage must be artificially aerated at the storage facility or evacuated for treatment in a timely manner to prevent septicity. To prevent the formation of odorous, noxious, explosive, and toxic gases, off-line storage facilities must maintain the stored combined sewage in an aerobic or facultative state.

2. CSO System Control

The control of combined sewer system operational components such as regulators and pumps to optimize system operation (minimize overflow, maximize capacity, minimize flooding) has been practiced for some time (Leiser 1974; Schilling 1986). Control schemes typically involve three levels of hierarchy:

(a) Local Control—which involves closed-loop control of regulators, gates or pumps.
(b) Regional Control—which provides coordinated operation of several local controllers.

(c) Global Control—which provides coordination of regional controllers to optimize system operation. Global control systems typically have the following components:
   (1) A data gathering system for rainfall, pumping rates, treatment capacity and regulator settings.
   (2) A telemetered control system for the manipulation of regulators, gates and pumps.
   (3) A computer system, with related mathematical models, for the centralized processing of system data and system control.

Local control is quite common, such as the use of a float controller to turn pumps on and off. Real-time control, on a regional or global basis, involves the use of **currently-monitored** process data (such as rainfall, flow, status of storage, wet well elevations, etc.) to operate regulators to achieve better (or optimal) systems performance **during** the actual process. Recent advances in microcomputers, weather radar, and control devices, as well as a growing body of experience in developing and operating real-time systems (Doering et al. 1987), have demonstrated that such control can provide significant improvements in system operation, although it can be difficult and costly to implement.

3. Treatment

Generally, interceptor capacities are inadequate to convey high combined sewage flows to the treatment facility, and even if one could convey the flows, it would be difficult to adapt the delicate microbial population associated with secondary treatment to the relatively short duration, high flow rates during runoff events. Treatment at individual CSO structures or for several consolidated overflows may be considered in conjunction with other CSO mitigation systems, when NPDES permit and state standards can be attained, and where the resulting head losses are acceptable.

The local treatment of CSOs has been primarily directed at the removal of settleable and suspended solids and floatable material, and disinfection. Reductions in suspended matter of 36–65% and in BOD of 25–40% are reported as achievable with primary treatment (Merritt 1968). Table 10.2 gives a comparison of typical physical treatment removal efficiencies for selected pollutant parameters (Lager et al. 1977). Disinfection is another local treatment process. A local treatment facility can use a combination of treatment processes.

(a) Sedimentation—The major objective of sedimentation is to produce a clarified effluent by gravitational settling. It is one of the most common and well established wastewater treatment unit operations. Sedimentation tanks provide some storage capacity, and disinfection can be concurrently effected. Sedimentation facilities can be used in conjunction with the addition of chemicals for improved removal of solids by coagulation (Graham 1978).

TABLE 10.2. Comparison of Typical Physical Treatment Removal Efficiencies for Selected Pollutant Parameters.

Physical Unit Process	Percent Reduction					
	Suspended Solids	BOD$_5$	COD	Settleable Solids	Total Phosphorus	Total Kjeldahl Nitrogen
Sedimentation						
Without chemicals	20-60	30	34	30-90	20	38
Chemically assisted	68	68	45	.....	..	..
Swirl concentrator/ flow regulator	40-60	25-60	..	50-90	..	..
Screening						
Microscreens	50-95	10-50	35	.....	20	30
Drum screen	30-55	10-40	25	60	10	17
Rotary screens	20-35	1-30	15	70-95	12	10
Disc strainers	10-45	5-20	15	.....	..	..
Static screens	5-25	0-20	13	10-60	10	8
Dissolved air flotation[a]	45-85	30-80	55	93[b]	10	35
High rate filtration[a]	50-80	20-55	40	55-95	55	21
High gradient magnetic separation[d]	92-98	90-98	75	99	50	..

[a]Process efficiencies include both prescreening and dissolved air flotation with chemical addition.
[b]From pilot plant analysis.
[c]Includes chemical addition.
[d]From bench scale and small scale pilot plant operation, 1 to 4 L/min (0.26 to 1.06 gal/min).

Sedimentation facilities have a relatively high land requirement, the captured primary sludge must be appropriately handled, and the solids removal efficiency is a function of the combined sewage influent flow rate, which can vary significantly.

(b) Swirl and helical concentrators—Swirl and helical concentrators regulate both the quantity and quality of combined sewage. Solids separation is caused by the inertia differential resulting from a nonlinear path of flow travel. The flow is separated into an overflow, and a concentrated low volume of wastewater that is intercepted for treatment at the treatment plant. Interceptor or storage capacity must be available for the concentrated effluent.

(c) Screening—Screening provides high rate separation of particulate matter from combined sewage. Pretreatment screening devices are most appropriately used for CSO applications. These include drum, rotary, and static screens. Removal efficiencies range from 5–20% of the suspended matter. Removal efficiency tends to increase as influent suspended solid concentrations increase. Only small removal increases are effected by the use of additional screens of the same size in series.

(d) Dissolved air floatation—Dissolved air floatation uses the small air bubbles that form on suspended particulate matter to float the particulate matter for removal. After the combined sewage is pressurized with air, depressurization releases the air as small bubbles. Oil, grease, and other floatables can also be removed. Small and light suspended matter can be removed more efficiently and quickly than by sedimentation. Chemical addition is sometimes used to improve removal efficiency. Operating costs are relatively high due to the compressed air requirement, and the process is sensitive to operational control.

(e) High rate filtration—High rate filtration has been used to capture suspended solids on a fixed bed for anthracite coal and sand. High rate filtration was developed for the treatment of industrial wastes. Backwashing is used to prevent the clogging of the filter. Pretreatment may be required to remove coarse solids.

(f) Disinfection—Disinfection does not quantitatively remove any of the organic or inorganic combined sewage pollution. Disinfection is used to control pathogenic organisms, and to prevent significant populations of living pathogenic organisms from entering receiving waters. It is often required as minimum treatment for CSOs. Generally, disinfection is effected by adding chemical oxidizers to the overflow. Chlorine and sodium hypochlorite are among the chemicals used. Adequate mixing and detention time must be provided, and dechlorination may be required.

4. Sewer Separation

Sewer separation can be defined as the division of an existing combined sewer system into non-interconnected sanitary and storm sewer systems. The sanitary sewer system is tributary to a wastewater treatment facility, and the storm sewer system discharges directly to receiving waters. Complete sewer separation will eliminate CSOs. However, heavily urbanized areas can generate significantly polluted stormwater runoff, and the negative impacts on a receiving stream may be significant. In fact, some European countries which have made major investments in sewer separation over the past decade have now decided that this practice is not environmentally acceptable. These countries have prohibited further separation, concentrating instead on source control and treatment of combined sewage.

Generally, sewer separation is effected by maintaining the existing combined sewers as either the storm or sanitary component of the new separated system. CSO mitigation may also be achieved by partial separation. The construction of new storm sewers to relieve selected portions of the combined system can reduce CSOs. Disconnection of rain leaders should also be considered if sewer separation is a viable alternative. Flow from rain leaders can be a significant portion of the stormwater entering a combined system.

F. Selection of CSO Mitigation Techniques

Mitigation techniques for CSOs include a number of possible technologies (discussed previously), as well as a range of degrees of control. Technologies include structural solutions—storage, treatment or conveyance facilities, and nonstructural or minimal structural solutions—source control programs, combined sewer operation and maintenance practices. Controls can be implemented to protect the receiving water quality for a design runoff return period, such as 1, 2, or 10 years. The objective of evaluations of CSO mitigation techniques is to determine the most practical and effective combination of technology and degree of control to achieve the water quality goals.

A cost-effectiveness analysis is usually performed as a means of collectively evaluating all of the factors of CSO control technologies, degree of control, and receiving water impacts. In such an analysis, the costs for various CSO mitigation alternatives are compared to the resulting benefits of water quality protection. Typically, a cost and benefit analysis will demonstrate that significant improvements in water quality protection can be attained through relatively low cost maintenance practices, especially if dry-weather overflows are a problem. Benefits to receiving water quality are estimated by using land-based and receiving water models discussed earlier.

After water quality improvements resulting from improved maintenance are achieved, the cost to provide higher degrees of protection generally increases geometrically. Typically, optimal CSO mitigation alternatives lie in the area of the "knee" of a cost-benefit curve where the marginal cost of CSO mitigation is small compared to the resulting benefit.

Once the range of solutions has been narrowed, detailed evaluation of a small number of CSO mitigation alternatives should be performed at the level of preliminary designs for specific sites. Given specific sites, subsurface soil conditions, depth of construction, constraints on use and access to the site, and environmental impact, mitigation features must be considered.

During the detailed evaluation of alternatives, the means of disposing of screenings and residual solids and handling and storing chemicals should be addressed. In storage and pump-back schemes, it is necessary to determine the capacity and routing of a force main and pumping station to de-water the storage tank.

V. REFERENCES

American Society of Civil Engineers. (1969). "Design and construction of sanitary and storm sewers." *Manuals and Reports on Engineering Practice No. 37*, ASCE, New York.

American Public Works Association. (1970). *Combined sewer regulator facilities.* APWA, Chicago, IL.

Bedient, P.B., Harned, D.A., and Characklis, W.G. (1978). "Stormwater analysis and prediction in Houston." *Journal of the Environmental Engineering Division*, 104(EED).

Binder, J.J., Katz, P.B., and Ripp, J.A. (1979). "The use of simplified modeling and monitoring as a screening technique to quantify urban runoff." *Proc of the international symp on urban storm runoff*, University of Kentucky, Lexington, KY.

Bradford, W.L. (1979). "Urban stormwater pollutant loadings: A statistical summary through 1972." *Journal of the Water Pollution Control Federation*, 49(4).

Brombach, H. (1990). "Equipment and instrumentation for CSO control."
Urban stormwater quality enhancement—source control, retrofitting, and combined sewer technology, Proceedings of An Engineering Foundation Conference, ASCE, New York.

Collins, P.G., and Ridgway J.W. (1980). "Urban storm runoff quality in southeast Michigan." *Journal of the Environmental Engineering Division*, 106(EED), 153–162.

Davis, W.K., and Browne, F.X. (1977). "Planning methodologies for analysis of land use/water quality relationships: Case study application." *EPA-440/3-77-025*, U.S. Environmental Protection Agency, Washington, D.C.

Doering, R. et al. (1987). "Real-time control of urban drainage systems, the state-of-the-art." *IAWPRC/IAHR* Joint Committee on Urban Drainage, Task Group on Real-Time Control of Urban Drainage Systems, International Association on Water Pollution Research and Control, London, England.

Donigian, A.S. and Crawford, N. H. (1976). "Modeling nonpoint pollution from the land surface." *EPA-600/3-76-033*, U.S. Environmental Protection Agency, Cincinnati, OH.

Driscoll, E.D., DiToro, D.M., and Thomann, R.V. (1979)."A statistical method for assessment of urban runoff." *EPA-440/3-79-023*, U.S. Environmental Protection Agency, Washington, D.C.

Driver, N.E., and Tasker, G.D. (1988). "Nationwide regression models for predicting urban runoff water quality at unmonitored sites." *Water Resources Bulletin*, 24(5).

"Experts Discuss Private Sector I/I" *Water Engineering and Management*, September, 1983.

Fair, G.M., Geyer, J.C., and Okun, D.A. (1966). "Wastewater systems." *Water and Wastewater Engineering, Volume I, Water Supply and Wastewater Removal*; John Wiley & Sons, Inc., New York, N.Y.

Graham, P.H. (Project Officer). (1978). "Report to Congress on control of combined sewer overflow in the United States." *EPA-430/9-78-006*, USEPA, Washington, D.C.

Griffin, D.M., Jr., et al. (1978). "An examination of nonpoint pollution export from various land use types." *Proc. of the international symp. on urban stormwater management*, University of Kentucky, Lexington, KY.

Hartigan, J.P., et al. (1978). "Calibration of urban nonpoint pollution loading models." *Proc. of the American Society of Civil Engineers Hydraulics Division specialty conf. on verification of mathematical and physical models in hydraulic engineering*, ASCE, New York.

Heaney, J.P., Huber, W.C. and Nix, S.J. (1976). "Storm water management model, level I preliminary screening procedures." *EPA-600/2-76-275*, U.S. Environmental Protection Agency, Cincinnati, OH.

Huber, W.C., et al. (1980). "Urban rainfall-runoff-quality data base: Update with statistical analysis." *Contract No. 68-03-0496*, U.S. Environmental Protection Agency, Cincinnati, OH.

Kibler, D.F., ed. (1982). "Urban stormwater hydrology." *Water Resources Monograph 7*, American Geophysical Union, Washington, D. C.

Lager, J.A., et al. (1977). "Urban stormwater management and technology: Update and users guide." *EPA-600/8-77-014*, Cincinnati, OH.

Leiser, C. (1974). "Computer management of a combined sewer system." *EPA 670/2-74-022*, U.S. Environmental Protection Agency, Washington, D. C.

Manning, M.J., Sullivan, R. H., and Kipp, T.M. (1977). "Nationwide evaluation of combined sewer overflows and urban stormwater discharges, volume III: Characterization of discharges." *EPA-600/2-77-064c*, U.S. Environmental Protection Agency, Cincinnati, OH.

McKee, J.E. (1947). "Loss of sanitary sewage through storm water overflows." *Journal Boston Society of Civil Engineers*, 34(55).

McPherson, M.B. (1980). *Study of Integrated Control of Combined Sewer Regulators*. Municipal Environmental Research Laboratory, USEPA, Cincinnati, OH.

Merritt, F.S. (1968). *Sanitary engineering*. Standard Handbook for Civil Engineers, McGraw-Hill, New York, NY.

Metcalf, L., and Eddy, H.P. (1914). "Design of sewers." *American Sewerage Practice*, Volume 1, McGraw-Hill, New York, NY.

Metcalf and Eddy, Inc. (1981). *Wastewater engineering, collection and pumping of wastewater*. McGraw-Hill, New York, NY.

Polls, I. and Lanyon, R. (1980). "Pollutant concentrations from homogeneous land uses." *Journal of the Environmental Engineering Division*, 106(EE1), 69–80.

Randall, C.W., Grizzard, T.J., and Hoehn, R.C. (1977). "Runoff sediment loads from urban areas." *Proc. of the international conference on urban hydrology, hydraulics, and sediment control*. University of Kentucky, Lexington, KY.

Sartor, J.D. and Boyd, G.B. (1972). "Water pollution aspects of street surface contaminants." *EPA-RZ-72-081*, U.S. Environmental Protection Agency, Washington, D.C.

Schilling, W. (1986). "Urban runoff quality management by real-time control." *Urban runoff pollution, Proc. of a NATO advanced research workshop, NATO ASI Series G, Vol. 10*, Springer-Verlag, Heidelberg, Germany.

Sutherland, R.C. (1980). "An overview of stormwater quality modeling." *Proc of the international symposium on urban storm runoff*, University of Kentucky, Lexington, Kentucky.

Sutherland, R.C. and McCuen, R.H. (1978), "Simulation of urban nonpoint pollution," *Water Resources Bulletin*, 14 (2).

U.S. Environmental Protection Agency. (1976). "Procedures for assessment of urban pollutant sources and loadings." *Areawide Assessment Procedures Manual*, Vol. 1, EPA-600/9-76-014, U.S. Environmental Protection Agency, Cincinnati, OH.

U.S. Environmental Protection Agency. (1983). *Results of the nationwide urban runoff program (NURP)*. Water Planning Division, Washington, D.C.

Water Pollution Control Federation. (1989). "Combined sewer overflow pollution abatement." *Manual of Practice FD-17*, WPCF, Alexandria, VA.

Wischmeier, W.H. and Smith, D.D. (1965). "Predicting rainfall-erosion losses from cropland east of the Rocky Mountains." *Agriculture Handbook No. 282*, United States Department of Agriculture, Agricultural Research Service, Washington, D.C.

Wu, J.S. and Ahlert, R.C. (1978a). "Assessment of Methods for Computing Storm Runoff Loads." *Water Resources Bulletin*, 14(2), 429–439.

Wu, J.S. and Ahlert, R.C. (1978b). "Prediction and analysis of stormwater pollution." *Proc. of the International Sym on Urban Stormwater Management*, University of Kentucky, Lexington, KY.

Wycoff, R.L. and Mara, M.J. (1978), *1978 needs survey—continuous stormwater pollution simulation system—users manual*. EPA-430/9-79-004, U.S. Environmental Protection Agency, Washington, D.C.

Young, G.K., Bondelid, T.R., and Athayde, D.N. (1978). "Simplified urban NPS estimation." *Proc. of the International Symp on Urban Stormwater Management*, University of Kentucky, Lexington, Kentucky.

Young, G.K., Bondelid, T.R., and Athayde, D.N. (1979). "Urban runoff pollution method." *Journal of Water Resources Planning and Management Division*, 105 (2).

Zison, S.W., et al. (1977). *Water quality assessment—a screening method for non-designated 208 areas*. EPA-600/9-77-023, U.S. Environmental Protection Agency, Athens, GA.

CHAPTER 11

DESIGN OF STORMWATER IMPOUNDMENTS

I. INTRODUCTION

Stormwater impoundments are flood flow detention or retention facilities. The larger ones are similar to the small dams that the Soil Conservation Service has built all over the country for the last half century. However, the majority of stormwater impoundments are smaller and have been designed for individual developments. They are intended to control peak flood flows in accordance with some local criteria (as opposed to being part of some overall drainage plan).

The design characteristics depend upon the local objectives, which may differ across the country. In addition to peak flow reduction, these objectives can include water quality enhancement, creation of open space (in urban areas), public recreation, creation of wildlife habitat, groundwater interception or recharge, and enhancement of property values.

Stahre and Urbonas (1990) list a total of nine different categories of stormwater detention that can be used to meet these various local objectives. For reasons explained later in the chapter, a system of impoundments designed for flood control can also provide a substantial degree of water quality control at little or no increase in capital cost. Stormwater management criteria therefore should normally include water quality control provisions. For best results, such facilities should be planned on a regional basis.

A. Environmental Considerations

Stormwater impoundments must be designed with adequate consideration of environmental impact. Either wet or dry basins, unless properly maintained, may become breeding grounds for mosquitoes or other insects. In some cases, destruction of large trees can cause adverse repercussions.

One particularly serious environmental issue is preservation of wet-lands. Fresh and saltwater wetlands are protected by federal law, and sometimes by state or local jurisdictions (see Chapter 2). In evaluating stormwater management interfaces with a wetlands program, it is usu-ally not clear what specific wetlands attributes are being protected, or whether the need for protection is absolute or can be traded off for other values.

All surface impoundments can act as settling basins for suspended solids. During construction, such impoundments can often be deepened to temporarily trap sediments from the construction site. They can be restored to their permanent design function after construction has been completed.

B. Water Quality Provisions

As previously discussed in Chapter 5, stormwater from developing and developed areas characteristically contains various contaminants (heavy metals, hydrocarbons, nutrients, and bacterial contamination) in amounts that are environmentally unacceptable.

In stormwater management, water quality control may be obtained through dual-purpose basins designed, first, to reduce flood damages downstream, and second, to reduce nonpoint source pollution from storm runoff. Although similar in concept to the much earlier sediment control and flood retention programs of the U.S. Soil Conservation Service, the idea of using stormwater detention basins to reduce en-vironmental pollution first gained currency through the Section 208 (of the Federal Water Quality Act) water quality planning studies started in 1975.

The underlying principle of dual-purpose detention is that the de-tention of flood flows for reduction of damages downstream, and the retardation of flood flows for settlement of particulates, can advanta-geously be combined in the same structure. Flood damages are almost entirely due to floods with return periods of more than two years, whereas the harmful pollution impacts occur mainly as the cumulative effect of a large number of small storms (see also Chapter 12). Storage of the runoff from storms of up to two-year frequency, for periods of 24–36 hours, in either wet ponds or extended dry detention basins, can reduce total contaminants by approximately 60% for lead and hy-drocarbons, and can achieve somewhat lower removal efficiencies for phosphates and other contaminants (Stahre and Urbonas 1990). By means of longer detention times, more complex forms of dual pond detention, or special wet basins, even higher water quality control efficiencies can be obtained (Hartigan 1989).

The design of dual-purpose detention basins is in most respects the same as that of other detention basins. The main differences are with respect to outlets. The lower portions of a detention basin, uses for storing the water quality design storm, should, wherever possible, be designed to maximize flow length from inlets to discharge points and avoid short circuiting. To keep metals bound to sediment particles, it is desirable that sediment pH be kept near 7 and that sediments remain aerobic. This more apt to be a problem in deep wet ponds.

Infiltration basins are sometimes used effectively for water quality improvement. Infiltration basins are usually built to recharge ground water, and care must be taken to avoid introducing unacceptable concentrations of contaminants. Infiltration basins are excellent from the viewpoint of instream water quality, since no sediments are released downstream.

Finally, there are circumstances in which stormwater detention basins may be built for water quality control only. Water quality needs are especially acute in coastal areas. Many small watersheds draining into tidal waters are not greatly affected by fluvial floods. Storm tides may be of predominant concern. In such cases, it may be unnecessary to require developers to provide detention storage to reduce peak storm flows. However, the seacoast, bays, and estuaries into which these streams drain are often environmentally fragile, and may be seriously affected by bacterial pollution, hydrocarbons, heavy metals, nutrients and other contaminants which the stormwater from developed areas may contain. The prolonged retention and sedimentation of runoff from small storms, such as the one-year frequency storm, will greatly reduce the adverse impact of storm runoff in such cases. The proper hydraulic design of the principal spillway can also provide temporary sediment storage. See Chapter 12 for further discussion.

II. TYPES OF IMPOUNDMENTS

The form that impoundments or detention facilities can take is almost limitless, as are their distribution, number, and size (APWA 1981; Taggart et al. 1982). However, there are two basic approaches to designing detention storage. When facilities are planned on an individual basis for the site being considered (rather than as a part of some overall regional plan), the structure may be referred to as "on-site" detention storage. The others are "regional" detention facilities (note that Stahre and Urbonas (1990) identified the two types as "source control" and "downstream storage" facilities, and that other designations have also been used).

On-site facilities usually are designed to control the short, intense storms that produce the highest peak flow, and are usually not located most advantageously for reduction of downstream flood damages. The total volume of runoff of such storms is quite small, and the detention time is relatively short. Therefore, unless design criteria require very low release rates, the effect upon flood peaks disappears rapidly downstream. Under some conditions, detention only in the lower portion of a watershed may actually increase the peak flow at the outlet of the watershed. Under other conditions, detention storage in the upper portion of a basin appears to be less effective.

The principal advantage of on-site facilities is that developers can be required to build them as a condition of site approval. Major disadvantages include the larger total land area required compared to regional basins, and, above all, the nuisance (odor, appearance) potential due to the difficulty of ensuring maintenance of all of these small facilities.

Facilities designed as part of a watershed planning process, in which the stormwater management needs for the watershed as a whole are developed in a staged "regional" plan, are called regional facilities. They provide more storage, and are more apt to be designed for longer release periods. The main disadvantages of the regional or watershed planning approach are the complex arrangements required to collect funds from developers and to use those funds efficiently for the intended stormwater management structures. Another disadvantage of regional basins is that they can leave substantial portions of the stream network unprotected, and plans must avoid this condition.

Generally, a stormwater management plan for a watershed incorporating a regional facility can produce more economical and effective results than the numerous small detention basins that usually result when each developer of a site provides his own detention facilities (Dendrou and Delleur 1982). As noted earlier, regional facilities also offer potential benefits, such as recreation and water quality enhancement. Despite this, the practical and institutional factors are so strong that by far the greater number of stormwater impoundments built in North America are on-site, rather than regional, facilities.

The major types of stormwater impoundments are:

(a) Detention—The temporary storage of flood water which is usually released by a measured but uncontrolled outlet. Detention facilities typically flatten and spread the inflow hydrograph, lowering the peak. Structures that release storage over a period of 12–36 (or more) hours may also serve water quality purposes (State of New Jersey 1986).
(b) Retention—Storage provided in a facility without a positive outlet, or with a specially regulated outlet, where all or a portion of the inflow is stored for a prolonged period. Infiltration basins are a common type of retention facility. Ponds that maintain water permanently, with freeboard provided for flood storage, are probably the most common type retention facility.

Detention and retention facilities can be further subdivided into:

(a) On-Stream Storage—A facility that intercepts the streamflow directly. On-stream storage occasionally is provided as an on-site facility, though it is more often an integral part of a watershed or regional stormwater plan.
(b) On-site Facilities—Special attention must be given to the design of outlet structures for controlling runoff from rooftops, parking lots, and

swales. Because runoff volumes from such areas are small, the required outlets are also small, which increases the potential for plugging by debris. Also, the outlet must release temporarily-stored water in a reasonable amount of time. As an example, parking lots must drain relatively fast in order not to be a nuisance. Roof top storage must be designed so as to provide safety of the structure if outlets are plugged.

(c) Off-Stream Storage—Diversion of flow out of the stream into a separate storage facility. A typical example is a side channel spillway that diverts storm flows from the stream into a storage impoundment (or a structure that can divert and store the "first flush" of particularly contaminated runoff).

(d) Conveyance Storage—Conveyance storage is an often neglected form of storage, because it is dynamic and requires channel storage routing analysis to identify. Slower-flowing conveyance caused by flatter slopes or rougher channels can markedly retard the buildup of flood peaks and alter the time response of the tributaries in a watershed.

(e) Wet Basins and Infiltration Basins—Wet basins are detention basins designed to maintain a permanent pool of water. In most aspects their design is similar to other detention basins (dry basins), except for the permanent pool. Wet basins are used for aesthetic or water quality enhancement, or for the maintenance of fish or wildlife. All outlets are above the normal level of the pool. Infiltration basins resemble other detention basins in most respects, though they may be built without

Copper Mountain, Colorado—Wet retention pond in a mountain resort area. A log wall has been provided for splash protection.

Denver, Colorado—Wet detention pond on a site with limited space. Excellent example of what is possible if a landscape architect is involved in the facility design.

outlets. They may retain flood flows for a prolonged period of time, for the purpose of encouraging infiltration into the groundwater.

When evaluating the relative merits of extended dry detention versus wet detention ponds, there are several factors to consider (see also Chapter 12 for further discussion on this subject). Extended dry detention generally requires much less storage volume (land area) than wet ponds (Hartigan 1988), yet despite this, wet ponds have other offsetting advantages, which include:

(a) Wet ponds generally provide more pollutant removal.
(b) Wet ponds are usually viewed as an amenity, which will enhance property values (and offset their higher cost).

Since the water in a detention structure may cause an inconvenience to those using the land for other purposes, the temptation is great to modify or eliminate the on-site detention after construction by changing the outlet. For example, perimeter curbing that serves as the outlet control for parking lot storage is frequently damaged by owners of the lot to release water at a faster rate. This is an institutional problem that needs to be addressed locally to ensure continuous and reliable functioning of such outlet facilities. Proper location and protection of such outlets can reduce these problems. Also, public education programs may be helpful. Manufactured outlets that are more difficult to alter can be used to protect roof drains.

III. THE TWO BASIC APPROACHES

A. On-site Impoundments

The typical regulatory approach to stormwater management is to require each developer to create on-site detention storage to reduce the peak outflow from his site after development, to the outflow that would occur from the same storm under pre-development conditions. Some jurisdictions require that the total post-development outflow volume not exceed the total pre-development outflow (however this is impracticable to achieve unless site conditions permit inflitration). More realistically, the criteria may specify that post-development **peak** outflow from a given storm may need to be reduced from its pre-development **peak** to compensate for the increase in volume.

When stormwater management first became common, some impoundments were designed for control of runoff from only a single return period, the 100-year storm. This sounds like an impressive criterion, but is in actuality rather ineffective, since such a design may provide little control of smaller floods, may actually increase downstream erosion, and may decrease the effect on control of larger floods downstream.

To illustrate the value of multiple-storm regulatory criteria, a comparative analysis was made of an assumed 8000 acre drainage basin in New Jersey, controlled by an on-site detention facility on each of 400 identical 20-acre watersheds (Whipple et al. 1983). The results are shown in Table 11.1.

It will be seen that Design 1, which requires only on-site control of a 100-year design flood, has virtually no effect on floods at a point downstream where the stream drains the entire 8000 acres. Design 2, which controls both 10-year and 100-year storm on-site, is somewhat better downstream, but only for large storms. The flood peaks below the frequency of 10 years, which largely determine channel erosion tendencies, are affected very little. Design 3, however, which requires on-site control of 2, 10, and 100-year floods, as well as retention of particulates, is most effective in reducing all types of floods downstream; however it also falls short in total control.

It is apparent that, in such a watershed, the requirement to control only the 100-year flood is completely ineffective a few miles downstream. On the other hand, three times as much storage is required to control peak flows for the 2, 10, and 100 year floods, and to obtain retention and settling of particulates. It is noteworthy that even with

TABLE 11.1.   Comparison of Effectiveness Downstream of Various Designs.

Flood Design & Criterion (1)	Storage, Per Basin, ft³ (2)	% Reduction in Peak Flow for 8000 Acre Watershed		
		2-yr (3)	10-yr (4)	100-yr (5)
1. Single outlet, control 100-yr flood to pre-D peak	38,000	0	0	2
2. Single outlet, control 10 and 100-yr floods to pre-D peak	89,000	1	7	30
3. Triple outlet, particulate retention and control 2, 10, and 100-yr storms to pre-D peak	113,000	44	24	20

these relatively strict criteria, the post-development peak flows at the bottom of the basin are larger than those occurring pre-development, the total volume of flood flow was considerably increased, and downstream landowners obviously do not receive adequate protection.

In another example, an investigation was made of three different design criteria; (a) regulating the 10-year flood to pre-development levels, (b) regulating the 100-year flood to pre-development levels and (c) limiting both the 10-year and 100-year floods both to pre-development levels. This analysis demonstrated that design for a single frequency, such as the 100-year flood event, has only a marginal effect on lesser events such as the 10-year, particularly as the area of the watershed increases. Design criteria addressing both the 10- and 100-year runoff can result in facilities that significantly control those events and also control increases in flood peaks further downstream (Urbonas and Glidden 1982). The engineer is strongly urged to examine a range of frequencies when undertaking any impoundment design.

The testing of criteria for on-site detention usually requires what has been called the "micro-macro" approach. In this approach, several small test areas (micro areas) representing typical subdivision and on-site detention designs are modeled. Runoff hydrographs for both pre- and post-development conditions are prepared for each micro area for the precipitation events desired. These micro hydrographs are then input to the macro model, which represents the appropriate major stream network. The advantage of the technique is that only one or a few discrete micro areas need to be modeled to identify the typical response from subdivision-type developments, with sufficient accuracy for basin-wide policy determination.

B. Watershed Planning and Regional Detention Basins

The regional approach to stormwater planning and management is generally preferable to the piecemeal site-by-site approach, though design objectives are usually not as clear-cut for watershed planning as for on-site developments (where the primary purpose usually is outflow regulation, rather than downstream flood control). It is usually economically more effective to develop a watershed or regional plan which will result in the control of storm flows in a manner equivalent to that achieved if the flows were controlled on a site-by-site basis.

Regional plans usually provide for a relatively small number of regional impoundments for flood control only, supplemented by a large number of small on-site facilities designed to retain the small floods in the interest of water quality and erosion control. The action of the larger regional impoundments makes it unnecessary to provide control of the larger floods at individual development sites. Figure 11.1 shows a watershed with three regional basins with smaller on-site facilities for water quality control. The alternative would be to require flood control storage at each of the sites shown.

The interrelated flood routings of multiple reservoirs within a watershed should be done iteratively, evaluating the relative merit (or impact) of various combinations. Deletion or addition of impoundments may point out the optimal combination. The watershed stormwater management plan may require different design criteria for different facilities. Unless an entire watershed is to be developed at one time there is no necessity to build all the master plan detention basins at once. They must, however, be carefully designed for staged construction.

Example 11-1: Tulsa, Oklahoma (Onsite vs. Regional Storage) As part of a master plan for the 3,000 acre Vensel Creek watershed prepared for the City Engineering Department of Tulsa, Oklahoma, various options for storage were evaluated (Taggart 1978). Vensel Creek (see Figure 11.2), is a tributary of the Arkansas River, with six upper tributaries which are well-defined streams in hilly wooded bluffs overlooking the Arkansas River Valley. The lower stream collects flows from the tributaries as it travels south roughly parallel to the Arkansas River. This lower stream reach is in the floodplain of the Arkansas. Its hydraulic capacity is quite limited, with overflows occurring at many points during heavy precipitation events. The overflows travel generally west to the Arkansas.

An important part of this study was to investigate regional and on-site storage options. Tulsa criteria called for regulation of developed flows from both the 5-year and 100-year design rainstorms. Thus, it was also important to test this ordinance within the context of the probable watershed response, the desired objective of limiting peak flows to historic levels, and identification of any potential for aggravating the problem.

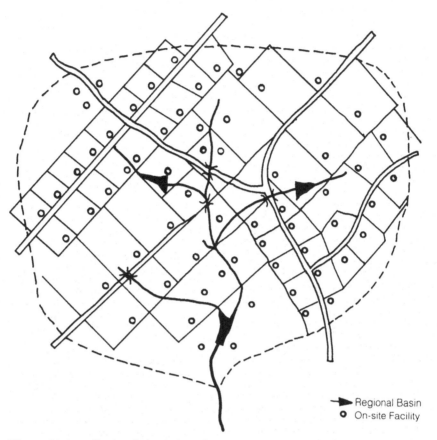

Figure 11.1.—Regional detention plan.

Micro models were prepared for all the types of development proposed. These micro watersheds varied in size from approximately 8–15 acres. Detailed modeling on tributaries *E* and *H* (see Figure 11.3) allowed comparisons in basins that were largely undeveloped. Proposed development in Sub-basin *E* is largely to be conventional residential zoning with a mixture of commercial and open space, and in Sub-basin *H*, residential with lots ½ to ¾ acre in size. Detailed hydrologic analysis was performed by means of a kinematic wave model (MITCAT) that can reflect both existing and proposed conditions, including typical subdivision layout, streets, and conveyance components. For the on-site detention option, the micro model was expanded to include impoundment facilities that meet Tulsa Design Criteria.

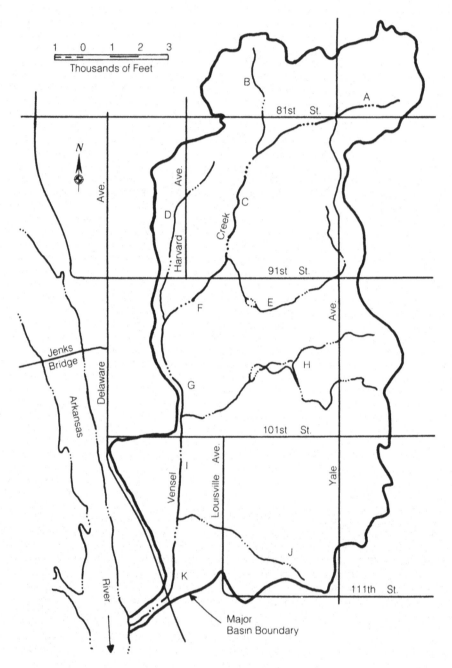

Figure 11.2. — Vensel Creek drainage basin.

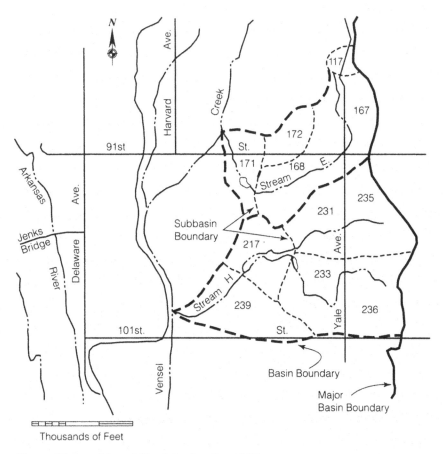

Thousands of Feet

Figure 11.3.—Vensel Creek basins E and H.

To evaluate the overall effectiveness of the on-site detention systems, the routed hydrographs from the micro models were entered into the macro watershed model by, in effect, entering multiple micro hydrographs which would correctly reflect the total watershed area and runoff volume. Care was taken to disperse the micro hydrographs throughout the basin. For regional storage facilities, the evaluation of impoundments was simpler, reflecting storage facilities along each of the tributaries.

Table 11.2 summarizes the results for the test areas. The analysis illustrated that, in this particular case, similar storage volumes would be required for either on-site or regional systems, and that peak flow reductions are roughly comparable (note that this is **not** always the case). This particular watershed could not easily accommodate larger storage sites, so on-site facilities were favored. The reader, however,

TABLE 11.2.   Summary of Vensel Creek On-site and Sub-regional
Storage Test

	Sub-basin E (1)	Sub-basin H (2)
Test Areas, Acres	287	472
100-Year Undeveloped Peak Flow (cfs)	697	780
100-Year Developed Flow Without Impoundments (cfs)	1062	1361
Sub-regional Detention System		
Resulting Test Area Peak Flow, cfs	678	726
Storage Required Acre Feet	35	47
On-Site Detention System		
Resulting Test Area Peak Flow, cfs	675	816
Storage Required Acre Feet	33	52

should not conclude that on-site facilities are more likely to provide
the best solution in other cases.

IV. DETERMINING STORAGE AND OUTLET
CHARACTERISTICS

While "simplified techniques" (rational method, etc.) have been widely
used in the analysis and design of storage, they have many drawbacks,
related primarily to the fact that the intensity-duration-frequency relationships upon which they are based were originally intended for the
calculation of peak flow rates and **not** for the calculation of either the
runoff hydrographs or flow volumes. The wide variety of computer
models now available for calculating runoff hydrographs and storage
volumes using actual rainfall distributions makes their use the method
of choice for the analysis and design of detention/retention storage.

A. Rational Method

For certain small drainage areas, a detention basin sizing procedure
based on the rational method may be acceptable. A simplified technique
has been developed for computing the detention volume of stormwater
runoff (Federal Aviation Agency 1966).

The following example, borrowed from the Urban Drainage and Flood
Control Manual (Urban Drainage and Flood Control District 1984), is

based on an analysis of storm runoff at an airport where a storage cell is created by taxiways. The tributary area to the inlet is 49.5 acres, and the analysis is shown for the storms having both 5-and 10-year frequencies of occurrence.

The cumulative rainfall for the 5- and 10-year frequency was used as the rate of supply. Table 11.3 is a tabulation of the rainfall intensity in inches per hour for various storm durations, as well as the data for the cumulative runoff and the discharge for a 33-inch diameter pipe. These

TABLE 11.3.   Computations for Ponding Example in Figure 11.4.

Hourly Intensities for Various Time Intervals (in/hr)

Time	5 yr. Frequency	10 yr. Frequency
5 min.	5.76	6.48
10 min.	4.92	5.70
15 min.	4.24	4.76
20 min.	3.72	4.19
30 min.	2.92	3.38
60 min.	1.87	2.28
90 min.	1.36	1.73
120 min.	1.09	1.40
180 min.	0.81	1.02

$Q = CIA$

A = 4.48 Acres, Pavement

= 45.04 Acres, Turf

= 49.52 Acres, Total

C = 0.90 For Pavement

= 0.30 For Turf

I = 2.00 in.

Q = 0.354 × 2.00 × 49.52 = 35.06 c.f.s.

n = 0.015   S = 0.7%

Distance most remote point—1600'

120' across pavement, 1480' across turf

Concentration Time: 4.5 + 50.5 = 55 minutes

$$\text{Average } C = \frac{4.48 \times 0.90}{49.52} + \frac{45.05 \times 0.30}{49.52} = 0.354$$

$CA = 49.52 \times 0.354 = 17.53$

Runoff rate when all areas contributing

33" pipe will carry 38 c.f.s.   1 hr. = 3600 × 38 = 136,800 c.f

Cumulative Runoff in cu. ft.

I = 5.76 (From above)

$Q = CIA$ $CA = 17.53$

$Q = 17.53 \times 5.76 = 100.97$ c.f.s.

For 5 min. for 5 yr. frequency

5 min. = 300 seconds

100.97 × 300 = 30292 cu. ft.

Thus:

Minutes	5 yr. Frequency	10 yr. Frequency
5	17.53 × 5.76 × 300 = 30292	17.53 × 6.48 × 300 = 34078
10	17.53 × 4.92 × 600 = 51749	17.53 × 5.70 × 600 = 59953
15	17.53 × 4.24 × 900 = 66894	17.53 × 4.76 × 900 = 75098
20	17.53 × 3.72 × 1200 = 78254	17.53 × 4.19 × 1200 = 88141
30	17.53 × 2.92 × 1800 = 92138	17.53 × 3.38 × 1800 = 106653
60	17.53 × 1.87 × 3600 = 118012	17.53 × 2.28 × 3600 = 143886
90	17.53 × 1.36 × 5400 = 128740	17.53 × 1.73 × 5400 = 163765
120	17.53 × 1.09 × 7200 = 137575	17.53 × 1.40 × 7200 = 176702
180	17.53 × 0.81 × 10800 = 153352	17.53 × 1.02 × 10800 = 193110

(Urban Drainage and Flood Control District 1969).

data have been plotted in Figure 11.4. The discharge capacities for 21-, 24-, and 30-inch pipes are also plotted for comparison purposes.

Computations indicated that if the outlet was placed slightly more than 2 feet below the maximum pond elevation, the storage capacity would be 243,300 cubic feet. The 21-inch pipe would provide sufficient discharge to keep the maximum ponding down to 102,500 cubic feet 60 minutes after the start of the runoff for the 10-year frequency storm; however, this pipe would not empty the pond for perhaps an additional three hours.

In view of these considerations, the 33-inch pipe could be reduced in size because the smaller pipe can dispose of the ponded volume. Without the pond, the diameter of the outlet pipe would be 33 inches for the 5-year storm, and 36 inches for the 10-year storm. The results of this procedure are considered adequate for the objectives involved.

Where control of several storms of different frequency, say, 2, 10, and 100 year, is required, and for larger areas, an iterative approach is required. The designer assumes sizes for three outlets, intended to have a specified aggregate discharge. The smallest of the outlets is the lowest. Hydrographs are then prepared for post-development flows of the more frequent (smaller) storms, and a test is made of the required degree of control of each storm, starting with the smallest. A few trials will determine which combination of outlets will provide the desired degree of control of each storm. The Colorado Urban Hydrograph Procedure (CUHP), described in Chapter 5, is a good procedure for determining the effect of storage on storm runoff. The Modified Puls method described in Chapter 6 may be used for storage routing.

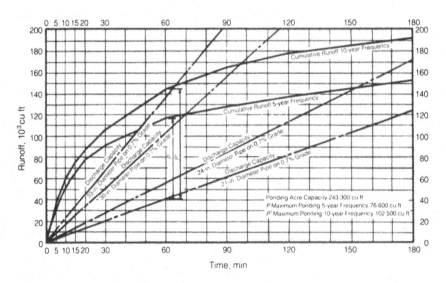

Figure 11.4.—Cumulative runoff for ponding in Table 11.3.

Where prolonged retention of runoff of small storms is required in the interest of water quality, a similar process is used. However, in this case, for small watersheds, the outflow from the bottom outlet is usually so small that it may be neglected. The detention basin capacity required for settlement of particulates is used to establish the elevation of the invert of the lowest flood control outlet, and design for control of floods continues as indicated above.

The above process is extremely simple, but reasonably reliable, if care is taken in estimating the coefficient C, and if good rainfall-frequency data are available.

B. Soil Conservation Service Methods

The Soil Conservation Service (SCS) has adapted the methods in Section 4 of the National Engineering Handbook (USDA 1985) to urban areas. The methods are applicable to small and large catchments. Their applicability to large catchments is based on a summation of small basin hydrographs. They are well adapted to showing differences in runoff due to differences in land use, soil type, and soil cover. For most of the United States, they are based upon use of a Type II storm of 24-hour duration, which contains within it shorter periods of more intense rainfall. These SCS methods are far more comprehensive, and cover a far wider range of conditions, than the rational method. In particular, they give volumes of runoff and complete hydrographs, rather than just peak flows (see Chapters 5 and 7 for further discussion of SCS methods).

When the entire hydrograph at a downstream location is required, it is necessary to perform a watershed routing, using the sub-catchment's runoff histogram and unit hydrograph. Space does not permit showing this calculation here. However, the method includes hydrographs derived from computer calculated watershed routings. These tabular method hydrographs can be used to develop hydrographs for a variety of commonly-occurring situations.

For various sub-catchment times of concentration, a family of hydrographs is presented. Each hydrograph represents a particular reach travel time (T_t). The hydrograph for $T_t = 0$ represents a point at the mouth of the sub-catchment. The other hydrographs are for point T_t-hours downstream from the sub-catchment mouth. This allows one to add the separate effects of each sub-catchment to obtain the total hydrograph at a particular point of interest. More specifically the procedure is:

(a) Determine the drainage area, time of concentration, and 24-hour depth of runoff for each sub-catchment (sub-basin).
(b) Calculate the travel time (T_t) from each sub-catchment to the mouth of the watershed by adding the travel times of the reaches through which it travels.
(c) For each sub-catchment, select the tabular hydrograph for the appropriate combination of t_c and T_t.

(d) Multiply these routed sub-catchment hydrographs by the sub-catchment's drainage area (in mi^2) and 24-hour runoff (inches).

(e) Add the sub-catchment hydrographs of step 4 to obtain the total watershed direct runoff hydrograph.

For single detention basins, once the flood hydrographs have been determined for pre-and post-development conditions, the pre-development flows are used to derive the allowable outflows from the proposed impoundment. The outlet characteristics then can be determined, by an approach similar to that described above for the rational method.

C. Other Modeling Procedures

There are many computer-based models (some of which are discussed in Chapter 7) suitable for the modeling of runoff and storage. The designer must choose one that is compatible with his or her capabilities and equipment, and use it to develop the inflow hydrograph(s) for the impoundment facility. The risks of working without historic precipitation and runoff data are substantial, and the reader is referred to Chapters 5 and 7 for further discussion of this important subject.

D. Provisions to Bypass Flows from Upstream

Stormwater facilities must provide for the conveyance of upstream runoff (i.e. bypass flows) while providing the necessary storage. Regulations in some areas specify a minimal bypass rate to be provided. The designer must consider the impacts of upstream runoff in both the existing and future states of watershed development. The requirement to bypass flood flows should be based on existing conditions upstream, and on the assumption that future increases in maximum discharge due to development will be controlled by the developer at that time. Provisions for bypassing of low flows through a detention facility may complicate the retention of small flood flows for water quality improvement. Spillway designs should consider full development conditions upstream in the future, since stormwater management upstream will not prevent the passage of the largest potential flood flows. If, because of particular site conditions, it is impractical for the detention basin to control all of the runoff from a development, it may be possible to control an equivalent amount of flows that not originate on-site, and that otherwise would be bypassed.

V. OUTLETS AND TRASH RACKS

A. Outlet Types

Outlets are designed for the planned release of water from the impoundment (this does not include the emergency spillway described in the following section). The Soil Conservation Service uses the term principal spillway for the main outlet of its small dams; but the usual practice for detention basins is to refer to outlets. The outlets for detention basins are ordinarily uncontrolled (i.e. without gates or valves), and there are usually several of them. The outlets may consist of separate conduits, or several outlets may enter a chamber or manifold that leads to a single pipe or conduit (preferably from a dam safety standpoint). Outlets for detention basins must be protected by trash racks. Two excellent general references on impoundment outlet design are ASCE (1985) and Schueler (1987). The latter provides many examples of outlets that enhance the ability of impoundments to trap pollutants.

Figure 11.5 shows outlets in a dual purpose detention basin, providing water quality control features. Figure 11.6 shows a single outlet for wet basins. To provide capacity at lower elevations, holes can be cut in the riser at appropriate elevations. Figure 11.7 shows a multiple stage

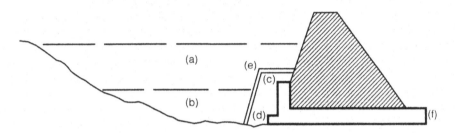

Figure 11.5.—Dual-purpose detention basin outlets.

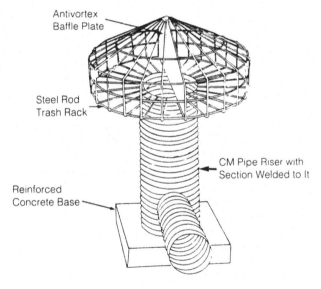

Figure 11.6.—Typical wet basin outlet.

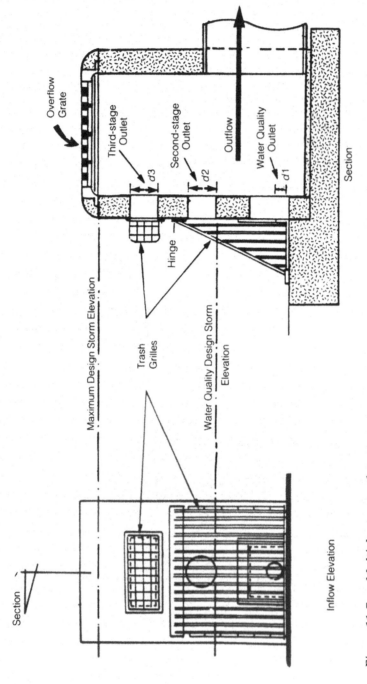

Figure 11.7. — Multiple-stage outlet structure.

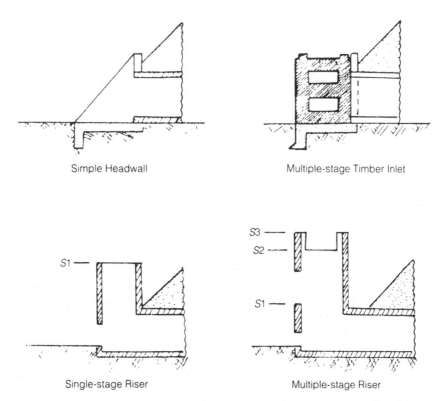

Figure 11.8. — *Typical inlet arrangements (for pond outlet structures).*

outlet structure for a dual-purpose detention basin. Figure 11.8 (ASCE 1985) shows inlets for several types of outlet structures.

The water quality outlet will usually be small in comparison with other outlets, and a restrictor plate containing an orifice designed to provide the required outflow capacity should be used in place of a small pipe in the structure. The restrictor plate can be removed to facilitate maintenance. The capacity computation is made by the usual orifice formula.

The capacity of other outlets is determined by weir, orifice, or pipe formulas, depending upon the design. In a design such as that of Figure 11.7, the discharge of the second stage outlet would be determined by an orifice formula, and that of the third stage by a weir formula.[1] The outlet conduit is sized to carry all discharges. In this design, the overflow grate is designed primarily as a relief outlet, in the event of

[1]Rossmiller (University of Wisconsin 1990) has determined through laboratory studies that complex outlet structures (consisting of multiple sizes and shapes of orifices and weirs) do not perform in accordance with hydraulic theory, and caution is urged in the design of such outlets.

clogging of the others, although it would also serve to add somewhat to spillway capacity.

Where conduits are intended to function under hydraulic head for prolonged periods of time, they should have seepage-drainage diaphragms, or geotextiles should be used to control erosion of fines. Where outlet conduits discharge into easily eroded materials, stilling basins or other energy-absorbing devices should be used. In the case of on-stream overflow dams, a notched spillway may function as an outlet as well as an emergency spillway (see also Chapter 14 for other dam safety considerations).

B. Trash Racks

The susceptibility of inlets to clogging by debris and trash needs to be considered when estimating their hydraulic capacities. In most instances, trash racks will be needed. Trash racks must be large enough that their partial plugging will not adversely restrict flows reaching the control outlet. No universal guidelines exist for the design of trash racks to protect urban stormwater detention control outlets, although a commonly used "rule-of-thumb" is to have the trash rack area at least ten times larger than the control outlet orifice. For very small outlets, an even larger ratio is usually necessary to control the onrush of debris at the onset of a storm, and a high degree of maintenance is required.

Examples of trash racks are shown in Figures 11.6 and 11.7. The inclined vertical bar rack is most effective for the lower stage outlets. Debris will ride up the trash rack as water levels rise. This design also allows for removal of accumulated debris with a rake while standing on top of the structure. Cage type racks or racks with horizontal members inhibit this type of debris removal.

The surface areas of all trash racks should be maximized and the trash racks should be designed to be as far away from the protected outlet as possible to avoid interference with the hydraulic capacity of the outlet. The spacing of trash rack bars must be proportioned to the size of the smallest outlet protected. Where a small water quality outlet orifice is involved, a separate trash rack for that outlet is frequently used, so that a simpler, sturdier trash rack with more widely spaced members can be used for the other outlets. Spacing of the rack bars should be wide enough to avoid interference, but close enough to provide the level of clogging protection required.

To facilitate removal of accumulated debris and sediment from around the outlet structure, the racks should have hinged connections. If the rack is bolted or set in concrete it will preclude removal of accumulated material and will eventually adversely affect the outlet hydraulics. Maintenance access, including access for heavy equipment, should be provided, as well as a means to drain the pond, if necessary.

Since sediment will tend to accumulate around the lowest stage outlet, the inside of the outlet structure should be depressed below the water quality outlet to minimize clogging due to sedimentation. Depressing the outlet bottom to a depth below the water quality outlet and equal to the diameter of the outlet is recommended (Figure 11.7).

C. Outlet Safety

Outlet safety considerations include both the safety of the structure and safety to the public. The outlet works create a potential hazard when in operation due to the possibility of a person being carried into the opening. Gratings or trash racks are often used; however, with substantial pressure a person can be forced against the grate or trash rack which can, in some cases, be worse than being carried through the conduit. To mitigate this, low entrance velocities at the trash rack are recommended. Fencing or other effective measures also should be provided to exclude people from potentially hazardous areas. Such measures include site grading, planting of thorny shrubs, or grading to assure "safety ledges" along the pond perimeter.

Outlet works can be designed to reduce the hazard to the public where heavy recreational use is anticipated. For instance, a vertical riser of concrete, timber, or steel can have a series of small openings (12" or less) from top to bottom with sufficient total area to cause low velocity at the entrances. The top of such risers can be grated, or even closed. In some instances the outlet works can be fenced. Fences are not universally recommended because of maintenance and operational needs, and because most fences do not fully prevent access. Signs are sometimes used to warn the public of the safety hazards involved at the outlet works.

During the periods of no operation, there is little hazard at most outlet works, although they can be attractive nuisances during operation. The designers need to be aware that an owner can be held liable, if an accident occurs, for having created an attractive nuisance. Design of outlets for which the orifices or weirs are not accessible from the embankment or shore when functioning is one method of reducing the hazard to playing children or curious adults, but this may complicate maintenance. Pipe openings on the upstream face of the embankment, where the pipe is only partially submerged, may be fenced on three sides to inhibit access from the embankment.

Finally, although not related to failure from overtopping, designers must recognize that a substantial percentage of embankment failures are due to inadequate outlet works design and construction. Chapter 14 discusses the subject in detail. The designer is reminded to direct special attention to the following:

(a) Avoid potential piping of water along the outside of the outlet conduits by using drainage-seepage diaphragms, or by careful material selection and good compaction around the conduit.
(b) Minimize the number of conduits through the embankment.
(c) Ensure against leaky joints within the embankment.

(d) Do not use thin-walled conduit through the embankment without a protective exterior encasement.

(e) Where reasonable, design the pipe to operate under little or no internal water pressure.

(f) Provide a safety factor in outlet works openings to account for debris collection. Spillway and outlet entrances are natural locations for debris buildup. Design the pond to minimize debris migration to both. Provide for vehicle and crane access so that debris can be removed from a basin when in operation.

(g) Do not depend upon human intervention to operate gates or other controls during a storm runoff event. Gates can jam or become inoperable when needed during emergencies. People can be unavailable or diverted by other activities.

VI. SPILLWAYS, EMBANKMENTS, AND UNDERDRAINAGE

A. Emergency Spillways

In many cases, stormwater detention structures do not warrant elaborate studies to determine spillway capacity. While the risk of damage due to failure is a real one, it normally does not approach the catastrophic risk involved in the overtopping or breaching of a major reservoir. The drainage areas of many sites are so small that only very short, sharp thunderstorms are apt to threaten overtopping or dam failure, and such storms are very localized. Also, capacities of the impoundments are usually too small to create a flood wave.

By contrast, regional, on-line facilities or smaller, onsite dams with homes immediately downstream, may pose a significant hazard if failure were to occur, in which case emergency spillway considerations are a major design factor. The engineer must characterize the potential for loss of life or property damage early in the design effort.

Dam safety awareness has improved nationwide as the result of analyses and surveys by the U.S. Army Corps of Engineers. In most states spillways of even small dams must provide a specific degree of protection by safe passage over the spillway of a specific frequency flood, usually some fraction of the Probable Maximum Flood (PMF), or the 100-year flood for small structures in undeveloped localities. Criteria for undeveloped localities should be applied with prudence, as flood plains uninhabited today may be largely developed in a few years time.

The Soil Conservation Service has provided general guidance on minimum spillway capacity for non-agricultural impoundments as shown in Table 11.4 (U.S. Soil Conservation Service, undated). It should be noted that this is **not** a national standard, it has been adjusted for New Jersey conditions and is provided simply to indicate that such tables exist. It is not recommended by ASCE, and the authors of this manual suggest that the criteria not be used without careful analysis.

TABLE 11.4.   Emergency Spillway *Minimum* Design Storms (Non-
Agricultural) (From U.S. Soil Conservation Service,
undated)—Adjusted for New Jersey Conditions

Raised Water Height (ft)[1] (1)	Total Storage (acre-ft)[2] (2)	Drainage Area (acres) (3)	Emergency Spillway Design Storm[3] (4)
≤5	≤50	≤20	10 yr 24 hr Type II or III
5.1–20	≤50	≤20	100 yr 24 hr Type II or III
≥20	≤50	≤20	100 yr 24 hr Type II or III
≤5	≥50	≤20	50 yr 24 hr Type II or III
5.1–20	≥50	≤20	100 yr 24 hr Type II or III
≥20	≥50	≥20	100 yr 24 hr Type II or III
≤5	≤50	21–320	25 yr 24 hr Type II or III
5.1–20	≤50	21–320	100 yr 24 hr Type II or III
≤20	≤50	21–320	100 yr 24 hr Type II or III
≤5	≥50	21–320	50 yr 24 hr Type II or III
5.1–20	≥50	21–320	100 yr 24 hr Type II or III
≥20	≥50	21–320	100 yr 24 hr Type II or III
≤5	all	≥320	50 yr 24 hr Type II or III
≥5	all	≥320	100 yr 24 hr Type II or III

(1) Measured from the lowest point on the downstream toe of the dam to the emergency
spillway crest or, in the absence of an emergency spillway, the top of the dam.
(2) Measured below the crest of the emergency spillway.
(3) Any pond for which a state stream encroachment permit is required must use a
minimum design storm of 100 yr 24 hr Type II or III for the emergency spillway.

In many instances, the small size of a detention pond virtually pre-
cludes a PMF spillway. In those cases, means to mitigate the adverse
effects of overtopping of the embankment can include:

(a) Flattening the downstream embankment face.
(b) Armoring the dam crest and downstream embankment face.
(c) Using regulated floodplain delineation and occupancy restrictions
downstream representative of conditions without the detention storage.
(d) Providing extra channel capacity downstream.
(e) Using a wide embankment crest, such as is common with urban roads
and streets (where rapid failure seldom occurs due to modest overtopping depths).
(f) Using non-eroding dam material such as rolled soil cement or concrete.
(g) Using small tributary basins, where the rate and volume of discharge
involved are limited, resulting in overtopping flows of short duration
and non-hazardous proportions.

If the engineer does not adopt the PMF as the spillway design flow
for a sizeable structure, he should so inform the client and be cognizant
of the increased liability that he and the facility's owner may be incur-

ring. The safest method available for justifying less than the PMF is an "incremental drainage analysis," which demonstrates that during the selected design, failure of the detention embankment will not create unacceptable increases in flood stage and velocity at critical downstream cross sections.

The emergency spillway is proportioned to pass flows in excess of the design flood without allowing overtopping of the embankment. Flow in the emergency spillway is open channel flow. Normally, it is assumed that critical depth occurs at the control section. For the larger structures, to avoid the possibility of an eroded channel developing in the spillway when the spillway is not lined, it is good practice to put a small concrete curb and cutoff wall in the throat as a point of hydraulic control.

Soil Conservation Service manuals (USDA 1982) provide guidance for the selection of emergency spillway characteristics for different soil conditions and different types of vegetation. The selection of degree of retardance for a given spillway depends on the vegetation. Knowing the retardance factor and the estimated discharge rate, the emergency spillway bottom width can be determined. For erosion protection during the first year assume minimum retardance.

B. Embankments

An embankment that raises the water level a specified amount as defined by the appropriate state dam safety group (generally 5–10 feet or more above the usual mean low water height, when measured along the downstream toe of the dam to the emergency spillway crest), is classified as a dam. Such embankments must be designed, constructed and maintained in accordance with state dam safety standards. All other detention basins with embankments should be designed in accordance with the following criteria (which are not intended as a substitute for a thorough, site-specific engineering evaluation).

(a) Side Slopes—For ease of maintenance, the side slopes of the settled embankment should not be steeper than 4 horizontal to 1 vertical (See Table 11.5).

(b) Freeboard—The elevation of the top of the settled embankment shall be a minimum of one foot above the water surface in the detention basin with the emergency spillway at the maximum design flow. All state dam safety criteria must be carefully considered when determining the freeboard capacity of an impoundment.

(c) Settlement—The design height of the basin embankment should be increased by 10% where hauling equipment is used for compaction and 5% where compaction equipment is used. All earth fill should be free from brush, roots, and other organic material subject to decomposition. The fill material in all earth dams and embankments should be compacted to at least 95% of the maximum density obtained from compaction tests performed by the Modified Proctor method of ASTM D698.

Boulder, Colorado—Dam safety is enhanced by flattening upstream and downstream side slopes of the dam and maximizing the width of the dam crest.

Table 11.5 gives minimum recommended side slopes for earth embankments, from a stability point of view (these must be confirmed by thorough geotechnical evaluation). To provide a safety factor and facilitate maintenance, however, side slopes of 4:1 (or flatter) are recommended.

Typical dam top widths are provided in Table 11.6. The embankment, emergency spillway, spoil and borrow areas, and other disturbed areas should be stabilized and planted with appropriate vegetation. Structural analysis of earthen embankments is covered in more detail in Chapter 14.

TABLE 11.5.   Generalized Side Slopes for Earth Embankments for Stability (USDA 1982).

Fill Material (1)	Side Slope Horizontal to Vertical*	
	Upstream (2)	Downstream (3)
Clay CH Clayey sand SC Sandy clay CL Silty clay CL Silty sand SS or	3–1	2–1
Clayey gravel GC Silty gravel GM	2.5–1	2.5–1
Silt ML or MH Clayey silt ML	3–1	3–1

TABLE 11.6.   Recommended Top Width for Earth
Embankments (modified for urban
settings) (USDA 1982)

Height of Dam (feet) (1)	Top Width (feet) (2)
Under 10	8
10 to 15	10
15 to 20	12
20 to 25	14

C. Underdrainage of Impoundment Areas

The feasibility, community acceptability, and maintenance costs of impoundments are significantly related to the facility drainage, both surface and subsurface, and the general groundwater situation in the area of the impoundment. There are many examples of facilities that can only be maintained at great expense because equipment cannot operate in soft bottom conditions, which cannot be used for recreation because the ground remains too wet, or which are generally aesthetically unacceptable.

Underdrainage or subsurface drainage of impoundments should be considered in each design case. The extent of effort and required facilities varies greatly. At one end of the spectrum, no special underdrainage improvements are necessary because depth to groundwater is large, water movement through the subsoils is fairly rapid and surface drainage in the impoundment is good. At the other extreme, however, any one of many variables can cause significant soil moisture and groundwater level problems.

1. Water Sources Contributing to Underdrainage

There are many potential sources of water that may create the need for underdrainage facilities. They include precipitation/ infiltration, irrigation, utility leakage, trapped runoff due to poor surface drainage, clogged trash racks, normal streamflow via groundwater and capillary movement, flood flows via groundwater movement, general groundwater lateral movement, and general groundwater upflow. Most facility designs must consider the first five of these sources. Frequently more sources can exist, however, which can create complex interactions that require significant effort to manage. The relative magnitude of the flows involved is usually small, but if not handled properly they can effectively ruin an impoundment. The following paragraphs will discuss some typical situations and suggested management actions. For further information see U.S. Bureau of Reclamation (1978) and U.S. Department of Agriculture (1962).

2. Surface Drainage Management

The usual surface drainage network includes graded areas and conveyances such as swales or conduits. For general surface grading, slopes of 2% or greater are recommended, with a minimum of 1% with special subsoils and underdrainage, or where the site is acceptable as open space. For conveyance grading, a 0.75–1.0% slope is recommended. As a minimum, a 0.5% slope is acceptable with special sub-drainage facilities or when acceptable as open space, as long as adequate relief is provided.

Generally speaking, when surface drainage is less than 2%, one finds that puddling, trapped water, damaged vegetation and conversion to nuisance plants occurs. The more frequent precipitation events and irrigation flows can leave the conveyance and large adjacent areas wet, unusable, and unmaintainable for long periods of time.

There is also water that will, regardless of the surface drainage system, continue to move by underdrainage, such as infiltrated precipitation and irrigation flow. Once these are identified they can be combined with the other subsurface sources for groundwater movement and management analysis.

3. Groundwater Management

It is most desirable to have the invert of the basin well above the groundwater table, and separated from it by permeable material. This is frequently not the case, and if the invert of the basin is within 3–5 feet of the groundwater level, sub-drainage facilities will be required unless the designs specify a permanently or periodically inundated pond bottom. Figure 11.9 illustrates a situation where an impoundment is to be constructed within a few feet of the existing groundwater table. At the uppermost end of the basin it may be desirable to lower the groundwater table sufficiently to prevent wet and unstable soils. A sub-drainage line placed above the barrier layer can intercept a portion of the groundwater flow.

Irrigation, which is common in the western states, can contribute significantly to groundwater problems. Numerous techniques are available to help manage this source (USBR 1978).

Frequently, impoundments have streams that flow through or along them. These flows can contribute to or create high groundwater table conditions and/or wet unstable soils. Figure 11.10 illustrates a valley cross section showing a stream, the invert of the adjacent impoundment and an intermediate embankment which might be used in an off-stream storage scheme. In this particular situation, the basin invert is 2' above the typical stream flow water level. With the highest groundwater table shown, the groundwater will migrate toward the stream. Obviously, the depth to the groundwater table will be unacceptable. Agronomists and agricultural engineers regard 3–5 feet as a minimum depth to

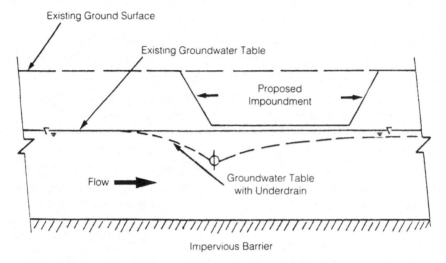

Figure 11.9.—*Impoundment close to groundwater table.*

groundwater for reasonable vegetation conditions, and soils engineers expect unstable conditions (pumping will occur) with less than 3 feet. Even if the stream loses flow to the groundwater system (shown by the middle groundwater profile in Figure 11.10), it is possible for the depth to the groundwater table to be too shallow, especially when considering local sources such as irrigation, precipitation or infiltration. If the impoundment invert cannot be raised, an underdrain system may be required.

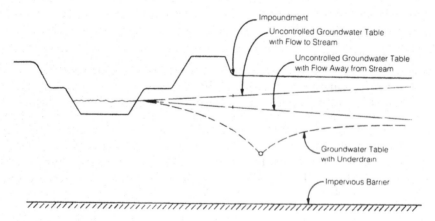

Figure 11.10.—*Impoundment near stream.*

Finally, it is advisable to consider long-term effects of the impoundment on the sub-drainage system. Long-term sedimentation within the impoundment can form an impervious layer which could hinder the sub-drainage and surface drainage system in the impoundment.

A riser and cap may be installed at the time of initial construction for maintenance of underdrainage. This will allow attachment of the underdrain lines to a pressurized water line to flush out the underdrain lines and the drainage field. This technique, similar to back pumping on a well screen, can extend the effective life of an underdrainage system, particularly in silty soils.

VII. SPECIAL APPLICATIONS

A. Detention Basins in Flood Plains

Stormwater management and floodplain management are separate and distinct programs (see Chapter 1), each with its own rationale, but there are some unavoidable interfaces. Though desirable, it is not always possible to locate all stormwater detention structures entirely above the floodplain. Much otherwise valuable land is located within the flood hazard area, and unless it lies within the floodway, it would be inequitable to preclude its development. Therefore, the conditions must be examined under which floodplain and stormwater management objectives may be reconciled.

When land is developed in the floodplain, or only slightly above it, it is almost impossible to avoid locating the detention basin itself in the floodplain. In such a case, when the detention basin is needed to store stormwater runoff, there is a chance that the floodplain will already be flooded, and that the detention basin will already be filled by floodwater from the main valley and rendered ineffective.

Obviously, the size of the drainage area of the main channel is relevant. If it drains several hundred square miles or more, the chance of interference is rather slight, since the local flooding is more apt to come from short storms than from the prolonged or extensive general rains required to bring the main stem to flood stage. On the other hand, if the drainage area is small, the same storms will probably affect the development site and the floodplain.

In the case of the floodplain of a minor stream, a computation showing the probable effect of detention storage can be made on the assumption that the design storm, or storms, will occur simultaneously on the entire watershed. Where a design storm of 100-year frequency is used, such a computation will presumably show detention storage within the floodplain to be virtually useless for controlling that storm. Where the design storm is of higher frequency, such as 10-years, storage provided at elevations lower than those defining the flood hazard area may be effective.

B. Wet Basins

Wet basins are usually built for one of two reasons. They are frequently provided to enhance the value of adjacent properties fronting on the resulting lake, or for the improvement of water quality. They are designed to be aesthetically pleasing, with curving shapes and even islands. The engineering design of embankments, outlets and spillways of wet basins is similar to that of any other detention basin, except that the bottom of the basin is below the level of the lowest outlet, with appropriate modifications of the sides and bottom. Usually the banks require rip-rap, masonry, or gravel protection at the low water line to prevent erosion, though vegetation stabilization is sometimes adequate. The depth should be at least six feet, if aquatic plants are to be discouraged. If aquatic plants are to be encouraged, depths of less than 2–4 feet are required. Goldfish or other small fish have been used for mosquito control.

The removal of stormwater pollutants in a wet basin is accomplished by a number of physical, chemical, and biological processes. Gravity settling removes particles through sedimentation with the removal rates directly related to the pond's geometry, volume, residence time and the size of the particles. Flocculation occurs when heavier sediment

Dallas, Texas—The "hard edge" approach to wet detention pond design.
Safety of small children is an obvious concern, although the wall is attractive.

Orlando, Florida—Good wet pond design, with flat slides slopes and a safety ledge at the interface between the ground and the water. Emerging vegetation in the littoral zone will be attractive and will enhance water quality.

In addition to being aesthetically pleasing, fountains can aid in maintaining suitable dissolved oxygen levels and in reducing algae blooms.

particles overtake and coalesce with smaller, lighter particles. The opportunity for particle contact increases as the depth of the permanent pool increases. Biological removal of dissolved stormwater pollutants includes metabolism by microorganisms that inhabit the bottom sediments, and by aquatic plants, if these are allowed to grow.

Additional pollutant removal occurs during the relatively long quiescent period between storms. The permanent water pool reduces runoff energy and provides a habitat for aquatic plants and algae—the biological filter that removes dissolved nutrients and metals. Aerobic conditions at the bottom of the permanent pool will maximize the uptake of dissolved pollutants (nutrients, metals) by bottom sediments and minimize release from the sediments into the water column (Yousef et al. 1985).

Designing wet detention systems to achieve a specified level of pollutant reduction is very difficult. Two approaches typically have been used to develop design criteria. One method relies upon solids settling theory and assumes that all pollutant removal is due to sedimentation (Driscoll 1983). The other approach views the wet detention system as a "lake" which achieves a controlled level of eutrophication, thereby causing biological, physical, and chemical assimilation of stormwater pollutants in addition to sedimentation (Hartigan and Quasebarth 1985). The basis for this design method is that stormwater is detained within the permanent pool long enough to produce adequate levels of nutrient uptake by algae and aquatic plants, but the hydraulic residence time is not so long as to induce stagnation, thermal stratification, or anaerobic bottom sediments.

Wet ponds should be shallow enough to minimize the risk of thermal stratification but deep enough to assure that algal blooms are not excessive. A mean depth of 3–10' normally achieves these conditions and has been shown to be effective in reducing stormwater pollutants (USEPA 1983).

A shallow littoral zone is an important component of the wet detention system, since aquatic plants within this zone provide biological assimilation of dissolved stormwater pollutants. The littoral zone should cover at least 30% of the pond's surface area and slope gently (6:1 or flatter) to a depth of 2' below the control elevation (Schueler 1987; ASCE 1990).

The littoral zone should include a variety of native aquatic plant species suitable for various depth ranges, nutrient assimilation or aesthetic purposes. The zone must either be planted with appropriate aquatic plants or covered with a 4–6 inch layer of topsoil containing the viable seeds of wetland plants. A combination of mulching plus planting one-third of the littoral zone is very cost effective and will provide a vigorous biological filter fairly quickly.

The long term viability of the biological filter is essential for good pollutant removal. Routine maintenance will include mowing along the pond perimeter, transplanting desirable wetland species into areas needing plants, and removal of undesirable aquatic plants such as cattails. Cattails will crowd out other more desirable aquatic plants, and also deposit a large amount of vegetative matter, which decays and creates anoxic conditions and poor water quality.

Pond geometry has a very strong influence on how effectively the detention system will remove pollutants, especially in systems with a short hydraulic residence time. Little or no pollutant removal occurs in dead storage areas where the inflow is bypassed without mixing. To avoid dead storage areas, the length to width ratio should be at least 3:1. In addition, the outlet structure should be located to maximize travel time from the inlet to the outlet. The effective length of a pond can be increased by the use of diversion barriers such as baffles, islands, or a peninsula within the pond.

C. Infiltration Basins

An infiltration basin needs to be shallow with a low rate of surface application. The seasonally high groundwater table should be located at least 4' below the bottom of the basin. Similarly, bedrock should also be located at least 4' below the bottom of the basin. Otherwise, unacceptable ponding may take place, resulting in grass kill and insect nuisances.

The soil permeability, or final infiltration rate, will determine how rapidly the stormwater will infiltrate the ground. Soil classes with final infiltration rates of at least 0.52" per hour (New Jersey Department of Environmental Protection 1989) allow for acceptable drain times, provided that the criteria for depth to high water table and bedrock are satisfied. In addition, soil porosity above the seasonal high water table needs to be considered. Design for a two-year runoff to fill less than one-half of the unsaturated soil mantle should provide a reasonable chance of keeping the basin from backing up and failing. Note that one foot of water depth in the infiltration basin becomes more than three feet in the soil mantle when the soil has a porosity ratio of 0.33. The water that infiltrates through the basin develops a temporary groundwater mound which drains slowly in the horizontal direction due to the very small available hydraulic gradient. This is the reason that a large storage depth in an infiltration basin has often been associated with its failure.

The soil textural class with a final infiltration rate of 0.27" per hour (silt loam) may have some limited suitability for a very shallow infiltration basin. Soils with infiltration capacities that allow them to infiltrate 36" of stored runoff in less than a 3-day period are particularly well suited for an infiltration basin.

Infiltration basin design should avoid the introduction of pollutants that violate groundwater quality criteria or standards.

1. Water Table, Bedrock, and Groundwater Conditions

Concerns related to the development of a groundwater mound below the wet pond or infiltration facility, as well as the potential for polluting down-gradient groundwater supplies, often arise when infiltration facilities are considered. Based on a limited data base for stormwater impoundments and infiltration facilities, groundwater pollution does not appear to be a problem with most residential and commercial land uses. Under many conditions, the addition of groundwater by means of detention basins is highly desirable.

2. Runoff Filtering

Grease, oil, floatable organic materials, and settleable solids should be removed from runoff water before it enters the infiltration basin. These materials take up storage capacity and reduce infiltration rates. Runoff filtering devices such as vegetative filters, sediment traps, and grease traps can be used to remove objectionable materials. A modified basin design such as that illustrated in Figure 11.11 can be used to enhance and prolong the infiltration capacity of the basin. When a runoff filtering system or structure is included, its design must allow adequate maintenance.

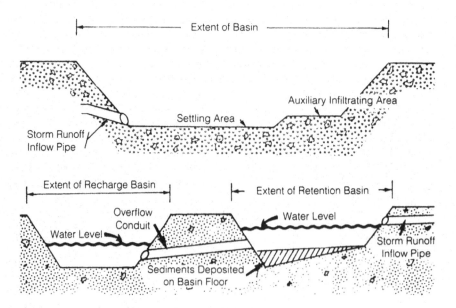

Figure 11.11.—Basins to enhance infiltration.

3. Excavation

Initial basin excavation should be carried to within 1' of the final elevation of the basin floor. Final excavation to the finished grade should be deferred until disturbed areas in the watershed have been stabilized or protected. The final excavation should remove all accumulated sediment. Relatively light tracked equipment is recommended for this operation to avoid compaction of the basin floor. After the final grading is completed, the basin floor should be deep-tilled using rotary tillers or disc harrows to provide a well-aerated, highly porous surface texture.

4. Sediment Control—Vegetated Basins

The cleanout frequency of infiltration basins will depend on whether they are vegetated or non-vegetated, and will be a function of their storage capacity, recharge characteristics, volume of inflow, and sediment load. Infiltration basins should be inspected at least once a year. Sedimentation basins and traps may require more frequent inspection and cleanout.

Grass bottoms on infiltration basins serve as a good filter material, although they may need occasional replacement. Use grass species that are most likely to work for the region of the country in which the facility is located that can withstand several days of submergence. Well-established turf on a basin floor will grow up through sediment deposits, forming a porous turf and retarding the formation of an impermeable layer. Grass filtration would work well with long, narrow, shoulder-type depressions (swales, ditches, etc.) where highway runoff flows down a grassy slope between the roadway and the basin. Grass planted on basin side slopes will help to prevent erosion.

5. Sediment Removal From Non-vegetated Basin

Sediment should be removed only when the basin floor is completely dry, after the silt layer has mud-cracked and separated from the basin floor. Equipment maneuverability and precise blade control are essential, and can greatly reduce the quantity of material to be removed. All sediment must be removed prior to tilling, which should be done at least once annually.

6. Side Slope Maintenance

Side slopes should have a dense turf with extensive root growth, which enhances infiltration through the slope surface and prevents weeds from gradually taking over. Grasses of the fescue family are recommended, primarily due to their adaptability to dry sandy soils, drought resistance, hardiness, and ability to withstand brief inundation. The use of fescues will also permit long intervals between mowings. This is important due to the relatively steep slopes which make mowing difficult. Mowing two to three times a year is generally satisfactory.

D. On-stream Impoundments

On-stream impoundments involve the use of the natural valley as the storage basin and the stream channel as the inflow-outflow conduit. Generally, an earth embankment is built to store the flood volume, although overflow structures are also used. Multiple-outlet spillways may be used to meet requirements for control of flows with different return frequencies. Open channels may also be enlarged to serve as stormwater impoundments when enough land is available and the channel has a relatively flat gradient.

On-stream impoundments are usually built as regional detention basins, but may also be on-site facilities built for a single development. In some respects, their design is similar to that of other impoundments. There are, however, some significant points of difference, which include:

(a) The on-stream impoundment must always provide for passing low stream flows. To avoid clogging and to enhance safety, the lowest outlet should be at the level of the stream bed.
(b) As a result of the need to pass low flows, it will usually be impractical to retain small storm flows in the interest of water quality.
(c) Unless the entire watershed above the impoundment is to be controlled, the emergency spillway capacity will be greater, sometimes many times greater, than would be required for an impoundment of similar storage capacity elsewhere. In such cases, an overflow dam may be used to avoid excessive spillway costs.
(d) In areas of erodible soils, sedimentation may be excessive.
(e) Backwater effects may require acquisition of rights-of-way higher than the structure itself.
(f) For on-stream impoundments, vertical drop outlets (such as those used in detention basins) must be designed with caution, because of the need to pass low flows, and because of potential sedimentation.

It should be obvious that on-stream impoundments must be designed with care and put in a proper regional context. **It is very important to note that simplified approaches, such as are often applied for small on-site detention basins, *cannot* be used to design on-stream impoundments.**

E. Oversizing Storm Sewers to serve as Stormwater Impoundments

The oversizing of storm sewers for stormwater impoundments has a narrow area of application (though it is often used in combined sewer systems to provide in-system storage). Such facilities are normally provided when land costs are extraordinarily high. For example, oversized storm sewers serving as stormwater impoundments have been used in connection with commercial developments where parking lot and roof storage availability were inadequate or considered undesirable. Oversizing storm sewers has also been used as a means of mitigating flooding problems in urbanized areas where land was not economically available.

Other limitations may include the oversized sewers' reduced ability to transport sediments, or the lack of a definitive site plan which provides for future stormwater access to the oversized sewer.

1. Sewer and Intake Sizing

The required storage volume can be determined by the methods discussed previously and in Chapter 5. The ultimate sewer size will be determined after considering upstream bypass flows, available head, physical constraints, and the overall hydraulics of the stormwater system.

Particular attention must be given to the location and types of intakes (i.e. inlets, catch basins, open grate manholes, and special structures), if the sewer's storage volume is to be fully utilized. In addition to evaluating the hydraulic characteristics of the intake structures, the designer should recognize that the effectiveness of intakes during major storm events can be significantly reduced by debris.

2. Bypass Considerations

Upstream bypass flows can rarely be economically conveyed through an oversized sewer impoundment facility, and overland conveyance should be considered.

3. Sediment Control

The low velocities that normally characterize oversized sewers require a careful evaluation of sediment deposition and removal. Estimates of the types and quantities of sediment accumulation are required to ascertain future maintenance requirements, and access for periodic sediment removal must be provided.

F. Recreation and Aesthetic Uses

Impoundment areas are often designed to improve the aesthetic quality of developments. Home sites are more valuable if adjacent to a lake; and corporate headquarters and industrial facilities often feature a small lake, which can function as a detention basin or as a water source for firefighting. In favorable climates, ducks and geese or other birds will occupy even a small permanent impoundment, and fish can provide a focus of interest. While the aesthetic qualities associated with any body of water provide benefits to urban and suburban developments, they also involve certain design constraints and maintenance responsibilities. If the impoundment area is to be an aesthetic focal point, the designer should consider the water quality that can be achieved. Runoff from nearby impervious areas could be channeled around the pond to avoid having greases and oils enter the impoundment area. If upstream areas will undergo development, control measures must be installed to prevent large amounts of sediment from entering the impoundment area.

If a permanent pool is to be established it should be as large and deep as possible.

Maintenance of any impoundment area designed to enhance the aesthetic quality of a developed area is extremely important. The designer should provide for easy maintenance of any area adjacent to the pool, of the area within the normal pool elevation including the outlet structure, and of any areas that will contribute runoff to the impoundment area.

When they are not storing water, detention basins should accommodate other uses such as parking; recreation sites ranging from soccer and baseball fields to handball and volley ball courts; and other more passive recreation uses such as shaded picnic sites, trails, and park benches. Some communities allow the impoundment area to be included in the open space areas of a development when calculating the density for zoning purposes. Thus the developer does not lose the total value of this land, though he may have to reorient the site development to accommodate the impoundment area.

When using impoundment areas for other uses it should be remembered that the design must accommodate both uses. Parking areas that are used for impoundments should be designed so the water never reaches depths that would damage parked cars. If some impoundment areas are to be used for soccer and baseball fields, the soils must be such that they will quickly drain to produce a usable playing surface. Any vegetation that is planted in the impoundment area must be able to withstand the expected frequency and depth of inundation.

G. Underground Impoundments

Underground impoundments should generally be avoided, except where site conditions preclude any alternative. They are ordinarily used only for on-site facilities, though there have been some very large systems built (the Chicago TARP project, Milwaukee, San Francisco, etc.). They may consist of one or more parallel tunnels, or a deep basin decked over for some particular purpose. The primary problem is one of maintenance, including the removal of both trash and accumulated sediment. Underground impoundments should be built with adequate maintenance access, and sufficient headroom for the operation of small mechanized equipment. Stahre and Urbonas (1990) provide many practical recommendations concerning the design of underground impoundments, and the reader is encouraged to study this reference before proceeding with the selection or design of such facilities.

H. Pump-Evacuated Impoundments

The control of stormwater may require that runoff be conveyed to retention basins or temporary storage impoundments. If the required storage volume and available surface area are such that the bottom of the required impoundment is below the grade of the conveyance channel, pumping may be required (see also Chapter 9 for a discussion of pump stations). In some instances, it may be possible to drain the upper portion of the impoundment over a control structure near the invert of the conveyance channel. Pumping would then begin sometime after the flood has subsided and the conveyance channel has receded below flood levels. Factors to evaluate during the design of a pump-evacuated impoundment include:

(a) Flood damage reduction versus construction and O & M costs.
(b) Required storage volume versus available surface area.
(c) Depth limitations due to adverse soil conditions.
(d) Groundwater recharge and pollution control.

At least two pumps should be installed to provide redundancy and reduce the probability of failure. An additional small pump is sometimes necessary to handle groundwater seepage. These pumps should be of the submersible type, and of a non-clogging design to allow passage of mud and small debris (see Chapter 9). A substantial concrete structure or pump wet well is usually preferred for support of the pumps. All pumps should be designed to start or stop automatically according to the wet well level or the receiving capability of the downstream channel.

The required storage volume can be determined by developing hydrographs for selected recurrence intervals and simulating the operation of the impoundment. The difference between the inflow and the pumped and gravity outflows is the necessary storage. Several iterations may be required to determine the optimum inlet and outlet configurations. The following examples illustrate the design considerations for pumped evacuated reservoirs.

Example 11-2: Chicago, Illinois—Retention Basin to Eliminate Overbank Flooding   This example is from a project constructed near Chicago, Illinois to reduce flooding from a small river in an urbanized region. The objective was to retain flood waters near their point of origin. The project, which was developed to eliminate overbank flooding caused by the 100-year storm was adopted for this comprehensive flood control plan to comply with the standards adopted by federal and local government agencies.

The recommended plan includes upstream channel improvements to increase the river's discharge capacity. A retention basin covering seven acres and with a storage volume of 110 acre feet was planned to intercept and retain a portion of the flood flows. The basin, shown in Figure 11.12, was located adjacent to the river and within the floodplain to minimize costs. By widening and deepening the channel, removing flow obstructions, and constructing a covered concrete conduit where space was restricted, the ability of the river to carry flood flows was increased. The basin invert is well below the conveyance channel, and pump-out facilities were required.

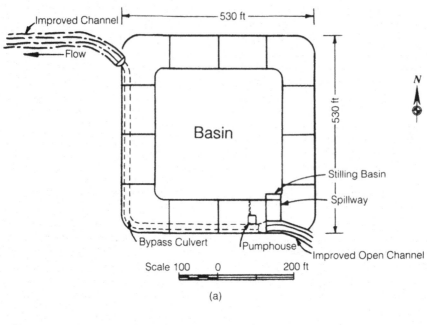

(a)

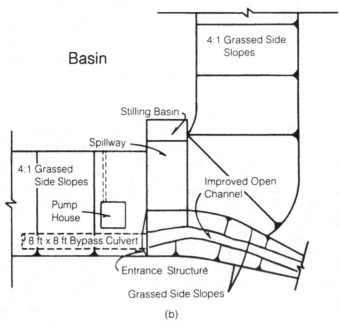

(b)

Figure 11.12a.—Retention basin (Chicago).

Figure 11.12b.—Inlet detail.

Flood water will flow into the basin over a concrete spillway and down a concrete lined channel to a stilling basin. For storms with less than a five-year return period, the river storage will not exceed the spillway crest elevation, and the entire flood will bypass the retention basin through an eight-foot square bypass culvert. As the river depth increases, flow in the bypass culvert will increase, up to a maximum of 390 cfs, and flows into the retention basin will be 450 cfs. During the 100-year storm 110 acre-feet of water will be stored in the basin with a maximum depth of 20'. The basin bank slopes are four horizontal to one vertical, which will allow easy access for grass cutting and other maintenance.

Evacuation of water stored in the basin for the 100-year storm will be by gravity backflow over the spillway from elevation 793 to the spillway crest elevation 791. Two manually-started pumps will empty the basin from elevation 791 to the basin bottom elevation 773 in about three days. A smaller, automatically operated sump pump will handle seepage.

The benefits from this project included a 50% reduction in the 100-year flow rate, and increased conveyance.

Example 11-3: Los Angeles, California—Pump-Evacuated Storage Facility The second example illustrates the use of pump-evacuated storage by the Los Angeles County Flood Control District for the Walteria Lake Project. In the design of such facilities, the District employs a 50-year storm which has the maximum rainfall intensity occurring on day four.

The design hydrograph for the Walteria Lake Project is shown on Figure 11.13. It has a peak flow rate of 3,000 cfs and a required storage volume of 1,057 acre-ft. A pump station with four main pumps, each of 55 cfs capacity, is used to drain the basin. After a 50-year design storm, nearly sixty hours of continuous pumping is required to empty the basin.

Figure 11.14 shows the method used to optimize pumping and storage costs. It is usually desirable to thoroughly study the costs of storage, pumping equipment, pump station construction, and operation and maintenance, to balance costs with allowable discharge requirements.

Other inflow hydrographs for different conditions may be found to have different shapes, peaking sooner and with less pronounced maximum inflow than the Walteria Lake example. The pumping rate or outflow is plotted on the inflow hydrograph and the excess of the curve about the pumping rate represents the necessary storage (see Figure 11.14). In any such installation, consideration must be given to safety of the structure in the event of flows in excess of the design flow.

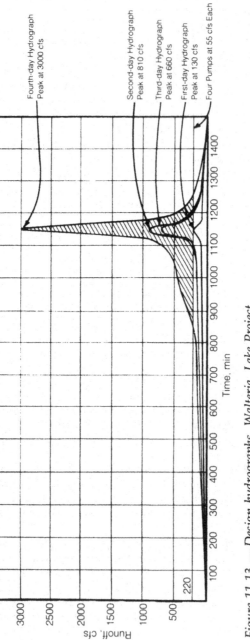

Figure 11.13. — Design hydrographs, Walteria, Lake Project.

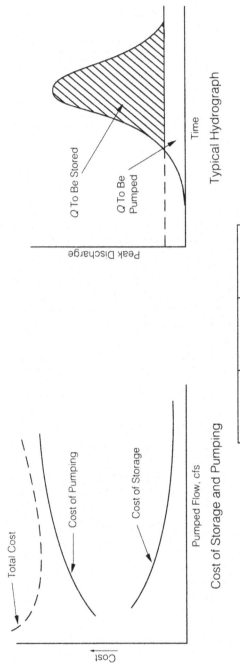

Storm Day	Rainfall, in.	Runoff, in.
First	0.65	0.19
Second	2.60	0.83
Third	2.28	0.71
Fourth	6.50	2.47

50-year Storm Rainfall and Runoff

Figure 11.14.—*Optimization of storage and pumping costs—Walteria Lake Project.*

IX. OPERATION AND MAINTENANCE CONSIDERATIONS

Economy and ease of maintenance are important considerations in regional planning, as well as in design of individual basins. Provisions to ensure reliable maintenance should be a key part of municipal regulations. Maintenance criteria for impoundment facilities include:

(a) Proper hydraulic functioning of the structure and its physical integrity.
(b) Avoidance of insect infestation or other nuisance.
(c) Safety and convenience of the visiting public, including children.
(d) Appearance of the facility.
(e) Utility for auxiliary functions such as recreation.
(f) Maintenance costs.

The responsibility for maintenance rests with the owner of the facility, however the local government should have the authority to inspect or review the facility to ensure that maintenance is being provided in accordance with municipal regulations.

Funding for maintenance, whether public or private, should be assured. In the case of private maintenance, two legal provisions should be made: (1) for public maintenance if the private organization fails in this responsibility, and (2) recorded easement to allow public access for maintenance. Situations should be avoided where developers sell off their building lots and depart, leaving no assurance that systems will be maintained. See Chapter 2 for more discussion of financing, legal, and regulatory issues, and Chapter 13 for further discussion on maintenance.

A. Maintenance Considerations In The Design Stage

Maintenance considerations during design include (ASCE 1985):

(a) If the detention period is long, and especially if children are apt to play in the vicinity of the impoundment, installation of an attractive fence or preferably, landscaping that will discourage entry (thick, thorny shrubs) along the periphery may be advisable. If the impoundment is situated at a lower grade and adjacent to a highway, installation of a guard rail would be in order. Access to the pond bottom and outlet is especially important.
(b) The impoundment should be accessible to maintenance equipment for removal of silt and debris, and for repair of inevitable minor erosion problems. Easements and/or rights-of-way are required to allow access to the impoundment by the owner or agency responsible for maintenance.
(c) In view of aesthetic requirements and the need for frequent mowing, bank slope, bank protection, and vegetation type are important design criteria.
(d) Permanent ponds should have provisions for complete drainage for

silt removal. The frequency of sediment removal will vary among facilities, depending on the original volume set aside for sediment, the rate of accumulation, drainage area erosion control measures and the aesthetic appearance of the pond.

(e) For basins designed for multiple-purpose use, the basin bottom needs special consideration to minimize periods of wetness. It may be advisable to provide an underground tile drainage system, if active recreation is contemplated and a flat bottom is required.

(f) Adequate dissolved oxygen supply in ponds (to minimize deterioration of water quality) can be maintained by artificial aeration. Use of fertilizer and pesticides adjacent to the pond should be carefully controlled.

(g) Secondary uses that would be incompatible with sediment deposits should not be planned unless a high level of maintenance will be provided. For example, planning a combination tennis court/detention basin may not be advisable downstream from an area prone to soil erosion.

(h) French drains and other small underground detention facilities are almost impossible to maintain, and should not be used where sediment loads are apt to be high.

(i) Underground tanks or conduits designed for detention should be sized and designed to permit entrance of mechanized equipment to remove accumulated sediment.

(j) All detention basins should be designed with sufficient depth to allow accumulation of sediment for several years prior to its removal.

(k) Wet basins should be of sufficient depth to discourage excessive aquatic vegetation on the bottom of the basin.

(l) Designing an outlet to minimize hazards is often "ensured" through construction of trash racks or fences, which may become eyesores, trap debris, impede flows, hinder maintenance, and, ironically, fail to prevent access to the outlet. On the other hand, desirable conditions can be achieved through careful design and positioning of the structure, as well as through landscaping that will discourage access (steep slopes, positioning the outlet away from the embankment, etc). Creative designs, integrated with innovative landscaping, can be safe and can also enhance the appearance of the outlet and pond. Such designs often are less expensive initially.

(m) To reduce maintenance, outlet structures should be designed with no moving parts (i.e., pipes, box culverts, orifices, and weirs). Manually and/or electrically operated gates should be avoided. The only exceptions are excavated storage or surface storage affected by tides or other high water that must be pumped dry, and structures with water-actuated flap gates. To reduce maintenance, outlets should be designed with openings as large as possible, compatible with the depth-outflow curve desired and with water quality, safety, and aesthetic objectives. One way of doing this is to use a larger outlet pipe and construct an orifice or a V-notch weir in the headwall to reduce outflow rates. Outlets should be robustly designed to lessen the chances of damage from debris or vandalism. The use of thin steel plates as sharp crested weirs is best avoided because of potential accidents, especially with children. Thin plate orifices must be protected by trash racks.

B. Maintenance Operations, General

The most important routine functions are cutting grass and weeds, removing sediment, repairing any erosion, and cleaning out debris. These operations must be planned and adjusted to local conditions, such as erodibility of upstream soils, prevalence of insect vectors, and the type of outlets. Small water quality outlets require more maintenance.

To maintain aesthetic appeal, floating debris must be removed from the pool surface after a storm. Therefore, to remove the debris, access must be provided for vehicles and boats. Accumulated sediment and weed growth can be removed by dredging or by excavating equipment after de-watering.

Where pumping facilities are required, the pump house should be designed to provide security and resistance to vandalism, which could include fencing and vandal-resistant doors and locks. Between storm events, maintenance should include lubrication and operation of the pumps on a regular schedule to ensure they will function when needed.

C. Outlet Maintenance

Outlets that are protected by a trash rack will accumulate trash during and between storm events. To facilitate outlet operation and maintenance, trash racks should be curved or inclined so that debris tends to ride up as the water level rises. Such a design leaves the rack clear and allows for easier cleaning during a storm event. The periodic removal of debris from orifice trash racks is the single most important maintenance aspect of any effective stormwater detention program.

Outlet structures may be partially or completely plugged by a build-up of deposited sediment, by floating plant growth such as water hyacinths, and by vegetation growing in the sediment. Sediment deposition is a natural occurrence in basins and periodic removal of vegetation and sediment is necessary to ensure that the intended hydraulic function of the outlet is not impeded. Such activities should be anticipated during design so that both maintenance access and a nearby waste disposal site are provided.

Proper design of the outlet structure can minimize the need for maintenance of the discharge end. High velocity outflow can erode the downstream outlet, foreslope, and channel. A well-designed and constructed energy dissipator, surrounded by large, well-graded riprap on the foreslope and downstream channel can do much to reduce the maintenance needed. Deep toe walls to resist scour (undercutting) should also be provided.

All portions of the outlet structure must be accessible to vehicles, equipment, and personnel between and during storm events. This includes the floor of the basin as well as ramps to points above the upstream and downstream sides of the outlet structure.

Provisions to remove particulate pollution are closely related to the desirability of prolonged retention of flood flows for flood damage reduction. From the viewpoint of capital investment, the additional water quality control function is obtained virtually without cost. However, prolonged retention of storm flows for either purpose requires a small outlet, which entails additional maintenance. The grass is cut periodically as a part of general grounds maintenance. At the same time, debris should be removed from in front of the trash racks protecting small orifices. Owners of detention basins understandably prefer large orifices which impound little or no water except during exceptionally large floods. Care must be taken to ensure that maintenance personnel do not remove orifice plates in order to simplify debris removal.

X. REFERENCES

Aaronson, D.A. and Seaburn, G.E. (1974). "Appraisal of Operating Efficiency of Recharge Basins on Long Island, N.Y., in 1969." *U.S. Geological Survey Water Supply Paper 2001-D*, U.S.G.S., Reston, VA.

American Public Works Association. (1981). "Urban Stormwater Management." *Special Report No. 49*, Chicago, IL.

American Society of Civil Engineers. (1985). "Final committee report, task committee on stormwater detention outlet structures (Urbonas, B., chairman)." Hydraulics Division, ASCE, New York.

American Society of Civil Engineers, "Urban stormwater quality management." Continuing education course notes, ASCE, New York. 1990.

Brater, E.F. and King, H.W. (1976). *Handbook of hydraulics.* 6th ed., McGraw-Hill, New York, NY.

Debo, T.N. and Williams, J.T. (1979). "Voter reaction to multiple-use drainage projects." *Journal of the Water Resources Planning and Management Division*, 105 (WR2), 295–304.

DeGroot, W., ed. (1982). "Stormwater detention facilities." *Proc. of an engineering foundation conference on stormwater detention facilities*, ASCE, New York, NY, 72–85.

Dendrou, S.A., and Delleur, J.W. (1982). "Watershed wide planning and detention basins." *Proc. of an engineering foundation conference on stormwater detention facilities*, ASCE, New York, NY, 72–85.

Driscoll, E.D. (1983). "Performance of detention basins for control of urban runoff." *Proc of the int'l symp on urban hydrology, hydraulics and sediment control*, University of Kentucky, Lexington, KY.

Federal Aviation Agency. (1966). *Airport drainage*, FAA, Washington, D. C.

Hartigan, J.P. and Quasebarth, T.F. (1985). "Urban nonpoint pollution management for water supply protection: Regional vs. onsite BMP plans." *Proc. of the intl symp on urban hydrology, hydraulics and sediment control*, University of Kentucky, Lexington, KY, 297–302.

Hartigan, J.P. (1989). "Basis for design of wet detention basin BMPs." *Design of urban runoff quality controls, Proc of an engineering foundation conference, Potosi, MO, July 10–15, 1988,* ASCE, New York, NY.

Lakatos, D.F., and Kropp, R.H. (1982). "Stormwater detention-downstream effects on peak flow rates." *Stormwater Detention Facilities, Proceedings of an Engineering Foundation Conference,* ASCE, New York, NY, 105–120.

Linsley, R.D., Kohler, M.A., and Paulhus, J.L.H. (1975). *Hydrology for engineers,* 2nd edition, McGraw-Hill, New York, NY.

New Jersey Department of Environmental Protection. (1986). *South branch Rockaway Creek stormwater management study.* Division of Water Resources, NJEPA, Trenton, NJ.

New Jersey Department of Environmental Protection. (1989). *Stormwater management facilities maintenance manual.* Division of Water Resources, NJEPA, Trenton, NJ.

Portland Cement Association. (1964). *Handbook of concrete culvert pipe hydraulics.* Skokie, IL.

Schueler, T.R. (1987). *Controlling urban runoff: A practical manual for planning and designing urban BMPs.* Metropolitan Washington Council of Governments, Washington, D.C.

Stahre, P. and Urbonas, B. (1990). *Stormwater detention for drainage, water quality, and CSO management.* Prentice-Hall Inc., Englewood Cliffs, NJ.

Taggart, W.C. (1978). *Vensel Creek master drainage plan.* University of Tulsa, OK.

Taggart, W.C. et al., (1982). "Man-made storage concepts." *Urban storm drainage management,* Marcell Dekker, Inc., New York, N.Y., 237–252.

University of Wisconsin. (1990). "Urban stormwater management." Professional Development Short Course Notes, Madison, WI.

Urban Drainage and Flood Control District. (1984). *Urban Storm Drainage Criteria Manual,* Denver Regional Council of Governments, Denver, CO.

Urban Land Institute. (1975). *Residential stormwater management,* published jointly with the American Society of Civil Engineers and the National Association of Home Builders, Washington, D. C.

Urbonas, B. and Glidden, M. (1982). "Development of simplified detention sizing relationship." *Stormwater detention facilities, Proc of an engineering foundation conference,* ASCE, New York, NY, 186–195.

U.S. Department of Agriculture. (1962). *National engineering handbook.* Soil Conservation Service, Washington, D.C.

U.S. Department of Agriculture. (1969). "Gated outlet appurtenances—earth dams." *Technical Release Number 46,* Soil Conservation Service, Washington, D.C.

U.S. Department of Agriculture. (1982). "Ponds—planning, design, construction." *Agriculture Handbook Number 590,* Soil Conservation Service, Washington, D.C.

U.S. Department of Agriculture. (1983). "Computer program for project formulation—hydrology." *Technical Release 20*, 2nd edition, Soil Conservation Service, U.S. Dept. of Agriculture, Available from NTIS, Springfield, VA.

U.S. Department of Agriculture. (1985). *National engineering handbook.* Section 4, Hydrology, Soil Conservation Service, Washington, D.C.

U.S. Department of Agriculture. (1985). *Engineering field manual for conservation practices.* U.S. Soil Conservation Service, Washington, D.C.

U.S. Department of Agriculture. (1986). "Urban hydrology for small watersheds." *Technical Release No. 55*, 2nd Edition, Soil Conservation Service, Engineering Division, Washington, D. C.

U.S. Department of the Interior. (1984). *Drainage manual.* Bureau of Reclamation, U.S. Government Printing Office, Washington, D.C.

U.S. Department of the Interior. (1981). *Water measurement manual.* Bureau of Reclamation, U.S. Government Printing Office, Washington, D.C.

U.S. Environmental Protection Agency. (1983). *Results of the nationwide urban runoff program: volume 2: Final report.* USEPA, Washington, D.C.

Whipple, W. Jr., Kropp, R., and Burke, S. (1987). "Implementing dual purpose stormwater detention programs." *Journal of Planning and Management*, 113 (6), 779–792.

Whipple, W. et al. (1983). *Stormwater management in urbanizing areas.* Prentice Hall, Englewood Cliffs, N.J.

Whipple, W. Jr. and DiLouie, J. (1981). "Coping with increased stream erosion in urbanizing areas." *Water Resources Research*, 17(5) 1561–4.

Whipple, W. Jr. (1979). "Dual purpose detention basins." *Journal of the Water Resources Planning and Management Division*, 105 (2), 403–412.

Yousef, Y.A., et al. (1985). "Fate of pollutants in retention/detention ponds." *Stormwater Management: An Update*, Publication 85-1, University of Central Florida, Orlando, FL, 259–275.

CHAPTER 12

STORMWATER MANAGEMENT PRACTICES
FOR WATER QUALITY ENHANCEMENT

I. INTRODUCTION

The control of water quality in urban runoff is in its technical infancy, however there are methods and techniques to insure that the levels of pollutant reduction achieved are indeed the maximum practicable. These techniques include the control of pollutants at their source, treatment of the polluted stream, or a combination of both.

Much can be done to improve the quality of the stormwater that runs off our urban developments if the designer is simply cognizant of possibilities for maximizing water quality. Moreover, many of these controls can be worked into the project in ways that can often enhance the aesthetic value of the project rather than detract from it (Roesner 1988).

II. HYDROLOGY FOR RUNOFF QUALITY CONTROL

The design runoff (flow rates and volumes) selected for sizing water quality controls is considerably different from that used for the design of drainage facilities. The damage done to a receiving water ecosystem by uncontrolled pollutant wash-off in the 50-year event is inconsequential compared to the hydraulic damage that results naturally to aquatic habitats from such an event.

(a) Drainage systems are designed for **large infrequent** runoff events (10-, 25-, 50-, or 100-year).
(b) Design events for runoff quality control are **small frequent** events (smaller than the 1-year runoff event).

To demonstrate this, consider the graphs shown in Figures 12.1a and 12.b. The curves were produced from an analysis of 39 years of sequential hourly rainfall data in Cincinnati, using the computer program STORM (Roesner et al. 1991). Detention basins of various volumes were simulated, and the frequency and volume of overflow were computed for each size basin, assuming the basin would be drained within 24 hours after a storm event. The upper curve shows the percentage of

486

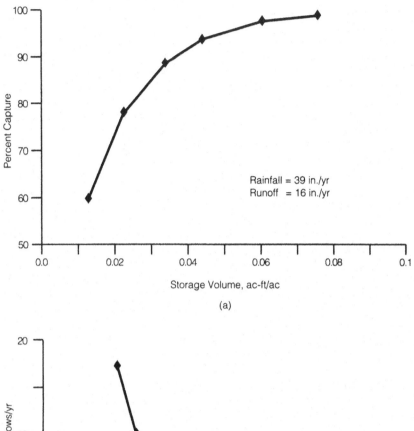

Figure 12.1.—*Effectiveness of surface detention on capture of runoff.*

the annual runoff that will be captured and detained for detention
facilities of various sizes. Capturing the first 0.25 inch of runoff (0.02
ac-ft/ac) results in a capture of 75% of the runoff on an annual basis,
while increasing that runoff capture to 0.5 inch (0.04 ac-ft/ac) allows
capture of 93% of the annual runoff.

To see what runoff volume must be captured to attain these per-centage captures of annual runoff, Figure 12.1b is used. This figure shows that, if 0.02 ac-ft/ac of storage is provided for the capture of runoff, only about 16 of the 80 events that occur per year (i.e., slightly more than one per month) overtop the detention facility. If that storage is increased to 0.5 inch (0.04 ac-ft/ac), the detention facility is overtopped only five times per year, or less than once every two months.

Put another way, capturing the 1-month runoff event (12 overflows per year) will result in capturing 80% of the runoff on an annual basis, while capturing the 2-month event (6 overflows per year) will result in the capture of more than 90% of the annual runoff.

The slope of the curve in Figure 12.1a indicates that, above a capture efficiency of about 90%, the marginal increase in storage volume re-quired to capture another 1% of the annual runoff is very large, and thus increasing storage capacity is probably not cost-effective above this level. Application of this analysis in a number of different locations across the United States has led to similar conclusions; i.e., capturing and/or treating the first 0.25–0.5 inch of runoff (0.02–0.04 ac-ft/ac) will result in 80–95% capture of the runoff volume. However, before de-veloping detention criteria for a given geographic area, local rainfall and reservoir routing analyses should be performed.

III. AXIOMS FOR THE DESIGN OF URBAN RUNOFF QUALITY CONTROLS

While runoff quality control is more an engineering art than a science with few established design criteria for pollutant removal, some em-pirical rules have been developed. They include:

(a) The most effective runoff quality controls reduce the runoff peak **and** volume (these are generally infiltration controls).
(b) The next most effective controls reduce the runoff peak (these controls generally involve storage).
(c) For small runoff events (those with return intervals of less than two years), the runoff should be retarded by detention in order to control downstream erosion (note, however, that the resulting duration of flows is longer, which can aggravate bank/channel erosion problems).
(d) Most obnoxious pollutants in urban runoff are settleable; however, appreciable amounts of nutrients and some heavy metals are dissolved and require treatment.

If these axioms are borne in mind while the designer develops the drainage plan for a development, the pollution load from the resulting system can be **significantly** reduced.

IV. SOURCE CONTROLS

Source control is a difficult matter. Homeowners generally are not restricted in the application of nutrients, herbicides, and pesticides to their yards. Chemical spillage on roadways, dustfall, etc., are also difficult to control. Nonetheless, attention to good housekeeping practices while designing the layout of the surface drainage system can significantly reduce the inevitable sources, and best management practice (BMP) programs now being introduced promise to reduce or eliminate some sources and their resultant impacts on water quality.

Source controls can also be effective at commercial/industrial sites. For example, chemical loading/unloading and storage areas could be covered or diked to capture 0.5 inch of runoff, with a closable outlet so that, if spillage occurs, it will not be washed into the drainage system before it can be cleaned up. Runoff from wash-down areas could be discharged to a sanitary sewer (though it may not be desirable if toxic chemicals are present) or to some treatment device before being discharged to a storm drain. Soil slopes should be mild (or otherwise terraced) and planted with a ground cover that minimizes erosion.

Being alert to the presence of chemicals and pollutants in areas subjected to rainfall or runoff is the first step in source control. Modifying the site plan and/or drainage plan is the second step toward eliminating these sources.

V. SITE CONTROLS FOR STORMWATER QUALITY MANAGEMENT

Site controls are generally those controls that attempt to reduce runoff rate and volume at or near the point where the rainfall hits the ground surface. The following types of site controls are common:

(a) Minimization of directly connected impervious area.
(b) Swales and filter strips.
(c) Porous pavement and parking blocks.
(d) Infiltration devices, such as trenches and basins.

Table 12.1, developed by Schueler (1987), shows the relative efficiency of various urban runoff quality controls in removing pollutants.

A. Minimization of Directly Connected Impervious Area

Directly connected impervious area (DCIA) is defined as the impermeable area that drains directly to the improved drainage system, i.e., paved gutter, improved ditch, or pipe. The minimization of DCIA is by far the most effective method of runoff quality control because it delays the concentration of flows into the improved drainage system and maximizes the opportunity for rainfall to infiltrate at or near the point at which it strikes the ground. Figure 12.2 illustrates the difference between an area where the DCIA is extensive and one where DCIA has been minimized. The residential lot on the north side of the street has all impervious areas on the lot draining directly to the gutter. This drainage plan allows no opportunity for water falling on the impervious surfaces to infiltrate into the ground; in fact, the system is laid out so that the rain falling on the impervious areas is quickly concentrated and drained to the gutter. The result is a greatly increased peak runoff

TABLE 12.1.   Comparative Pollutant Removal of Urban Runoff Quality Controls (Schueler 1987).

BMP/design		SUSPENDED SEDIMENT	TOTAL PHOSPHORUS	TOTAL NITROGEN	OXYGEN DEMAND	TRACE METALS	BACTERIA	OVERALL REMOVAL CAPABILITY
EXTENDED DETENTION POND	DESIGN 1	●	○	○	●	⊗		MODERATE
	DESIGN 2	●	●	●	●	⊗		MODERATE
	DESIGN 3	●	●	●	●	⊗		HIGH
WET POND	DESIGN 4	●	●	●	○	⊗		MODERATE
	DESIGN 5	●	●	●	●	⊗		MODERATE
	DESIGN 6	●	●	●	●	⊗		HIGH
INFILTRATION TRENCH	DESIGN 7	●	●	●	●	●		MODERATE
	DESIGN 8	●	●	●	●	●		HIGH
	DESIGN 9	●	●	●	●	●		HIGH
INFILTRATION BASIN	DESIGN 7	●	●	●	●	●		MODERATE
	DESIGN 8	●	●	●	●	●		HIGH
	DESIGN 9	●	●	●	●	●		HIGH

KEY:
○ 0 TO 20% REMOVAL
◑ 20 TO 40% REMOVAL
◐ 40 TO 60% REMOVAL
● 60 TO 80% REMOVAL

● 80 TO 100% REMOVAL
⊗ INSUFFICIENT KNOWLEDGE

POROUS PAVEMENT							
DESIGN 7	◐	●	◐	●	●	●	MODERATE
DESIGN 8	●	●	●	●	●	●	HIGH
DESIGN 9	●	●	●	●	●	●	HIGH
WATER QUALITY INLET							
DESIGN 10	○	⊗	⊗	⊗	⊗	⊗	LOW
FILTER STRIP							
DESIGN 11	◐	○	○	○	◐	⊗	LOW
DESIGN 12	●	◐	◐	◐	●	⊗	MODERATE
GRASSED SWALE							
DESIGN 13	○	◐	○	○	○	⊗	LOW
DESIGN 14	◐	◐	◐	○	○	⊗	LOW

Design 1: First-flush runoff volume detained for 6–12 hours.
Design 2: Runoff volume produced by 1.0 inch, detained 24 hours.
Design 3: As in Design 2, but with shallow marsh in bottom stage.
Design 4: Permanent pool equal to 0.5 inch storage per impervious acre.
Design 5: Permanent pool equal to 2.5 (Vr); where Vr = mean storm runoff.
Design 6: Permanent pool equal to 4.0 (Vr); approx. 2 weeks retention.
Design 7: Facility exfiltrates first-flush; 0.5 inch runoff/imper. acre.
Design 8: Facility exfiltrates one inch runoff volume per imper. acre.
Design 9: Facility exfiltrates all runoff, up to the 2 year design storm.
Design 10: 400 cubic feet wet storage per impervious acre.
Design 11: 20 foot wide turf strip.
Design 12: 100 foot wide forested strip, with level spreader.
Design 13: High slope swales, with no check dams.
Design 14: Low gradient swales which check dams.

rate and runoff volume compared to the pre-development condition. The pollutants contained in the runoff from the rooftop, driveway, sidewalk and street are simply collected in the gutter and must be dealt with at some location further down in the drainage system.

In contrast, the layout for the lot on the south side of the street has been designed to minimize DCIA. **All** impervious areas drain to a pervious area before they reach the grassed swale that serves as the primary conveyance facility for runoff from the lot. The roof runoff drains to the lawn and flows (sheet flow) across it, the driveway is sloped to drain to the lawn instead of the street, and runoff from the sidewalk and the street flow across a grass filter strip before reaching the water in the grassed swale. All of these techniques combine to provide maximum opportunity for infiltration and for retardation of the runoff rate. This approach to drainage system layout, which emphasizes peak flow reduction and pollutant capture, is called **stormwater management**, as contrasted with the north lot design, which is simply a drainage plan.

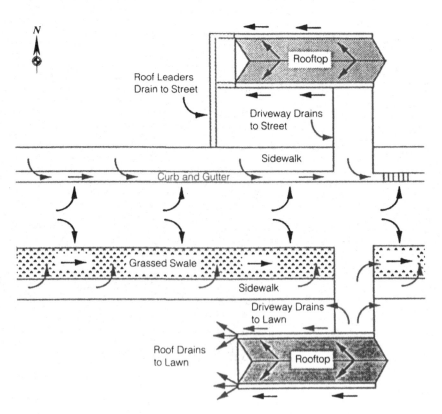

Figure 12.2. — Examples of maximizing (north lot) and minimizing (south lot) directly connected impervious areas.

Bear in mind that the aim of this design is minimization of the runoff peak and volume for small storms (the one- to two-month storm). The grassed swale, of course, will have been designed to transport the five- or ten-year storm through the area, so during the small events it will not be flowing full and the unsubmerged area of the swale will serve as a filter strip for the runoff from the street and sidewalks.

In the case of streets without curbs, an additional maintenance problem arises. While runoff from the street can easily flow onto bordering grassy areas, over time, as cars drive or park off of the asphalt pavement the edges begin to break up, and the grass either pushes out into the asphalt or is destroyed by the weight of the vehicles driving off the pavement. One way to minimize these problems without inhibiting runoff is to provide a one-foot-wide concrete border along the edge of the street. At intersections where cars tend to drive onto the grass to go around another car turning left, a curb can be installed that turns the radius of the corner. Figure 12.3 shows such a curbless street and intersection.

B. Swales and Filter Strips

Swales, or grassed waterways, and filter strips are among the oldest stormwater control measures, having been used alongside streets and highways for many years. A swale is a shallow trench which has the following characteristics:

Seattle, Washington—Draining parking lots through grassed swales and into grassed depressions can aid in removing pollutants from parking lot runoff.

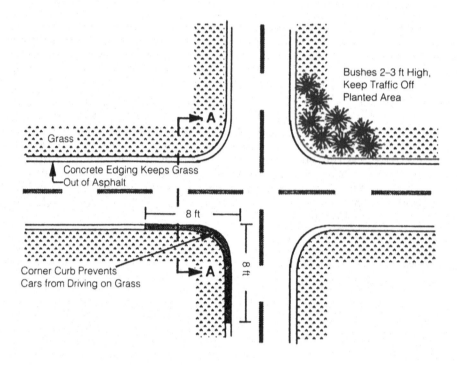

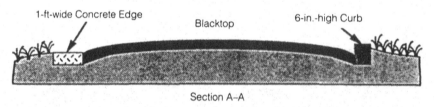

Section A–A

Figure 12.3. — Example of concrete road edging and corner curb.

(a) The side slopes are flatter than three feet horizontally to one foot vertically.

(b) It contains contiguous areas of standing or flowing water **only** following rainfall.

(c) It is planted with or contains vegetation suitable for soil stabilization, stormwater treatment, and nutrient uptake.

A filter strip is simply a strip of land across which stormwater from a street, parking lot, rooftop, etc., flows before entering adjacent re-

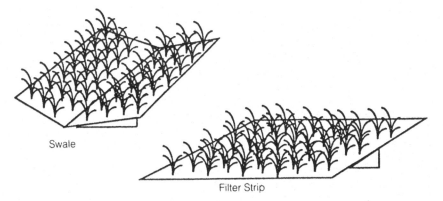

Figure 12.4.—Swales and filter strips as controls.

ceiving waters. A summary of the characteristics of swales and filter strips is presented in Figure 12.4.

1. Design Considerations

For small storms, both swales and filter strips remove pollutants from stormwater by slowing the water and settling or filtering out solids as the water travels over the grassed area, and by allowing infiltration into the underlying soil. In general, the higher the flow rate, the lower the efficiency. Thus, low velocity and shallow depth are key design criteria. A swale designed with a low bottom slope and check dams will perform much more efficiently than one without check dams. Raised driveway culverts are also very effective as check dams. For maximum efficiency of pollutant removal during small storms, a trapezoidal swale with as large bottom width as can be fitted into the site plan is desirable, since this will maximize the amount of runoff in contact with the vegetation and soil.

Design flows are calculated using equations for open channel flow (and small, frequent runoff events) with a roughness coefficient suitably adjusted for the grass or vegetation on the channel cross section. If the soil is sufficiently permeable that infiltration through the bottom is significant, this may be taken into account in the channel design. Avellaneda (1985) and Wanielista et al. (1986) provide some guidance in this area. For filter strips, there are two primary design considerations. The first is to minimize the grade in the direction of the flow. One effective technique for this is terracing. The second is to make certain that the water flowing across the strip is introduced at the upstream side of the strip in such a way that it flows across the strip in sheet flow and does not channelize.

2. Maintenance

Maintenance of swales and filter strips is an important consideration, for aesthetic reasons and for hydraulic efficiency. In the case of the swale, care must be taken to insure that flows through a swale used for drainage purposes during large storms are not impeded by an overgrowth of vegetation. To prevent this, the vegetation planted in the channel should be suitable for mowing, and the channel designed so that mowing machines can be easily and efficiently operated along the swale. The swale should be mowed on a regular basis. For filter strips that are not part of the drainageway during large storms, maintenance is purely an aesthetic matter. These strips can be planted in grass and mowed, or natural vegetation can be used. Any ground cover, however, must be sufficiently dense to keep the overland flow from channelizing and eroding rivulets through the filter strip.

C. Porous Pavement and Parking Blocks

Porous pavement has excellent potential for use on streets and in parking areas (Niemczynowicz 1990). When properly designed and carefully installed and maintained, porous pavement can have load-bearing strength and longevity similar to conventional pavement. In addition, porous pavement can help to reduce the amount of land needed for stormwater management to preserve the natural water balance at a given site, and to provide a safer driving surface that offers better skid resistance and reduced hydroplaning.

However, porous pavement is only feasible on sites with permeable soils, fairly flat slopes, and relatively deep water-table and bedrock levels. In addition, batching and placement of the material require special expertise to avoid clogging, which is the principal concern associated with porous pavement. The risk of clogging is high, and, once it has occurred, it is difficult and costly to correct. The chief means of preventing the problem is to keep sediment off the underlying soil before construction and off the pavement during and after construction.

Porous pavement is being used fairly extensively as a viable alternative in Florida, and a design manual has recently become available (Florida Concrete and Products Association 1989). However, where winter conditions are severe, additional consideration should be given to the structural integrity of porous pavement under freeze-thaw conditions. Detailed design information on porous asphalt is available in Schueler (1987) and Maryland Water Resources Administration (1984).

Another very effective site-control device is parking blocks (Pratt 1990). These are hollow concrete blocks similar to but smaller than those used in construction. They are an excellent site control that is unfortunately rarely used. In parking lots for retail stores, sports arenas, civic theaters, and the like, where more than half of the parking area is used less than 20% of the time, the use of parking blocks in the less-used portions of such lots give them a much more attractive appearance and will considerably reduce runoff quantity, flow rates, and pollution from these areas.

Figure 12.5a shows typical blocks, which are placed in rows with soil surrounding each one and planted with vegetation. Runoff quantity reduction occurs as infiltration takes place in the planted areas. More-

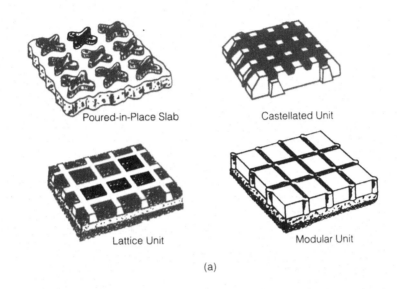

Poured-in-Place Slab

Castellated Unit

Lattice Unit

Modular Unit

(a)

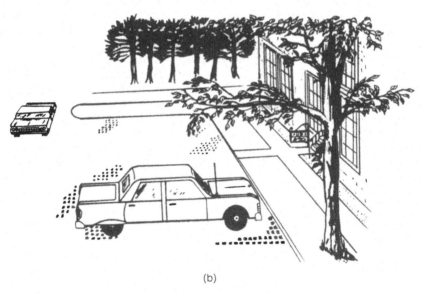

(b)

Figure 12.5.—Types of grid and modular pavement. Typical parking lot layout using grid or modular pavement.

over, the greater flow resistance of the grassed areas retards the runoff rate, especially during small storms. Most importantly, the quality of the runoff is greatly enhanced over that from a normal parking lot because the pollutants, being restrained by the vegetation matrix, are more difficult to wash off.

In designing a parking block area, the block manufacturer should be consulted to determine the most suitable sub-base to use. Also, it is suggested that only the actual parking spaces be paved with the blocks, since they do not hold up well under traffic. Traffic lanes through the lot should be paved in the normal fashion. A typical layout is illustrated in Figure 12.5b.

Finally, it should be noted that parking blocks are best suited for use where the rainfall is frequent enough to keep the planted grass alive. In arid areas, this control should not be used unless provision for watering is made. When the grassed areas are not well-maintained and the grass has been allowed to die, soil erosion inevitably occurs and the parking lot can be muddy.

D. Infiltration Devices

Infiltration devices are those stormwater quality control measures that completely capture runoff from the water quality design storm and allow it to infiltrate into the ground. They are the most effective stormwater quality control device that can be implemented, but they can only be used in situations where the captured volume of water can infiltrate into the ground before the next storm, and where they will not cause structural or groundwater pollution problems (particularly if groundwater is used as a potable water source).

Infiltration devices can be either above-ground infiltration basins or buried infiltration trenches. Among their advantages are that they can help to minimize alterations to the natural water balance of a site, can be integrated into a site's landscaped and open areas, and, if carefully designed, can serve larger developments. Disadvantages can include a fairly high rate of failure due to unsuitable soils, the need for frequent maintenance, and possible nuisance factors, such as odors, mosquitos, or soggy ground. The general performance characteristics of these two types of infiltration device are shown in Figure 12.6.

1. Infiltration Basins

Infiltration basins temporarily store stormwater until it infiltrates the surrounding ground through the bottom and sides of the basin. Such basins are made by constructing an embankment or by excavating in or down to relatively permeable soils. They also can be a natural depression within an open area or a recreational area such as a soccer field. Infiltration basins generally serve areas ranging in size from a front yard to 50 acres.

Infiltration basins can be constructed on-line or off-line with respect to the normal drainage path. When a basin is located on-line, it is

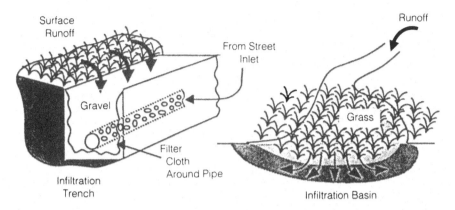

Figure 12.6. — Infiltration devices.

designed to capture the water quality design storm entirely. When a larger storm occurs, it overflows the basin, which then serves as a detention pond for those larger events. Experience in Florida indicates that these on-line basins are not as efficient in the removal of stormwater as off-line infiltration basins.

Off-line infiltration basins are designed to divert the more polluted first flush of stormwater out of the normal path and hold it for later water quality treatment. They also help to reduce stormwater volume and to recharge the groundwater, and should be used wherever site conditions allow. When the basin reaches capacity, the flow path for any additional stormwater returns to normal. The diverted first flush is not discharged to surface water but is stored until it is gradually removed by infiltration, evaporation, and evapo-transpiration.

Off-line infiltration areas can often be easily incorporated into the landscaped/open areas of a site. These can include natural or excavated grassed depressions, recreational areas, and even landscape islands in parking lots. However, if site conditions prevent the exclusive use of infiltration, then small off-line retention areas scattered over the development or site should be used as pre-treatment devices for the runoff prior to subjecting it to the primary runoff treatment device for the development. This is especially necessary if detention lakes are the primary component of the stormwater system and the lakes are intended to serve as an aesthetic focal point of the development (see Section VI).

Off-line infiltration devices are designed to store a selected volume of stormwater for a specified period of time with a predetermined infiltration rate. The success of such practices will depend on the careful evaluation of a potential site for suitability. Special attention must be given to such aspects as soils, bedrock, setbacks, vegetation, and maintenance, as discussed in the following paragraphs.

Infiltration basins must be located in soils with hydraulic conductivity rates that will allow the diverted volume to infiltrate within 72 hours or within 24–36 hours for infiltration areas that are planted with grasses. The seasonal high-water table should be at least three feet beneath the bottom of the infiltration area to assure that stormwater pollutants are removed by the vegetation, soil, and microbes before coming into contact with the groundwater.

Bedrock also should be at least four feet beneath the bottom of the infiltration area. In those parts of the country where limestone is at or near the land surface, the potential for groundwater contamination is quite high, especially in Karst-sensitive areas, where sinkhole formation is common. Similar concerns exist in mountainous areas where thin soil mantles overlay fractured bedrock aquifers that serve as water supplies.

Site selection and planning also must address several other important issues. For example, infiltration basins should not be located on areas with slopes greater than 20%, to minimize the chance of downstream water seepage from the subgrade. Care should be taken to avoid areas with expansive soils, or soils subject to substantial consolidation or settlement upon the introduction of water. In addition, infiltration areas should be set back from water-supply wells or septic systems (distance will vary with site conditions), and at least 10 feet down gradient from any building foundations. Local groundwater quality standards, if any, must be considered, and land used for such purposes as gasoline stations, or industrial lands susceptible to spills of potentially hazardous materials should not use infiltration basins at all unless precautions are taken to prevent such materials or spills from mixing with stormwater.

Once areas with suitable characteristics have been selected for infiltration, the sites should be well marked during surveying and protected during construction. Heavy equipment, vehicles, and sediment-laden runoff should be kept out of infiltration areas to prevent compaction and loss of infiltration capacity. Public safety and owner liability must also be carefully considered during site planning. Additional discussions of the concerns involved in the design and siting of infiltration basins are addressed by Schueler (1987) and Maryland Water Resources Administration (1986).

The side slopes and bottoms of infiltration areas should be planted with a dense turf of native vegetation immediately after construction. This grass should be water-tolerant. Depending on local hydrology. supplemental irrigation at these areas may be required during dry periods. Similarly, inlet channels into and within the basin should be planted with stabilizing vegetation to prevent inflows from reaching erosive velocities and scouring the bottom. Stabilizing these detention areas through the use of vegetation not only helps to eliminate erosion and scouring but also filters stormwater pollutants, reduces maintenance needs, and even maintains or improves infiltration rates. Finally, a maximum slope of 4:1 ensures that maintenance equipment can easily be brought into the site and moved around within it.

2. Infiltration Trenches

In general, site planning considerations for infiltration trenches are the same as for infiltration basins. As with basins, trenches must be very carefully designed, installed, and maintained, because they are very susceptible to clogging, and, once clogged, rehabilitation is a major effort. If properly constructed, with pretreatment practices in place to prevent heavy sediment loading, infiltration trenches can provide stormwater benefits without tremendous maintenance needs. Since trenches are usually "out-of-sight, out-of-mind," getting property owners to maintain them can be difficult. Accordingly, a public commitment for regular inspection of privately owned trenches is essential, as is a legally binding maintenance agreement and owner education regarding the function and maintenance needs of trenches.

Infiltration trenches, which can be located on the surface of the ground or buried beneath the surface, are usually designed to serve areas ranging up to 5–10 acres in size and are especially appropriate in urban areas, where land costs are very high. An infiltration trench generally consists of a long, narrow excavation, ranging from 3–12 feet in depth, backfilled with stone aggregate to allow for the temporary storage of the first-flush stormwater in the voids between the aggregate material (See Figure 12.6). Stored runoff then infiltrates into the surrounding soil, through either the trench bottom or the sides, depending on the elevation of the water table and the soil properties. In addition to the infiltration rate of the subsoil, the hydraulic conductivity of the surrounding filter fabric is crucial and, in fact, can become a limiting factor. The use of fabrics designed as landfill liners should be avoided.

Trench bottoms should be at least four feet above the seasonal high water table level to prevent mounding of groundwater, consequent loss of infiltration capacity, and insect problems.

There are two major types of trenches, surface trenches and underground trenches. The differences between the two involve the amount of stormwater that can be handled and the ease of maintenance. Surface trenches receive sheet-flow runoff directly from adjacent areas after it has been filtered by a grass buffer. They are typically used in residential areas where relatively small loads of sediment and oil can be trapped in grass filters strips at least 20' wide. While surface trenches may be more susceptible to sediment accumulation, their accessibility makes them easier to maintain. In addition, they are very appropriate for use in highway medians, parking lots, and narrow landscape areas.

Underground trenches can potentially be used in many development situations, although discretion must be exercised. For instance, while underground trenches can accept runoff from storm sewers, they require installation of special inlets to prevent coarse sediment and oils and greases from clogging the stone reservoir. These inlets should include oil/grease traps, catch basins, and baffles to reduce sediment, leaves, and debris (such measures, however, cannot remove petroleum hydrocarbons normally found in urban runoff). In addition, pretreatment by routing the flow over grassed filter strips or vegetated swales is essential to protect the infiltration trench.

The most commonly used underground trench is an exfiltration system in which the stormwater treatment volume is diverted into an oversized perforated pipe placed within an aggregate envelope. The first flush stormwater is stored in the pipe and exfiltrates out of the holes through the gravel and filter fabric and into the surrounding soil. Because of the lack of accessibility, maintenance of these underground trenches can be very difficult and expensive, especially if they are placed beneath parking areas or pavement.

One other type of infiltration trench is the dry well. This device is used extensively in Maryland to store and infiltrate runoff from rooftops. In a dry well, the downspout from the roof gutter is extended into an underground trench which is constructed at least ten feet away from the building foundation. Again, clogging is a concern, and rooftop gutter screens must be used to trap particles, leaves, and other debris. Additional design information on dry wells is available from the Maryland Water Resources Administration (1986).

Trenches should be inspected frequently within the first few months of operation and regularly thereafter. Such inspections should be done after large storms to check for ponding, with water levels in the observation well recorded over several days to check drawdown. In addition, grass buffer strips should maintain a dense, vigorous growth of vegetation; they should receive regular mowing (with bagging of grass clippings) as needed. Finally, pretreatment devices should be checked periodically and cleaned when the sediment reduces available capacity by more than 10% (Schueler 1987).

Design, construction, and maintenance procedures for infiltration trenches of all types can be found in references such as Schueler (1987), Maryland Water Resources Administration (1986), and Stahre (1988), which should be carefully reviewed when use of these devices is planned.

VI. DETENTION PRACTICES

The site controls described above are the most effective and efficient for reducing the pollutant load in urban runoff. Unfortunately, variations in soil, water table, and geologic conditions throughout the country preclude the use of infiltration practices in many locations. These locales are characterized by slowly percolating soils, subsurface hardpans that restrict infiltration, flat terrain, controls on groundwater quality, and/or high water tables. In such areas detention, either wet or dry, serves as an alternative or supplementary stormwater quality control device.

Detention basins, in fact, are more widely used for stormwater management than any other type of control. However, their primary application to date has been for drainage control, i.e. peak-flow atten-uation, rather than for water quality control. The state of practice in the design of detention basins for water quality management is still in its formative stage. Detention basins designed for peak flow attenuation can be given an effective water quality control function at little added cost (see Chapter 11).

The design of both wet and dry detention basins is considered in Chapter 11. The following two sections contain supplemental infor-mation that can be incorporated into the design to maximize the pol-lutant removal efficiency of both types of detention basins.

A. Dry Detention

Detention basins designed for flood peak reduction only are the most common type of detention basin used around the country. These basins generally are not as effective in removing pollutants as wet basins. This is primarily due to the short residence time. For basins with detention times of less than twelve hours, no more than 10% of stormwater pollutants are removed, and, in fact, there may even be a **negative** pollutant removal rate due to the washout of pollutants captured in earlier small storms (Camp Dresser & McKee 1985; Jellerson 1981). Performance can be improved when dry detention basins are designed to include water quality features, as explained in Chapter 11.

Figure 12.7 presents a schematic of a dry detention pond and sum-marizes typical performance characteristics. There are several design modifications that can be made to a dry detention basin to enhance overall system performance. For example, the use of swales (as pre-treatment conveyances) and landscape infiltration will greatly improve the pollutant removal effectiveness and life of a dry detention system. Another possibility is to incorporate a shallow marsh around the outlet, as is shown in Figure 12.8. It should be noted, however, that the addition of a marsh makes this control device no longer a **dry** detention

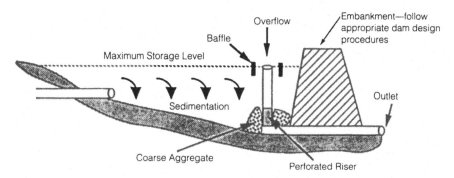

Figure 12.7.—Dry detention ponds.

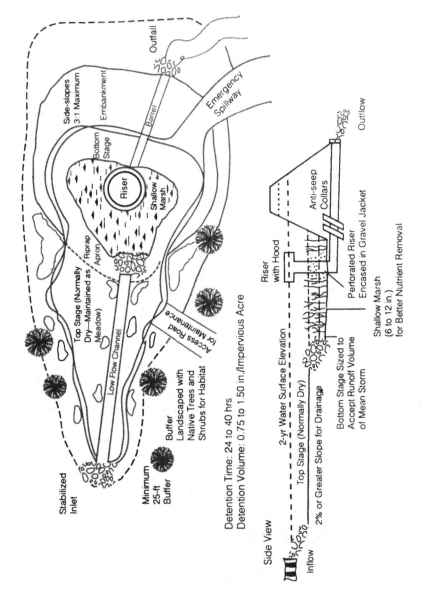

Figure 12.8.—Extended detention pond design features.

basin but rather a **transitional** device between dry and wet detention. Further discussion of these and other improvements to significantly enhance the pollutant removal performance of dry detention basins can be found in Schueler (1987).

One of the treatment practices that has been commonly used in the southeastern United States for stormwater treatment is detention with filtration, in which stored stormwater is discharged through a filter. Typical filtration systems have included bottom or side-bank sand or natural soil filters, along with multi-media filters composed of materials such as alum sludge or activated charcoal. Except for the latter two filters, these systems only remove particulate pollutants, which limits their overall benefit.

In addition, difficulties associated with the design, construction, and, most importantly, maintenance of stormwater filters suggests that they should be used only as a last resort. Experience has shown that it is not a question of **if** a filter will clog but **when** (and then who will maintain the filter).

B. Wet Detention Ponds

Figure 12.9 illustrates the basic components of a wet detention system that is used for both flood control and water quality enhancement. The figure also lists the performance characteristics of a typical wet detention basin. Essentially, a wet-detention lake consists of 1) a permanent water pool, 2) an overlying zone in which the design runoff volume temporarily increases the depth of the pool while it is stored and released at the allowed peak discharge rate, and 3) a shallow littoral zone (the biological filter). During storms, runoff replaces treated waters detained within the permanent pool after the previous storm, thus making the permanent water pool volume and the vegetated littoral zone of utmost importance for water quality enhancement. Wet detention ponds are often used in series with swale interconnectors. If properly designed

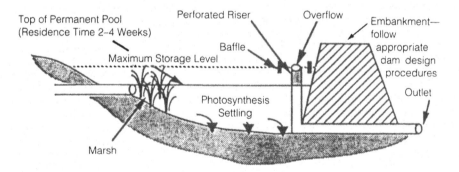

Figure 12.9. — Wet detention ponds.

and maintained, they can provide effective flood and water quality protection, as well as an aesthetic wildlife habitat.

The removal of stormwater pollutants in a wet-detention system is accomplished by a number of physical, chemical, and biological processes. Gravity settling removes particles. Chemical flocculation occurs when heavier sediment particles overtake and coalesce with smaller, lighter particles to form still larger particles. Biological removal of dissolved stormwater pollutants includes uptake by aquatic plants and metabolism by phytoplankton and micro-organisms that inhabit the bottom sediments.

Removal of dissolved pollutants primarily occurs during the relatively long quiescent period between storms. Accordingly, the permanent water pool is especially vital, since it permits treatment between storms, reduces runoff energy, and provides a habitat for aquatic plants and algae (the biological filter that removes dissolved nutrients and metals). Aerobic conditions at the bottom of the permanent pool will maximize the uptake of dissolved pollutants (nutrients, metals) by bottom sediments and will minimize release from the sediments into the water (Yousef 1985).

Designing wet detention systems to achieve a specified level of reduction of dissolved pollutants is very difficult. Two approaches have typically been used to develop design criteria for such systems. One method relies upon solids settling theory and assumes that all pollutant removal is due to sedimentation (Driscoll 1983). The other approach views the wet detention system as a lake that achieves a controlled level of eutrophication, thereby accounting for biological, physical, and chemical assimilation of stormwater pollutants in addition to sedimentation (Hartigan 1985). The basis for this second design method is that stormwater is detained within the permanent pool long enough to produce adequate levels of nutrient uptake by algae and aquatic plants, but the hydraulic residence time is not so long as to induce stagnation, thermal stratification, or anaerobic bottom sediments. Specific design criteria for wet-detention systems using this controlled level of eutrophication approach are described by Hartigan (1988).

There are three critical criteria in determining how efficiently the wet-detention type of device will work. The first is the volume of the permanent pool, which should be sufficient to provide 2-4 weeks of detention time so that algae can grow. Second is the depth of the permanent pool, which should be greater than 4-6 feet but less than 10-15 feet, so that the water remains wind-mixed and the bottom sediment stays aerobic. If the bottom sediment turns anaerobic, it will **release** nutrients into the overlying pool, and these nutrients will be washed out in the next rainstorm.

The third criterion involves the presence of a shallow littoral zone, typically concentrated around inflow points and the outflow (If vegetation is planted continuously around the perimeter, experience shows that homeowners tend to remove it). This is one of the very important components of the system, since it is the aquatic plants within this zone that provide much of the biological assimilation of dissolved stormwater pollutants. The littoral zone should cover at least 30% of the pond's surface area and should have a gentle slope (6:1 or shallower) to a depth of two feet below the control elevation. Another benefit of flat side slopes is enhanced public safety, especially for children. Figure 12.10 by Scheuler (1987) shows an example of a wet-detention pond sized for 2- to 4-week residence time.

A detention system with a permanent water pool must receive and keep enough water to maintain the pool. In some arid parts of the United States, this is not possible. The volume of stormwater that will be detained can be determined from the pond's drainage area, the amount of imperviousness, soil types, and water table elevation. Wet ponds should not be considered in areas where the underlying soil infiltration rate will not allow water to remain in the pond on a permanent basis or where the rainfall is not sufficient to sustain the necessary volume in the pool.

The use of several detention lakes in series or the separation of a pond into multiple cells will enhance pollutant removal and lessen maintenance tasks. In fact, the configuration of the components of a wet-detention system is limited only by the designer's imagination and site constraints. Figure 12.11 illustrates a multi-cell design in which the relatively small deep pool serves as the primary sedimentation area (which can easily be dredged) and the larger shallower pool provides for flood protection and pollutant attenuation. The vegetated littoral area is concentrated in two spots—a shallow ledge between the two pools and near the outlet structure.

VII. USING WETLANDS FOR STORMWATER QUALITY ENHANCEMENT

Wetlands can provide water quality enhancement through sedimentation, filtration, absorption, and biological processes, as well as natural flood protection. Further, the incorporation of wetlands into a comprehensive stormwater management system achieves several additional objectives, including reduced operation/ maintenance and wetland preservation and revitalization.

However, the use of wetlands for stormwater management should not be considered a panacea, nor should it be considered applicable in much of the United States. While much research has been completed on the ability of wetlands to remove wastewater pollutants (USEPA 1985), many questions still remain. For example, how long can a wetland continue to remove stormwater pollutants effectively? What type of maintenance (harvesting, sediment banking, or other) should be planned for a given wetland area? How frequently should an area be scheduled for this maintenance? These and other such questions must

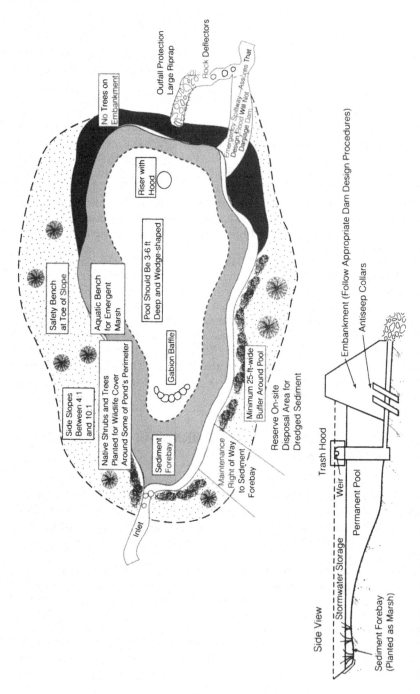

Figure 12.10.—Design schematic of a wet pond.

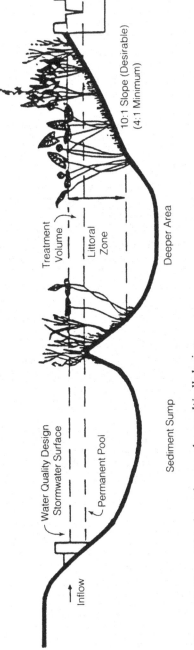

Figure 12.11. — Wet detention pond — multicell design.

be addressed in order to design the most effective wetland stormwater management systems. In addition, for both fresh and saltwater wetlands there are laws and regulations (see Chapter 2) severely limiting such possibilities.

Given the many unknowns still associated with this treatment practice, only relatively isolated wetlands or constructed wetlands should be used for stormwater management. Wetlands that have been previously ditched and drained can be revitalized by blocking the drains and incorporating them into the stormwater system. Wetlands intermittently connected to other waters through flows that result when ground water rises to ground level should also be considered for stormwater management.

In designing a stormwater management system that incorporates a wetland, care should be taken to minimize changes to the hydro-period (the time that water remains at a particular level in the wetland—which determines the form, function, and nature of the wetland) caused by the addition of stormwater. Finally, pretreatment practices are needed to reduce oil, grease, and sediment levels to protect the wetland. Design guidance in these and other matters relative to constructed wetlands can be found in Maryland Water Resources Administration (1987) and Livingston (1989).

VIII. ALUM TREATMENT OF STORMWATER

Alum (aluminum sulfate) has been used extensively to clarify potable water supplies, to remove phosphorus from wastewater and to inactivate phosphorus in lake sediments. Injection of liquid alum inside storm sewers to treat stormwater is innovative and has potential for application in reducing the water quality impact of stormwater discharges on lakes.

Alum forms harmless precipitates of $Al(PO_4)$ and $Al(OH_3)$ that combine with phosphorus and heavy metals, causing them to be deposited into the sediments in a stable inactive state. The insoluble precipitates formed between alum and phosphorus in the sediments are exceptionally stable, since they are immune to changes in sediment redox potential and, to a lesser degree, to pH changes.

Among the advantages of injecting alum into stormwater are the excellent reduction of the amounts of most stormwater pollutants (to more than 80%), the relatively low construction and operation costs, the removal of nutrients from the receiving water column, and the reduction of nutrient recycling from bottom sediments.

On the other hand, the disadvantages of using alum include the experimental nature of the technology, the need for careful evaluation of alum dosage levels, the potential for pH changes in the receiving water, and the possible toxicity of the aluminum compound. However, it is known that if alum injections into a particular body of water do affect the benthic organisms residing in bottom sediments, those effects are not long-lasting. For more information on this type of treatment, see Harper, et al. (1986).

IX. IMPLEMENTING STORMWATER QUALITY CONTROLS

Integrating urban runoff quality controls into a drainage plan requires creativity. During the site-planning phase, the designer/planner must constantly have in mind the four objectives of the stormwater management. These are:

(a) Surface drainage.
(b) Flood control.
(c) Erosion and sediment control.
(d) Control of pollutants in the runoff.

The approach to runoff water quality control is to minimize the adverse impacts of stormwater through a coordinated system of source controls and site controls. Source controls emphasize the prevention and reduction of nonpoint pollution by eliminating the opportunity for pollutants on the land surface to be entrained into surface runoff. They are fundamental to the various management methods. After that, a coordinated set of site controls, detention devices, and other practices is necessary to achieve the desired level of runoff quality control. In order for this second step to be both aesthetically acceptable and cost-effective, the stormwater management system must be an integral part of the site-planning process for every project.

Each site will contain natural attributes that should influence the type and configuration of the stormwater system. for example, sandy soils imply the possibility to integrate infiltration practices such as retention areas into the development's open space and landscaping, while natural low areas and isolated wetlands may offer opportunities for detention/ wetland treatment. The variety of features contained on a site will often suggest which particular combination of runoff quality controls can best be integrated into an effective system.

A stormwater management system can be viewed as a treatment train in which the individual treatment practices are the cars. The first car is source controls, followed by site controls, such as minimizing DCIA, filter strips, and infiltration, that capture and remove pollutants near their point of entrainment in the runoff. Grassed swales, detention ponds, wetlands, and other devices add to the train. The more controls incorporated into the system, the better the performance of the train.

It should be noted that the preceding discussion applies primarily to the establishment of new regulatory criteria, and to governmental programs such as best management practices and nonpoint source control programs. The engineer charged with developing designs for a particular client must of course conform to regulatory criteria, including any requirements resulting from regional planning of stormwater management systems, requirements in the interest of flood peak reduction and erosion control, and any specific environmental constraints. He must include design provisions to satisfy these constraints, and also any additional water quality control provisions (such as those in this chapter) that can be included at nominal cost. He can thus satisfy governmental requirements, protect the interests of his client, and produce the most environmentally favorable design possible.

If the design is developed with the following concepts in mind, a good water quality management system will result.

(a) Design runoff quality controls to capture small storms.
(b) Design to maximize sediment removal, and removal of other pollutants.
(c) The most effective method for reducing urban runoff pollution is to minimize directly connected impervious area (DCIA).
(d) Infiltration devices are most efficient but most difficult to maintain (and can only be used where groundwater pollution is not a problem).
(e) Dry detention is easiest to design and operate, and efficiency is satisfactory if properly designed.
(f) Wet detention is more difficult to design and maintain but more efficient than dry detention, and often more aesthetically desirable.

It can be extremely difficult (and expensive) to insert detention facilities into areas that are already intensively developed—the NIMBY (not in my back yard) syndrome is strong. It is possible, however, with thoughtful planning and careful design, to integrate cost-effective runoff quality controls into urban development plans to achieve the required level of pollutant reduction with no negative impact on aesthetics. When all is said and done, the incorporation of runoff quality controls into urban landscape design is more an art than a science. If properly done, the aesthetic character of the development is actually enhanced by integration of runoff quality controls into the site plan.

XI. REFERENCES

Avellaneda, E. (1985). "Hydrologic design of swales." Master's thesis, University of Central Florida, Orlando, FL.

Camp Dresser & McKee. (1985). "Final report on an assessment of stormwater management programs." Prepared for Florida Department of Environmental Regulation, Tallahassee, FL.

Driscoll, E.D. (1983). "Performance of Detention Basins for Control of Urban Runoff." *Proc of the intl symp on urban hydrology, hydraulics and sediment control*, University of Kentucky, Lexington, KY.

Florida Concrete and Products Association, *Pervious Pavement Manual*.

Harper, H.H. (1988). "Effects of stormwater management systems on groundwater quality." Interim report for Florida Department of Environmental Regulation, Orlando, FL.

Harper, H.H. (1985). Fate of heavy metals from highway runoff in stormwater management systems, Ph.D. diss., University of Central Florida, Orlando, FL.

Harper, H.H., Murphy, M.P., and Livingston, E.H. (1986). "Inactivation and Precipitation of Urban Runoff by Alum Injection in Storm Sewers," Paper presented at sixth symposium on lake and reservoir management, North American Lake Management Society, Portland, OR.

Hartigan, J.P. (1986). "Regional BMP master plans," *Urban runoff quality—impact and quality enhancement technology; Proc of an engineering foundation conf.*, ASCE, New York, NY.

Hartigan, J.P. and Quasebarth, T.F. (1985). "Urban nonpoint pollution management of water supply protection: Regional vs. onsite BMP plans." *Proc of an intl symp on urban hydrology, hydraulics and sediment control*, University of Kentucky, Lexington, KY.

Jellerson, D. (1981). "Impacts of alum sludge on lake sediment phosphorus release and benthic communities." Master's thesis, University of Central Florida, Orlando, FL.

Livingston, E. et al. (1988). *The Florida Development Manual: A Guide to Sound Land and Water Management.* Florida Department of Environmental Regulation, Tallahassee, FL.

Maryland Water Resources Administration. (1987) *Guidelines for constructing wetland stormwater basins.* Annapolis, MD

Maryland Water Resources Administration. (1986). *Maintenance of stormwater management structures: A departmental summary.* Annapolis, MD.

Maryland Water Resources Administration. (1984). *Standards and specifications for infiltration practices.* Annapolis, MD.

Miller, R.A. (1985). *Percentage entrainment of constituent loads in urban runoff, South Florida.* U.S. Geological Survey, USGS WRI 84-4329.

Niemczynowicz, J. (1990). "Swedish way to stormwater enhancement by source control." *Urban Stormwater Quality Enhancement—Source Control, Retrofitting, and Combined Sewer Technology; Proc of an Engineering Foundation Conf.* ASCE, New York, NY.

Pratt, C.J. (1990). "Permeable Pavements for Stormwater Quality Enhancement." *Urban Stormwater Quality Enhancement—Source Control, Retrofitting, and Combined Sewer Technology, Proc of an Engineering Foundation Conf.* ASCE, New York, NY.

Rast, W., Jones, R., and Lee, G.F. (1983). "Predictive capability of US OECD phosphorus loading eutrophication response models." *Journal of the Water Pollution Control Federation*, 55 (7).

Roesner, L.A. (1988). "Aesthetic implementation of nonpoint source controls." *Proc. of the symp on nonpoint pollution: 1988 policy, economy, management, and appropriate technology*, American Water Resources Association, Bethesda, MD.

Roesner, L.A., Urbonas, B., and Sonnen, M.B., eds. (1989). "Design of urban runoff quality controls." *Proc of an engineering foundation conf.*, ASCE, New York.

Roesner, L.A., Burgess, E.H. and Aldrich, J.A. (1991). "The hydrology of urban runoff quality management." *Water resources planning and management and urban water resources, Proc of ASCE specialty conf.*, ASCE, New York, NY.

Saint John's River Water Management District. (1986). *Applicants handbook.* Palatka, FL

Schillinger, J.E. and Gannon, J.J. (1985). "Bacterial adsorption and suspended particles in urban stormwater." *Journal of the Water Pollution Control Federation,* 57 (5).

Schueler, T.R. (1987). *Controlling urban runoff: A practical manual for planning and designing urban BMPs.* Metropolitan Washington Council of Governments, Washington, D.C.

Shaver, H.E. (1986). "Infiltration as a stormwater management component." *Urban runoff quality—impact and quality enhancement technology,* Proc of an engineering foundation conf., ASCE, New York, NY.

Stenstrom, M.K., Herman, G.S., and Burstynsky, T.A. (1984). "Oil and grease in urban stormwater." *Journal of Environmental Engineering,* 110 (1).

Urbonas, B. and Roesner, L.A., eds. (1986). "Urban runoff quality—impact and quality enhancement technology." *Proc of an engineering foundation conf., Henniker, New Hampshire, June, 1986,* ASCE, New York, NY.

U. S. Army Corps of Engineers. (1976). *Storage, treatment, overflow, runoff model (STORM).* Hydrologic Engineering Center, Davis CA.

U.S. Environmental Protection Agency. (1985). *Freshwater wetlands for wastewater management handbook.* EPA 904/9-85-135, Washington, D. C.

U.S. Environmental Protection Agency (1983). "Results of the nationwide urban runoff program: volume 2: Final report." U.S.E.P.A., Washington, D. C.

Wanielista, M.P. (1977). "Quality considerations in the design of holding ponds." *Stormwater retention/detention basins seminar,* University of Central Florida, Orlando, FL

Wanielista, M.P. et al. (1986). "Best management practices enhanced erosion and sediment control using swale blocks." Florida Department of Transportation *Report FL-ER-35-87,* Orlando, FL.

Wanielista, M.P., and Yousef, Y.A. (1986). "Best management practices overview." *Urban runoff quality—impact and quality enhancement technology. Proc of an Engrg Foundation Conf.,* ASCE, New York, NY.

Wiegand, C. et al. (1986). "Cost of urban runoff controls." *Urban runoff quality—impact and quality enhancement technology, Proc of an Engrg Foundation Conf.,* ASCE, New York, NY.

Yousef, Y.A. et al. (1985). "Fate of pollutants in retention/detention ponds." Stormwater management: An update, *Publication 85-1,* University of Central Florida, Orlando, FL.

CHAPTER 13

MATERIALS OF CONSTRUCTION AND MAINTENANCE

I. INTRODUCTION

A. General

Most man-made drainage works throughout recorded history have served the field of agriculture. The age of the Pharaohs was made possible by the Egyptians' ingenious use of the Nile floods, and man has continually sought to understand and to utilize nature's way of draining the earth of storm waters. Numerous examples, however, have shown that nature will return storm drainage patterns to their natural state unless the man-made improvements are maintained continuously. The concept of forever is not a possibility available to the engineer. All drainage structures have a usable life, and we must design and utilize materials to handle drainage for specific lifetimes.

B. Environmental Considerations

There are many materials available to the engineer designing a storm water conveyance system. The environment in which a given material will be installed will greatly influence its useful service life. Soil and water Ph, soil resistivity, stream silt load, stream geology, and other environmental factors combine to limit or extend the usable life of the planned storm drainage system.

If possible, the designer should analyze the chemical characteristics of the liquid that is being transported by the system as well as the soil characteristics in which the system is to be installed. These analyses will be helpful in selecting the appropriate material and ensuring a more durable system.

C. Economic Considerations

The drainage system's initial cost and estimated usable life are most important to the engineer. The concept of life cycle cost analysis for engineering systems was introduced in Chapter 8. Knowing the cost of materials and installation, the cost of money over the projected life cycle of the facility, and the cost of maintaining the facility, a total comparative cost in terms of present dollars can be derived which allows the engineer to compare alternative systems in a clear and objective manner.

Different sections of the country have different needs that strongly influence the type of materials appropriate for a specific project. The cost of materials in various regions of the United States strongly influence the life cycle cost of the facility. The engineer should therefore spend a significant amount of design time and study in unit-cost comparison of available materials, prior to actual specification. Various sources of information are listed in the reference section of this chapter.

D. Materials

The evaluation and selection of materials for sewer construction depend on the anticipated conditions of service. Consider the following factors:

(a) Intended use.
(b) Scour or abrasion conditions.
(c) Installation requirements—pipe characteristics and sensitivities.
(d) Corrosion conditions—chemical and biological factors both within the pipe and in the surrounding soil.
(e) Flow requirements—pipe size, velocity, slope and friction coefficient.
(f) Product characteristics—cross sectional shapes, fitting and connection requirements, laying strength, supplementary protective coating systems.
(g) Cost effectiveness—materials, installation, maintenance, life expectancy.
(h) Physical properties—crush strength for rigid pipe, pipe stiffness or stiffness factor for flexible pipe, soil conditions, pipe loading strength, pipe shear loading strength, pipe flexural strength.
(i) Handling requirements—weight, impact resistance.

No single product will provide optimum capability in every characteristic for all design conditions. Specific application requirements should be evaluated prior to selecting or specifying materials. Many studies have been conducted concerning durability and are available to the designer and/or owner to facilitate design and construction of a relatively maintenance-free storm sewer system. New materials are continually being offered for use in storm drainage construction. The discussion has been limited to the commonly accepted materials currently available today.

II. MATERIALS FOR OPEN, LINED CHANNELS

Open, lined channels are used extensively throughout the world as water conduits. Their performance depends largely on the suitability of the channel lining. For a comprehensive discussion of channel lining materials and the design criteria for their use, see Chapter 9.

III. CONDUIT MATERIALS

Because of the maintenance costs involved with open channel drainage systems, the drainage design engineer should always investigate the feasibility of an enclosed conduit system in each design situation. This section deals with the materials available for the construction of closed conduits. Details on the structural aspects of design will be found in Chapter 14.

A. Rigid Pipe

Pipe materials in this classification derive a substantial part of their basic earth load carrying capacity from the structural strength inherent in the rigid pipe wall. Commonly specified rigid pipe materials include:

1. Asbestos Cement Pipe (ACP)

Asbestos-cement pipe has been used for both gravity and pressure sanitary sewers but has now become less common. The product, produced from asbestos fiber and cement, is available in nominal diameters from 4–36" and in some areas up to 42 inches. A full range of compatible fittings is available. Jointing is accomplished by compressing elastomeric rings between pipe ends and sleeves or couplings.

ACP manufactured for gravity drain applications is available in seven strength classifications. The class designation represents a minimum crushing strength of the pipe expressed in pounds per linear foot of pipe. Potential advantages of asbestos-cement pipe include:

(a) Long laying lengths (in some situations).
(b) Wide range of strength classifications.
(c) Wide range of fittings available.

Potential disadvantages of asbestos-cement pipe include:

(a) Subject to corrosion where acids are present.
(b) Subject to shear and beam breakage when improperly bedded.
(c) Low beam strength.

Asbestos-cement pipe is specified by pipe diameter and class or strength. The product should be manufactured in accordance with one or more of the following specifications:

(a) "Standard Specification for Asbestos-Cement Non-Pressure Pipe," ASTM C 428.
(b) "Standard Specification for Asbestos-Cement Non-Pressure Small Diameter Sewer Pipe," ASTM C 644.

2. Cast Iron Pipe (CIP)

CIP (gray iron) has been used for both gravity and pressure drainage systems, although recently ductile iron pipe has generally been specified in its place. Standards specify the product in nominal diameters from 2–48" inches with a variety of joints. Cast iron fittings and appurtenances are generally available. Product availability is limited, as manufacturers are converting to ductile iron production.

CIP is manufactured in a number of thicknesses, classes, and strengths. A cement mortar lining with an asphaltic seal coating may be specified on the interior of the pipe. An exterior asphaltic coating is also commonly specified. Other linings and coatings may be specified. Potential advantages of cast iron pipe (gray iron) include:

(a) Long laying lengths (in some situations).
(b) High pressure and load bearing capacity.

Potential disadvantages of cast iron pipe (gray iron) include:

(a) Subject to corrosion where acids are present.
(b) Subject to chemical attack in corrosive soils.
(c) Subject to shear and beam breakage when improperly bedded.
(d) High weight per length ratio.
(e) Product availability.

CIP is specified by nominal diameter, class, lining, and type of joint. CIP is manufactured in accordance with one or more of the following standard specifications:

(a) "Cast Iron Pipe Centrifugally Cast in Metal Molds, for Water or Other Liquids," ANSI A 21.6 (AWWA C 106).
(b) "Gray-Iron and Ductile Iron Fittings, 2 through 48–inch, for Water and Other Liquids," ANSI/AWWA C 110.
(c) "Polyethylene Encasement for Gray and Ductile Iron Piping for Water and Other Liquids," ANSI/AWWA C-105/A 21.5.
(d) "Flanged Cast-Iron and Ductile-Iron Pipe with Threaded Flanges," ANSI A 21.15 (AWWA C 115).
(e) "Cement Mortar Lining for Cast-Iron and Ductile Iron Pipe and Fittings for Water," ANSI A 21.4 (AWWA C-104).

Additional information relative to the selection and design of CIP may be obtained from the Ductile Iron Pipe Research Association (1984).

3. Concrete Pipe

Reinforced and non-reinforced concrete pipe are used for gravity storm drainage. Reinforced concrete pressure pipe and prestressed concrete pipes are used for pressure as well as gravity drains. Non-reinforced concrete pipe is available in nominal diameters from 4–36". Reinforced concrete pipe is available in nominal diameters from 12–200". Pressure pipe is available in diameters from 12–120". Concrete fittings and appurtenances such as wyes, tees, and manhole sections are generally available. A number of jointing methods are available depending on the tightness required and the operating pressure, and various linings and coatings are available. A number of mechanical processes are used in the manufacture of concrete pipe, including centrifugation, vibration, and packing and tamping. Gravity and pressure concrete pipe may be manufactured to any reasonable strength requirement by varying the wall thickness, concrete strength, quantity and configuration of reinforcing steel or prestressing elements. Potential advantages of concrete pipe include:

(a) Wide range of structural and pressure strengths.
(b) Wide range of nominal diameters.
(c) Wide range of standard lengths (generally 4–24').
(d) Moderately low friction losses.
(e) Resistance to galvanic corrosion and abrasion.

Potential disadvantages of concrete pipe include:

(a) High weight.
(b) Subject to corrosion where acids are present.

Concrete pipe is normally specified by nominal diameter, class or D-load strength and type of joint. The product should be manufactured in accordance with one or more of the following standard specifications.

(a) "Concrete Sewer, Storm Drain, and Culvert Pipe," ANSI/ASTM C 14.
(b) "Reinforced Concrete Culvert, Storm Drain, and Sewer Pipe," ANSI/ASTM C 76.
(c) "Reinforced Concrete Arch Culvert, Storm Drain, and Sewer Pipe," ANSI/ASTM C 655.
(d) "Reinforced Concrete Elliptical Culvert, Storm Drain, and Sewer Pipe," ANSI/ASTM C 507.
(e) "Reinforced Concrete Box Culverts," ANSI/ASTM C250, C789M, and C789.
(f) "Reinforced Concrete Low-Head Pressure Pipe," ANSI/ASTM C 361.
(g) "Joints for Circular Concrete Sewer and Culvert Pipe, Using Rubber Gaskets," ANSI/ASTM C 443.
(h) "External Sealing Bands for Non-Circular Concrete Sewer, Storm Drain, and Culvert Pipe," ANSI/ASTM C 877.

Additional information relative to the selection and design of concrete pipe may be obtained from the American Concrete Pipe Association (1970, 1988).

4. Vitrified Clay Pipe (VCP)

VCP, manufactured from clay and shales, is used for gravity storm drainage. The pipe is vitrified at a temperature at which the clay mineral particles become fused. The product is available in diameters from 3–36″ and in some areas up to 42″. Clay fittings are available to meet most requirements, with special fittings manufactured upon request. A number of jointing methods are available.

VCP is manufactured in standard and extra-strength classifications, although in some areas the manufacture of standard-strength pipe is not common in sizes 12″ and smaller. The strength of vitrified clay pipe varies with the diameter and strength classification. The pipe is manufactured in lengths up to 10′. Potential advantages of vitrified clay pipe include:

(a) High resistance to chemical corrosion.
(b) High resistance to abrasion.
(c) Wide range of fittings available.
(d) Low friction losses.

Potential disadvantages of vitrified clay pipe include:

(a) Limited range of sizes available.
(b) High weight.
(c) Subject to shear and beam breakage when improperly bedded.
(d) Low beam strength.

VCP is specified by nominal pipe diameter, strength, and type of joint. The product should be manufactured in accordance with one or more of the following standard specifications:

(a) "Standard Specification for Vitrified Clay Pipe, Extra Strength, Standard Strength and Perforated," ANSI/ASTM C 700.
(b) "Compression Joints for Vitrified Clay Pipe and Fittings," ASTM C 425.
(c) "Pipe, Clay, Sewer," Federal Specification SS-P361d, Standard Methods of Testing Vitrified Clay Pipe, ANSI/ASTM 301.
(d) Crushing Strength for Pipe and Fittings for Perforated VCP in Accordance with NCPI ER4–67.

Additional information relative to the selection and design of vitrified clay pipe may be obtained from the National Clay Pipe Institute (1978).

B. Flexible Pipe

Storm drainage pipe materials in this classification derive their load carrying capacity from a combination of the inherent strength of the pipe and of the interaction of the flexible pipe and the embedment soils (affected by the deflection of the pipe under load). Commonly specified flexible pipe materials are discussed below:

1. Ductile Iron Pipe (DIP)

DIP is used for both gravity and pressure drains. DIP is manufactured by adding cerium or magnesium to cast (gray) iron just prior to the casting process. The product is available in nominal diameters from 3–54" and in lengths to 20'. Cast iron (gray iron) or ductile iron fittings are used with ductile iron pipe. Various jointing methods for the product are available.

Applications of DIP generally involve one or more of the following conditions, high impact and/or loading, minimal cover, and long service life with minimal maintenance.

DIP is manufactured in various thicknesses, classes and strengths. Pipe linings such as cement mortar lining with asphaltic coating, coal tar epoxies, epoxies, and polyethylene may be specified. An exterior asphaltic coating and polyethylene exterior wrapping are also commonly specified. Potential advantages of DIP include:

(a) Long laying lengths (in some situations).
(b) High impact strength.
(c) High pressure and load bearing capacity.
(d) High beam strength.
(e) Low friction losses.

Potential disadvantages of DIP include:

(a) Subject to corrosion where acids are present.
(b) Subject to chemical attack in corrosive soils.
(c) High weight.

DIP is specified by nominal diameter, class, lining, and type of joint. DIP should be manufactured in accordance with one or more of the following standards specifications:

(a) "Polyethylene Encasement for Gray and Ductile Cast-Iron Piping for Water and Other Liquids," ANSI A 21.5 (AWWA C 105).
(b) "Ductile Iron Gravity Sewer Pipe," ASTM A 746.
(c) "Gray-Iron and Ductile Iron Fittings, 3 inch through 48 inch, for Water and Other Liquids," ANSI/AWWA C 110.
(d) "Cement Mortar Lining for Cast-Iron and Ductile-Iron Pipe and Fittings for Water," ANSI A 21.4 (AWWA C 104).

Additional information relative to the selection and design of DIP may be obtained from the Ductile Iron Pipe Research Association (1984).

2. Fabricated Steel Pipe

Corrugated Steel Pipe, Arches, and Pipe Arches—galvanized corrugated steel is fabricated in a variety of conduit shapes with a choice of additional protective coatings when deemed necessary. Available sizes and shapes include: circular, in diameters from 12–144"; pipe arches manufactured from circular pipe from 15–120" in diameter; structural plate structures of 60–312" in diameter; structural plate arches from 6–25' in span with concrete base; and circular, arch, or horseshoe in shape. Strengths to meet a variety of design loads may be obtained by specifying from a range of gages and joint types.

Pipe sections are furnished in standard lengths of 20, 30, or 40' in multiples of 2 or 4'. Shorter or longer lengths can be provided by the fabricator, as lock seam or welded seam pipe is manufactured from a continuous coil and can be cut to any length. The sections are joined by coupling bands which may be single piece, two piece, or of an internal expanding type used in lining work. Large sizes (structural plate conduits) are field-bolted. CMP is also available in helical corrugations with improved flow characteristics.

Appurtenances include tees, wyes, elbows, and manholes fabricated from the same corrugated material as well as end sections. Corrugated pipe may be designed specially for jacking purposes. Advantages of corrugated steel pipe, arches, and pipe arches include:

(a) Light weight.
(b) Long laying lengths (in some situations).
(c) Flexibility.
(d) Usefulness as a lining for the repair of existing structures.
(e) Wide range of coatings available.

The principal disadvantages are:

(a) Relatively poor hydraulic coefficient unless fully lined with bituminous materials.
(b) Subject to corrosion in aggressive environments.
(c) The requirement for structurally satisfactory horizontal support of the spring-line of the upper semi-circular arch of the pipe or culvert (special attention must be given to proper bedding).

Bituminous linings, which cover the crests of the corrugations and form a smooth surface to improve flow characteristics, sometimes are used to improve hydraulic performance (the invert may also be paved), and to impart corrosion resistance. An inherent problem with coated corrugated steel pipe is the achievement of an effective bond between the coating and pipe. Durability is increased with coating, but the coated material should not be expected to remain intact for the full life of the pipe. Smooth-coated corrugated steel pipe should not be used in outfall situations where the materials are subject to freezing and thawing conditions. External corrosion protection may be necessary depending on soil conditions. Bituminous coatings are flammable and may be damaged or destroyed by petroleum wastes or solvents. Care should also be taken in handling coated pipe during installation to prevent damage to the coating. Continuous adequate lateral support is critical to obtain structural stability of the pipe.

Corrugated steel pipe, arches, or pipe arches are specified by size (nominal diameter, span and rise, or arc length), shape (full circle, elliptical, arch, or segmental plate arch), gage of metal (depending on strength requirements), assembly of sections with bands or bolts, coatings or linings, and couplings (width of single piece and two-piece).

Additional information relative to the selection and design of steel pipe may be obtained from the American Iron and Steel Institute (1971), and a number of books published by the National Corrugated Steel Pipe Association.

3. Corrugated Aluminum Pipe

Corrugated aluminum pipe, arches, and box culverts are available in many sizes and shapes. Circular pipe is available from 6–180 in. Arches can be specified in a variety of sizes up to 14 foot rise by 30 foot span. Box culverts are available up to 10 feet-2 inch rise by 25 feet-5 inch span. The strength of pipe is specified by a range of gages, joint types (welded, mechanical coupling, bolted), and method of bedding or backfilling. Pipe sections can be furnished in lengths up to 40 ft. and are easy to handle due to their light weight. Appurtenances include tees, wyes, elbows, manholes and end sections. Large size arches and box culverts are generally field bolted. Smaller sizes can be cut and welded in the field and are extremely versatile for on-site fabrication. Advantages of corrugated aluminum pipe, arches and box culverts include:

(a) Light weight.
(b) Long laying lengths.
(c) Ease of field connection and fabrication.
(d) Flexibility.
(e) Durability and resistance to corrosive environments such as saltwater.

Disadvantages include:

(a) Relatively poor hydraulic coefficient unless lined with bituminous material.
(b) The requirement for structurally satisfactory horizontal support (see corrugated steel pipe).

The aluminum pipe should conform to one or more of the following standard specifications:

(a) Standard Specification for Corrugated Aluminum Alloy Culverts and Underdrains, AASHTO Designation M-196.

(b) Standard Specification for Aluminum Structural Plate for Pipe, Pipe Arches, and Arches, AASHTO, Designation M-219.
(c) Standard Specification—For Clad Aluminum Alloy Sheets for Culverts and Underdrains, AASHTO, Designation M-197.
(d) Federal Specification—Pipe, Corrugated (Aluminum Alloy) WW-P-402.

Additional information relative to selection and design of corrugated aluminum pipe may be obtained from the Aluminum Association.

4. Thermoplastic Pipe

Thermoplastic materials include a broad variety of plastics that can be repeatedly softened by heating and hardened by cooling through a temperature range characteristic for each specific plastic. Generally, thermoplastic materials used in sanitary sewers and drainage are limited to acrylonitrile-butadiene-styrene (ABS), polyethylene (PE), and polyvinyl chloride (PVC).

Acrylonitrile-Butadiene-Styrene (ABS) Pipe ABS pipe is used for both gravity and pressure drains. Non-pressure rated ABS sewer pipe is available in nominal diameters from 3–12" and in lengths up to 35'. A variety of ABS fittings and several jointing systems are available.

ABS pipe is manufactured by extrusion of ABS plastic material, and is available in three dimension ratio (DR) classifications (23.5, 35, and 42) depending on nominal diameter selected. The classifications relate to three pipe stiffness values, PS 150, 45 and 20 psi, respectively. The DR is the ratio of the average outside diameter to the minimum wall thickness of the pipe. Potential advantages of ABS pipe include:

(a) Light weight.
(b) Long laying lengths (in some situations).
(c) High impact strength.
(d) Ease in field cutting and tapping.

Potential disadvantages of ABS pipe include:

(a) Limited range of sizes available.
(b) Subject to environmental stress cracking.
(c) Subject to excessive deflection when improperly bedded and haunched.
(d) Subject to attack by certain organic chemicals.
(e) Subject to surface change affected by long-term ultra-violet exposure.

ABS pipe is specified by nominal diameter, dimension ratio, pipe stiffness and type of joint. ABS pipe should be manufactured in accordance with one or more of the following specifications:

(a) "Acrylonitrile-Butadiene-Styrene (ABS) Sewer Pipe and Fittings," ANSI/ASTM D 2751.
(b) "Solvent Cement for Acrylonitrile-Butadiene-Styrene (ABS) Plastic Pipe and Fittings," ANSI/ASTM D 2235.

(c) "Elastomeric Seals (Gaskets) for Jointing Plastic Pipe," ANSI/ASTM F 477.
(d) "Joints for Drain and Sewer Plastic Pipes Using Flexible Elastomeric Seals," ANSI/ASTM D 3212.
(e) "PVC and ABS Injected Solvent Cemented Pipe Joints," ANSI/ASTM F 545.

Acrylonitrile-Butadiene-Styrene (ABS) Composite Pipe ABS composite pipe can be used for gravity drainage pipes. The product is available in nominal diameters from 8–15" and in lengths from 6.25–12.5'. ABS fittings are available for the product. The jointing systems available include elastomeric gasket joints and solvent cemented joints.

ABS composite pipe is manufactured by extrusion of ABS plastic material with a series of truss annuli which are filled with filter material such as light-weight portland cement concrete. Potential advantages of ABS composite pipe include:

(a) Light weight.
(b) Long laying lengths (in some situations).
(c) Ease in field cutting.

Potential disadvantages of ABS composite pipe include:

(a) Limited range of sizes available.
(b) Subject to environmental stress cracking.
(c) Subject to rupture when improperly bedded.
(d) Subject to attack by certain organic chemicals.
(e) Subject to surface change affected by long-term ultraviolet exposure.

ABS composite pipe should be manufactured in accordance with one or more of the following standard specifications:

(a) "Acrylonitrile-Butadiene-Styrene (ABS) Composite Sewer Piping," ANSI/ASTM D 2680.
(b) "Solvent Cement for Acrylonitrile-Butadiene-Styrene (ABS) Plastic Pipe and Fittings," ANSI/ASTM D 2235.
(c) "Joints for Drain and Sewer Plastic Pipes Using Flexible Elastomeric Seals," ANSI/ASTM D 3212.
(d) "Elastomeric Seals (Gaskets) for Joining Plastic Pipe," ANSI/ASTM F477.

Polyethylene (PE) Pipe PE pipe is used for both gravity and pressure drains. Non-pressure PE pipe, primarily used for sewer relining, is available in nominal diameters from 4–48". PE fittings are available. Jointing is primarily accomplished by butt fusion or flanged adapters.

PE pipe is manufactured by extrusion of PE plastic material. Non-pressure PE is produced at this time in accordance with individual manufacturer's product standards. Potential advantages of PE pipe include:

(a) Long laying lengths (in some situations).
(b) Light weight.

(c) High impact strength.
(d) Ease in field cutting.

Potential disadvantages of PE pipe include:

(a) Relatively low tensile strength and pipe stiffness.
(b) Limited range of sizes available.
(c) Subject to excessive deflection when improperly bedded and haunched.
(d) Subject to attack by certain organic chemicals.
(e) Subject to surface change affected by long-term ultraviolet exposure.
(f) Special tooling required for fusing joints.

PE pipe is specified by material designation, nominal diameter (inside or outside), standard dimension ratios and type of joint. PE pipe should be manufactured in accordance with one or more of the following specifications:

(a) "Butt Heat Fusion Polyethylene (PE) Plastic Fittings for Polyethylene (PE) Plastic Fittings for Polyethylene (PE) Pipe and Tubing," ANSI/ASTM D 3261.
(b) "Polyethylene (PE) Pipe and Tubing," ANSI/ASTM D 2239.
(c) "Polyethylene (PE) Plastic Pipe (SDR-PR)," ANSI/ASTM D 3261.
(d) "Polyethylene (PE) Plastic Pipe (SDR-PR)," ANSI/ASTM D 2239.
(e) "Polyethylene (PE) Plastic Pipe (SDR-PR), Based on Controlled Outside Diameter," ANSI/ASTM D 3035.

Polyvinyl Chloride (PVC) Pipe PVC pipe is used for both gravity drains and storm sanitary sewers. Non-pressure PVC sewer pipe is available in nominal diameters from 4–27". PVC pressure and non-pressure fittings are available. PVC pipe is generally available in lengths up to 20'. Jointing is primarily accomplished with elastomeric seal gasket joints, although solvent cement joints for special applications are available.

PVC pipe is manufactured by extrusion of the plastic material. A wide variety of appurtenances are available including wyes, tees, saddles, elbows, increasers, plugs, couplings, manhole adapters, and adapters to other pipe material.

Non-pressure PVC sanitary sewer pipe is provided in three dimension ratios, DR 35, 41, and 51, which relate to pipe stiffness values, PS 46, 28, and 80 psi, respectively. Potential advantages of PVC pipe include:

(a) Light weight.
(b) Long laying length (in some situations).
(c) High impact strength.
(d) Ease in field cutting and tapping.

Potential disadvantages of PVC pipe include:

(a) Subject to attack by certain organic chemicals.
(b) Subject to excessive deflection when improperly bedded and haunched.

(c) Limited range of sizes available.
(d) Subject to surface changes affected by long-term ultraviolet exposure.

PVC pipe is specified by nominal diameter, dimension ratio, pipe stiffness and type of joint. PVC pipe should be manufactured in accordance with one or more of the following standard specifications:

(a) "Type PSM Polyvinyl Chloride (PVC) Sewer Pipe and Fittings," ANSI/ASTM D 3034.
(b) "Type PSP Polyvinyl Chloride (PVC) Sewer Pipe and Fittings," ANSI/ASTM D 3033.
(c) "Elastomeric Seals (Gaskets) for Joining Plastic Pipe," ANSI/ASTM F 477.
(d) "Joints for Drain and Sewer Plastic Pipes Using Flexible Elastomeric Seals," ANSI/ASTM D 3212.
(e) "PVC and ABS Injected Solvent Cemented Plastic Pipe Joints," ANSI/ASTM F 545.
(f) "Solvent Cements for Polyvinyl Chloride (PVC) Plastic Pipe and Fittings," ANSI/ASTM D 2564.
(g) "Standard Specification for Polyvinyl Chloride (PVC) Large Diameter Plastic Gravity Sewer Pipe and Fittings," ASTM F-679.

5. Thermoset Plastic Pipe

Thermoset plastic materials include a broad variety of plastics. These plastics, after having been cured by heat or other means, are substantially infusible and insoluble. Generally, thermoset plastic materials used in sewers are provided in two categories—reinforced thermosetting resin (RTR) and reinforced plastic mortar (RPM).

Reinforced Thermosetting Resin (RTR) Pipe RTR pipe is used for both gravity and pressure drains. RTR pipe is generally available in nominal diameters from 1—12", manufactured in accordance with ASTM standard specifications. The product is available in nominal diameters from 12–144" manufactured in accordance with individual manufacturer's specifications. In small diameters, RTR fittings are available. In large diameters, RTR fittings are manufactured as required. A number of jointing methods are available. Various methods of interior protection (e.g., thermoplastic or thermosetting liners or coatings) are available.

RTR pipe is manufactured using a number of methods including centrifugal casting, pressure laminating and filament winding. In general, the product contains fibrous reinforcement materials, such as fiberglass, embedded in or surrounded by cured thermosetting resin.

An example of an RTR pipe material is the "Insituform" process which is most commonly used for the rehabilitation of existing pipe materials. A polyester fiber felt tube with an impermeable layer on one side is impregnated with a liquid thermosetting resin. The tube is inserted through the existing pipe through an access point such as a manhole. Cold water is forced through the pipe, pressing the tube against the existing pipe. The water is then heated, curing the inserted tube into a pipe within a pipe. This "Insituform" process is available for sizes from 4" to several feet and can be used for various shaped conduits. The thermoset resins can be selected to meet the requirements for resistance to corrosion. There are no joints for infiltration to enter. The wall thickness is selected to meet structural requirements and is expressed as a standard dimension ratio, SDR. The existing waterway area is reduced by the tube wall thickness. The reduced area may be compensated in part by improved smoothness of the pipe. Potential advantages of RTR pipe include:

(a) Light weight.
(b) Long laying lengths (in some situations).

Potential disadvantages of RTR pipe include:

(a) Subject to strain corrosion in some environments.
(b) Subject to excessive deflection when improperly bedded and haunched.
(c) Subject to attack by certain organic chemicals.
(d) Subject to surface change affected by long-term ultraviolet exposure.

RTR pipe is specified by nominal diameter, pipe stiffness, lining and coating, method of manufacture, thermoset plastic material, and type of joints. RTR pipe should be manufactured in accordance with one or more of the following standard specifications:

(a) "Filament-Wound Reinforced Thermosetting Resin Pipe," ASTM D 2996.
(b) "Centrifugally-Cast Reinforced Thermosetting Resin Pipe," ANSI/ASTM D 2997.
(c) "Machine-Made Reinforced Thermosetting Resin Pipe," ASTM D 2310.

Reinforced Plastic Mortar (RPM) Pipe   RPM pipe is used for both gravity and pressure sewers. RPM pipe is available in nominal diameters from 8–144". In smaller diameters, RPM fittings are manufactured as required. A number of jointing methods are available. Various methods of interior protection (e.g., thermoplastic or thermosetting liners or coatings) are available.

RPM pipe is manufactured containing fibrous reinforcements such as fiberglass and aggregates such as sand embedded in or surrounded by cured thermosetting resin. Potential advantages of RPM pipe include:

(a) Light weight.
(b) Long laying lengths (in some situations).

Potential disadvantages of RPM pipe include:

(a) Subject to strain corrosion in some environments.
(b) Subject to excessive deflection when improperly bedded and haunched.
(c) Subject to attack by certain organic chemicals.
(d) Subject to surface change affected by long-term ultraviolet exposure.

RPM pipe is specified by nominal diameter, pipe stiffness, stiffness factor, beam strength, hoop tensile strength, lining or coating, thermoset plastic material, and type of joint. RPM pipe should be manufactured in accordance with one or more of the following standard specifications:

(a) "Reinforced Plastic Mortar Sewer Pipe," ANSI/ASTM D 3252.
(b) "Reinforced Plastic Mortar Sewer and Industrial Pressure Pipe," ASTM D 3754.

C. Pipe Joints

A substantial variety of pipe joints are available for the different pipe materials used in drainage construction. A common requirement for the design of all drainage systems, regardless of the type of pipe specified, is the use of reliable, tight pipe joints. A good pipe joint must be watertight, flexible and durable. Currently, various forms of gasket (elastomeric seal) pipe joints are used with many pipe materials. They generally can be assembled by unskilled labor in a broad range of weather conditions and environments with good assurance of a reliable, tight seal.

1. Gasket Pipe Joints

Gasket joints seal against leakage through compression of an elastomeric seal or ring. Gasketed pipe joint design is generally divided into two types; push-on joint and mechanical compression pipe joint.

2. Push-on Pipe Joint

This type of pipe joint uses a continuous elastomeric ring gasket compressed into an annular space formed by the pipe, fitting, or coupler socket, producing a positive seal when the pipe spigot is pushed into the socket. When using this type of pipe joint in pressure sewers, thrust restraint may be required to prevent joint separation under pressure. Push-on pipe joints (fittings, couplers, or integral bells) are available on nearly all pipe products mentioned.

3. Mechanical Compression Pipe Joint

This type of pipe joint uses a continuous elastomeric ring gasket which provides a positive seal when the gasket is compressed by means of a mechanical device. When using this type of pipe joint in pressure sewers, thrust restraint may be required to prevent joint separation under pressure. This type of pipe joint may be provided as an integral part of cast iron or ductile iron pipe. When incorporated into a coupler, this type of pipe joint may be used to join two similarly sized plain spigot ends of commonly used sewer pipe materials.

4. Bituminous Pipe Joints

This type of pipe joint involves use of hot-poured or cold-packed bituminous material forced into a bell-and-spigot pipe joint to provide a seal. The use of this joint is discouraged, since reliable, watertight joints are not assured.

5. Cement Mortar Pipe Joint

This type of pipe joint involves use of shrink-compensating cement mortar placed into a bell-and-spigot pipe joint to provide a seal. The use of this joint is discouraged for pressure drains because reliable, watertight joints are not assured. Cement mortar joints are not flexible and may crack if there is any pipe movement.

6. Elastomeric Sealing Compound Pipe Joints

Elastomeric sealing compound may be used in jointing properly prepared concrete gravity pipe. Pipe ends must be sandblasted and primed for elastomeric sealant application. The sealant, a thixotropic, two-compound elastomer is mixed on the job site and applied with a caulking gun and spatula. The pipe joint, when assembled with proper materials and procedures, provides a positive seal against leakage.

7. Solvent Cement Pipe Joints

Solvent cement pipe joints may be used in jointing thermoplastic pipe materials such as ABS, ABS composite, and PVC pipe. This type of pipe joint involves bonding a sewer pipe spigot into a sewer pipe bell or coupler using a solvent cement. Solvent cement joints can provide a positive seal provided the proper cement is applied under proper ambient conditions with proper techniques. Refer to ASTM D 402 for safe handling procedures. Precautions must be taken to ensure adequate trench ventilation and protection for workers installing the pipe. Solvent cement pipe joints may be desired in special situations and with some plastic fittings.

8. Heat Fusion Pipe Joints

Heat fusion pipe joints are commonly specified for PE pipe. The general method of jointing PE pipe involves butt fusion of the pipe lengths. After the ends of two lengths of PE pipe are trimmed and softened to a melted state with heated metal plates, the pipe ends are forced together until they fuse, providing a positive seal. The pipe joint does not require thrust restraint in pressure applications. Trained technicians with special apparatus are required to achieve reliable watertight pipe joints.

9. Mastic Pipe Joints

Mastic pipe joints are frequently used for special non-round shapes of concrete pipe that are not adaptable for gasketed pipe joints. The mastic material is placed into the annular space to provide a positive seal. Application may be by trowelling, caulking, or by the use of preformed segments of mastic material in a manner similar to gaskets. Satisfactory performance of the pipe joints depends upon the proper selection of primer and mastic material, and on good workmanship.

10. Sealing Band Joints

External sealing bands of rubber made in conformance with ASTM C 877 are also used on non-circular concrete pipe. These elastomeric bands are wrapped tightly around the exterior of the pipe at the joint and extend several centimeters (inches) on each side of the joint. Sealing against the concrete is achieved by a mastic applied to one side of the band.

IV. MAINTENANCE

A. Introduction

All storm drainage systems must be maintained. A well-maintained storm drainage system will be ready to convey the runoff from the next storm with minimal damage to the storm drainage facilities. A poorly maintained drainage system may not be able to function at its design conveyance and could be damaged by the runoff. The increases in potential repair costs and liability exposure are less obvious, but no less serious. Minor repairs can often prolong the service life of the facility, and can reduce the costs of future major repairs and/or replacement.

Because storm drainage systems function intermittently and seldom at full capacity, it is all too easy to defer maintenance activities. The storm drainage system must be actively maintained—it is too late to repair the damage from the last storm or to do preventive maintenance if storm clouds are again gathering over the basin. It is especially important to recognize that maintenance includes both scheduled (mowing, trash pickup, etc.) and unscheduled (erosion damage repair, etc.), and funds must be provided for both. Excellent discussions of the subject can be found in ASCE (1983) and WPCF (1989).

The owner of storm drainage facilities should establish a routine maintenance inspection program once the facility has been completed and placed in service. The inspections should be conducted on an annual or semi-annual basis, as well as following major storms. The inspections may be accomplished by visual means or by using a television camera, where applicable.

The inspection should be documented. Items to be recorded should include size and type of facility, date of inspection, location of facility, minor deficiencies, major deficiencies, and areas of possible future problems. The documentation should be kept current, and when any repair work has been accomplished, it should be recorded.

B. MAINTENANCE OBJECTIVES

A thorough drainage system maintenance program will provide for scheduled maintenance activities and will also accommodate necessary unscheduled work. Staff from the maintenance department should be involved in all aspects of a drainage system, from planning and design through to construction, if maintenance considerations are to receive adequate attention. The general goals of a maintenance program should include the following:

(a) Participate in drainage project planning and in design review to facilitate maintenance activities.
(b) Participate in construction progress meetings to determine if maintenance-oriented facilities are being built as called for in the design.
(c) Inspect facilities regularly to monitor their effectiveness and need for repairs.
(d) Prevent drainage systems from falling into visual disrepair. Aesthetic considerations are important within the community. A respectable-looking drainage facility is less likely to attract vandalism and garbage dumping, and will help maintain adjacent property values.
(e) Reduce life-cycle costs through effective design review, timely maintenance activities, documentation of crew and equipment productivity, and analysis of repair costs and longevity.
(f) Repair deteriorated facilities before major damage or failure occurs.
(g) Have drainage systems repaired, cleaned, and ready to function before the next rainy season arrives.
(h) Repair and maintain facilities as necessary to insure they are capable of operating at full design conveyance.

C. Life-cycle Stages of a Storm Drainage System

All improved storm drainage systems pass through three stages in their life-cycle. Those stages are:

(a) Design and Construction.
(b) Drainage Service.
(c) Rehabilitation/Replacement.

If a drainage system includes many structures and man-made features, the stages are quite separate and distinct. The stages are less obvious and less important for streams that have been only slightly modified by man. The higher the degree of improvement to be done to a drainage channel, the more imperative it is that maintenance personnel are involved in every stage.

The design and construction stage is short but it is a period of much activity with far-reaching affects. It is the beginning of the life-cycle for the drainage system. The maintenance-oriented decisions made during this stage will dictate much of what happens during the other two stages.

The drainage service stage will be long and uneventful if maintenance concerns are given full consideration during design. During the drainage service stage the drainage system should function as designed. Maintenance activities are directed at extending the service life as much as possible.

All drainage systems will eventually pass through a rehabilitation stage, or will have features that are better replaced than rehabilitated. Facilities that are storm-damaged or simply worn out will need repair. As long as the hydrology and hydraulics are still valid the system can be rehabilitated and returned to the drainage service stage. The more comprehensive and maintenance-oriented the original design the more likely it is that the system will need rehabilitation rather than complete re-design and reconstruction. The life-cycle stages will be discussed in each of the following sections regarding maintenance of different drainage systems.

D. Maintenance of Open Channel Drainage Systems

Many factors make open channel drainage systems more desirable than other systems. Those same factors need careful attention from maintenance personnel during the three life-cycle stages.

1. Design and Construction Stage

(a) Access—Vehicle access is vital to the maintainability of a drainage system. Ramps leading into channels and/or all-weather trails paralleling the system are frequently multi-purpose designs. Be sure access ramps and trails have traffic control barriers to keep unwanted traffic from using the trails while still allowing pedestrian movement.

(b) Side slopes—Grass-lined channels must have slopes that are steep enough to drain toward the channel and yet are gentle enough to allow vegetation to establish and to permit mowing and clean-up activities.

(c) Vandalism—Drainage facilities can be attractive nuisances and can be damaged by those who use the area. Preventive measures may be necessary to keep graffiti off walls, to keep rock riprap from being relocated, or to keep gabion baskets from being cut open.

(d) Trickle channels—Base flow erosion damage continues day after day. The cumulative effect can be dramatic. If the soils are erodible it may be necessary to install a trickle channel to halt the erosion. Pay attention to the potential for erosion immediately outside the trickle channel during intermediate runoff events. Reach agreement, in advance of construction, with pertinent regulators that channel maintenance will not require a special permit, such as a "404" permit under the Clean Water Act.

(e) Localized erosion—There are several locations that can suffer erosion and subsequently need increased maintenance work. Proper design will reduce the likelihood of erosion problems. A practical review of the project plans may reveal the need for additional erosion protection in the following places:
   (1) All transitions, such as changes in cross-section or changes in channel lining material.
   (2) At the outside of curves where flow velocities are higher.
   (3) At the outlet of all tributary storm sewer pipes.
   (4) On the bank opposite all tributary pipes and channels.
   (5) Downstream from drop structure energy dissipation basins.
   (6) Downstream of bridges and box culverts.
(f) Toe protection—Localized scour or general degradation can quickly lower the bottom of a channel. Erosion protection facilities must have deep toe protection or they can fail by being undermined.
(g) Rundowns—All drainage systems have many small capacity tributaries. Runoff events can damage these tributary connections as well as the main channel if the connections are not built to withstand the erosion impact.
(h) Trash racks—Normally, these structures do exactly what they are designed to do—catch debris. The bars should be spaced to allow small debris to pass through yet catch large material. Arrange the bars to facilitate cleaning and to allow the debris to float out of the way as the water level rises.
(i) Sediment traps—If called for in the design, they will certainly need regular silt removal to protect the downstream facilities. Sediment traps effectively reduce downstream maintenance needs.

2. Drainage Service Stage

(a) Mowing—In urbanized areas the drainage channel should be mowed often enough to control weeds and to show community responsibility. For native grass vegetation in a semi-arid climate three to six cuttings per year is satisfactory.
(b) Debris control—Debris blockage at drainage structures often contributes to flooding problems. Trash racks and debris traps help reduce the problem if they function properly and are regularly cleaned. Regular debris removal along the length of the drainage system also helps. This should include trimming and thinning of trees if they encroach on the drainage channel or if they have become overgrown.
(c) Inspection—An inspection at least annually of drainage facilities will detail long-term changes in the system and will highlight needed maintenance work. Inspections should also be done following major storm events.
(d) Silt removal—Some silt accumulation in stilling basins and around channel obstructions is inevitable, and is harmless in limited amounts. Silt should be removed if it is severe enough to alter the water surface or affect the function of drainage facilities such as drop structures. Silt accumulations can also cause trouble by supporting undesirable or obstructive vegetation. Remember that suitable silt disposal sites can

be a problem (due to contaminants) and should be evaluated during design.
(e) Trail repair—An annual effort to repair damaged trail sections will result in guaranteed maintenance access and better pedestrian use. The best time for repairs is right after the cold/rainy season.

3. Rehabilitation Stage

With regular inspection reports a drainage maintenance department will know when a drainage system is in need of repairs. If the problems are repaired promptly the facility can be returned to service with little threat of further damage or failure. Listed below are many of the typical problem areas that signal the need for rehabilitation.

(a) Hard-lined trickle/low flow channels—Local undermining of the structure or secondary channel erosion parallel to the main channel.
(b) Soft-lined trickle/low flow channels—Random bank failure and bottom degradation that is unsafe or threatens other improvements.
(c) Tributary channels and pipe outlets—Erosion from the receiving channel leading back to the tributary outlet and/or erosion under the outlet structure.
(d) Drop structures and grade control structures—Frequent problems include erosion damage in and around the energy dissipation basin and around the outside edges of the structure. Physical damage can occur to the structure in the form of uplifted or depressed concrete, broken gabion baskets, and displaced riprap.
(e) Channel banks—Channel bottom and sides should be maintained to their original slopes. Bank protection such as riprap, slope paving, or retaining walls can be undermined by local scour or general degradation if not "toed-in" deep enough. Grass-lined banks can lose vegetative cover and can suffer spot erosion that may quickly worsen.

E. Maintenance of Piped Drainage Systems

Right-of-way constraints frequently dictate use of a piped drainage system, which in turn create particular maintenance constraints.

1. Design and Construction Stage

(a) Access—Manpower and equipment access for the length of the piped system must be available. Publicly owned right-of-way or easements are normally sufficient (be sure the easement connects to a public right-of-way and is wide enough for maintenance activities). The easement language must be restrictive enough to prohibit undesirable activities on the land surface.
(b) Erosion protection—The inlets and outlets to piped systems are subject to high velocities. Adequate protection will usually take the form of riprap aprons, energy dissipation structures, or concrete headwalls, wingwalls and aprons. Steep earth slopes at inlet and outlet transitions frequently need short walls to hold the soil in place.

(c) Trash Racks—This is one of the most frequent problem areas. The design should consider the potential debris sources upstream. Designers should assume at least 50% blockage of a trash rack when designing for the maximum storm runoff. Another rule of thumb requires a trash rack to have four times the clear opening area of the pipe being protected.

(d) Manholes—Most local governments have their own requirements for spacing. Manholes need to be accessible in all weather conditions. Be sure access is available to all pipes of a multi-barrel system. Drop manholes can be especially difficult when designing for adequate access and safety.

(e) As-built drawings—It is an absolute necessity to obtain as-built drawings of the completed project.

2. Drainage Service Stage

(a) Curb inlet cleaning—Because of their location and shape inlets often trap sediment and debris. They should be cleaned twice a year to insure their proper function. If only one cleaning is possible it should occur prior to the rainy season.

(b) Debris control—Trash racks should be cleaned regularly to keep accumulations from forming. In-pipe debris should be removed if it is large enough to create a flow obstruction.

(c) Overflow channel maintenance—If the pipe system was designed with a surcharge or overflow channel it deserves occasional attention. It must be kept clear of debris and excessive vegetation. In general, it should be maintained as an open channel to be ready to function when called upon.

(d) Inspection—A regular in-pipe inspection of piped drainage systems will detail long term changes and will point out needed maintenance work such as debris removal or joint patching. Special attention is necessary to insure the safety of the inspection team if the pipe is long. Small pipes and pipes that carry continuous flow can be viewed with automated equipment. Inspections should be done following major runoff events. Inlet grates should be checked for clogging, and catch basins and pipes for sediment/waste blockage.

3. Rehabilitation Stage

Typical problem areas that can signal the need for rehabilitation of a piped system include:

(a) Inlet and outlet structures—Local erosion due to high velocities, lack of protection, or transition turbulence.

(b) Trench backfill—Subsidence of the trench, which can result from poor initial compaction or from pipe or joint failure. Earth settlement around manholes is a frequent indicator of compaction problems.

(c) Pipe joints—The first sign of problems in the system shows at the pipe joints. Spalled concrete, cracks, distorted pipe geometry, backfill movement, and water inflow occur at the joints and are precursors of greater problems to come.

V. SUMMARY

The choice of a particular material depends upon a number of variables. The best choice is the one that yields the best performance over the project life cycle. Some products have a much longer useful life in a particular environment than others. The engineer must evaluate material longevity based on a realistic appraisal of how long the stormwater conveyance or system will be necessary with respect to projected development growth rates and master planning priorities established by the community. The life-cycle cost analysis will insure the proper material specification after the factors of capital cost, annual maintenance costs, and life expectancy are applied. Regionally specific concerns, such as material availability, will influence the selection. Finally, the forces that degrade storm drainage systems vary from one location to another, and prudent evaluation of these factors is advised.

VI. REFERENCES

American Concrete Pipe Association. (1970). *Concrete pipe design manual*. 7th printing (Revised), ACPA Arlington, VA.

American Concrete Pipe Association. (1988). *Concrete pipe handbook*. 3rd printing, ACPA, Arlington, VA.

American Iron and Steel Institute. (1971). *Handbook of steel drainage and highway construction products*. 1st ed., AISI.

Ductile Iron Pipe Research Association. (1984). *Handbook of ductile iron pipe*. DIPRA, Birmingham, AL.

National Clay Pipe Institute. (1978). *Clay pipe engineering manual*. NCPI, Washington, D. C.

CHAPTER 14
STRUCTURAL REQUIREMENTS

I. INTRODUCTION

Storm drainage structures vary from relatively conventional storm sewers and culverts to unique facilities designed for specific purposes. Structural details of all of the facilities that might be included in drainage and flood control structures are beyond the scope of this Manual, though the design engineer must, nevertheless, be cognizant of them.

It is assumed that the design of the more conventional structures will be completed by the hydraulic engineer, after appropriate consultation with a geotechnical engineer. Specific guidelines or references are provided to facilitate this effort. It is also assumed that unique, large, or complex structures will probably be designed by structural and/or geotechnical engineers. In such cases, stormwater facility designers must take an active role, providing guidance or input in the structural design of all hydraulic facilities. The hydraulic engineer should be responsible for identifying the critical design conditions and computing the loads on the various members of the structure.

II. STRUCTURAL DESIGN PROCESS

The structural design process consists of the following steps:

(a) Identify critical design condition(s).
(b) Determine loadings in the critical design condition(s).
(c) Perform a stability analysis on the overall structure.
(d) Select and design structural members.
(e) Prepare a structural plan and details.

A structural engineer unfamiliar with drainage facilities can be of significant assistance only in the final two steps. The first four steps of the structural design process are addressed in detail in the remainder

538

of this chapter. The fifth step, the preparation of plans and details, is covered in numerous texts and manuals and is not addressed here.

III. PROJECT LIFETIME FOR STRUCTURAL DESIGN

It should not be assumed that the structural elements of a facility need to be designed for a service life equal to the design return period for the facility itself, and it is important that a designer distinguish between several different "time intervals" frequently associated with a project.

"Project life" or "design life," also discussed in Chapter 13, are terms that refer to the length of time a facility will last, after which it may require replacement. This lifetime is generally based upon factors that occur gradually, such as wearing of equipment or linings, corrosion of metals, and deterioration of concrete.

Since the lengths of time within which a facility may need repair due to these processes are all different, they are frequently aggregated into the terms "life cycle" or "economic life." These refer to a time period after which it would be more economical to replace than repair the facility. Commonly selected economic lives for drainage and flood control projects are 35, 50, or 75 years, despite the fact that many projects provide much longer service. It is important to note that the project life is essentially independent of a project's design return period. Projects designed to convey a five-year flood or a 100-year flood may each have a project life of 35 years.

The project design return period (for hydraulic performance) will establish the flow conditions—and therefore the loading conditions— for which the structures will be designed. When greater loads occur, it must be assumed that the structure could fail, unless specifically designed for overload conditions. Note, however, that structural failure is generally unacceptable in any but the most rare events. This is particularly true if a failure would result in damage, injury, or death, as in the case of a failure of a dam or levee.

The project design life (for structural design) must be selected carefully for each project, and an analysis must be made of the critical design conditions and the failure mechanisms for specific structures. The following guidelines should be considered:

(a) The project design life should, in many cases, be at least as great as the project's economic life. A facility that an owner would plan to replace at fifty-year intervals should not experience structural failure more frequently (Owners should, however, be reminded that failure due to hydraulic events is based upon probability and not upon elapsed time).

(b) If structural failure could be catastrophic, such as the failure of a large dam, then the structures should be designed to survive very rare events, such as the Probable Maximum Flood. It should be noted that completely unsatisfactory hydraulic performance can occasionally be tolerated, if the structural performance remains satisfactory.

IV. ESTABLISHMENT OF DESIGN CONDITIONS

A. General

The hydraulic selection and sizing of a drainage facility requires the identification of at least one "critical design condition." For purposes of this discussion, a "design condition" is a combination of loads that could be exerted simultaneously on a structure or any of its members. A "critical design condition" is one in which the loads create a maximum net force in a single direction on a structure or any of its members. Any structure will have a large number of design conditions, from which must be selected the few critical design conditions to be addressed in the design process. The engineer must consider all conditions that could occur during the life of the structure. He must identify conditions that could or could not occur simultaneously, and he must consider which simultaneous conditions might cause increased loads and which loads would tend to reduce or cancel other loads.

As part of the process of identifying critical design conditions, the engineer should consider mechanisms through which failure might occur, such as rare flow events or excessive scour.

The following sections describe some conditions that could constitute critical design conditions (it should not be assumed that all possible critical design conditions are included). While each possible condition should be considered for each structure, the designer's experience should allow him to quickly discard certain possibilities as inappropriate for specific situations.

B. Flow Conditions

Since most drainage and flood control structures are designed to convey or control water, flow conditions are frequently the most readily predictable design conditions. Each of the following should be considered:

(a) No Flow—The absence of flow is a relatively common condition that may not, in itself, create excessive forces, but that may be the condition under which other loads reach their maximum values.
(b) Low Flow—Such a condition would rarely create a critical design condition, but it should be considered.
(c) Design Flow—Design flow generally is a large flow with high velocities and, in most cases, will represent a portion of one or more critical design conditions.
(d) Rare Flow—Most drainage and flood control structures have design flows with return periods that may range up to 100 years. However, many could also be subject to much larger flood events. The engineer

should consider the potential for structural failure as a result of such events. If they could cause significant additional damage or loss of life, the facility should be designed accordingly.

C. Groundwater Conditions

If all or part of any structure or member will be below groundwater level, hydrostatic pressures will be exerted, as if the structure were partially or completely submerged in a body of water. This can result in uplift forces that tend to lift the structure out of the ground and reduce the sliding friction resistance at boundaries between the structure and the soil or rock. The groundwater pressures also produce loads within the structural elements, such as bending moments and shear stresses within a retaining wall that has groundwater behind it.

When a designer selects the groundwater level for the design of a particular structure, he must consider both the typical groundwater conditions and the fluctuations in level that might be anticipated due to seasonal variations, flood events, and other applicable factors. The permeability of the soil and rock strata will affect the magnitude and timing of the expected fluctuations.

Careful consideration should be given to the adequacy and effectiveness of design details intended to either encourage groundwater flow through a structure or to prohibit groundwater flow into a structure. For example, weep holes and other drainage behind retaining walls should have adequate capacity to maintain the groundwater levels assumed in the design. Water stops and sealed joints used to prevent seepage through structural elements should be strong enough to withstand the expected hydrostatic pressures, and should be able to accommodate any expected differential movements caused by settlement or loads.

D. Adjacent Earth Conditions

All structures are affected to varying degrees by earth loadings. The loadings to impose on a structure after backfilling are generally determined by a geotechnical engineer. His recommendations must be based upon analyses of the soils to be used for backfilling, which may be either earth existing at the site or imported material.

When a hydraulic structure reaches various stages of partial completion, it could have loading conditions imposed upon it that stress its members either to a greater degree or in a different direction than would occur after backfilling. In particular, flexible conduits and sloping (or warped) wing walls may require special supports until backfilling is completed.

Many hydraulic structures can be subjected to critical design conditions resulting from either excessive erosion or scour, or from the deposition of sediment. While such conditions generally last for only a short time, they are usually associated with large flows or rare flood events, with the critical hydraulic loads occurring simultaneously with a removal of resisting or supporting soil.

It is important that the hydraulic engineer make a careful evaluation of two scour conditions. The first is scour that could reasonably be anticipated during high flow conditions. It is generally prudent to assume some level of erosion in scour-prone areas adjacent to a structure, even when great care is given to erosion prevention.

The second condition to evaluate is one that might occur when flows exceed the design flow. The engineer should consider whether extreme scour could lead to a complete structural failure, and whether it would cause damage or loss of life. In such cases, it may be appropriate to insure that a structural failure would not occur.

E. Superimposed Loadings

Superimposed loadings result from movable objects, including pedestrians, automobiles, trucks, railroad locomotives and aircraft. Maintenance equipment, such as mowers or dump trucks, also create superimposed loads. In unusual conditions, unexpected construction of future roadways, buildings or other facilities adjacent to a structure could create superimposed loads.

F. Construction Conditions

Hydraulic structures are frequently exposed to critical design conditions during construction. The hydraulic and geotechnical engineers must advise the contractor when this is likely to occur, so that damage to a partially completed structure can be minimized. The standard contract language that makes a contractor responsible for protection of the site during construction does not eliminate the engineer's responsibility to insure that a structure's integrity can be maintained during construction. If a specific structure is uniquely vulnerable to damage or failure in specific stages of construction, the designer should consider ways to reduce the vulnerability and should advise the contractor accordingly. Loadings of particular concern include hydrostatic loads behind walls (where soil backfill would eventually be placed), earth loadings due to excessive sediment deposition or sliding banks, and uplift created before backfill is placed.

Finally, any member that is to be moved into place must be designed and constructed so that it can be lifted and transported without damage. Supporting requirements should be identified for items such as pipes, which might be lifted by cables or slings.

G. Documentation of Critical Design Conditions

Once the critical design conditions are established, the hydraulic engineer should identify and record the following information for each condition:

(a) Finished earth grades adjacent to the structure (including scour predictions).
(b) Design water surfaces.
(c) Design flow velocities.
(d) Ground water levels.
(e) Seepage conditions.
(f) Superimposed loads.
(g) Construction stages and loadings.

V. DETERMINATION OF LOADS

A. General

The design of a drainage or flood control structure requires the determination of the composite forces on the overall structure and on each individual member of the structure. These composite forces result primarily from fluid pressures, earth pressures, and superimposed loads. Miscellaneous loadings such as wind or earthquakes can have effects on certain types of structures. Such loads are beyond the scope of this Manual, and may be determined with the aid of manuals or textbooks on fluid and soil mechanics or structural design.

B. Hydraulic Loads

Hydraulic loads on structures are characterized as either hydrostatic, resulting from fluid pressures, or hydrodynamic, resulting from fluid movement. Most structures described in this Manual are affected by both hydrostatic and hydrodynamic forces. In many cases the hydrostatic forces will represent the dominant load on a structure.

Hydrodynamic forces may be significant if flow through or adjacent to the structure experiences an abrupt change in elevation, velocity, or direction. Hydrostatic pressures act equally in all directions in both static and moving water. Hydrostatic pressure may be determined by:

$$P = \gamma h \qquad (14\text{-}1)$$

where:

P = pressure, in pounds per square foot
γ = unit weight of water, 62.4 pounds per cubic foot
h = vertical distance to the water surface, in feet

Hydrostatic pressures act perpendicular to the surface in contact with the water, and increase in proportion to the water depth. Structures designed to produce significant changes in water surface elevations or pressures may experience significant differential hydrostatic loads.

Hydrostatic loads that tend to reduce the composite loads on structures should be considered only if it is certain that those forces will exist in the design condition. Proper structural design practice requires that intermittent forces, or forces that may cease to exist, must be neglected to the extent that they would tend to reduce or offset other loads.

The hydrodynamic forces most commonly of concern in structural design result from friction and momentum. Water flowing at a uniform depth (or pressure, in a closed conduit) and at a uniform velocity exerts a force on the walls of the conveying facility or on obstructions to the flow. The force acts in the direction of the flow and is frequently negligible in comparison to other forces on a structure. Conversely, water flow that changes in depth, velocity or direction can exert a relatively greater force, which is commonly referred to as an impact force or momentum force. The magnitude of a momentum force is determined by use of the momentum equation. Newton's second law of motion states that force equals the time rate of change of momentum. The resultant of forces on an element of water is determined by Equation 14-2:

$$F = \rho Q \Delta V \qquad\qquad 14\text{-}2$$

where

F = the net force on an element of water
ρ = the density of water, 1.94 slugs per cubic foot
Q = the flow rate, in cubic feet per second
ΔV = change in water velocity in the direction of the force, in feet per second

Since F is the force acting on the water element that causes the momentum change, its reaction is the force exerted on the structure by the moving water. Equation 14-2 is generally solved in vector form and can be simplified for one dimensional flow as follows:

$$F_x = \rho Q \, (V_{1x} - V_{2x}) \qquad\qquad (14\text{-}3)$$

where the subscript x implies a directional vector, and where V_{1x} and V_{2x} are the velocities on each side of the object subjected to the hydrodynamic force.

If the flow is two dimensional or three dimensional, Equation 14-3 can be written to determine forces parallel to the other axes. The determination of hydrodynamic forces begins with the computation of flow depths, velocities, and directions. Complex flow patterns in a hydraulic structure may create complex forces on the structural members.

C. Earth Loads

Hydraulic structures generally are affected by earth loads and, in turn, impose a load on the soil or rock adjacent to the structure. Determinations of earth loads should be based on a soils analysis made by a geotechnical engineer. The hydraulic engineer and the geotechnical engineer should together determine the soil parameters and loads required for a particular situation. These typically include the unit weight of the soil or soils involved; active, at-rest, and passive earth pressures; design groundwater levels; allowable bearing capacity; and coefficients of friction.

In general, earth loads on structures are the sum of the effective soil stress (the pressure exerted on the structure by the soil particles) and the groundwater pressures (the hydrostatic pressures exerted on the structure by the pore water in the soil).

The vertical effective soil stress (the soil stress acting downward on a horizontal plane) is the total downward stress minus the groundwater pressure acting upward at that point. For the simple case of a point at depth z_1 below a horizontal soil surface, and at a depth z_2 below the groundwater level, with z_2 less than z_1, the vertical effective stress is given by:

$$\bar{\sigma}_z = \gamma_t Z_1 - \gamma_w Z_2 \qquad (14\text{-}4)$$

where

$\bar{\sigma}_z$ = the vertical effective stress, lbs/ft^2
γ_t = the total unit weight of the soil, lbs/ft^3
γ_w = the unit weight of water, lbs/ft^3

For cases where building or wall foundations or underground structures such as buried tanks are involved, the calculations are somewhat more complicated and distributions of stresses within the soil must be determined. However, the principle is still the same.

The horizontal effective soil stress (the soil stress acting laterally on a vertical plane, such as the back of a retaining wall or the vertical wall of an underground structure) is normally estimated by multiplying the vertical effective soil stress by an earth pressure coefficient, as in Equation 14-5.

$$\bar{\sigma}_h = K\sigma_v \qquad (14\text{-}5)$$

where

$\bar{\sigma}_H$ = the horizontal effective soil stress, lb/ft^2
K = the earth pressure coefficient
$\bar{\sigma}_v$ = the vertical effective soil stress, lb/ft^2

The three commonly used earth pressure coefficients are the active, at-rest, and passive coefficients.

The active earth pressure is appropriate for estimating the earth pressure on a wall that moves away from the soil mass, and corresponds to a state of failure in the soil. The active earth pressure is often used for the pressure behind a retaining wall. However, the factors of safety must be high enough to insure that a structure is safe. The design of a structure with active earth pressures and factors of safety near unity will result in a design very close to a state of failure.

The at-rest earth pressure would result if there were no horizontal movement of the surface. It is the appropriate pressure for rigid walls, such as the walls of underground concrete tanks. Some designers also use at-rest pressures to design retaining walls, since it is likely that, unless the wall moves repeatedly over its lifetime, the earth pressure behind the wall will build up to active earth pressure. If at-rest pressures are used in the design of a retaining wall, it is reasonable to use lower factors of safety, because the at-rest pressures are reasonable estimates of the maximum pressures the wall might experience.

The passive pressure is the maximum pressure on a surface that is pushed or moved into a soil mass in a horizontal direction. Passive pressure is often used as a component of resistance to sliding of retaining walls or other structures. Some cautions are appropriate relative to the use of passive pressure. First, it typically takes several inches of wall movement into the soil to fully develop the passive pressure, while it takes only a fraction of an inch of wall movement away from the soil to reduce pressure to active conditions. Unless several inches of movement can be tolerated, full passive pressure should not be included as a resisting force in design calculations, and full passive pressure values should be divided by factors of between two and three to arrive at design values. Second, the soil surface at the toe of the retaining wall must extend a significant distance outward from the wall for the passive pressure to be fully developed.

Some designers use "equivalent fluid pressures" to estimate horizontal earth pressures rather than using earth pressure coefficients applied to the vertical effective stress. In this method, the soil is assigned an "equivalent unit weight" and the horizontal pressure at any point is calculated as the hydrostatic pressure that would exist at a point below the surface of a fluid with that unit weight. Equivalent fluid pressures can be selected to represent either the effective soil stress or groundwater pressure. In the former case, any existing groundwater pressure would be added to the equivalent fluid pressure. When the equivalent fluid pressure method is applied, care must be taken so that the resulting pressures reasonably approximate the actual combination of effective soil stress and groundwater pressure.

Values of bearing capacity are provided by geotechnical engineers in two different formats. The designer must understand whether the value provided is an ultimate bearing capacity, which would not include a factor of safety, or an allowable or design bearing capacity, which has been reduced by an appropriate factor of safety or by consideration of an allowable settlement, if appropriate.

When coefficients of friction are used in an evaluation of sliding, it is important that consideration be given to both the potential for sliding within the soil itself and the potential for sliding along the boundary between the soil and the structure (e.g., concrete or steel). The coefficients of friction are different for the two cases, and the more critical should be determined. It should also be noted that the coefficients of friction apply only to forces resulting from effective soil pressures, and not to those resulting from groundwater pressures. The effects of groundwater pressures, however, must be considered when sliding friction forces are calculated, since they would have an uplift component that would reduce the structure weight. This would, in turn, reduce the friction force that would resist sliding.

D. Groundwater Loads

Groundwater pressures on drainage structures are frequently overlooked, but can produce significant and sometimes unexpected forces. For the purposes of this discussion, groundwater is defined as water able to move freely between soil particles, and the groundwater level is defined as the elevation to which water would rise in an open, uncased hole. These definitions differentiate between groundwater and the normal soil moisture present in almost all soils.

Since drainage structures are frequently built in low-lying or channel areas, they may extend underground into the natural groundwater. The natural soils in valley areas are frequently coarse and permeable, thus allowing free movement of groundwater. In addition, drainage facilities built in excavations are frequently backfilled with relatively coarse, freely draining material.

Pressures can not only be produced by the natural groundwater in the vicinity of a structure, but also from the sudden flooding or saturation of adjacent soil as a result of high flow through (or overtopping) a structure.

Since groundwater pressures are related to soil pressures, they are sometimes assumed to be part of the soil pressures in computations. This practice may be an oversimplification. When excessive soil pressures cause a deflection in structural members, the interlocking of particles within the soil matrix will frequently result in a "relaxation" of the load, limiting further deflections. Groundwater pressures will not relax, since the water can flow into the void created by a structural deflection and maintain the loading.

Groundwater pressures are identical to the hydrostatic pressures described in the previous subsection. When the groundwater conditions are relatively static and the level is approximately horizontal, groundwater pressures may be computed using Equation 14.1, which implies that equal pressures occur at equal depths. When a structure is immersed completely or partly in groundwater, it may act as a floating vessel, with the upward force equal to the weight of the displaced fluid. If the upward buoyant force exceeds the weight of the structure and the weight of any soil, water, or other material, as may occur in the case of an empty tank or pipe, flotation can occur.

Groundwater moving around or beneath a structure, such as a dam or drop structure, moves from an area of relatively higher to lower ground water level. In such a case, the groundwater level cannot be assumed to be horizontal, and groundwater pressures along seepage paths are reduced in the direction of the seepage. "Lane's Weighted Creep" procedure, (U.S. Department of Interior 1960), provides a simplified method of computing groundwater pressures along a seepage path. The procedure also provides a mechanism for evaluating various methods of reducing groundwater pressures and seepage by adding cut-off walls or modifying the structure's dimensions. If significant groundwater pressures along a seepage path are anticipated, more detailed analyses are merited.

Accurate determination of groundwater pressures for structural design requires a careful evaluation of the extent to which the pressure can build up outside of a structure. In some structures, full groundwater pressures are allowed to develop and the structures are designed to resist them. In other structures, exterior drainage facilities are constructed to insure that the groundwater pressures cannot exceed the levels for which the structures are designed.

E. Superimposed Loads

Superimposed loads result from the weight and impact of vehicles over and adjacent to structures. The vehicles that could create the loads have legal limits that establish maximum wheel or axle loads. When the vehicle comes in contact with the structure, such as a bridge, inlet, or manhole vault, the load is equal to the maximum number of wheels, axles, or vehicles that could simultaneously be located over the structure. Loads are generally increased to represent impact if the vehicles are moving. The designer must determine the location of the load or combination of loads that would create the critical design condition for the structural member. For example, it is common to compare a single wheel in the center of a span to two adjacent wheels located as close to the center of the span as possible. The wheel load could be assumed to be a point load in a relatively large structure, or to be spread over the area of the wheel on smaller structures. For a further discussion, see Section VIII.

VI. STABILITY ANALYSIS

A. General

The initial step in the structural design of a drainage facility is the analysis of the overall structure's stability under the critical design conditions. The structure itself, and the resisting forces acting on it, must be sufficient to insure that the structure can resist overturning, sliding and uplift. In addition, the structure's load on the underlying soil must not exceed the allowable bearing pressure.

B. Overturning

Many structures are exposed to loads that exert little or no overturning forces. However, in structures such as retaining walls, drop structures, or dams, overturning is frequently a principal concern.

Overturning is evaluated by computing the forces that would tend to overturn the structure, and the forces that would resist overturning. The point at which overturning would occur is determined, and moments are computed about that point. Moments that resist overturning should exceed those causing overturning by a factor of safety of 1.5 or more. Figures 14.1 and 14.2 illustrate the typical forces that would act on a small drop structure in an overturning evaluation, under two different conditions.

The designer who has difficulty providing an adequate factor of safety against overturning for a specific structure has two principal alternatives available. He may modify the structure to increase the magnitude of the resisting moments, or he may increase the distance, and thus the moment arm, to the point about which overturning would occur.

Resisting moments are increased through an addition to the structure weight or an extension of underground footings, which would mobilize a greater weight of resisting soil. Anchors into the soil or rock can also be constructed to add resisting moments. The resisting moment arm is generally increased by an extension of the structure length or foundation.

C. Sliding

Structures subjected to large horizontal forces, including dams, drop structures, and retaining walls, may slide along the earth foundation. The sliding could occur along a plane between adjacent layers of soil or rock, or along a plane between the structure and the underlying soil. The determination of a structure's stability in sliding requires an evaluation of the forces causing sliding and those resisting sliding in the critical design conditions.

Forces causing sliding are generally hydrostatic forces, hydrodynamic forces, and active earth pressures. Forces that resist sliding usually include the passive earth pressures of the downstream soil and the friction force developed between the structure and its soil or rock foundation. Figures 14.1 and 14.2 also illustrate the forces acting on a small drop structure subjected to sliding. The friction force is computed as follows:

$$F = \mu_F W \qquad (14\text{-}6)$$

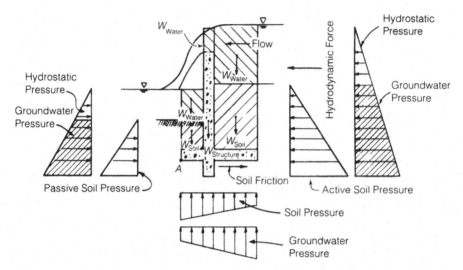

Figure 14.1.—*Evaluation of drop structure for overturning about toe (point A) and sliding along structure base during high flow.*

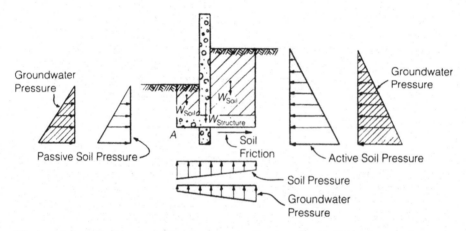

Figure 14.2.—*Evaluation of drop structure for overturning about toe (point A) and sliding along structure base after passage of high flow.*

where

F = force, in pounds, required to initiate sliding

μ_F = coefficient of friction between soil layers or between the structure and the soil

W = the downward force, in pounds, the structure exerts on the soil

Values of μ_F must be determined based upon soils investigations. The downward force, W, includes the weight of the structure as well as any material, including earth, water, or vehicles in or above the structure that tend to increase the downward force on the soil foundation. The weight of water or vehicles should be included only when it is certain that they would be present during the design condition under consideration.

The critical design conditions under which sliding should be evaluated require careful identification. As described earlier, the critical design condition with respect to sliding may occur immediately after a high flow, as illustrated in Figure 14.2, since downstream erosion may tend to reduce resisting forces and high groundwater pressures may tend to reduce the value of W.

The earth pressures tending to create sliding are assumed to be active earth pressures, since the structure would tend to slide away from the earth. Conversely, earth pressures tending to resist sliding are assumed to be passive earth pressures, since the structure would be forced into the earth. Resisting passive pressures may be reduced by a factor of safety to allow for failure that might result from an inadequate extent of resisting soil. After the forces are determined, a structure's stability is computed by determining its safety factor against sliding using equation (14-7):

$$F.S._S = \frac{F_R}{F_S} \qquad\qquad (14\text{-}7)$$

where

$F.S._S$ = factor of safety against sliding
F_R = summation of forces resisting sliding including the friction force
          of Equation (14-6)
F_S = summation of forces creating sliding

Generally, a safety factor against sliding of 1.5 or greater is considered adequate, although values as low as 1.25 may be satisfactory in some circumstances. A geotechnical engineer should be consulted before a safety factor of less than 1.5 is adopted.

When a designer has difficulty achieving satisfactory stability against sliding, his options are usually limited to increasing the resisting forces. This can be accomplished by a vertical keyway into the foundation, a greater depth of soil at the downstream toe of the structure, or an increase in the structure's weight. Anchors constructed in underlying soil or rock will also provide resistance to sliding. A structure's weight may sometimes be increased by increasing the lateral extent of footings, which would mobilize additional earth loading to increase the value of W. If the assumption of erosion at the structure's toe results in the loss of too great a resisting force, it could be beneficial to provide more extensive erosion protection to insure that the resisting soil remains in place.

If extensive uplift from groundwater forces reduces W substantially, the exterior drainage and seepage conditions can be modified to some degree. Filter materials, crushed rock, or PVC liners can be installed outside of a structure to convey groundwater away from the structure more quickly than it can move through the soil to the structure. Uplift pressures due to seepage can be reduced to some degree by an increase in the seepage path under the structure by use of cut-off walls or curtains. As noted in Section V. D., Lane's Weighted Creep procedure can be used for this analysis.

Inspection of Equations (14-6) and (14-7) indicates that greater benefit will be derived from increasing resisting forces acting on the structure than by increasing the downward force, W.

D. Uplift and Flotation

The pressures of groundwater under a hydraulic structure exert a vertical force referred to as an uplift force. The uplift force is counteracted by downward forces, which include the structure's weight, soil loads that bear downward on the structure's base or footings, friction of the soil along structure walls, and, in some cases, water within or flowing through the structure. When the uplift force exceeds the sum of the downward forces, flotation of either an entire structure or a part of a structure can occur. Figure 14.3 illustrates forces on a typical energy dissipator in an uplift evaluation (after the passage of a design flow), and on an outlet structure in a detention pond.

Flotation failures have occurred in several types of drainage structures. Pipes constructed of lightweight materials, such as corrugated steel or PVC, which extend into flood control ponds or reservoirs, can become plugged with debris to the extent that little water is in the pipe while the water continues to exert an uplift force. Lengths of pipe have been forced out of the ground under these conditions. Similarly, structures designed to be normally filled with a fluid have floated upward when they were drained for cleaning or repairs. While some flotation failures have been spectacular, most result in only a relatively small vertical displacement. Even small movements, however, can cause total failure of a structure, if slabs or walls crack or buckle.

The possibility of flotation of a structure is evaluated using Equation 14-8:

$$F.S._u = \frac{F_d}{F_u} \tag{14-8}$$

where

$F.S._u$ = factor of safety against uplift
F_d = the sum of the downward forces on the structure
F_u = the sum of the upward forces on the structure

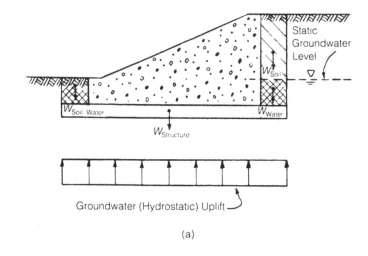

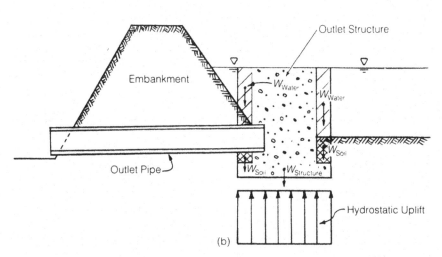

Figure 14.3.—Structures subject to uplift.

Design engineers have used values of $F.S._u$ as low as 1.0. However, because many of the forces are difficult to predict with accuracy, a value of $F.S._u$ of 1.25 is recommended. When a designer finds that the value of $F.S._u$ is smaller than he considers appropriate, he must generally find ways to increase F_d, such as by adding dead weight to the structure, or by extending a footing outward from the structure to mobilize additional downward soil weight, providing the footing is structurally designed to handle the loads. Uplift forces may also be resisted by anchors. In some structures it may be appropriate to include the weight of water contained in the structure, but only if its presence is assured during events which would create the uplift. A final approach which could apply to some specific structures is a system of weep holes or similar devices which would allow groundwater to flow into and fill a structure if it reached a critical elevation. Some underground lift stations and wet wells are designed in this manner.

In some cases, F_u could be reduced by lengthening a seepage path, constructing cutoff walls or membranes, or modifying drainage under the structure. However, caution must be exercised to insure that such devices will work under all conditions.

E. Soil Bearing Pressures

Drainage or flood control structures, which have weight and may support a combination of other loads, exert a force on the underlying soil. The downward force must not exceed the capability of the soil to support the load, or excessive settlement or consolidation may occur. The downward pressure the structure may be allowed to exert on the soil is limited by two parameters. The pressure must not create a failure of the soil and must not cause settlement that would exceed the adjusting capabilities of the structure. The allowable soil bearing capacity is a parameter that must be determined by a geotechnical engineer, based on both concerns. The value is normally determined as a safe bearing capacity, and therefore includes a safety factor. As long as the downward pressure of the structure's foundation does not exceed the allowable bearing capacity, no additional safety factor is applied.

Small drainage and flood control structures seldom cause significant soil pressures by virtue of their own weight. However, some structures, such as retaining walls and dams, are designed in a way that concentrates the load (as a result of an overturning tendency) in a specific part of the footing, usually the toe. The resultant soil pressures under points of load concentration must be evaluated to insure that they will not exceed the allowable pressure. When a designer finds that the concentrated pressures exceed allowable soil capacities, he generally must increase the footing dimensions to spread the load over a larger area of the foundation.

Large drainage structures, such as dams and levees, may create loads that greatly exceed the capacity of the underlying soil. In such cases, several options exist. The load may be spread out over a larger area; the weaker underlying soil may be removed to allow the facility to bear upon deeper strata with greater capacity; or deep foundations such as piles or caissons may be employed to transfer the load to strata with adequate load-bearing capacity.

VII. DESIGN OF STRUCTURAL MEMBERS

A. General

The stability analyses described in the previous section frequently result in a refinement of the overall size of the structure. Structural modifications to achieve stability supplement the dimensions first established to achieve the desired hydraulic function. The structural design of individual members involves the selection of the materials to be used, the sizing of the members, and the preparation of the plans and specifications.

B. Selection of Materials

The selection of the materials from which the structure will be built is frequently done prior to the structural design stage, for reasons such as aesthetics or the availability of materials. The two materials used most commonly are earth (including soil cement) and concrete (reinforced, non-reinforced, and roller-compacted). In addition, many precast or pre-formed units are used, primarily pipe. Less commonly used materials include steel sheet piling, timber, gabions, and synthetic and organic erosion protection devices.

C. Sizing of Members

Each member must be designed to withstand the loads imposed on it. In addition, each member must be connected to the adjacent members in such a way that its loads are properly transferred to the overall structure. Finally, each member must be detailed, in both contract drawings and specifications, so that it can be properly constructed.

VIII. CONDUIT STRUCTURAL REQUIREMENTS

A. Introduction

The structural design of a storm sewer requires that the supporting strength of the installed sewer pipe, divided by a suitable factor of safety, equal or exceed the loads imposed on it by the combined weight of soil and any superimposed loads.

The following are generally accepted criteria and methods for determining combined loads and supporting strength of the sewer pipe, as well as procedures for combining these elements with the application of a factor of safety to produce a safe and economical design.

Methods are presented for estimating probable maximum loads caused by soil forces and for both static and moving superimposed loads (note also that the determination of loads on rigid conduits is presented in great detail in ASCE Manual 60, *Gravity Sanitary Sewers*, Chapter 9, "Structural Requirements" (ASCE 1982)). Where so noted, the methods apply to rigid and flexible conduits in the three most common conditions of installation: in a trench in natural ground; in an embankment; and in a tunnel.

The design of rigid and flexible pipes is treated separately. There are no specific design procedures given for flexible pipes of intermediate stiffness. For such cases, design procedures such as computer analysis based on soil-structure interaction or the designs for rigid or flexible pipes may be used (not interchangeably) for conservative results.

The supporting strength of a buried sewer pipe is a function of installation conditions as well as the strength of the sewer pipe itself. Structural analysis and design of the sewer line are problems of soil-structure interaction. This chapter presents procedures for determining the field or installed supporting strength of rigid sewer pipe based on its established relationship to the laboratory test strength, commonly called the Indirect Design Method. It also presents methods of predicting approximate field deflections for flexible pipe, based on empirical methods. Since installation conditions have such an important effect on both load and supporting strength, a satisfactory sewer construction project requires that assumed design conditions be adhered to on the job site.

This chapter does not include information on reinforced concrete design of rigid sewer pipe sections. Reference should be made to standard textbooks and to ACI/ASTM/AASHTO Specifications or FHWA or industry handbooks for such design data.

B. Loads on Sewers Caused by Gravity Earth Forces

1. General Method—Marston Theory

Marston's Theory, which is widely accepted, is used for determining the vertical load on buried conduits caused by soil forces in all of the most commonly encountered construction conditions (Marston and Anderson 1913; Marston 1930). Recent analysis and actual observation of field performance have shown that designs based on the Marston Theory yield satisfactory results, especially for small diameter conduits in narrow trenches. For larger diameter conduits, the results are conservative.

In general, the theory states that the load on a buried pipe is equal to the weight of the prism of soil directly over it, called the interior prism, plus or minus the frictional shearing forces transferred to that prism by the adjacent prisms of soil. The magnitude and direction of these frictional forces are a function of the relative settlement between the interior and adjacent soil prisms. The theory makes the following assumptions:

(a) The calculated load is the load that will develop when ultimate settlement has taken place.
(b) The magnitude of the lateral pressures that induce the shearing forces between the interior and adjacent soil prisms is computed in accordance with Rankine's theory.
(c) Cohesion is negligible except for tunnel conditions.

The general form of Marston's equation is:

$$W = CwB^2 \qquad (14\text{-}9)$$

in which W is the vertical load per unit length action on the sewer pipe because of gravity soil loads; w is the unit weight of soil; B is the trench width or sewer pipe width, depending on installation conditions; and C is a dimensionless coefficient that measures the effect of the following variables:

(a) The ratio of height of fill to width of trench or sewer pipe.
(b) The shearing forces between interior and adjacent soil prisms.
(c) The direction and amount of relative settlement between interior and adjacent soil prisms for embankment conditions.

2. Types of Loading Conditions

Although the general form of Marston's equation includes all the factors necessary to analyze all types of installation conditions, it is convenient to classify these conditions, write a specialized form of equation, and prepare separate graphs and tables of coefficients for each.

The accepted system of classification is shown diagrammatically in Figure 14.4 and is described here briefly:

Trench conditions are defined as those in which the sewer pipe is installed in a relatively narrow trench cut in undisturbed ground and covered with soil backfill to the original ground surface.

Embankment conditions are defined as those in which the sewer pipe is covered above the original ground surface or when a trench in un-

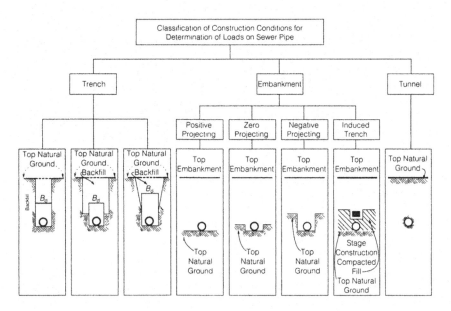

Figure 14.4.—Classification of construction conditions.

disturbed soil is so wide that trench wall friction does not affect the load on the sewer pipe. The embankment classification is further sub-divided into two major subclassifications—positive projecting and neg-ative projecting. Sewer pipe is defined as positive projecting when the top of the sewer pipe is above the adjacent original ground surface. Negative projecting sewer pipe is that installed with the top of the sewer pipe below the adjacent original ground surface in a trench that is narrow with respect to the size of pipe and the depth of cover (Figure 14.4), and when the native material is of sufficient strength that the trench shape can be maintained dependably during placement of the embankment.

A special case, called the induced trench condition, may be employed to minimize the load on a conduit under an embankment of unusual height.

3. Loads for Trench Conditions

Sewers usually are constructed in ditches or trenches excavated in natural or undisturbed soil, and then covered by refilling the trench to the original ground line. This procedure often is referred to as "cut and cover," "cut and fill," or "open cut."

The vertical soil load to which a sewer pipe in a trench is subjected is the result of two major forces. The first is produced by the mass of the prism of soil within the trench and above the top of the sewer pipe. The second is the friction or shearing forces generated between the prism of soil in the trench and the sides of the trench.

The backfill soil has a tendency to settle in relation to the undisturbed soil in which the trench is excavated. This downward movement or tendency for movement induces upward shearing forces that support a part of the weight of the backfill. Thus, the resultant load on the horizontal plane at the top of the sewer pipe within the trench is equal to the weight of the backfill minus these upward shearing forces (Figure 14.5) (ASCE 1969).

Unusual conditions may be encountered in which poor natural soils may effect a change from trench to embankment conditions with con-siderably increased load on the sewer pipe. This is covered in the next subsection.

Use of Marston's Formula   Marston's formula for loads on rigid sewer pipe in trench condition is:

$$W_c = C_d w B_d^2 \qquad (14\text{-}10)$$

in which W_c is the load on the sewer pipe, in pounds per foot; w is the density of backfill soil, in pounds per cubic foot; B_d is the width of trench at the top of the sewer pipe, in feet; C_d is a dimensionless load coefficient, which is a function of the ratio of height of fill to width of

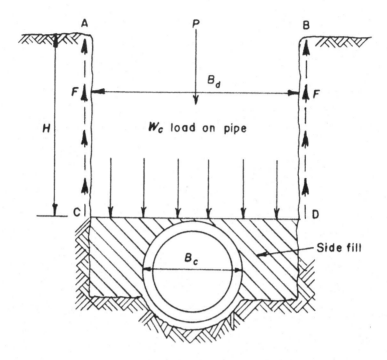

Figure 14.5.—Load-producing forces.

trench and of the friction coefficient between the backfill and the sides of the trench. The load coefficient, C_d, is computed as follows:

$$C_d = \frac{1 - e^{-2\kappa\mu'\frac{H}{B_d}}}{2\kappa\mu'} \tag{14-11}$$

in which e is the base of natural logarithms and κ is Rankine's ratio of lateral pressure to vertical pressure:

$$\kappa = \frac{\sqrt{\mu^2 + 1} - \mu}{\sqrt{\mu^2 + 1} + \mu} = \frac{1 - \sin\phi}{1 + \sin\phi} \tag{14-12}$$

The other terms are: $\mu = \tan\phi$ = the coefficient of internal friction of backfill material; $\mu' = \tan\phi'$ = the coefficient of friction between backfill material and sides of trench (μ' may be equal to or less than, but never greater than μ); and H is the height of fill above top of pipe, in feet. The value of C_d for various ratios of H/B_d and various types of soil backfill may be obtained from Figure 14.6.

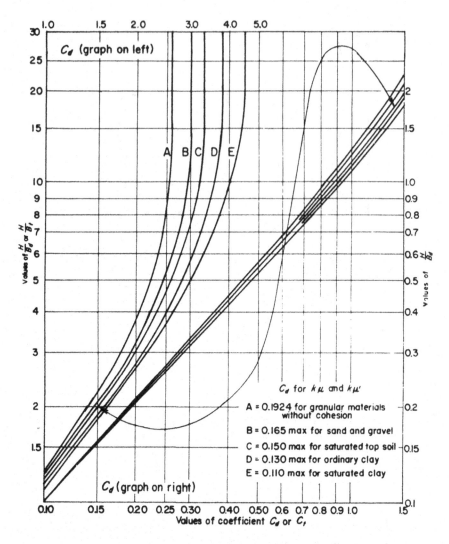

Figure 14.6.—Computation diagram for soil loads on trench installations (sewer pipe completely buried in trenches).

The trench load formula, Equation (14-10), gives the total vertical load on a horizontal plane at the top of the sewer pipe. If the sewer pipe is rigid, it will carry practically all this load. If the sewer pipe is flexible and the soil at the sides is compacted to the extent that it will deform under vertical load less than the sewer pipe itself will deform, the side fills may carry a proportional share of the total load. Under these circumstances the trench load formula may be modified to:

$$W_c = C_d w B_c B_d \qquad (14\text{-}13)$$

in which B_c is the outside width of pipe, in feet.

It is emphasized that Equation 14-13 is applicable only if the backfill is compacted as described above. The equation should not be used merely because the pipe is a flexible type.

The term "side fill" refers to the soil backfill placed between the sides of sewer pipe and the sides of the trench. The character of this material and the manner of its placement have two important influences on the structural behavior of a sewer pipe. First, the side fill may carry a part of the total vertical load on the horizontal plane at the elevation of the top of the sewer pipe. Second, the side fill plays an important role in helping the sewer pipe carry vertical load. Every pound of force that can be brought to bear against the sides of an elastic ring increases the ability of the ring to carry vertical load by nearly the same amount.

Examination of Equation 14-10 indicates the important influence the width of the trench exerts on the load as long as the trench condition formula applies. This influence has been extensively verified through experimentation. These experiments also have indicated that the width of trench at the top of the sewer pipe is the controlling factor.

The width of trench below the top of the sewer pipe also is important. It must not be permitted to exceed the safe limit for the strength of sewer pipe and class of bedding used. The minimum width must be consistent with the provision of sufficient working space at the sides of the sewer pipe to assemble joints properly, to insert and strip forms, and to compact backfill. The design engineer must allow reasonable tolerance in width for variations in field conditions and accepted construction practice.

Narrow trench construction can result in some degree of soil arching, thereby reducing the load on the pipe. Conservative design practice for pipe is based on the prism load. There are other approaches that may be considered for load calculation, some of which are discussed in later sections.

The position of the lower wale usually will determine the proper width of trench (from sheeting face to sheeting face of sheeting, where sheeting and bracing are required). A working-room allowance of 12″ from each side of the sewer pipe or sewer pipe cradle to the face of the sheeting is a practical minimum for small and medium-sized sewer pipe for trenches up to about 13′ deep.

At any given depth and for any given sewer pipe size there is a certain limiting value to the width of trench beyond which no additional load is transmitted to the sewer pipe. This limiting value is called the "transition width." There are sufficient experimental data to show that it is safe to calculate the imposed load by means of the trench-conduit formula, Equation (14-10), for all widths of trench less than that which gives a load equal to the load calculated by the projecting-conduit formula (see section 4a). In other words, as the width of the trench increases (other factors remaining constant), the load on a rigid sewer pipe increases in accordance with the theory for a trench sewer pipe until it equals the load determined by the theory for a projecting sewer pipe. The width of trench at which this transition occurs may be determined from Figure 14.7 (Schlick 1932). The curves in Figure 14.7 are calculated for sand and gravel, where $\kappa\mu = 0.165$, but can be used for other types of soil since the change with varying, values of $\kappa\mu$ is small. In any event, the design engineer can check by calculating loads for both trench and embankment conditions. There is little research on the appropriate value of $r_{sd}p$ (the projection ratio times the settlement ratio) to use in the application of the transition width concept. In the absence

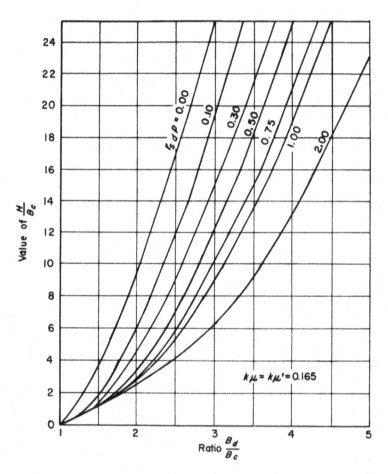

Figure 14.7. —Values of B_c/B_c at which pipe in trench and projecting pipe load formulas give equal loads.

of specific information, a value of 0.5 is suggested as a reasonably good working value. These quantities are defined in section 4a.

It is advisable, in the structural design of sewers, to evaluate the effect of the transition width on both the design criteria and the construction procedure. A contractor, for instance, may wish to place well points for drainage in the trench. If this requires a wider trench than usual, a stronger sewer pipe or higher class of bedding may be necessary.

It may be economical and proper to excavate the trench with sloping sides in undeveloped areas where no inconvenience to the public or danger to property, buildings, subsurface structures, or pavements will result. A subtrench (Figure 14.8) may be used in such cases to minimize the load on the pipe. When sheeting of the subtrench at the pipe is necessary, it should extend about 1.5' above the top of the pipe.

When sheeting of the trench is necessary, it should be driven at least to the bottom of the pipe bedding or foundation material, if used. In general, in a constantly wet or dry area, sheeting and bracing should be left in place to prevent reduction in lateral support at the sides of the pipe because of voids formed by removal of the sheeting. Sheeting left in place should be cut off as far below the surface as practicable, but in no case less than 3 feet below final ground elevation.

When wood sheeting is to be removed, the sheeting should be cut off 1.5 feet above the top of the pipe and the sheeting alongside the pipe left in place. Steel sheeting to be removed should be pulled in increments as the trench is backfilled, and the soil should be compacted to prevent formation of voids. The portion of wood sheeting to be removed should be handled similarly.

Loads on sewer pipe in sheeted trenches should be calculated from a trench width measured to the outside of the sheeting if it is pulled or to the inside if it is left in place. Voids created by removal of the sheeting should be backfilled with a flowable material such as pea gravel.

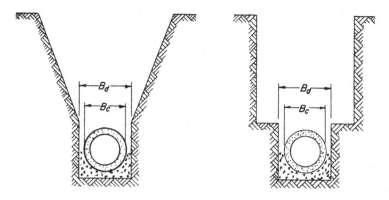

Figure 14.8.—Examples of subtrench.

If a shield is used in sewer pipe-laying operations, the shield width controls the width of the trench at the top of the sewer pipe. This width, with a small addition for the space needed to advance the shield without a large friction loss, should be the width factor used in computing loads on the sewer pipe. Extreme care must be taken when advancing the shield in the trench to prevent the pipe joints from pulling apart or to avoid disturbance of the pipe bedding.

Sewers to be constructed in sloping-sided trenches with the slopes extending to the invert, or to any place above the invert but below the top of the sewer, should be designed for loads computed by using the actual width of the trench at the top of the sewer pipe, or by the projecting-sewer formula, whichever gives the smallest load on the sewer pipe.

If for any reason the trench becomes wider than that specified and for which the sewer pipe was designed, the load on the sewer pipe should be checked and a stronger sewer pipe or higher class of bedding used if necessary.

Soil Characteristics   The load on a sewer pipe is influenced directly by the density of the soil backfill. This value varies widely for different soils, from a minimum of about 100 pounds per cubic foot to a maximum of about 135 pounds per cubic foot. The average maximum unit weight of the soil that will constitute the backfill over the sewer pipe may be determined by density measurements in advance of the structural design of the sewer pipe. A design value of not less than about 125 pounds per cubic foot is recommended if such measurements are not made.

The load is also influenced by the coefficient of friction between the backfill and the sides of the trench, and by the coefficient of internal friction of the backfill soil. Ordinarily these two values will be nearly the same and may be so considered for design purposes, as in Figure 14.6. However, in special cases this may not be true. For example, if the backfill is sharp sand and the sides of the trench are sheeted with finished lumber, μ may be substantially greater than μ'. Unless specific information to the contrary is available, values of the products $\kappa\mu$ and $\kappa\mu'$ may be assumed to be the same and equal to 0.130. If the backfill soil is a "slippery" clay and there is a possibility that it will become very wet after being placed, $\kappa\mu$ and $\kappa\mu'$ equal to 0.110 (maximum for saturated clay, Figure 14.6) should be used.

4. Loads for Embankment Conditions

A sewer pipe is described as a projecting sewer pipe when installed in a wide trench in such a manner that the top of the sewer pipe is at or near the natural ground surface or the surface of thoroughly compacted soil and subsequently is covered with an embankment. If the top of the sewer pipe projects some distance above the natural ground surface, or if it is installed in a wide trench, it is a positive projecting sewer pipe. Other methods of installing sewer pipe under embankments, however, have the favorable effect of minimizing the load on the pipe. In these cases, the installation is classified as a negative projecting sewer pipe or an induced trench sewer pipe (Figure 14.4).

Positive Projecting Sewer Pipe   The load on a positive projecting sewer pipe is equal to the weight of the prism of soil directly above the structure, plus (or minus) vertical shearing forces that act on vertical planes extending upward into the embankment from the sides of the sewer pipe. For an embankment installation of sufficient height, these vertical shearing forces may not extend to the top of the embankment, but terminate in a horizontal plane at some elevation above the top of the sewer pipe known as the "plane of equal settlement" (see Figure 14.9). The shear increment acts downward when $(s_m + s_g) > (s_f + d_c)$, and upward when $(s_f + d_c) > (s_m + s_g)$, where s_m is the compression of the columns of soil of height pB_c; s_g is the settlement of the natural ground adjacent to the sewer pipe; s_f is the settlement of the bottom of the sewer pipe; and d_c is the deflection of the sewer pipe.

The location of the plane of equal settlement is determined by equating the total strain in the soil above the pipe to that in the side fill plus the settlement of the critical plane. When the plane of equal settlement is an imaginary plane above the top of the embankment (i.e., shear forces extend to the top of the embankment), the installation is called

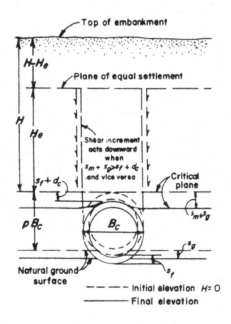

Figure 14.9. — *Settlements that influence loads on positive projecting sewer pipe.*

either "complete trench condition" or "complete projection condition" depending on the direction of the shear forces. When the plane of equal settlement is located within the embankment (Figure 14.9) (ASCE 1969), the installation is called "incomplete trench condition," or "incomplete projection condition."

In computing the settlement values, the effect of differential settlement caused by any compressible layers below the natural ground surface also must be considered. An exceptional situation for a sewer pipe in a trench can be encountered where the natural soil settles more than the trench backfill, such as where the natural soils are organic or peat, and the trench backfill is relatively incompressible compacted fill. A more common situation is where the sewer pipe is pile-supported in organic soils. In such cases, the load on the sewer pipe is greater than that of the prism above the pipe, and downward drag loads should be considered in the design of the piles.

Marston's Formula   Marston's formula for loads on rigid positive projecting sewer pipe is:

$$W_c = C_c w B_c^2 \tag{14-14}$$

in which W_c is the load on the sewer pipe, in pounds per foot; B_c is the outside width of the sewer pipe, in feet; and C_c is the load coefficient. Values of C_c may be obtained from Figure 14.10. In this diagram, H is the height of fill above the top of the sewer pipe, in feet; B_c is the outside width of sewer pipe, in feet; p is the projection ratio; and r_{sd} is the settlement ratio (the last two terms are defined in the next subsection).

Influence of Environmental Factors   The shear component of the total load on a sewer pipe under an embankment depends on two factors associated with the conditions under which the sewer pipe is installed. These are the projection ratio and the settlement ratio.

The projection ratio, p, is defined as the ratio of the distance that the top of the sewer pipe projects above the adjacent natural ground surface, or the top of thoroughly compacted fill, or the bottom of a wide trench, to the vertical outside height of the sewer pipe. It is a physical factor that can be determined in advanced stages of planning when the size of the sewer pipe and its elevation have been established.

The settlement ratio, r_{sd}, indicates the direction and magnitude of the relative settlements of the prism of soil directly above the sewer pipe and of the prisms of soil adjacent to it. In computing the settlement, the influence of any compressible layers below the sewer pipe also must be considered.

These relative settlements generate the shearing forces that combine algebraically with the weight of the central prism of soil to produce the resultant load on the sewer pipe. The settlement ratio is the quotient obtained by taking the difference between the settlement of the hori-

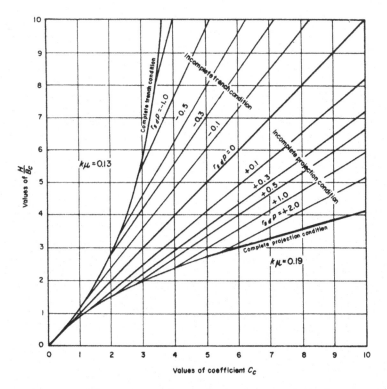

Figure 14.10.—Diagram for coefficient C_c *for positive projecting sewer pipes.*

zontal plane in the adjacent soil which was originally level with the top of the sewer pipe (the critical plane) and the settlement of the top of the sewer pipe, and dividing the difference by the compression of the columns of soil between the natural ground surface and the level of the top of the sewer pipe. The formula for the settlement ratio is:

$$r_{sd} = \frac{(s_m + s_g) - (s_f + d_c)}{s_m} \qquad (14\text{-}15)$$

in which r_{sd} is the settlement ratio; s_g is the settlement of the natural ground adjacent to the sewer pipe; s_m is the compression of the columns of soil of height Pb_c; $(s_m + s_g)$ is the settlement of the critical plane; d_c is the deflection of the sewer pipe, that is, the shortening of its vertical dimension; s_f is the settlement of the bottom of the sewer pipe; and $(s_f + d_c)$ is the settlement of the top of the sewer pipe.

The elements of the settlement ratio are shown in Figure 14.9. When the settlement ratio is positive, the shearing forces induced along the sides of the central prism of soil are directed downward, and the load on the sewer pipe is greater than the weight of the central prism. When the settlement ratio is negative, the shearing forces act upward and the load is less than the weight of the central prism.

The numerical magnitude of the product of the projection ratio and the settlement ratio, $r_{sd}p$, is an indicator of the relative height of the plane of equal settlement and, therefore, of the magnitude of the shear component of the load. The plane of equal settlement is at the top of the sewer pipe when this product is equal to zero. There are no induced shearing forces in this case, and the load is equal to the weight of the central prism (the "prism load").

It is not practical to predetermine a value of the settlement ratio by estimating the magnitude of its various elements except in very general terms. Rather, it should be treated as an empirical factor. Recommended design values of r_{sd}, based on measured settlements of a number of actual installations, are given in Table 14.1. The last three cases in Table 14.1 presume soil conditions immediately under the sewer pipe to be the same as those in the adjacent areas outside the trench. In these cases, the settlement ratio may be conservatively assumed as zero in locations with highly fluctuating water tables above the pipe or plastic native trench soils. This results in designing for the "prism load," or the weight of the prism of soil above the pipe. In such cases, C_c is equal to H/B_c and Marston's formula for the prism load becomes:

$$W_c = HwB_c \qquad (14\text{-}16)$$

The prism load may also be expressed in terms of soil pressure, P, in pounds per square foot at depth H as:

$$P = wH = \frac{W_c}{B_c} \qquad (14\text{-}17)$$

TABLE 14.1.   Recommended Design Values of r_{sd}.

Type of Sewer Pipe (1)	Soil Conditions (2)	Settlement Ratio, r_{sd} (3)
Rigid	Rock or unyielding foundation	+1.0
Rigid	Ordinary foundation	+0.5 to +0.8
Rigid	Yielding foundation	0 to +0.5
Rigid	Negative projecting installation	−0.3 to −0.5
Flexible	Poorly compacted side fills	−0.4 to 0
Flexible	Well compacted side fills	0

(ASCE 1969).

Embankment Soil Characteristics   The load on a projecting sewer pipe is influenced directly by the density of the embankment soil. If the soil is to be compacted to a specified dry density, the corresponding wet density under normal moisture conditions should be used in calculating the load. A design value of not less than about 125 pounds per cubic foot is recommended if specific information relative to soil density is not available.

The load also is influenced by the coefficient of internal friction of the embankment soil. Recommended values of the product (Figure 14.10) are:

$$\text{for a positive settlement ratio, } \kappa\mu = 0.19$$

$$\text{for a negative settlement ratio, } \kappa\mu = 0.13$$

Negative Projecting and Induced Trench Sewer Pipes   A negative projecting sewer pipe (Figure 14.11) is one installed in a relatively shallow trench with its top at some elevation below the natural ground surface.

The trench above the sewer pipe is refilled with loose, compressible material, and the embankment is constructed to finished grade by ordinary methods. The greater the value of the negative projection ratio, p', and the more compressible the trench backfill over the sewer pipe, the greater will be the settlement of the interior prism of soil in relation to the adjacent fill material. In using this technique, the plane of equal settlement must fall below the top of the finished embankment. This

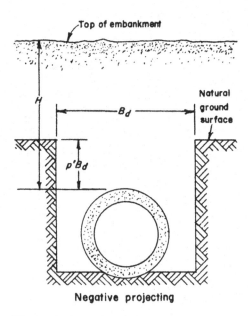

Figure 14.11.—Negative projecting sewer pipe.

action generates upward shearing forces that relieve the load on the sewer pipe.

An induced trench sewer pipe (Figure 14.12) first is installed as a positive projecting sewer pipe. The embankment then is built up to some height above the top and thoroughly compacted as it is placed. A trench of the same width as the sewer pipe next is excavated directly over the sewer pipe down to or near the top of the sewer pipe. This trench is refilled with loose, compressible material, and the balance of the embankment is completed in a normal manner. Sometimes straw, hay, cornstalks, sawdust or similar materials may be used in the trench backfill to augment the settlement of the interior prism.

The formula for loads on negative projecting sewer pipe is:

$$W_c = C_n w B_d^2 \qquad (14\text{-}18)$$

in which W_c is the load on the sewer pipe, in pounds per foot; w is the density of soil; B_d is the width of the trench; C_n is the load coefficient (Figure 14.13) (ASCE 1969), a function of H/B_d or H/B_c, p', and r_{sd}; p' is the projection ratio; and r_{sd} is the settlement ratio as defined below.

In the case of the induced trench sewer pipe, B_c is substituted for B_d in Equation (14.18), in which B_c is the width of the sewer pipe, assuming the trench in the fill is no wider than the sewer pipe.

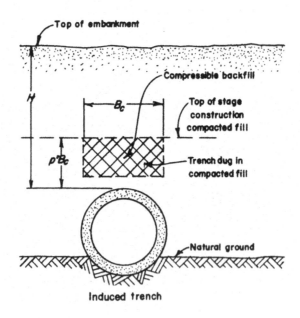

Figure 14.12.—Induced trench pipe.

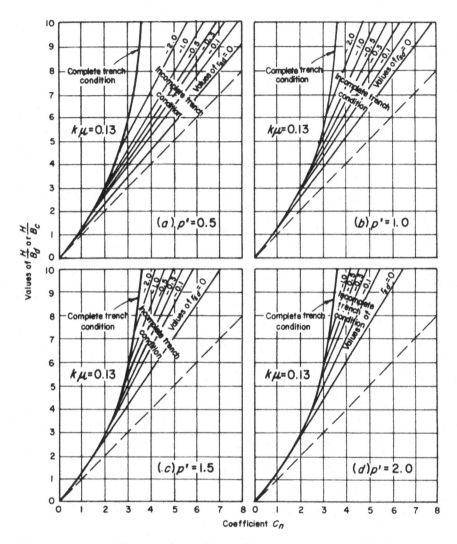

Figure 14.13.—*Diagrams for coefficient* C_n *for negative projecting and induced trench sewer pipes.*

The projection ratio, p', is equal to the vertical distance from the firm ground surface down to the top of the sewer pipe divided by the width of the trench, B_d in the case of negative projecting sewer pipe, or by the width of the sewer pipe, B_c, in the case of induced trench sewer pipe.

The settlement ratio, r_{sd}, for these cases is the quotient obtained by taking the difference between the settlement of the firm ground surface and the settlement of the plane in the trench backfill that was originally level with the ground surface (the critical plane), and dividing the difference by the compression of the column of soil in the trench. The formula for the settlement ratio is:

$$r_{sd} = \frac{s_g - (s_d + s_f + d_c)}{s_d} \qquad (14\text{-}19)$$

in which r_{sd} is the settlement ratio for negative projecting or induced trench sewer pipe; s_g is the settlement of the firm ground surface; s_d is the compression of trench backfill within the height $p'B_d$ or $p'B_c$; s_f is the settlement of the bottom of the sewer pipe; d_c is the deflection of the sewer pipe, that is, the shortening of its vertical dimension; and $(s_d + s_f + d_c)$ is the settlement of the critical plane. The elements of the settlement ratio are shown in Figure 14.14.

In the absence of extensive data on the probable values of the settlement ratio, it is tentatively recommended that this ratio be assumed to be between -0.3 and -0.5. Research (Taylor 1971) has indicated that the measured settlement ratio of 48-inch reinforced concrete pipe

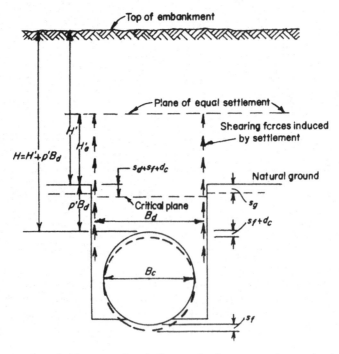

Figure 14.14. — Settlements that influence loads on negative projecting sewer pipes.

culvert, installed as an induced trench conduit under approximately 30' of fill, varied from -0.25 to -0.45.

Sewer Pipe Under Sloping Embankment Surfaces Sewer pipe can sometimes have different heights of fill on two sides because of the sloping surface of the embankment or when embankment exists on one side of the sewer pipe only. Such cases require special analysis. Design based on the larger fill height may not yield conservative results. Where yielding ground may surround the sewer pipe, a surcharge on one side of the sewer pipe may result in vertical displacement.

5. Loads for Jacked Sewer Pipe and Certain Tunnel Conditions

When the sewer is more than 30–40' deep or when surface obstructions are such that it is difficult to construct the sewer by trenching, it may be more economical to place the sewer by means of jacking or tunneling. The theories set forth in this Manual usually will be appropriate for materials where jacking of the sewer pipe is possible and for tunnels in homogeneous soils of low plasticity. Where a tunnel is to be constructed through materials subject to unusually high internal pressures and stresses, such as some types of clays or shales that tend to squeeze or swell, or through fissured and seamed rock, the loads on the sewer pipe cannot be determined from factors discussed here. Reference should be made to the following section on tunnels.

The methods of constructing sewers by tunneling and jacking are described in Chapter 16. Tunnel supports carry the earth load until the sewer pipe is constructed and the voids between the sewer pipe and tunnel supports are filled. Jacked sewer pipe is assumed to carry the earth load as it is pushed into place (Contractors and Engineers Monthly 1948; ACPA 1960).

Load-Producing Forces The vertical load acting on the jacked sewer pipe or tunnel supports, and eventually the sewer pipe in the tunnel, is the result of two major forces; the weight of the overhead prism of soil within the width of the jacked sewer pipe or tunnel excavation and the shearing forces generated between the interior prisms and the adjacent material.

During excavation of a tunnel, and varying somewhat with construction method, the soil directly above the face of the tunnel tends to settle slightly in relation to the soil adjacent to the tunnel because of the lack of support during the period immediately after excavation and prior to placement of the tunnel support. Also, the tunnel supports and the sewer pipe must deflect and settle slightly when the vertical load is applied. This downward movement or tendency for movement induces upward shearing forces that support a part of the weight of the earth prism above the tunnel. The cohesion of the material provides additional support. The resultant load on the horizontal plane on the top of the tunnel and within the width of the tunnel excavation is equal to the weight of the prism of earth above the tunnel minus the upward friction forces and cohesion of the soil along the limits of the prism of soil over the tunnel.

Hence, the forces involved with gravity earth loads on jacked sewer pipe or tunnels in such soils are similar to those discussed for loads on sewer pipe in trenches except for the cohesion of the material. Cohesion also exists in the case of loads in trenches and embankments, but is neglected because the cohesion of the disturbed soil is of minor consequence and may be absent altogether if the soil is saturated. However, in the case of jacked sewer pipe, or in tunnels where the soil is undisturbed, cohesion can reduce loads appreciably, and may be considered safe if reasonable coefficients are assumed.

Jacking stresses must be investigated in pipe that is to be jacked into place. The critical section is at the pipe joint where the transfer of stress from one pipe to the adjacent pipe occurs. Jointing materials should be used that will provide uniform bearing around the pipe circumference (ACPA 1960). Thrust at the joint is usually transmitted through the tongue or groove but not both. Concrete stress in the tongue or groove should be checked and additional reinforcement for both longitudinal and bursting stresses provided if required.

Marston's Formula   When modified to include cohesion, Marston's formula may be used to determine the gravity soil loads on jacked sewer pipe or sewer pipe in tunnels through undisturbed soil (Figure 14.15). The modified Marston formula is as follows:

$$W_t = C_t B_t (w B_t - 2c) \tag{14-20}$$

in which W_t is the load on the sewer pipe or tunnel support, in pounds per foot; w is the density of the soil above the tunnel; B_t is the maximum width of the tunnel excavation (B_c in the case of jacked sewer pipe); c is the cohesion coefficient, in pounds per square foot; and C_t is a load coefficient, which is a function of the ratio of the distance from the ground surface to the top of the tunnel to the width of the tunnel excavation and of the coefficient of internal friction of the natural material above the tunnel.

The formula for C_t is identical to that for C_d (Equation 14-11), except that H is the distance from the ground surface to the top of the tunnel and B_t is substituted for B_d. The values of the coefficient for C_t for various ratios of H/B_t and various types of materials may be obtained from Figure 14.16 (ASCE 1969) or Figure 14.6. Values of $\kappa\mu$ and $\kappa\mu'$ are the same as those noted in Figure 14.6.

An analysis of the formula for computing C_t indicates that for very high values of H/B_t the coefficient C_t approaches the limiting value of $1/(2\kappa\mu')$. Hence, where the tunnel is very deep, the load on the tunnel can be calculated readily by using the limiting value of C_t.

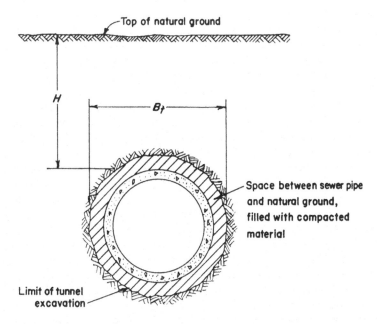

Figure 14.15.—Sewer pipe in tunnel.

Tunnel Soil Characteristics   The discussion regarding unit weight and coefficient of friction for sewers in trenches applies equally to the determination of earth loads on jacked sewer pipe or sewer pipe in tunnels through undisturbed soil.

The one additional factor that enters into the determination of loads on tunnels is c, the coefficient of cohesion. An examination of Equation (14-20) shows that the proper selection of c is very important; unfortunately, it can vary widely even for similar types of soils. It may be possible in some instances to obtain undisturbed samples of the material and to determine the value of c in the laboratory. Such testing should be done whenever possible. Conservative values of c should be used to allow for a saturated condition of the soil or for other unknown factors. Design values should probably be about 33% of the laboratory test value to allow for uncertainties.

For cases in which it is not practicable to determine c (the coefficient of cohesion) from laboratory tests, recommended safe values of c are listed in Table 14.2.

It is suggested that the value of c be taken as zero in the zone subject to seasonal frost and cracking because of dessication or loss of strength from saturation.

Effect of Excessive Excavation   Where the tunnel is constructed by a method that results in excessive excavation and where the voids above

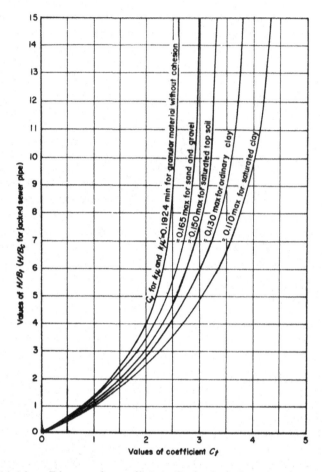

Figure 14.16. —*Diagram for coefficient* C_t *for jacked sewer pipe or tunnels in undisturbed soil.*

TABLE 14.2.   Recommended Safe Values of Cohesion, c.

Material (1)	Values of c	
	In Kilopascals (2)	In Pounds per Square Foot (3)
Clay, very soft	2	40
Clay, medium	12	250
Clay, hard	50	1,000
Sand, loose dry	0	0
Sand, silty	5	100
Sand, dense	15	300

(ASCE 1969).

the sewer pipe or tunnel lining are not backfilled carefully or packed with grout or other suitable backfill materials, saturation of the soil or vibration eventually may destroy the cohesion of the undisturbed material above the sewer pipe and result in loads in excess of those calculated using Equation (14-20). If this situation is anticipated, it is suggested that Equation (14-20) be modified by eliminating the cohesion term. The calculated loads then will be the same as those calculated for the trench condition from Equation (14-10).

6. Loads for Tunnels

When the sewer is to be constructed in a tunnel through homogeneous soils of low plasticity, the design should be based on the theories set forth in the previous section describing jacked sewer pipe. The design of tunnels through other types of materials is discussed in this section. The usual procedure in tunnel construction is to complete the excavation first, then place either a cast-in-place concrete liner or sewer pipe, and then grout or concrete the pipe in place. Additional strength in such a section can be obtained by means of pressure grouting to strengthen the surrounding material instead of relying totally on the liner or pipe itself. Tunnel loads, therefore, usually are determined for purposes of selecting supports to be used during excavation, and the sewer pipe or cast-in-place liner is designed primarily to withstand loads from pressure grouting.

Load-Producing Forces   When the tunnel is to be constructed through soils that tend to squeeze or swell (such as some types of clay or shale), or through fissured or seamed rock, the vertical load cannot be determined from a consideration of the factors discussed previously, and Equation (14-20) is not applicable.

The determination of rock pressures exerted against the tunnel lining is largely an estimate based on previous experience of the performance of linings in similar rock formations, although attempts at numerical analysis of stress conditions around a tunnel shaft have been made.

In the case of plastic clay, the full weight of the overburden is likely to come to rest on the tunnel lining some time after construction. The extent of lateral pressures to expect is yet to be determined fully, especially the passive resistance which will be maintained permanently by a plastic clay in the case of a flexible ring-shaped tunnel lining. For normally consolidated clays, suggested lateral pressures are on the order of $2/3$ to $7/8$ of vertical overburden pressures.

On the other hand, when tunneling through sand, only part of the weight of the overburden will come to rest on the tunnel lining at any time if adequate precautions are taken. The relief will be the result of the transfer of the soil weight immediately above the tunnel to the adjoining soil mass by shearing stresses along the vertical planes. In this case Marston's formula may be used for estimating the total load the tunnel lining may have to carry.

Great care must be taken to prevent any escape of sand into the tunnel during construction. Moist sand usually will arch over small openings and not cause trouble in this respect; however, entirely dry sand, which is sometimes encountered, is liable to trickle into the tunnel through gaps in the temporary lining. Wet sand or sand under the natural water table will flow readily through the smallest gaps. Sand movements of this kind destroy most if not all of the arching around the tunnel, resulting in a significant increase in both vertical and horizontal pressures on the supports of the lining. Such cases have been recorded and have caused considerable difficulty.

7. Alternate Design Method

For large diameter sewers, such as those greater than 48 inches in diameter, Indirect Designs based on the Marston method may yield conservative results. In such cases a more precise analysis can be made using a Direct Design method based on the principles of soil-structure interaction. Analysis should consider both the geometry of the system and material properties of the sewer pipe and the surrounding soil mass.

A simpler method based on arch analysis, which considers only the geometry of the sewer pipe and section properties of the sewer pipe material (Portland Cement Association 1975), can also be used for any specified loading condition.

In the method of soil-structure interaction analysis, loadings on the sewer pipe are automatically generated from the specified boundary conditions, the material properties, and the constitutive relationships of material behavior. Most solutions consider elastic behavior of the materials. Elastoplastic behavior and nonlinear analyses are also available.

The arch analysis method requires specification of vertical and lateral loads. The vertical loads can be determined by the Marston method and distributed uniformly over the full width of the sewer pipe. Lateral loads depend on the soil type and geologic history of the soil deposit. Design parameters should be obtained from a soils consultant knowledgeable in the subsurface conditions in the area. For sewer pipe installed in a tunnel or in a trench with properly compacted backfill, the recommended design lateral pressures are those corresponding to "at-rest" conditions. Where the backfill on the sides of the sewer may be loosely placed or insufficiently compacted, "active" pressure coefficients should be used to determine the lateral pressures. For preliminary analysis, the "at-rest" pressure coefficients in Table 14.3 are suggested.

Since active and passive earth pressures are the result of lateral strain in the soil mass, the at-rest condition refers to the lateral pressures existing in a large soil mass not subject to horizontal forces or strains except those resulting from its own weight.

TABLE 14.3.   "At-Rest" Pressure Coefficients.

Soil Type (1)	"At-Rest" Coefficient (2)
Granular soils	0.5 to 0.67
Cohesive soils, medium to hard	0.67 to 0.88
Cohesive soils, soft	0.75 to 1.0

C. SUPERIMPOSED LOADS ON SEWERS

1. General Method

Two types of superimposed loads are encountered commonly in the structural design of sewers; concentrated and distributed. Loads on sewer pipe caused by these superimposed loads can be determined by application of Boussinesq's solution for stresses in semi-infinite elastic medium (Spangler and Hennessy 1946).

Other methods, like the one given in the AASHTO Code, can be used to determine loads on sewer pipe from superimposed loads (AASHTO undated). The AASHTO method is intended for use with wheel loads, and may not be conservative or applicable for other types of loads, such as those from adjacent building foundations. Empirical studies indicate the difficulties of accurately predicting the actual loads on the pipe.

In the design of buried sewer pipe systems, proper consideration of construction loads is necessary. Loads resulting from heavy equipment and reduced backfill heights can produce loads on the sewer pipe that exceed final design loads.

2. Boussinesq Solution

Concentrated Loads   The formula for load caused by a superimposed concentrated load (Figure 14.17), is as follows:

$$W_{sc} = C_s \frac{PF}{L}$$ (14-21)

in which W_{sc} is the load on the sewer pipe, in pounds per unit length; P is the concentrated load, in pounds; F is the impact factor; C_s is the load coefficient (Table 14.4), a function of $B_c/2H$ and $L/2H$; H is the height of fill from the top of sewer pipe to ground surface; B_c is the width of sewer pipe; and L is the effective length of sewer pipe.

The effective length of a sewer pipe is defined as the length over which the average load caused by surface load produces nearly the

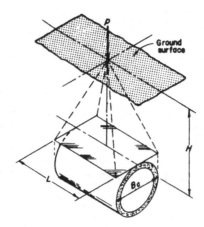

Figure 14.17.—Concentrated superimpopsed load vertically centered over sewer pipe.

same stress in the sewer pipe wall as does the actual load, which varies in intensity from point to point. Little information is available on this subject.

If the concentrated load is displaced laterally and longitudinally from a vertically centered location over the section of sewer pipe under construction, the load on the pipe can be computed by adding algebraically the effect of the concentrated load on various rectangles each with a corner centered under the concentrated load. Values of C_s in Table 14.4 divided by 4 equal the load coefficient for a rectangle the corner of which is vertically centered under the concentrated load.

Impact Factor   The impact factor, F, reflects the influence of dynamic loads caused by traffic at the ground surface. Suggested values for various kinds of traffic are shown in Table 14.5.

The impact effect decreases with increasing cover. The AASHTO (highway) Code (AASHTO undated) recommends a reduction to 1.0 where depth of cover exceeds 3' or the pipe outside diameter, whichever is larger. The AREA (railway) Code (AREA 1981–1982) recommends 10 feet of cover for the elimination of impact effect.

Distributed Loads   For the case of a superimposed load distributed over an area of considerable extent (Figure 14.18), the formula for load on the sewer pipe is:

$$W_{sd} = C_s p F B_c \tag{14-22}$$

in which W_{sd} is the load on the sewer pipe, in pounds per unit length; p is the intensity of distributed load, in pounds per square foot; F is the impact factor; B_c is the width of the sewer pipe; and C_s is the load

TABLE 14.4. Values of Load Coefficients, C_S, for Concentrated and Distributed Superimposed Loads Vertically Centered over Sewer Pipe[a]

$\dfrac{D}{2H}$ or $\dfrac{B_c}{2H}$ (1)	\multicolumn{14}{c}{$\dfrac{M}{2H}$ or $\dfrac{L}{2H}$}
	0.1 (2)	0.2 (3)	0.3 (4)	0.4 (5)	0.5 (6)	0.6 (7)	0.7 (8)	0.8 (9)	0.9 (10)	1.0 (11)	1.2 (12)	1.5 (13)	2.0 (14)	5.0 (15)
0.1	0.019	0.037	0.053	0.067	0.079	0.089	0.097	0.103	0.108	0.112	0.117	0.121	0.124	0.128
0.2	0.037	0.072	0.103	0.131	0.155	0.174	0.189	0.202	0.211	0.219	0.229	0.238	0.244	0.248
0.3	0.053	0.103	0.149	0.190	0.224	0.252	0.274	0.292	0.306	0.318	0.333	0.345	0.355	0.360
0.4	0.067	0.131	0.190	0.241	0.284	0.320	0.349	0.373	0.391	0.405	0.425	0.440	0.454	0.460
0.5	0.079	0.155	0.224	0.284	0.336	0.379	0.414	0.441	0.463	0.481	0.505	0.525	0.540	0.548
0.6	0.089	0.174	0.252	0.320	0.379	0.428	0.467	0.499	0.524	0.544	0.572	0.596	0.613	0.624
0.7	0.097	0.189	0.274	0.349	0.414	0.467	0.511	0.546	0.584	0.597	0.628	0.650	0.674	0.688
0.8	0.103	0.202	0.292	0.373	0.441	0.499	0.546	0.584	0.615	0.639	0.674	0.703	0.725	0.740
0.9	0.108	0.211	0.306	0.391	0.463	0.524	0.574	0.615	0.647	0.673	0.711	0.742	0.766	0.784
1.0	0.112	0.219	0.318	0.405	0.481	0.544	0.597	0.639	0.673	0.701	0.740	0.774	0.800	0.816
1.2	0.117	0.229	0.333	0.425	0.505	0.572	0.628	0.674	0.711	0.740	0.783	0.820	0.849	0.868
1.5	0.121	0.238	0.345	0.440	0.525	0.596	0.650	0.703	0.742	0.774	0.820	0.861	0.894	0.916
2.0	0.124	0.244	0.355	0.454	0.540	0.613	0.674	0.725	0.766	0.800	0.849	0.894	0.930	0.956

[a]Influence coefficients for solution of Holl's and Newmark's integration of the Boussinesq equation for vertical stress.

TABLE 14.5.   Suggested Values of Impact Factor, F.

Traffic Type (1)	F (2)
Highway	1.30
Railway	1.40
Airfield runways (for taxiways, consult FAA)	1.00

(ASCE 1969).

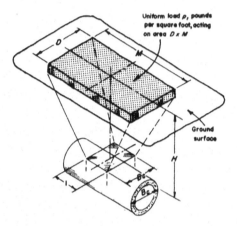

Figure 14.18.—Distributed superimposed load vertically centered over sewer pipe.

coefficient, a function of $D/(2H)$ and $(M/2H)$ from Table 14.4; H is the height from the top of the sewer pipe to the ground surface.

For the case of a uniform load offset from the center of the sewer pipe, the loads per unit length of the sewer pipe may be determined by a combination of rectangles. For determination of the stress below a point such as A in Figure 14.19, as a result of the loading in the rectangle BCDE, the area may be considered to consist of four rectangles: (AJDF) − (AJCG) − (AHEF) + (AHBG). Each of these four rectangles has a corner at point A. By computing $D/2H$ and $M/2H$ for each rectangle, the load coefficient for each rectangle can be taken from Table 14.4. Since point A is at the corner of each rectangle, the load coefficients from Table 14.4 should be divided by 4. A combination of the stresses from the four rectangles, with signs as indicated above, gives the desired stress.

Values of C_s can be read directly from Table 14.4 if the area of the distributed superimposed load is not centered over the sewer pipe under consideration.

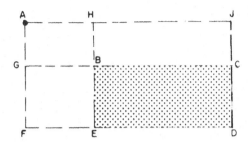

Figure 14.19.—Diagram for obtaining stress at point A caused by load in shaded area BCDE.

The load on the sewer pipe can be computed by adding algebraically the effect of various rectangles of loaded area if the area of the distributed superimposed load is not centered over the sewer pipe, but is displaced laterally and longitudinally. It is more convenient to work in terms of load under one corner of a rectangular loaded area rather than at the center. Dividing the tabular values of C_s by 4 will give the effect for this condition. Stresses from various types of surcharge loadings can be computed using computer solutions (Jumikis 1969, 1971).

3. Highway Loads

Pavements designed for heavy truck traffic substantially reduce the pressure transmitted through a wheel to the subgrade, and consequently to underlying sewer pipe. The pressure reduction is so great that generally the live load can be neglected. For heavy duty asphalt or flexible pavements, the reduction will be comparable to that for concrete pavements. The cost savings as a result of the resulting reduction in required pipe strength can be substantial.

For intermediate thicknesses of asphalt or flexible pavements, there is no generally accepted theory for estimating load distribution effects.

Relatively thin pavements do not reduce the pressure transmitted from the wheel to the subgrade to any significant degree. Such pavements are generally considered as unsurfaced roadways for determination of the effect of live loads on buried sewer pipe. A practical method, accepted by AASHTO, is developed in the following paragraphs.

Load Assumptions   The maximum highway wheel loads generally considered for design purposes are those specified by AASHTO for HS20 truck and alternate load configurations (Figure 14.20). The critical axle loads for these configurations are carried on dual wheels. The contact area of the dual wheels is assumed to be oval in shape and is approximately equal to the wheel load divided by the tire pressure. Several assumptions are normally made to simplify the evaluation process. For an HS20 wheel load, the contact pressure is assumed equal to the tire

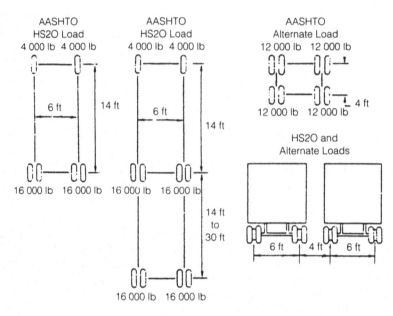

Figure 14.20.—Live load spacing.

pressure. The contact area can be approximated by a rectangle, as shown in Figure 14.21. The possible combinations of load applications are a single dual wheel load, two HS20 trucks passing, and the alternate load configuration in the passing mode. For highway loads, AASHTO recommends that the live load be increased by an impact factor when the pipe is less than three feet beneath the pavement surface. The recommended impact factors are presented in Table 14.6.

Load Distribution Assumptions At depths H below the surface, the length and the width of the rectangle are assumed to increase by the value of $1.75H$. The conditions for a single dual wheel, two HS20

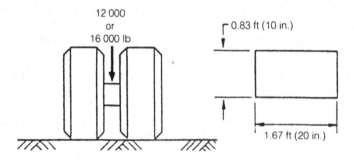

Figure 14.21.—Wheel load surface contact area.

TABLE 14.6. Impact Factors for Traffic Loads (AASHTO 1990).

Height of Cover	Impact Factor
0'-0" to 1'-0"	30%
1'-1" to 2'-0"	20%
2'-1" to 2'-11"	10%
3'-0" and Greater	0

Note: Impact factors recommended by the American Association of State Highway and Transportation Officials in "Standard Specifications for Highway Bridges," Twelfth Edition.

trucks passing, and alternate loads in the passing mode are shown in Figures 14.22, 14.23, and 14.24. The maximum value for average pressure intensity is obtained under various conditions of live load and depth of pipe. As shown in Figures 14.23 and 14.24, at greater depths the distributed loaded areas overlap, and the maximum pressure will develop under either the truck passing or the alternate loads in passing mode conditions. The depths for these transitions is shown in Table 14.7, along with the total loading, and rectangular areas that will produce the critical load effect.

Live Load on Pipe To determine the live load pressure on a buried sewer pipe, the average pressure intensity at the elevation of the outside top of the pipe is calculated. Then the total live load acting on the pipe is calculated based on the average pressure intensity. Finally, the live

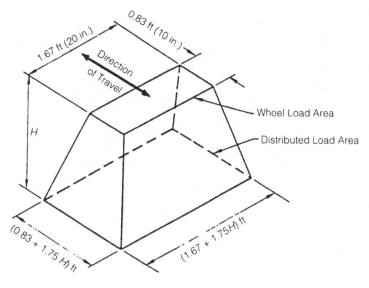

Figure 14.22. — Distributed load area — Single dual wheel.

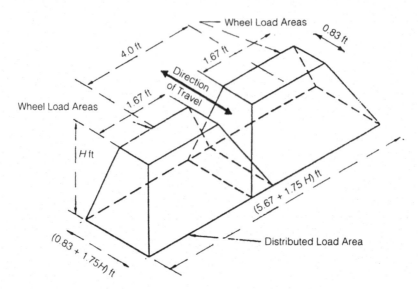

Figure 14.23.—*Distributed load area—Two HS 20 trucks passing.*

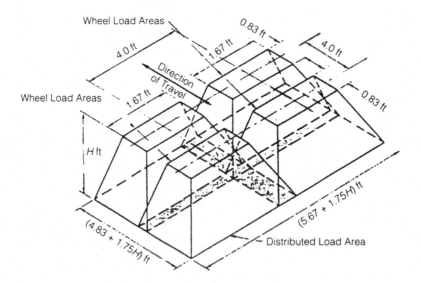

Figure 14.24.—*Distributed load area—Alternate loads in passing mode.*

TABLE 14.7. Critical Loading Configurations (ACPA 1988).

H, Feet	P, Pounds	A_{LL}, Square Feet
H < 1.33	16,000	(0.83 + 1.75H) (1.67 + 1.75H) (See Fig. 4.26)
1.33 ≤ H < 4.10	32,000	(0.83 + 1.75H) (5.67 + 1.75H) (See Fig. 4.27)
4.10 ≤ H	48,000	(4.83 + 1.75H) (5.67 + 1.75H) (See Fig. 4.28)

load pressure per linear foot of pipe is obtained by dividing the total live load on the pipe by the effective supporting length, as shown in Figure 14.25.

The average pressure intensity at the level of the outside top of the pipe is obtained by:

$$w_L = \frac{P(1 + I_f)}{A_{LL}}$$
(14-23)

in which w_L is the average pressure intensity, in pounds per square foot; P is the total applied surface wheel loads; A_{LL} is the distributed live load area on the subsoil plane at the outside top of the pipe; and I_f is the impact factor.

Given the average pressure intensity that is critical, the total live load, W_T, is obtained by the following equation:

$$W_T = w_L L S_L$$
(14-24)

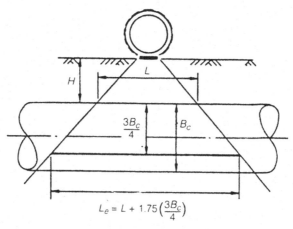

Figure 14.25. — Effective supporting length of pipe.

in which W_T is the total live load; L is the length of A_{LL} parallel to longitudinal axis of pipe; and S_L is the outside horizontal span of pipe or width of A_{LL} transverse to longitudinal axis of pipe, whichever is less.

A designer is primarily concerned with maximum loads. The most critical pipe location is centered under the distributed load area. The most critical loading can occur either when the longitudinal pipe axis is parallel or is transverse to the direction of truck travel, depending upon the diameter of pipe and depth of cover. Values for L and S_L to be used in Equation 14-24 are obtained from the orientation of the longitudinal pipe axis with respect to the distributed load area. The maximum value of W_T calculated for the two possible orientations, is used in Equation (14-25) to determine the live load, W_L:

$$W_L = \frac{W_T}{L_e} \tag{14-25}$$

in which W_L is the live load on pipe, and L_e is the effective supporting length of pipe.

The buried concrete pipe is assumed to be a beam on a continuous support, and L_e is at a level $B_c/4$ above the bottom of the outside of the pipe, as shown in Figure 14.25. Highway live load effects at depths greater than 10' below the pavement surface are insignificant, and, as a result, for pipes with cover exceeding that amount, live load effects can be neglected. The following equation is used to obtain L_e:

$$L_e = L + 1.75\left(\frac{3B_c}{4}\right) \tag{14-26}$$

4. Sewer Pipe Under Airport Pavements

An important factor in airport operations is the proper functioning of its drainage facilities. Because of relatively shallow covers associated with subsurface drainage for airfields, and the high wheel loads, the effect of aircraft loads on the structural design of underground conduits is more critical than most highway facilities.

The pressure distribution of aircraft wheel loads on any horizontal plane in the soil mass depends on the magnitude and characteristics of the aircraft loads, including tire pressures, the landing gear configuration, the pavement structure, and the subsoil conditions. Larger aircraft have multiple wheel assemblies and dual tandem assemblies to reduce the load concentrations. These cause a combination of loadings that impose overlapping pressures similar to, but greater than, highway loadings.

Major airports are paved with relatively strong surfaces of concrete or asphalt materials, which effectively reduce the load intensity on the subgrade. The method described in Section 6 can be used for estimating pressure intensities under concrete pavements. For asphalt surfaces, either local engineering practice or a nominal increase in pressure intensity over that estimated for concrete can be used.

There are a large number of small airports with no surfacing beyond a nominal amount to control dust and minimize the destructive effect of windblown aggregate particles. For these, the Boussinesq Solution, which assumes no pavement effect on load distribution, can be used to determine the subsoil pressure intensities from live loads.

5. Sewer Pipe Under Railway Tracks

To evaluate the live load effect for railroad loadings, the American Railway Engineering Association recommends the use of a combination of axle loads and axle spacing represented by the Cooper E 80 loading shown in Figure 14.26 and an impact factor ranging from 40% at zero cover to zero percent with 10' of cover. The series of axle loads and spacings is converted into a uniform load at the bottom of the railroad ties. The pressure intensity on a pipe at various depths and various offsets is then computed based upon the Boussinesq theory. The live load transmitted to a pipe underground is computed by:

$$W_L = Cp_o B_c(1 + I_f) \tag{14-27}$$

in which W_L is the live load transmitted to the pipe; C is the pressure coefficient; p_o is the intensity of the distributed load at the bottom of the ties; B_c is the outside horizontal span of the pipe; and I_f is the impact factor.

This equation is similar to the expressions for highway and airport live loadings in that the pressure intensity on the pipe is equal to the pressure on the surface multiplied by an appropriate coefficient.

A locomotive load is assumed uniformly distributed over an area 8' by 20', and the dead load weight of the track structure is assumed to be 200 lbs per linear foot. Live and dead load curves, including impact factors, are plotted in Figure 14.27 as determined from Equation (14-27) for a Cooper E 80 loading. For any given height of cover from the

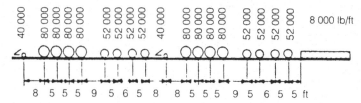

Figure 14.26.—Spacing of wheel loads per axle for a Cooper E 80 design loading.

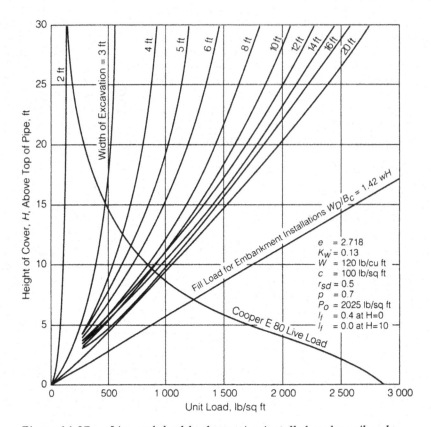

Figure 14.27.—Live and dead loads on pipe installed under railroads.

top of the pipe to the bottom of the ties, the live load can be read directly from Figure 14.27. To obtain the live load per linear foot, it is necessary to multiply the unit load from Figure 14.27 by the outside span of the pipe, B_c, in feet.

6. Sewer Pipe Under Rigid Pavement

Based upon the early work of Westergaard and others, the Portland Cement Association (PCA) developed a method to determine the vertical pressure on buried pipe due to wheel loads applied to concrete pavements (PCA 1951). The PCA approach assumes that both the slab and the earth are elastic materials, the compression of the soil is proportional to the deflection of the slab, and the effect of the load radiates equally in all directions. The following equation for the pressure intensity on concrete pipe under concrete pavements was developed:

$$p_{(H,X)} = \frac{CP}{R_s^2} \tag{14-28}$$

in which C is the pressure coefficient, dependent on H, X, and R_s; P is the wheel load; R_s is the radius of stiffness of the rigid pavement; and $p_{(H,X)}$ is the vertical pressure intensity at any horizontal distance, X, and any vertical distance, H, within the soil mass.
R_s is further defined as:

$$R_s = \sqrt[4]{\frac{Eh^3}{12(1 - \mu^2)k}} \tag{14-29}$$

in which E is the modulus of elasticity of concrete, in pounds per square inch; h is the thickness of concrete pavement; μ is the Poisson's ratio of concrete (assumed a constant 0.15); and κ is the modulus of subgrade reaction, pounds per cubic inch.

Equation (14-28) has been solved for various conditions to simplify the task of estimating the live load. The results are included in Tables 14.8 through 14.12 in the form of pressure coefficients for various values of H, X, and R_s. Equation (14-29) also has been solved for a range of conditions and R_s is tabulated for various values of h and κ in Table 14.13. Figure 14.28 provides a basis for estimating the modulus of subgrade reaction using as a basis the various soil classification systems.

The values given in the tables are used to determine the pressures on horizontal planes at depth H in a semi-infinite elastic body from a single wheel and two wheels at various spacings on the pavement surface. The coefficient values in Tables 14.9 through 14.11 are larger, for any given fill height, than the coefficient values presented for a single wheel (Table 14.8). Thus, the combined pressure on any horizontal plane for wheels at a spacing of $2.4R_s$ or less is greater than that for a single wheel. For the greater wheel spacing of $3.2R_s$, the coefficient values in Table 14.12 are greater than for a single load (Table 14.8), except close to the pavement. Therefore, for a wheel spacing of $3.2R_s$, the combined pressure should be used except where the pipe is very close to the pavement subgrade elevation, when the single wheel load pressure can be used. For wheel spacing greater than $3.2R_s$, the combined pressure of two wheels will never be more than the pressure from a single wheel.

The presence of a pipe introduces a boundary condition which, theoretically, creates a case simulating that of an elastic layer of depth H resting on a rigid base. Although pressures based on this concept would be slightly higher than those included in the tabular coefficients, variability of the elastic properties of the backfill over the pipe, the rigidity of the pipe, and the inadequacies of the theory are such that precise computations are not justified.

TABLE 14.8.   Pressure Coefficients for a Single Load (PCA 1951).

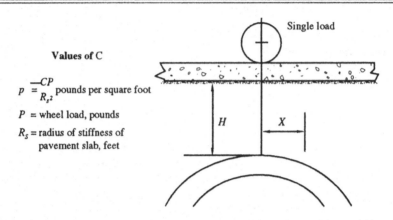

Values of C

$$p = \frac{CP}{R_s^2} \text{ pounds per square foot}$$

P = wheel load, pounds

R_s = radius of stiffness of pavement slab, feet

$\dfrac{H}{R_s}$	X/R_s										
	0.0	0.4	0.8	1.2	1.6	2.0	2.4	2.8	3.2	3.6	4.0
0.0	.113	.105	.089	.068	.048	.032	.020	.011	.006	.002	.000
0.4	.101	.095	.082	.065	.047	.033	.021	.011	.004	.001	.000
0.8	.089	.084	.074	.061	.045	.033	.022	.012	.005	.002	.001
1.2	.076	.072	.065	.054	.043	.032	.022	.014	.008	.005	.003
1.6	.062	.059	.054	.047	.039	.030	.022	.016	.011	.007	.005
2.0	.051	.049	.046	.042	.035	.028	.022	.016	.011	.008	.006
2.4	.043	.041	.039	.036	.030	.026	.021	.016	.011	.008	.006
2.8	.037	.036	.033	.031	.027	.023	.019	.015	.011	.009	.006
3.2	.032	.030	.029	.026	.024	.021	.018	.014	.011	.009	.007
3.6	.027	.026	.025	.023	.021	.019	.016	.014	.011	.009	.007
4.0	.024	.023	.022	.020	.019	.018	.015	.013	.011	.009	.007
4.4	.020	.020	.019	.018	.017	.015	.014	.012	.010	.009	.007
4.8	.018	.017	.017	.016	.015	.013	.012	.011	.009	.008	.007
5.2	.015	.015	.014	.014	.013	.012	.011	.010	.008	.007	.006
5.6	.014	.013	.013	.012	.011	.010	.010	.009	.008	.007	.006
6.0	.012	.012	.011	.011	.010	.009	.009	.008	.007	.007	.006
6.4	.011	.010	.010	.010	.009	.008	.008	.007	.007	.006	.005
6.8	.010	.009	.009	.009	.008	.008	.007	.007	.006	.006	.005
7.2	.009	.008	.008	.008	.008	.007	.007	.006	.006	.006	.005
7.6	.008	.008	.008	.007	.007	.007	.006	.006	.006	.005	.005
8.0	.007	.007	.007	.007	.006	.006	.006	.006	.005	.005	.005

TABLE 14.9.   Pressure Coefficients for Two Loads Spaced $0.8R_s$ Apart (PCA 1951).

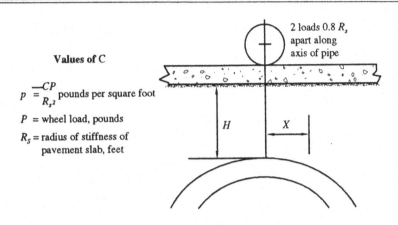

Values of C

$$p = \frac{CP}{R_s^2} \text{ pounds per square foot}$$

P = wheel load, pounds

R_s = radius of stiffness of pavement slab, feet

2 loads $0.8 R_s$ apart along axis of pipe

$\frac{H}{R_s}$	X/R_s										
	0.0	0.4	0.8	1.2	1.6	2.0	2.4	2.8	3.2	3.6	4.0
0.0	.210	.198	.168	.130	.092	.062	.038	.022	.011	.004	.000
0.4	.190	.181	.156	.126	.092	.064	.040	.023	.010	.002	.000
0.8	.168	.160	.140	.117	.088	.063	.042	.024	.010	.003	.001
1.2	.144	.139	.124	.106	.083	.062	.043	.027	.013	.007	.004
1.6	.118	.115	.105	.094	.076	.060	.044	.030	.020	.014	.009
2.0	.098	.095	.089	.081	.070	.056	.043	.032	.023	.017	.012
2.4	.083	.080	.076	.069	.061	.050	.040	.031	.023	.017	.012
2.8	.071	.069	.066	.060	.053	.045	.037	.029	.022	.017	.012
3.2	.061	.059	.057	.052	.046	.040	.034	.028	.022	.017	.013
3.6	.052	.051	.049	.046	.041	.036	.032	.027	.022	.018	.014
4.0	.045	.044	.042	.040	.037	.034	.030	.026	.022	.018	.015
4.4	.039	.038	.037	.035	.033	.030	.027	.024	.021	.017	.015
4.8	.034	.034	.033	.031	.029	.027	.024	.021	.019	.016	.014
5.2	.030	.029	.028	.027	.025	.023	.021	.019	.017	.015	.013
5.6	.026	.026	.025	.024	.022	.021	.019	.018	.016	.014	.012
6.0	.023	.023	.022	.021	.020	.019	.017	.016	.015	.013	.011
6.4	.021	.021	.020	.019	.018	.017	.016	.015	.014	.012	.011
6.8	.019	.019	.018	.018	.017	.016	.015	.014	.013	.012	.010
7.2	.017	.017	.016	.016	.015	.014	.013	.013	.012	.011	.010
7.6	.016	.015	.015	.015	.014	.013	.012	.012	.011	.010	.009
8.0	.014	.014	.014	.013	.013	.012	.012	.011	.010	.010	.009

TABLE 14.10.   Pressure Coefficients for Two Loads Spaced $1.6R_s$
Apart (PCA 1951).

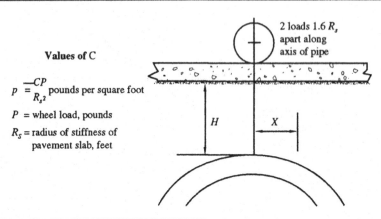

Values of C

$$p = \frac{-CP}{R_s^2} \text{ pounds per square foot}$$

P = wheel load, pounds

R_s = radius of stiffness of
pavement slab, feet

2 loads 1.6 R_s
apart along
axis of pipe

$\frac{H}{R_s}$	X/R_s										
	0.0	0.4	0.8	1.2	1.6	2.0	2.4	2.8	3.2	3.6	4.0
0.0	.178	.167	.142	.112	.080	.054	.034	.019	.009	.004	.000
0.4	.164	.156	.136	.109	.080	.056	.036	.019	.008	.002	.000
0.8	.147	.141	.126	.103	.078	.057	.037	.020	.008	.002	.001
1.2	.128	.124	.106	.094	.074	.056	.039	.023	.012	.006	.004
1.6	.108	.105	.097	.082	.070	.054	.040	.028	.019	.014	.009
2.0	.092	.090	.084	.075	.065	.052	.040	.030	.022	.017	.012
2.4	.079	.076	.072	.065	.056	.047	.038	.029	.022	.017	.012
2.8	.068	.066	.062	.058	.050	.043	.035	.028	.022	.017	.012
3.2	.058	.056	.054	.050	.044	.038	.032	.027	.022	.017	.012
3.6	.050	.049	.047	.044	.040	.035	.030	.026	.022	.017	.013
4.0	.043	.042	.041	.039	.036	.033	.030	.026	.022	.018	.015
4.4	.038	.037	.036	.034	.032	.029	.026	.023	.020	.016	.014
4.8	.033	.032	.031	.030	.028	.026	.024	.021	.018	.015	.013
5.2	.029	.028	.027	.026	.025	.023	.021	.019	.016	.014	.012
5.6	.025	.025	.024	.023	.022	.020	.019	.017	.015	.013	.012
6.0	.023	.022	.022	.021	.019	.018	.017	.016	.014	.013	.011
6.4	.020	.020	.019	.019	.018	.016	.015	.015	.013	.012	.011
6.8	.018	.018	.018	.017	.016	.015	.014	.013	.012	.011	.010
7.2	.017	.016	.016	.015	.015	.014	.013	.013	.012	.011	.010
7.6	.015	.015	.014	.014	.014	.013	.012	.012	.011	.010	.010
8.0	.014	.014	.013	.013	.013	.012	.011	.011	.010	.010	.009

TABLE 14.11. Pressure Coefficients for Two Loads Spaced 2.4 R_s Apart (PCA 1951).

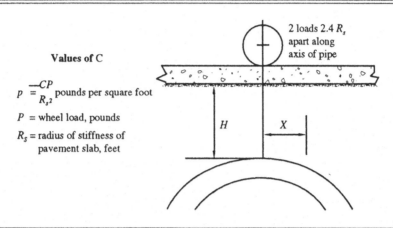

Values of C

$$p = \frac{-CP}{R_s^2} \text{ pounds per square foot}$$

P = wheel load, pounds

R_s = radius of stiffness of pavement slab, feet

$\dfrac{H}{R_s}$	X/R_s										
	0.0	0.4	0.8	1.2	1.6	2.0	2.4	2.8	3.2	3.6	4.0
0.0	.137	.130	.112	.088	.065	.044	.028	.014	.007	.003	.000
0.4	.130	.125	.109	.087	.066	.047	.028	.013	.005	.001	.000
0.8	.121	.117	.104	.085	.066	.048	.030	.014	.006	.002	.001
1.2	.109	.105	.096	.079	.064	.048	.033	.018	.012	.006	.005
1.6	.095	.092	.084	.072	.060	.047	.035	.025	.018	.012	.009
2.0	.083	.081	.077	.068	.057	.046	.035	.026	.020	.015	.010
2.4	.070	.069	.065	.059	.052	.044	.034	.026	.020	.015	.011
2.8	.062	.060	.058	.053	.046	.039	.033	.027	.020	.015	.011
3.2	.053	.052	.050	.046	.041	.035	.032	.026	.020	.016	.012
3.6	.046	.045	.044	.042	.038	.034	.030	.026	.021	.017	.013
4.0	.040	.040	.039	.037	.035	.032	.029	.025	.021	.017	.014
4.4	.036	.035	.034	.033	.031	.028	.025	.022	.019	.016	.013
4.8	.031	.031	.030	.029	.027	.025	.022	.020	.017	.015	.012
5.2	.027	.027	.026	.025	.024	.022	.020	.018	.016	.014	.012
5.6	.024	.023	.023	.022	.021	.020	.018	.017	.015	.013	.011
6.0	.022	.021	.021	.020	.019	.018	.017	.015	.014	.012	.011
6.4	.019	.019	.019	.018	.017	.016	.015	.014	.013	.012	.010
6.8	.018	.017	.017	.016	.016	.015	.014	.013	.012	.011	.010
7.2	.016	.016	.016	.015	.014	.014	.013	.012	.011	.010	.009
7.6	.015	.014	.014	.014	.013	.013	.012	.011	.011	.010	.009
8.0	.013	.013	.013	.013	.012	.012	.011	.011	.010	.009	.009

TABLE 14.12.   Pressure Coefficients for Two Loads Spaced 3.2 R_S Apart (PCA 1951).

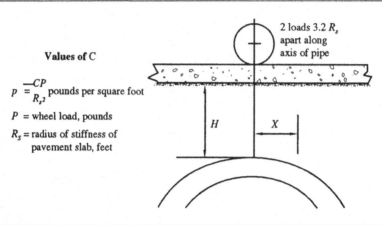

Values of C

$p = \dfrac{-CP}{R_s^2}$ pounds per square foot

P = wheel load, pounds

R_S = radius of stiffness of pavement slab, feet

2 loads 3.2 R_s apart along axis of pipe

H

X

$\dfrac{H}{R_s}$	X/R_s										
	0.0	0.4	0.8	1.2	1.6	2.0	2.4	2.8	3.2	3.6	4.0
0.0	.097	.093	.080	.065	.048	.032	.020	.011	.004	.000	.000
0.4	.096	.092	.079	.067	.050	.034	.020	.010	.003	.000	.000
0.8	.092	.088	.078	.066	.051	.036	.021	.010	.003	.000	.000
1.2	.086	.082	.074	.066	.050	.038	.025	.014	.007	.003	.001
1.6	.077	.075	.068	.060	.049	.039	.030	.021	.015	.011	.007
2.0	.070	.068	.063	.057	.048	.040	.031	.023	.017	.013	.009
2.4	.061	.060	.056	.051	.045	.038	.030	.023	.017	.013	.010
2.8	.056	.054	.052	.048	.042	.036	.029	.023	.018	.013	.010
3.2	.048	.046	.044	.041	.037	.032	.028	.023	.018	.014	.010
3.6	.043	.041	.040	.038	.034	.030	.027	.022	.019	.015	.012
4.0	.038	.037	.036	.035	.032	.029	.026	.022	.019	.016	.013
4.4	.033	.033	.032	.031	.029	.027	.024	.020	.018	.015	.013
4.8	.029	.029	.028	.027	.025	.023	.021	.018	.016	.014	.012
5.2	.025	.025	.025	.024	.022	.021	.019	.017	.015	.013	.012
5.6	.022	.022	.022	.021	.020	.018	.017	.016	.014	.012	.011
6.0	.020	.020	.020	.020	.020	.017	.016	.015	.013	.011	.011
6.4	.018	.018	.018	.018	.018	.016	.015	.014	.012	.011	.010
6.8	.016	.016	.016	.016	.016	.014	.014	.013	.012	.010	.010
7.2	.015	.015	.015	.015	.015	.013	.013	.012	.011	.010	.009
7.6	.014	.014	.013	.013	.013	.012	.012	.011	.010	.009	.009
8.0	.013	.013	.012	.012	.012	.011	.011	.010	.010	.009	.008

TABLE 14.13. Values of Radius of Stiffness, R_s (PCA 1951).

h (in.)	Values of k								
	50	100	150	200	250	300	350	400	500
6	34.84	29.30	26.47	24.63	23.30	22.26	21.42	20.72	19.59
6.5	36.99	31.11	28.11	26.16	24.74	23.64	22.74	22.00	20.80
7	39.11	32.89	29.72	27.65	26.15	24.99	24.04	23.25	21.99
7.5	41.19	34.63	31.29	29.12	27.54	26.32	25.32	24.49	23.16
8	43.23	36.35	32.85	30.57	28.91	27.62	26.58	25.70	24.31
8.5	45.24	38.04	34.37	31.99	30.25	28.91	27.81	26.90	25.44
9	47.22	39.71	35.88	33.39	31.58	30.17	29.03	28.08	26.55
9.5	49.17	41.35	37.36	34.77	32.89	31.42	30.23	29.24	27.65
10	51.10	42.97	38.83	36.14	34.17	32.65	31.42	30.39	28.74
10.5	53.01	44.57	40.28	37.48	35.45	33.87	32.59	31.52	29.81
11	54.89	46.16	41.71	38.81	36.71	35.07	33.75	32.64	30.87
11.5	56.75	47.72	43.12	40.13	37.95	36.26	34.89	33.74	31.91
12	58.59	49.27	44.52	41.43	39.18	37.44	36.02	34.84	32.95
12.5	60.41	50.80	45.90	42.72	40.40	38.60	37.14	35.92	33.97
13	62.22	52.32	47.27	43.99	41.61	39.75	38.25	36.99	34.99
13.5	64.00	53.82	48.63	45.26	42.80	40.89	39.35	38.06	35.99
14	65.77	55.31	49.98	46.51	43.98	42.02	40.44	39.11	36.99
14.5	67.53	56.78	51.31	47.75	45.16	43.15	41.51	40.15	37.97
15	69.27	58.25	52.63	48.98	46.32	44.26	42.58	41.19	38.95
15.5	70.99	59.70	53.94	50.20	47.47	45.36	43.64	42.21	39.92
16	72.70	61.13	55.24	51.41	48.62	46.45	44.70	43.23	40.88
16.5	74.40	62.56	56.53	52.61	49.75	47.54	45.74	44.24	41.84
17	76.08	63.98	57.81	53.80	50.88	48.61	46.77	45.24	42.78
17.5	77.75	65.38	59.08	54.98	52.00	49.68	47.80	46.23	43.72
18	79.41	66.78	60.35	56.16	53.11	50.74	48.82	47.22	44.66
19	82.70	69.54	62.84	58.48	55.31	52.84	50.84	49.17	46.51
20	85.95	72.27	65.30	60.77	57.47	54.92	52.84	51.10	48.33
21	89.15	74.97	67.74	63.04	59.62	56.96	54.81	53.01	50.13
22	92.31	77.63	70.14	65.28	61.73	58.98	56.75	54.89	51.91
23	95.44	80.27	75.52	67.49	63.83	60.98	58.68	56.75	53.67
24	98.54	82.86	74.87	69.68	65.90	62.96	60.58	58.59	55.41

$$R_s = \sqrt[4]{\frac{Eh^3}{12\,(1 - \mu^2)\,k}}$$

where: $E = 4{,}000{,}000$ psi $\quad R_s = 24.1652 \sqrt[4]{\dfrac{h^3}{k}}$
$\mu = 0.15$

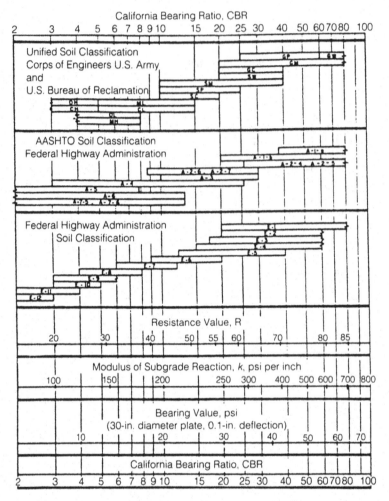

Figure 14.28.—Approximate interrelationships of soil classifications and bearing values.

D. Pipe Bedding and Backfilling

1. General Concepts

The ability of a sewer pipe to safely support the calculated soil load depends not only on its inherent strength but also on the distribution of the bedding reaction and on the lateral pressure acting against the sides of the sewer pipe.

Construction of the sewer pipe/soil system focuses attention on the pipe zone, which is made up of five specific areas; foundation, bedding, haunching, initial backfill, and final backfill (see Figure 14.29 for definitions and limits of these five areas). Note that all of these areas are not necessarily referred to in all pipe design standards. The discussion in this section is in general terms and is intended to describe the effect of the various areas on the pipe-soil system.

For all sewer pipe materials, the calculated vertical load is assumed to be uniformly distributed over the width of the pipe. This assumption originated in Marston's work; the assumption is part of the bedding factors developed by Spangler and presented in this chapter. Many years of field experience indicate that the assumption results in conservative designs.

The load-carrying capacity of sewer pipes of all materials is influenced by the sewer pipe/soil system, although the importance of the specific areas may vary with different pipe materials. Detailed information on pipe bedding classes is contained in the various ANSI and ASTM Specifications or industry literature for each material. The design engineer should consult the applicable specification or literature for information to be used in design.

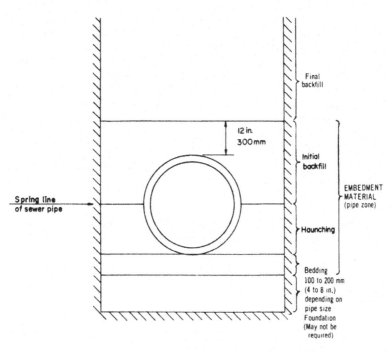

Figure 14.29.—Trench cross section illustrating terminology.

2. Foundation

The foundation provides the base for the sewer pipe-soil system. In trench conditions, the total weight of the pipe and soil backfill will normally be no more than the weight of the excavated soil. In this case, foundation pressures are not increased from the initial condition, and the designer should be concerned primarily with the presence of unsuitable soils, such as peat or other highly organic or compressible soils, and with maintaining a stable trench bottom. If the full benefit of the bedding is to be gained, the bottom of the trench or embankment must be stable.

3. Bedding

Bedding provides the interface between the pipe and its foundation, and has an important influence on the distribution of the reaction against the bottom of the sewer pipe. Bedding therefore influences the supporting strength of the pipe as installed. Some research (Griffith and Keeney 1967; Sikora 1980) has indicated that a well-graded stone is a suitable material for sewer pipe bedding, and that the most important property of any bedding material is the degree to which the pipe can settle into the bedding. A crushed material that is self-compacting and will interlock should be avoided.

Even though larger particle sizes give greater stability, the maximum size and shape of granular embedment should also be related to the pipe material and the recommendations of the manufacturer. For example, sharp angular embedment material larger than 0.5–0.75 inch should not be used against corrosion protection coatings. For small sewer pipes, the maximum size should be limited to about 10% of the pipe diameter.

Soil classifications under the Unified Soil Classification System, including manufactured materials, are grouped into five broad categories according to their ability to develop an interacting sewer pipe-soil system (Table 14.14). These soil classes are described in ASTM D 2321.

In general, crushed stone or gravel meeting the requirements of ASTM Designation C33, Gradation 67 [0.75 in. to No. 4)] will provide the most satisfactory sewer pipe bedding.

In some locations, the natural soils at the level of the bottom of the sewer pipe may be sands of suitable grain size and density to serve as both foundation and bedding for the pipe. In such situations it may not be necessary to remove and replace these soils with the special bedding materials described above.

4. Haunching

The soil placed at the sides of a pipe from the bedding up to the spring line is the haunching. The care with which this material is placed has a significant influence on the performance of the sewer pipe, particularly in the space just above the bedding. Poorly compacted material

TABLE 14.14. Soil Classifications* (ASCE 1982).

Soil Class (1)	Group Symbol (2)	Typical Names (3)	Comments (4)
I		Crushed rock	angular, 6–40 mm
II	GW GP SW SP	Well graded gravels Poorly graded gravels Well graded sands Poorly graded sands	40 mm maximum
III	GM GC SM SC	Silty gravels Clayey gravels Silty sands Clayey sands	
IV	MH, ML CH, CL	Inorganic silts Inorganic clays	Not recommended for bedding, haunching or initial backfill
V	OL, OH PT	Organic silts and clays Peat	

*For a more detailed description, see ASTM D2321—"Recommended Practice for Underground Installation of Flexible Thermoplastic Sewer Pipe."
Copyright, American Society for Testing and Materials, 1916 Race St., Philadelphia, PA 19103, reprinted with permission.
Note: 1mm = 0.039 in.

in this space will result in a concentration of reaction at the bottom of the pipe.

For flexible pipe, compaction of the haunching material is essential. For rigid pipe, compaction can ensure better distribution of the forces on the pipe. Material used for sewer pipe haunching should be shovel sliced or otherwise placed to provide uniform support for the pipe barrel and to fill completely all voids under the pipe. Because of space limitations, haunching material is often compacted manually. Results should be checked to verify that the bedding or installation criteria are achieved.

Material used for haunching may be crushed stone or sand, or a well-graded granular material of intermediate size. If crushed stone is used, it should be subject to the same size limitations and cautions regarding use against corrosion protection coatings. Sand should not be used if the pipe zone area is subject to a fluctuating groundwater table or where there is a possibility of the sand migrating into the pipe bedding or trench walls, unless a geotextile is incorporated in the installation.

5. Initial Backfill

Initial backfill is the material that covers the sewer pipe and extends from the haunching to some specific point (generally 6–12") above the top of the pipe, depending on the class of bedding and type of pipe material. Its function is to protect the pipe from damage by subsequent backfill, and to ensure the uniform distribution of load over the top of the pipe. For rigid pipe, the structural performance is primarily based on the haunching material; whereas for flexible pipe, compaction of the initial backfill is critical to performance.

The initial backfill is usually not mechanically tamped or compacted, since such work may damage the sewer pipe, particularly if done over the crown of the pipe. Therefore it should be a material that will develop a uniform and, for flexible sewer pipe, relatively high density with little compactive effort. Initial backfill should consist of suitable granular material, but for rigid pipe not necessarily as select a material as that used for bedding and haunching. Clay materials requiring mechanical compaction should not be used for initial backfill.

The fact that little compaction effort is used on the initial backfill should not lead to carelessness in choice or placement of material. Particularly for large sewer pipes, care should be taken in placing both the initial backfill and final backfill over the crown to avoid damage to the sewer pipe.

6. Final Backfill

The choice of material and placement methods for final backfill are related to the construction site and (generally) not related to the design of the sewer pipe. Under special embankment conditions or induced trench conditions, final backfill may play an important part in the sewer pipe design. However, for most trench installations, final backfill does not affect the pipe design.

The final backfill of trenches in traffic areas, such as under improved existing surfaces, is usually composed of material that is easily densified to minimize future settlement. In underdeveloped areas, final backfill often will consist of the excavated material placed with little compaction and left mounded over the trench to allow for future settlement. Studies have indicated that with some soils, this settlement may continue for over 10 years.

Trench backfilling should be done in such a way as to prevent dropping of material directly on top of a sewer pipe through any great vertical distance. When placing material with a bucket, the bucket should be lowered so that the shock of falling earth will not cause damage.

E. Design Safety Factor and Performance Limits

1. General Concepts

In the design of sewer pipes, the selection of a factor of safety is an essential element of the structural design requirements. When these requirements are defined for a given construction material, the most severe or maximum performance limits of that material as related to the proposed service or design application limit must be determined, and the ratio of those two limits is the "factor of safety." Maximum performance limits and design values may be defined in terms of initial or long-term strength, stress, strain, or product deformation depending on the characteristics of the material under consideration. Typical design values are presented in AASHTO standards (AASHTO undated). The selected factor of safety is applied to the maximum performance limit to calculate a lower value, which is then used as a design or service performance value.

The selection of a desired factor of safety is essentially based on a risk value assessment which must relate to the specific conditions anticipated in an application, the failure mode of the construction material, and the potential cost of system failure. Factors of safety compensate for unexpected construction deficiencies. They should not be relied on to compensate for poor construction practice or for inadequate inspection. Properly established design performance values, including adequate factors of safety, must be realized in installation and operation to provide reasonable assurance of acceptable long-term system performance.

The relationship between safety factors and design performance values is similar for rigid and flexible sewer pipe. However, an understanding of the differences between design requirements for rigid sewer pipe and the requirements for flexible sewer pipe system design is important.

2. Rigid Sewer Pipe

Design performance limits for rigid sewer pipes generally are expressed in terms of strength under load. There are two alternative methods of determining the service strength of reinforced concrete sewer pipe: by "strength and design" analysis, or by testing. The direct design method more accurately represents the relative performance of pipe in the field under the service load. However, testing is generally used for the indirect design types of precast or prefabricated rigid pipe.

Indirect design strengths of rigid sewer pipe usually are measured in terms of the ultimate three-edge bearing strength, and of ultimate and 0.01" crack, three-edge bearing strengths for reinforced concrete sewer pipe. A safety factor of 1.0 should be applied to the 0.01" crack load for reinforced concrete sewer pipe, and a safety factor of 1.25–1.50 should be applied to the specified minimum ultimate three-edge bearing strength to determine the working strength for other rigid pipes. Common practice is to use a factor of safety of 1.25 for the ultimate load of reinforced concrete sewer pipe, and up to 1.50 for vitrified clay tested in three-edge bearing. Such safety factors relate to test loads that concentrate maximum shear and flexure at the same point. They are very conservative and do not represent ultimate supporting strength in the buried condition.

3. Flexible Sewer Pipe

Design performance limits for flexible sewer pipe may be expressed in terms of stress or strain in the pipe wall, crushing or buckling in the pipe wall, or deflection. The most common limitation is deflection. The deflection limitation is established as a design performance limit to provide a factor of safety against structural failure or any type of distress that might tend to limit the service life of the pipe. This design limit will vary with different pipe materials and the pipe manufacturing process. Pipes must be able to deflect without cracking, liner failure, joint leaks, excessive strain or other distress, and they should be designed with a reasonable factor of safety.

Certain types of plastic pipes are subject to strain deterioration. There is a limiting strain which, if exceeded over a period of time, will eventually result in cracking of the pipe wall, which is called environmental stress cracking (ESC). AASHTO Standards (AASHTO undated) provide design values for this limiting strain.

4. Recommendations for Field Procedures

The factor of safety against ultimate collapse of sewer pipe is about the same as that used in the design of most engineered structures. However, the design of sewer pipe is based on calculated loads, bedding factors, and experimental factors, which are less well-defined than the dead and live loads used in building design. It is therefore important that the loads imposed on the sewer pipe not exceed the design loads.

To obtain this objective, the following procedures are recommended:

(a) Specifications. Construction specifications should set forth limits for the width of trench at the top of sewer pipe. The width limits should take into account the minimum allowable width for each class of sewer pipe and bedding to be used. Where the depth is such that a positive projecting condition will be obtained, maximum width should be specified as unlimited unless the width must be controlled for some reason other than to meet structural requirements of the sewer pipe. Appropriate corrective measures should be specified in the event the maximum allowable width is exceeded. These measures may include provision for a higher class of bedding or stronger pipe. Maximum allowable construction live loads should be specified for various depths of cover if appropriate.

(b) Inspection. Construction should be observed by an experienced engineer, or by an inspector who reports to a competent field engineer.

(c) Testing. Sewer pipe testing should be under the supervision of a reliable testing laboratory, and close liaison should be maintained between the laboratory and the field engineer.

(d) Field Conditions. The field engineer should be furnished with sufficient design data to enable him to evaluate unforeseen conditions intelligently. The field engineer should be instructed to confer with the design engineer if design changes appear advisable.

(e) Sheeting. Where sheeting is to be removed, pulling should be done in stages. The space formerly occupied by the sheeting must be backfilled completely, and field inspection should verify this.

Effect of Trench Sheeting Because of the various alternative methods employed in sheeting trenches, generalizations on the proper construction procedure follow to ensure that the design load is not exceeded are risky and dangerous. Each method of sheeting and bracing should be studied separately. The effect of a particular system on the sewer pipe load, as well as the consequences of removing the sheeting or the bracing, must be estimated.

It is difficult to obtain satisfactory filling and compaction of the void left when wood sheeting is pulled. Wood sheeting driven alongside the sewer pipe should be cut off and left in place to an elevation of 1.5' above the top of the sewer pipe.

If granular materials are used for backfill it is possible to fill and compact the voids left by the wood sheeting if the material is placed in lifts and jetted as the sheeting is pulled. If cohesive materials are used for backfill, a void will be left after pulling the wood sheets and the full weight of the prism of earth contained between the sheeting will come to bear on the sewer pipe.

Skeleton sheeting or bracing should be cut off and left in place to an elevation of 1.5' over the top of the sewer pipe if removal of the trench support might cause a collapse of the trench wall and a widening of the trench at the top of the conduit. Entire skeleton sheeting systems should be left in place if removal would cause collapse of the trench before backfill can be placed.

Where steel soldier beams with horizontal lagging between the beam flanges are used for sheeting trenches, efforts to reclaim the steel beams before the trench is backfilled may damage pipe joints. It is recommended that use of this type of sheeting be allowed only on stipulation that the beams be pulled after backfilling and the lagging be left in place.

Steel sheeting may be used and reused many times, and the relative economy of this type of sheeting compared with timber or timber and soldier beams should be explored. Because of the thinness of the sheeting, it is often feasible to achieve reasonable compaction of backfill so that the steel sheeting may be withdrawn with about the same factor of safety against settlement of the surfaces adjacent to the trench as that for other types of sheeting left in place.

Trench Boxes Several backfilling techniques are possible when a trench box is used. Granular material can be placed between the box and the trench wall immediately after placing the box, and the box is advanced by lifting slightly before moving forward. Boxes can be made with a step at the rear, which makes the trench wall accessible for compacting the embedment against the walls. Also, the box can be used where the sewer pipe is laid in a small sub-trench, and the box is used only in the trench above the top of the sewer pipe. Advancement of the box must be done carefully to avoid pulling pipe joints apart.

Pipe Bedding and Embedment To assure that the sewer pipe is properly bedded or embedded it is suggested that compaction tests be made at selected or critical locations, or that the method of material placement be observed and correlated with known results.

Where compaction measurement or control is desired or required, the recommended references are:

(a) ASTM D2049, "Standard Method for Test for Relative Density of Cohesionless Soils."

(b) ASTM D698, "Standard Method of Test for Moisture Density Relations of Soils Using 5.5-lb (2.5 kg) Rammer and 12-inch (204.8 mm) Drop."

(c) ASTM D2167, "Standard Method of Test for Density of Soil in Place by the Rubber-Balloon Method."

(d) ASTM D1556, "Standard Method of Test for Density of Soil in Place by the Sand-Cone Method."

(e) ASTM D2922, "Standard Method of Test of Density of Soil and Soil-Aggregate in Place by Nuclear Methods (Shallow Depth)."

It is recommended that the in-place density of Class I and Class II embedment materials be measured by ASTM D2049 by percentage of relative density, and Class III and Class IV measured by either ASTM D2167, D1556, or D2922, by percentage of Standard Proctor Density according to ASTM D698 or AASHTO T99.

F. Rigid Sewer Pipe Design

1. General Relationships

The inherent strength of a rigid sewer pipe usually is given by its strength in the three-edge bearing test. Although this test is both convenient and severe, it does not reproduce the actual field load conditions. Thus to select the most economical combination of bedding and sewer pipe strength, a relationship must be established between calculated load, laboratory strength, and field strength for various installation conditions (Figure 14.30).

Field strength, moreover, depends on the distribution of the reaction against the bottom of the sewer pipe and on the magnitude and distribution of the lateral pressure acting on the sides of the pipe. These factors, therefore, make it necessary to qualify the term "field strength" with a description of conditions of installation in a particular case, as they affect the distribution of the reaction and the magnitude and distribution of lateral pressure.

Just as for sewer pipe load computations, it is convenient when determining field strength to classify installation conditions as either trench or embankment.

2. Laboratory Strength

Rigid sewer pipe is tested for strength in the laboratory by the three-edge bearing test. Methods of testing are described in detail in ASTM Specification C 301 for vitrified clay sewer pipe and C 497 for concrete and reinforced concrete sewer pipe.

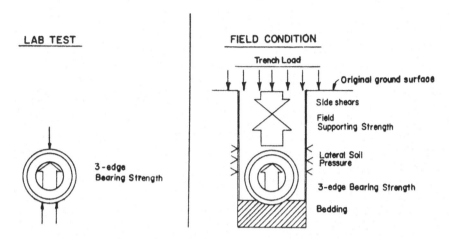

Figure 14.30.—*Both laboratory testing and field conditions should be used for rigid sewer pipe strength determination in the indirect design method.*

The minimum strengths required for the three-edge bearing tests of the various types of rigid sewer pipe are stated in the ASTM Specifications for the pipe.

In the case of reinforced concrete sewer pipe, laboratory strengths are divided into two categories: The load that will produce a 0.01″ crack, and the ultimate load the sewer pipe will withstand. The ultimate strength load in three-edge bearing is 1.50–1.25 times greater than the load that will produce a 0.01-in. crack, and the factor of safety used in design depends on which pipe class is being considered.

Non-reinforced concrete, and clay sewer pipe class designations indicate the ultimate strength in three-edge bearing directly in pounds per foot for any diameter.

3. *Design Relationships*

The structural design of rigid sewer pipe systems relates to the product's performance limit, expressed in terms of strength of the installed sewer pipe. Based on anticipated field loadings and concrete and reinforcing steel properties, calculations of shear, thrust, and moment can be made and compared to the ultimate strength of the section. This method is commonly used for reinforced concrete cast-in-place sewer pipes, and for precast concrete pipe using the direct design method.

For indirect design precast sewer pipes, the design strength is commonly related to a three-edge bearing test strength measured at the manufacturing plant, as described in the following text. The plant test is much more severe because maximum shearing forces occur at the point of maximum moment. Special reinforcement to resist shear in testing may be required by the more uniformly distributed field loading.

In either case, the design load is equal to the field strength divided by the factor of safety. The design load is the calculated load on the pipe, and the field strength is the ultimate load the pipe must support when installed under specified conditions of bedding and backfilling.

The field strength is equal to the three-edge bearing test strength of the pipe times the bedding factor. The ratio of the strength of a sewer pipe under any stated condition of loading and bedding to its strength measured by the three-edge bearing test is called the bedding factor.

Bedding factors for trenches and embankments were determined experimentally between the years 1925 and 1935 at Iowa State College (Spangler 1956). The required three-edge bearing strength for a given class of bedding can be calculated as follows:

$$\frac{\text{Required Three-Edge}}{\text{Bearing Strength}} = \frac{\text{Design Load} \times \text{Safety Factor}}{\text{Bedding Factor}}$$

The strength of reinforced concrete sewer pipe in pounds per foot at either the 0.01-inch crack or ultimate load, divided by the nominal internal diameter of the sewer pipe, is defined as the D-load strength: Different classes of reinforced concrete sewer pipe per ASTM C 76 and other reinforced concrete pipe specifications are defined by D-load strengths:

$$\text{D-Load} = \frac{\text{Design Load} \times \text{Safety Factor}}{\text{Bedding Factor} \times \text{Diameter}}$$

In considering the design requirements, the design engineer should evaluate the options of different types of rigid sewer pipe and different bedding classes, keeping in mind the differences in the relationship between test loadings and field loadings.

4. Rigid Sewer Pipe Installation—Classes of Bedding and Bedding Factors for Trench Conditions

Four classes of beddings (Figure 14.31) are used most often for sewer pipes in trenches (Moser et al. 1977; Sikora 1980). They are described in the following subsections.

Class A—Concrete Cradle   The sewer pipe is bedded in a cast-in-place cradle of plain or reinforced concrete having a thickness equal to one-fourth the inside pipe diameter, with a minimum of 4 inches and a maximum of 15 inches under the pipe barrel and extending up the sides for a distance equal to one-fourth the outside diameter. The cradle shall have a width at least equal to the outside diameter of the sewer pipe barrel plus 8 inches. Construction procedures must be executed carefully to prevent the sewer pipe from floating off line and grade during placement of the cradle concrete.

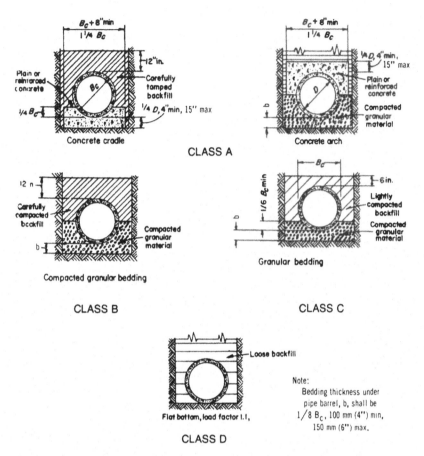

Figure 14.31.—Classes of bedding for rigid sewer pipes in trench.

If the cradle is made of reinforced concrete, the reinforcement is placed transverse to the pipe and 3 inches clear from the bottom of the cradle. The percentage of reinforcement, p, is the ratio of the area of transverse reinforcement to the area of concrete cradle at the pipe invert above the centerline of the reinforcement.

Consideration must be given to the points at which the cradle (or arch, in the next section) begins and terminates with respect to the pipe joints. In general, the concrete cradle, or envelope should start and terminate at the face of a pipe bell or collar to avoid shear cracks.

Haunching and initial backfill above the cradle to 12″ above the crown of the sewer pipe should be placed and compacted as described earlier in this chapter. The cradle must be cured sufficiently to develop full bedding prior to final backfilling.

The bedding factor for Class A concrete cradle bedding is 2.2 for plain concrete with lightly tamped backfill; 2.8 for plain concrete with carefully tamped backfill; up to 3.4 for reinforced concrete with the percentage of reinforcement, p, equal to 0.4%, and up to 4.8% with p equal to 1.0%.

Class A—Concrete Arch   The sewer pipe is bedded in carefully compacted granular material having a minimum thickness of one-eighth the outside diameter but not less than 4" or more than 6" between the sewer pipe barrel and bottom of the trench excavation. Granular material is then placed to the spring line of the sewer pipe and across the full breadth of the trench. The haunching material beneath the sides of the arch must be compacted so as to be unyielding. Crushed stone in the 0.25–0.75" size range is the preferred material. The top half of the sewer pipe is covered with a cast-in-place plain or reinforced concrete arch having a minimum thickness of 4" or one-fourth the inside pipe diameter (but not to exceed 15 inches), and having a minimum width equal to the outside sewer pipe diameter plus 8".

If the arch is made of reinforced concrete, the reinforcement is placed transverse to the pipe and 2" clear from the top of the arch. The percentage of reinforcement is the ratio of the transverse reinforcement to the area of concrete arch above the top of the pipe and below the centerline of the reinforcement.

Class B Bedding   The sewer pipe is bedded in carefully compacted granular material. The granular bedding has a minimum thickness of one-eighth the outside sewer pipe diameter, but not less than 4" or more than 6", between the barrel and the trench bottom, and covering the full width of the trench.

The haunch area of the sewer pipe must be fully supported; therefore, the granular material should be shovel sliced or otherwise compacted under the pipe haunch. Both haunching and initial backfill to a minimum depth of 12" over the top of the sewer pipe should be placed and compacted.

The bedding factor for Class B bedding is 1.9.

Class C Bedding   The sewer pipe is bedded on compacted granular material. The bedding has a minimum thickness of one-eighth the outside sewer pipe diameter, but not less than 4" or more than 6", and shall extend up the sides of the sewer pipe one-sixth of the pipe outside diameter. The remainder of the sidefills, to a minimum depth of 6" over the top of the pipe, consists of lightly compacted backfill.

The bedding factor for Class C bedding is 1.5.

Class D Bedding   For this class of bedding, the bottom of the trench is left flat, as cut by excavating equipment. Successful Class D installations of rigid conduits can be achieved in locations with appropriate soil conditions, trench width control, and light superimposed loads. Care must be taken to prevent point loading of sewer pipe bells; the excavation of bell holes will prevent such loading. Existing soil should be shovel sliced or otherwise compacted under the haunching of the sewer pipe to provide some uniform support. In poor soils, granular bedding material is generally a more practical and cost effective installation.

Since the field conditions with Class D bedding can approach the load conditions for the three-edge bearing test, the bedding factor for Class D beddings is 1.1.

5. Variable Bedding Factors for Trench Conditions

Both Spangler (1941) and Schlick (1932) postulate that some active lateral pressure is developed in trench installations before the transition width is reached. Experience indicates that the active lateral pressure increases as the trench width increases from a very narrow width to the transition width, provided the sidefill is compacted. Defining the narrow trench width as a trench having a width at the top of the pipe equal to or less than the outside horizontal span plus one foot, and assuming a conservative linear variation, the variable trench bedding factor can be determined by:

$$B_{fv} = (B_{fe} - B_{ft}) \left[\frac{B_d - (B_c + 1.0)}{B_{dt} - (B_c + 1.0)} \right] + B_{ft} \qquad (14\text{-}30)$$

in which B_c is the outside horizontal span of pipe, in feet; B_d is the trench width at top of pipe; B_{dt} is the transition width at top of pipe; B_{fe} is the bedding factor, embankment; B_{ft} is the fixed bedding factor, trench; and B_{fv} is the variable bedding factor, trench.

A six-step design procedure for determining the trench variable bedding factor is:

(a) Determine the trench fixed bedding factor, B_{ft}.
(b) Determine the trench width, B_d.
(c) Determine the transition width for the installation conditions, B_{dt}.
(d) Determine H/B_c ratio, settlement ratio, r_{sd}, projection ratio, p, and the product of the settlement and projection ratios, $r_{sd}p$.
(e) Determine positive projecting embankment bedding factor, B_{fe}.
(f) Calculate the trench variable bedding factor, B_{fv}.

6. Encased Pipe

Total encasement of rigid sewer pipe in concrete may be necessary where the required field strength cannot be obtained by other bedding or installation methods.

A typical concrete encasement detail is shown in Figure 14.32. The bedding factor for concrete encasement varies with the thickness of concrete and the use of reinforcement, and may be greater than that for a concrete cradle or arch. Concrete thickness and reinforcement should be determined by the application of conventional structural theory and analysis. The bedding factor for the encasement shown in Figure 14.32 is 4.5.

Concrete encasement also may be required for sewers built in deep trenches to ensure uniform support, or for sewer pipe built on comparatively steep grades where there is the possibility that earth beddings may be eroded by currents of water under and around the pipe. Flotation of the pipe during concrete placement should be prevented.

7. Field Strength in Embankments

The active soil pressure against the sides of rigid sewer pipe placed in an embankment can be a significant factor in the resistance of the structure to vertical load. This factor is important enough to justify a separate examination of the field strength of embankment sewer pipe.

The following discussion of field strength in embankments is based on a theory developed by Marston and Spangler. The design engineer, however, should approach design based on this theory with some caution. With time, lateral pressures for trench installation usually will approach "at rest" conditions, which correspond to vertical overburden times a factor (usually between 0.5 and 1.0). However, for a negative projecting conduit, a positive projecting conduit, and induced trench embankment conditions, the lateral pressure magnitude and distribu-

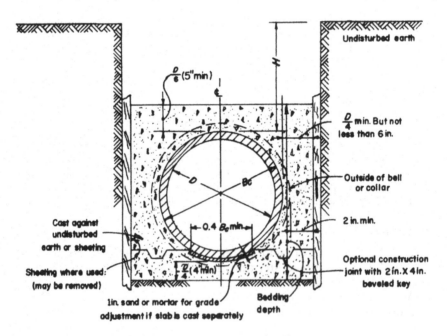

Figure 14.32.—Typical concrete encasement details.

tion may be much different, and these may control the structural design of the sewer pipe.

Positive Projecting Sewer Pipe   The bedding factor for rigid sewer pipes installed as projecting sewer pipe under embankments or in wide trenches depends on the bedding in which the sewer pipe is laid, the magnitude of the active lateral soil pressure against the sides of the sewer pipe, and the area of the sewer pipe over which the active lateral pressure is effective.

For projecting sewer pipe, the bedding factor B_f, is:

$$B_f = \frac{A}{N - xq} \tag{14-31}$$

in which A is a sewer pipe shape factor; N is a parameter that is a function of the bedding class; x is a parameter dependent on the area over which lateral pressure effectively acts; and q is the ratio of total lateral pressure to total vertical load on the sewer pipe.

Classes of bedding for projecting sewer pipe are shown in Figure 14.33. The values of A for circular, elliptical, and arch sewer pipe are shown in Table 14.15.

Values of N for various classes of bedding are given in Table 14.16. Values of x for circular, elliptical and arch sewer pipe are listed in Table 14.17. The projection ratio, m, refers to the fraction of the sewer pipe diameter over which lateral pressure is effective. For example, if lateral pressure acts on the top half of the sewer pipe above the horizontal diameter, m equals 0.5. The ratio of total lateral pressure to total vertical load, q, for positive projecting sewer pipe may be estimated by the formula:

$$q = \frac{m\kappa}{C_c}\left(\frac{H}{B_c} + \frac{m}{2}\right) \tag{14-32}$$

in which κ is the ratio of unit lateral pressure to unit vertical pressure (Rankine's ratio). A value of κ equal to 0.33 usually will be sufficiently accurate for use in Equation (14-32). Values of C_c are found in Figure 14.10.

Negative Projecting Sewer Pipe   The bedding factor for negative projecting sewer pipe may be the same as that for trench conditions corresponding to the various classes of bedding given in this chapter. The bedding factors for Class B, C, and D trench bedding do not take into account lateral pressures against the sides of the sewer pipe. However, in the case of negative projecting sewer pipe, it may be possible to compact the sidefill soils to the extent that some lateral pressure against the sewer pipe can be relied on. If such favorable conditions are anticipated, it is suggested that the bedding factor be computed by means

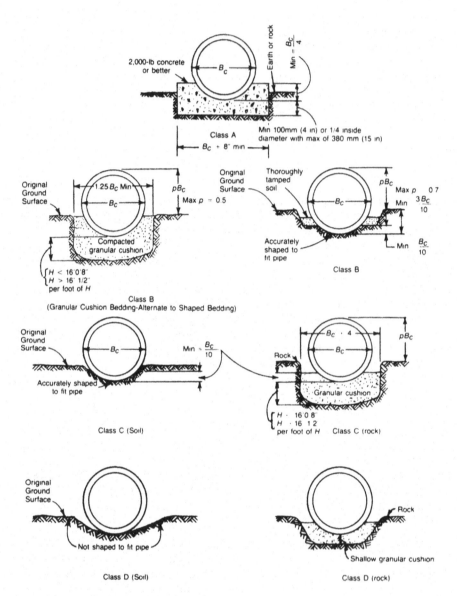

Figure 14.33.—Classes of bedding for projecting sewer pipes.

TABLE 14.15.   Values of A for Circular, Elliptical, and Arch Sewer
Pipe (ASCE 1982).

Sewer Pipe Shape (1)	A (2)
Circular	1.431
Elliptical:	
Horizontal elliptical and arch	1.337
Vertical elliptical	1.021

TABLE 14.16.   Values of N (ASCE 1982).

	Value of N		
	Sewer Pipe Shape		
Class of Bedding (1)	Circular (2)	Horizontal Elliptical (3)	Vertical Elliptical (4)
A (reinforced cradle)	0.421 to 0.505	—	—
A (unreinforced cradle)	0.505 to 0.636	—	—
B	0.707	0.630	0.516
C	0.840	0.763	0.615
D	1.310	—	—

TABLE 14.17.   Values of x (ASCE 1982).

	x			
	Class A Bedding	Other than Class A Bedding		
Fraction of Sewer Pipe Subjected to Lateral Pressure, m (1)	Circular Sewer Pipe (2)	Circular Sewer Pipe (3)	Horizontal Elliptical Sewer Pipe (4)	Vertical Elliptical Sewer Pipe (5)
0	0.150	0	0	0
0.3	0.743	0.217	0.146	0.238
0.5	0.856	0.423	0.268	0.457
0.7	0.811	0.594	0.369	0.639
0.9	0.678	0.655	0.421	0.718
1.0	0.638	0.638	—	—

of Equations (14-31) and (14-32), using a value of κ equal to 0.15 for estimating the lateral pressure on the sewer pipe.

Induced Trench Conditions   Induced trench sewer pipes usually are installed as positive projecting sewer pipes before the overlying soil is compacted and the induced trench is excavated. Therefore, lateral pressures are effective against the sides of the sewer pipe, and the bedding factor should be calculated using Equations (14-31) and (14-32).

G. Flexible Sewer Pipe Design

Several types of flexible pipe are available for use as sewer pipe material. Among the most common are coated corrugated metal pipe (CMP), acrylonitrile-butadiene-styrene (ABS) composite pipe and (ABS) solid wall pipe, polyvinyl chloride (PVC) pipe, polyethylene (PE) pipe, and fiberglass-reinforced plastics (FRP).

1. General Method

Flexible sewer pipes under earth fills and in trenches derive their ability to support load from their inherent strength plus the passive resistance of the soil as the pipe deflects and the sides of sewer pipe move outward against the soil sidefills. Proper compaction of the soil sidefills is important to the long term structural performance of flexible sewer pipe. The extent to which flexible pipe deflects is most commonly used to judge performance and as a basis for design. The amount of deflection considered permissible is dependent on physical properties of the pipe material used and project limitations.

The limiting buckling stress for flexible pipes takes into account the restraining effect of the soil structure around the pipe and the properties of the pipe wall. Equations for the critical stress in the pipe wall can be found in manufacturers' handbooks for the various types of pipe.

The approximate long-term deflection of flexible sewer pipe can be calculated using the Modified Iowa Formula developed by Spangler and Watkins, provided the pipe to soil stiffness relationship produces an elliptical deformation:

$$\Delta X = \frac{D_t K_b W_e r^3}{EI + 0.061\ E'r^3} \qquad (14\text{-}33)$$

where

ΔX = horizontal deflection, in inches;
K_b = bedding factor;
D_L = deflection lag factor;
W_e = load, in pounds per linear inch;
r = mean radius of pipe, in inches;
E = modulus of tensile elasticity, in pounds per square inch;

I = moment of inertia per length, in inches to the fourth power per inch; and

E' = modulus of soil reaction, in pounds per square inch.

For small deflections, the vertical deflection ΔY may be assumed to approximately equal the horizontal deflection ΔX in Equation (14-33). Research has been conducted on the application of Equation (14-33) to pipe having a low ratio of pipe stiffness to soil stiffness where the vertical deflection may not be assumed to equal the horizontal deflection. For low pipe to soil stiffness ratios, a correction factor must be applied to the calculated horizontal deflection to accurately predict the vertical deflection (Zicaro 1990).

Flexible sewer pipes that are to support a fill should not be placed directly on a cradle or pile bents. If such supports are necessary, they should have a flat top and be covered with a compressible earth cushion. In those instances where flexible pipe is to be encased in concrete, the pipe manufacturer should be consulted.

The deflection lag factor, empirically determined, compensates for the time consolidation characteristics of the soil, which may deform flexible sewer pipes for some period after installation. Long-term deflection will be greater with light or moderate degrees of compaction of sidefills when compared to values for heavy compaction. The better the compaction, the lower the initial deflection, and the greater the magnitude of the long-term lag factor. Lag factors over 2.5 have been recorded in dry soil. Recommended values of this factor range from 1.25–2.50. Bedding requirements for flexible sewer pipe installations are discussed below.

Values of the bedding constant K_b, depending on the width of the sewer pipe bedding, are shown in Table 14.18.

The passive resistance of the soil at the sides of the pipe greatly influences flexible pipe deflection. This passive resistance is expressed as the Modulus of Soil Reaction, E'. It is an empirical number related to the degree of compaction of the soil and to the type of soil. The U.S.

TABLE 14.18.   Values of Bedding Constant, K_b (ASCE 1982).

Bedding Angle, in degrees (1)	K_b (2)
0	0.110
30	0.108
45	0.105
60	0.102
90	0.096
120	0.090
180	0.083

Note: 1 deg = 0.017 rad.

Bureau of Reclamation (Howard 1977) has established an empirical relationship between Modulus of Soil Reaction values, degree of compaction of bedding, and type of bedding material. These values are shown in Table 14.19. (See Table 14.14 for soil symbol definition.)

In the deflection formula (Equation (14-33)) the first term in the denominator, EI (stiffness factor), reflects the influence of the inherent stiffness of the sewer pipe on deflection. The second term, $0.061\,E'r^3$, reflects the influence of the passive pressure on the sides of the pipe. The second term may predominate in the case of large-diameter pipes, with the result that a very lightweight pipe may appear to be satisfactory. Since the pipe wall must have sufficient local strength in bending and thrust to develop and utilize the passive resistance pressure on the sides of the pipe, it is recommended as a practical measure that the value of EI should never be less than about 10–15% of the term $0.061\,E'r^3$.

It should be noted that the E' values in Table 14.19 are average values and do not reflect field variables. A conservative approach would be to use 75% of the E' values given to calculate maximum deflections.

The ability of a flexible sewer pipe to retain its shape and integrity is largely dependent on the selection, placement, and compaction of the envelope of soil surrounding the structure. For this reason as much care should be taken in the design of the bedding and initial backfill as is used in the design of the sewer pipe. The backfill material selected preferably should be of a granular nature to provide good shear characteristics. Cohesive soils are generally less suitable because of the importance of proper moisture content and the difficulty of obtaining proper compaction in a limited work space.

If, under embankment conditions, the material placed around the sewer pipe is different from that used in the embankment, or if for construction reasons fill is placed around the sewer pipe before the embankment is built, the compacted backfill should cover the pipe by at least 1 foot.

Design Relationships   Structural design of flexible pipe requires definition of the critical deflection limit for the specific pipe considered. The critical deflection limit for flexible pipe is commonly based on structural performance characteristics and potential modes of failure. The requirements of serviceability, such as effect on carrying capacity, passage of cleaning equipment, velocity and differential deflection at fittings, manholes and joints, should be considered. Maximum long-term strain in the pipe wall, discussed above, also should be considered. The design basis is expressed as follows:

$$\text{Design Deflection Limit} = \frac{\text{Allowable Deflection Limit}}{\text{Factor of Safety}}$$

TABLE 14.19.   Bureau of Reclamation Average Values of E' for Iowa Formula (for Initial Flexible Pipe Deflection) (ASCE 1982).

Soil Type-Pipe Bedding Material (Unified Classification System[a]) (1)	E' for Degree of Compaction of Bedding, in pounds per square inch			
	Dumped (2)	Slight, <85%, Proctor, <40% Relative Density (3)	Moderate, 85%–95% Proctor, 40%–70% Relative Density (4)	High >95% Proctor, >70% Relative Density (5)
Fine-grained Soils (LL > 50)[b] Soils with medium to high plasticity CH, MH, CH-MH	No data available; consult a competent soils engineer: Otherwise use $E' = 0$			
Fine-grained Soils (LL < 50) Soils with medium to no plasticity CL, ML, ML-CL, with less than 25% coarse-grained particles	50	200	400	1,000
Fine-grained Soils (LL < 50) Soils with medium to no plasticity CL, ML, ML-CL, with more than 25% coarse-grained particles Coarse-grained Soils with Fines GM, GC, SM, SC[c] contains more than 12% fines	100	400	1,000	2,000
Coarse-grained Soils with Little or No Fines GW, GP,SW,SP[c] contains less than 12% fines	200	1,000	2,000	3,000
Crushed Rock	1,000	3,000	3,000	3,000
Accuracy in Terms of Percentage Deflection[d]	±2	±2	±1	±0.5

[a]ASTM Designation D-2487, USBR Designation E-3.

[b]LL = Liquid limit.

[c]Or any borderline soil beginning with one of these symbols (i.e., GM-GC, GC-SC).

[d]For ±1% accuracy and predicted deflection of 3%, actual deflection would be between 2% and 4%.

Note: Values applicable only for fills less than 50 ft (15 m). Table does not include any safety factor. For use in predicting initial deflections only, appropriate Deflection Lag Factor must be applied for long-term deflections. If bedding falls on the borderline between two compaction categories, select lower E' value or average the two values. Percentage Proctor based on laboratory maximum dry density from test standards using about 12,500 ft-lb/cu ft (598,000 J/m^3) (ASTM D-698, AASHO T-99, USBR Designation E-11). 1 psi = 6.9 KPa.

In structural design for a flexible sewer pipe system, the design deflection limit should not be more than the maximum long-term deflection anticipated under load. In the determination of projected long-term deflection, the design element considerations are load (soil, superimposed dead load, and superimposed live load), pipe stiffness and soil stiffness. The primary controlling elements of deflection are the load, the pipe stiffness, the soil stiffness and the ratio of pipe to soil stiffness. Typical allowable long-term deflection limits are:

- Factory cement-mortar-lined and/or cement-mortar-coated steel pipe   3%
- Flexible lined and coated steel pipe   5%
- Flexible coated and cement-mortar-lined in-place steel pipe   5%
- Fiberglass pressure pipe   5%
- Thermoplastic pipe (PVC)   5%

The soil stiffness depends on the type of soil and its density as placed in the sewer pipe embedment zone. In consideration of design requirements, the engineer should evaluate the options of different types of flexible sewer pipes, different embedment soils, and different embedment material compaction and placement requirements.

The choice of factor of safety should be influenced by soil characteristics, the degree of compaction likely to be obtained, and available field tests, and practical experience.

Loads on Flexible Pipe The load carried by a buried flexible pipe in a narrow trench may be calculated using the Marston Formula, Equation (14-13). A conservative design approach may be used by assuming the dead load carried by a flexible pipe-soil system in any installation to be the prism load. For normal installations the prism load is the maximum load that can be developed.

The load on a projecting flexible sewer pipe is calculated using Equation (14-14). As before, the load coefficient (C_c) depends on the projection ratio (p), settlement ratio r_{sd}, and the ratio of fill height to pipe width (H/B_c), and can be determined from Figure 14.10. For flexible projecting pipe, the product $r_{sd}p$ is negative or zero. As shown in Figure 14.10, when the product is zero, the load coefficient, C_c, equals H/B_c. Equation (14-14) then becomes Equation (14-16), which gives the prism load.

2. Design of Plastic Sewer Pipe

Thermoplastic pipe materials (ABS, PE, and PVC) are all affected by temperature. Tests are usually specified to be conducted at 23° C (73.4° F). At higher temperatures the pipe stiffness is decreased, and lower temperatures result in greater pipe stiffness. Pipe being installed at high and low temperatures requires careful handling.

Laboratory Load Test The standard test to determine "pipe stiffness" or load deflection characteristics of plastic pipe is the parallel-plate loading test. The test is conducted in accordance with ASTM C 2412, "Standard Test Method for External Loading Properties of Plastic Pipe by Parallel-Plate Loading."

In the test, a short length of pipe is loaded between two rigid parallel flat plates that are moved together at a controlled rate. Load and deflection are noted.

The parallel-plate loading test determines the pipe stiffness (PS) at a prescribed deflection (ΔY), which for convenience in testing is arbitrarily set at 5%. This is not to be considered the field deflection limitation. The pipe stiffness is defined as the value obtained by dividing the force (F) per unit length by the resulting deflection in the same units at the prescribed percentage deflection, and is expressed in pounds per inch:

$$PS = \frac{F}{\Delta Y} = \frac{EI}{0.149r^3} \qquad (14\text{-}34)$$

where

E = modulus of elasticity in pounds per square foot;
I = $t^3/12$;
r = mean radius of pipe; and
t = wall thickness.

Minimum required pipe stiffness values are stated in plastic sewer pipe specifications. Table 14.20 lists the ASTM Specifications for various types of plastic pipe and the corresponding pipe stiffness values.

The stiffness factor (SF) is the pipe stiffness multiplied by the quantity $0.149r^3$:

$$SF = EI = \frac{F}{\Delta Y} 0.149r^3 = 0.149r^3 (PS) \qquad (14\text{-}35)$$

The stiffness factor or EI is used in the Modified Iowa Formula (Equation 14-33) to determine approximate field deflections under earth loads. It is the engineer's responsibility to establish the acceptable field deflection limit, and to design the installation accordingly. The manufacturer should be consulted for recommended field installation deflection limits.

Field Deflection of Flexible Plastic Pipe As previously discussed, the pipe stiffness is determined in a standard parallel-plate loading test for an arbitrary deflection of the nominal diameter. This is a measure of the inherent strength of plastic pipe.

The pipe stiffness and soil stiffness as measured by E' must develop sufficient field strength that the deflection of the pipe under load will not exceed the acceptable deflection.

Approximate values of horizontal deflections for field installations can be calculated using the Modified Iowa Formula (Equation (14-33)).

TABLE 14.20.   Stiffness Requirements for Plastic Sewer Pipe Parallel-Plate Loading* (ASCE 1982).

Material** (1)	ASTM Specification (2)	Nominal Diameter, d, in inches (3)	Required Stiffness at 5% Deflection, in pounds per square inch (4)
ABS Composite	D 2680	8–15	200
ABS Plain	D 2751		
	SDR 23.5	4 & 6	150
	SDR 35	3	50
		4 & 6	45
	SDR 42	8, 10 & 12	20
RPM	D 3262	8–18	Varies (99–17)
		20–108	10
PVC	D 2729	2	59
	(PVC-12454)	3	19
		4	11
		5	9
		6	8
	D 2729	2	74
	(PVC-13364)	3	24
		4	13
		5	12
		6	10
	D 3033		
	SDR 41	6–15	28
	SDR 35	4–15	46
	D 3034		
	SDR 41	6–15	28
	SDR 35	4–15	46

*ASTM 2412
**Other plastic pipe materials are not listed, in that insufficient data is currently available.
Note: 1 in. = 25.4 mm; 1 psi = 6.89 kPa.

A correction factor must be applied to the calculated horizontal deflection to accurately predict the vertical deflection for low pipe to soil stiffness ratios (Zicaro 1990). This formula can be simplified to permit a calculation of approximate deflection based on pipe stiffness as follows:

$$\Delta X = \frac{D_L K_b W_c}{0.149PS + 0.061E'} \qquad (14\text{-}36)$$

Values of the bedding constant for use in this formula were presented earlier in the "General Methods" subsection of this section.

The solution of Equations (14-33) and (14-36) requires that various factors be determined. It is desirable, where possible, to establish anticipated long-term field deflections based on well-documented empirical data.

3. Soil Classification

For different categories of embedment materials, different construction procedures are specified. Soil classifications under the Unified Soil Classification System, including manufactured materials, are grouped into five broad categories according to their ability to develop an interacting sewer pipe-soil system (Table 14.14).

For the larger-grained soils (Class I and Class II gravels), compatibility with the existing subgrade and trench side soils should be considered. Particularly for uniformly graded or gap graded materials, the potential exists for the migration of the finer fraction of the existing soils into the embedment materials, with resultant settlement and loss of side support for the sewer pipe. Analytical techniques for assessing this possibility are similar to those for filter blankets and are well covered in the references. See Table 14.19 for the suggested E' values given for each soil class.

Class I This class includes angular, 0.25–1", graded stone, including a number of fill materials that have regional significance, such as coral, clay, cinders, crushed stone, and crushed shells.

Class I material provides the best material for the construction of a stable sewer pipe-soil system. When used for underdraining, Class I material should be placed to the top of the sewer pipe.

Class I material used for haunching and initial backfill, when dumped into place with little or no compaction, will produce an average E' value of 1,000 psi. Care must be taken to place material under the haunches and in contact with the sides of the pipe. Class I material compacted to 85% Standard Proctor Density or higher has an average E' value of 3,000 psi.

Class II This class comprises coarse sands and gravels with maximum particle size of 1.5", including variously graded sands and gravels containing small percentages of fines, generally granular and non-cohesive, either wet or dry. Soil types GW, GP, SW, and SP are included.

Use of sands of Type SP should be done with caution. Poorly graded fine sands with little material finer than that passing a 200-sieve have a tendency to flow when wet.

Class II material used for haunching and initial backfill, when dumped into place with little or no compaction, will product an average E' value of 200 psi. Compacted to 85% Standard Proctor Density, the average E' value is raised to 1,000 psi. Higher E' values are obtained with greater compaction.

Class III This class comprises fine sand and clayey gravels, including fine sands, sand-clay mixtures, and gravel-clay mixtures. Soil types GM, GC, SM and SC are included.

Class III materials dumped in place for haunching and initial backfill will produce an average E' of 100 psi. Careful placement and compaction to 90% Standard Proctor Density will produce an average E' of 1,000 psi.

Class IV and V   Class IV materials require special effort for compaction, thus may be suitable for sewer pipe foundation if special care is taken during excavation to provide a uniform, undisturbed trench bottom. Use of Class IV materials for bedding, haunching, or initial backfilling is not recommended.

Class V materials present special problems in providing an adequate foundation, and should not be used for any part of the sewer pipe envelope.

4. Design of Corrugated Metal Sewer Pipes

Corrugated metal pipe is manufactured in a variety of gages, corrugation depths, and corrugation spacings.

The longitudinal seam formed by bolting or riveting curved sheets together for corrugated metal pipe should be checked for crushing strength. Tables of seam strengths for various metal gages and bolt or rivet sizes and spacing can be found in manufacturers' handbooks. To improve hydraulic properties, the pipe may have a paved invert (25% of circumference) or may have the interior completely paved.

Corrugated metal pipe sewers may be designed for a limiting deflection using the Modified Iowa Formula (Equation (14-33)) or by the manufacturer's handbook figures. A design maximum deflection limitation of 5% is commonly used.

Corrugated metal sewers that are to support a fill should not be placed directly on a cradle or pile bents. If such supports are necessary, they should have a flat top and be covered with a compressible earth cushion. Corrugated metal should not be encased in concrete. For corrugated metal pipes installed in trenches, reference is made to manufacturers' handbooks for recommended gages and corrugations.

5. Flexible Sewer Pipe Installation

General bedding and backfilling concepts developed for rigid sewer pipe may also be applied to flexible pipe, with some exceptions. The general pipe terminology for installation was shown in Figure 14.29

Detailed information on bedding of flexible pipe is contained in the various ANSI and ASTM specifications or manufacturers' literature. The design engineer should consult the applicable specification or literature for design information.

Bedding, Haunching and Initial Backfill   The bedding requirements for flexible thermoplastic sewer pipe are given in ASTM D 2321 for

Class I, II, and III material (Table 14.14). Haunching and initial backfill requirements are also given, including minimum compaction recommendations. Similarly, bedding and initial backfill requirements for thermosetting reinforced plastic pipe are given in ASTM D 3839, and a guide for estimating both initial and long-term deflection is included in the appendices. Installation requirements for corrugated metal pipe are given in ASTM A 798, which describes procedures, soils and soil placement.

By referring to Tables 14.18 and 14.19, it is possible to estimate the E' value obtained for the expected compaction, or to determine the compaction requirements needed to develop the E' value that will keep the pipe deflection within allowable limits.

Final Backfill The procedure for installing final backfill for the remainder of the trench is the same as for rigid pipe.

IX. DESIGN OF OTHER SPECIFIC STRUCTURES

A. Open Channel Linings

Open channels are widely used in drainage and flood control projects, and often require linings that will remain stable during the flows the channels are designed to carry. Channel linings include bare earth, various forms of grasses and plants, riprap, grouted riprap, gabions, soil cement and concrete. Except for grouted rock, all were discussed in detail in Chapter 9.

1. Grouted Rock

Riprap channel linings can be grouted together to form a more rigid surface. The grouted riprap requires a lesser thickness of rock and reduces maintenance. However, the grouting results in the loss of several of the advantages of loose riprap. The lining's impermeability prevents groundwater pressures from being released between the rocks. The flow then concentrates at openings through the lining, with the potential of erosion of the underlying soil. The rigidity of the lining makes it more susceptible to failure if it is undermined by erosion or piping.

Grouted riprap channel linings form retaining walls and are loaded by lateral earth forces and groundwater pressures. As with riprap linings, a geotechnical engineer should be consulted regarding stable slopes which will not load the walls excessively.

2. Concrete

Channel linings are frequently constructed of reinforced concrete. The steeper the side slopes, the more the wall serves as a retaining wall. Reinforced concrete retaining walls can serve as channel lining in two different structural configurations—as independent retaining walls or attached to a floor slab. These typical designs are illustrated in Figure 14.34.

Each wall can be an independent cantilever retaining wall with its own footing. In such a case, the channel bottom would not be connected to the walls, and thus could be concrete, riprap, or earth. Each wall would be structurally designed to resist lateral earth and groundwater pressures, and would require analyses of stability against overturning and sliding, but would be little affected by uplift (except as it reduces the weight and thus friction resistance).

Reinforced concrete retaining walls can also be rigidly attached to the channel floor to form a rectangular or U-shaped channel. In this con-

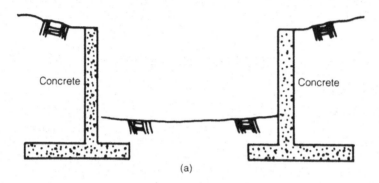

(a)

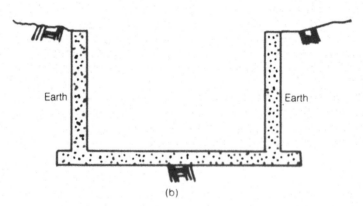

(b)

Figure 14.34. —*Alternate approaches to channel design with vertical concrete retaining walls.*

figuration, the same lateral soil and groundwater pressures are exerted on the walls, but the structure is subjected to little or no net overturning or sliding forces. The bending moment from the walls is transferred to the floor, which must also be a structural member. This increases the bending in the floor slab. These channels are subject to uplift forces and frequently require special attention to insure that flotation will not occur.

3. Soil Cement and Roller Compacted Concrete

These two materials are used for channel linings primarily to prevent bank erosion. They are generally designed to perform as gravity retaining walls to resist the lateral earth and groundwater pressures. The structures cannot be loaded in tension or bending, since the tensile strength, particularly in the vertical direction, is negligible. In a retaining wall configuration, the structures are subject to overturning and sliding (but have ample weight to resist both) in addition to uplift. Advantages of these products include relatively lower cost than concrete and a more attractive appearance (blends with natural soils).

4. Geotechnical Considerations for Channel Linings

For channels with vertical or near-vertical sides, the principal geotechnical considerations are stability, seepage control, and toe protection. For this type of channel, the side walls act as retaining structures, whether they are concrete walls, gabions, grouted riprap, or other types of materials, and should be evaluated as retaining walls. Evaluations of stability should include overturning and sliding. In addition, an evaluation should be made of the possibility of slope failure on a surface that passes completely beneath the structure. As for most retaining wall design, it is desirable to include seepage control measures to lower the water level behind the wall. In designing the seepage control measures, the engineer must provide adequate filters so that the facilities operate as intended. For a weep hole through a concrete wall, a two-stage filter consisting of concrete sand and a filter-compatible crushed stone should be used to keep the weep hole from plugging. Enough pervious soil must be placed behind the wall to allow the water to drain to the weep holes. All structural walls should be adequately embedded into the bottom of the channel, so that they are not undermined by scour or erosion.

For channels with steeply sloping sides requiring slope protection, the principal geotechnical considerations are slope stability, seepage control, adequate filter systems and toe protection. An analysis of the stability of the slope, including the protection system, should be made and the design slope angle selected accordingly. Seepage control systems should be included to keep water levels within limits assumed in the stability analyses. In addition, for impervious forms of slope protection, such as concrete, fabriform and grouted riprap, the seepage control systems must be sufficient to prevent build-up of enough water pressure beneath the protection to lift and damage it. Any open protection, such as riprap or gabions, should be underlain by filter layers, which prevent the underlying soils from being washed out either by groundwater or channel flow. For example, riprap should be underlain by successive layers of graded filters, such that the bottom layer is filter-compatible with the soils in the slope. The slope protection on steep slopes should be extended into the channel bottom so that it will not be undermined by erosion.

B. Open Channel Structures

The design of open channels usually requires the additional design of an assortment of associated structures, whose hydraulic functions and designs are described in Chapter 9. Following are structural comments on some of the more common open channel facilities.

1. Drop Structures

A drop structure is constructed to lower a channel abruptly in elevation. It usually serves as a combination of an energy dissipator and a retaining wall, and is commonly designed either as a cantilever retaining wall or a gravity retaining wall. Each structure must resist sliding in the direction of flow and overturning about the toe of the footing.

The design of the structure must also address the potential for development of groundwater pressures under the footing, with the resulting decrease in frictional resistance to sliding.

When a drop structure fails, it is frequently due to unexpected erosion at the toe or ends of the structure. The formation of a scour hole beneath the toe not only reduces the resistance to overturning and sliding, but also shortens the moment arm to the point about which the overturning could occur. For these reasons, it is important that riprap or other erosion protection be installed at the downstream edge of the structure.

2. Check Structures

A check structure differs from a drop structure in that there is no abrupt drop in the channel invert. A check structure forces the water upstream to flow at a higher elevation than would otherwise occur. In addition to hydraulic turbulence, the principal forces on a check structure are the hydrostatic and hydrodynamic forces of the water. The structure must be anchored in the earth by a foundation and must be evaluated for stability against sliding and overturning.

3. Energy Dissipators

An energy dissipator is constructed where the energy in flowing water must be reduced. The structural elements that create the reduction in energy may be integral with a conveying facility, such as a pipe or channel, or may be elements of a separate structure between two conveying facilities. Energy dissipators are frequently located at the bases of spillways, the ends of steep channels, or culvert outlets. The dissipation of energy generally results in complex hydraulic flow patterns that require simplification in order to compute the hydrostatic and hydrodynamic forces. The mechanism by which the energy is dissipated determines the manner in which the hydrodynamic forces of the water are transferred to the structure walls and interior elements, such as baffles.

The hydrodynamic loads tend to force the structure to slide downstream and to overturn about the toe. Stability against sliding and overturning must be evaluated. The turbulent flow in and around an energy dissipator can result in erosion of earthen areas. Care must be taken to insure that erosion is either prevented or is not critical to the structure's stability. Many energy dissipators are constructed with rigid floors and partly enclosed walls. In such cases, an evaluation of uplift is required.

C. Closed Conduit Appurtenances

Storm sewer systems and appurtenances represent a significant proportion of drainage and flood control structures. The appurtenant facilities that merit structural consideration include manholes, junction vaults, and inlets. Each is discussed briefly below:

1. Manholes

Under most circumstances, manholes present no stability concerns, because they are completely surrounded by soil. However, manholes are frequently subjected to significant superimposed loads from vehicular traffic. Manholes are designed in two different manners to resist the surface loads.

In one approach, a concrete footing is constructed around the manhole ring and cover, so that the vehicle load is immediately transferred to the soil adjacent to the manhole. This requires the ring and cover to be disconnected from the manhole ring, so that the load is not transmitted downward to the storm sewer.

In the second approach, the manhole ring is connected directly to the manhole cone. In this configuration, a vehicle's load is transmitted directly to the base of the manhole, with a small (and usually neglected) resistance provided by the soil friction on the outside of the manhole. If the manhole has a horizontal base, the load is transferred past the pipe to the earth, provided the manhole does not bear directly on the pipe. If the manhole is a "tee" manhole, directly connected to the pipe, or if it sits on a large storm sewer (either circular or rectangular in section), then the load is transferred directly to the pipe. The tee manhole connection makes the pipe inherently stronger, because of the stiffening of the structure at the opening. In addition, the bedding and soil support under the pipe section must spread the load sufficiently so that the soil bearing pressure is not exceeded.

Differential settlement between the manhole and pipe may sometimes occur because of differences in bedding, vertical load and backfill settlement. If the pipe is rigidly connected to the manhole, differential settlement can induce excessive shear forces, resulting in circumferential cracks in the pipe, or shear failure. Resilient connectors accommodate differential settlements and minimize the effect of shear forces, as well as minimizing leakage by providing a positive seal between the manhole wall and the connector, and between the connector and the pipe. Resilient connectors are available throughout North America, and are manufactured in accordance with ASTM C 923.

2. Junction Boxes

Junction boxes or vaults are constructed in storm sewer systems when large pipes converge. Such vaults are subjected to the same concentrated loads as manholes described previously. In addition, the vaults experience lateral hydrodynamic forces if the flow changes direction or velocity. Under most circumstances, a lateral force in a storm sewer is easily resisted by the soil outside the storm sewer. However, small openings in pipe joints or between the vault and the pipe can allow jets of water to erode the soil outside of the pipe. The erosion, coupled with the hydrodynamic forces, can cause shifting of pipes, increased erosion, and eventual collapse of the backfill or the pipe itself.

Designers must insure that pipes adjacent to junctions have tight, impermeable joints and that the trench bedding and backfill are firmly compacted. In cases where lateral flows have high velocity, the designer should determine whether or not the lateral force would exceed the allowable passive resistance of the soil.

3. Inlets

Storm sewer inlets are generally simple structures but are subjected to numerous superimposed loads from motor vehicles. Any inlet, even if constructed primarily behind a curb line, can be crossed by wheels of heavily loaded trucks. The exposed parts of the inlet can also experience severe impact loads. The engineer cannot design an inlet that will survive all possible abuse by motor vehicles, but he should be certain to include vehicle loads in the design.

4. Internal Energy Dissipators

Research conducted in the 1960s at the Virginia Polytechnic Institute and State University (VPI) established that excess energy in storm water flowing down steep drainage channels could be dissipated by constructing roughness elements within the channel.

Initial tests were conducted on partially-full pipes, and subsequent tests were conducted for full-flow conditions occurring near the outlet end of the pipe at maximum discharge. By allowing the culvert to approach full flow, it was found that velocity reductions could be effected with the roughness elements without increasing pipe diameter (American Concrete Pipe Association 1988).

D. Detention/Retention Dams and Ponds

The design of detention/retention dams and ponds probably requires more geotechnical input than any other aspect of storm drainage design. Geotechnical considerations for design of these facilities include embankment stability, pond-related seepage, foundations of appurtenant structures, sources of embankment materials, and embankment placement specifications.

The stability of the embankment should be evaluated for all reasonable loading conditions. Those conditions include normal static loading, loading during flood flows, rapid drawdown loading (if the pond drains quickly after each filling), and earthquake loading, if applicable for the project site. If accommodating an earthquake loading is very expensive, the engineer and the owner may elect to evaluate the risk of eliminating the load from the design conditions. In typical embankment dam design, the simultaneous occurrence of an earthquake and a flood is not normally considered. For a detention dam that is normally dry, the principal risks associated with earthquake loading are slope failure, cracking of the dam, or damage to appurtenant structures. If such events occur when the pond is dry, they do not pose any downstream threats and repairs can be made before the next flood occurs. In such circumstances, it may be reasonable for the owner to elect not to have a design that can withstand earthquake loading.

Any evaluation should consider seepage through the embankment, under the embankment through the foundation, and into or through the reservoir bottom or abutments. The significance of any of these potential seepage paths depends on the expected duration of pond storage. If the duration of pond storage is long enough for seepage to occur through either the embankment or its foundations, then the design must include provisions to prevent failure of the embankment by internal erosion or "piping" of the embankment or foundation materials. Such a failure could result in a rapid, uncontrolled release of the stored water, which could flood downstream areas. Measures to prevent piping include impervious upstream blankets or impervious core zones to prevent seepage; or internal filters or internal drains and toe drains to safely control and collect the seepage.

Depending on the composition of the materials in the abutments and the reservoir bottom, seepage through those areas can also cause piping failures, which could result in the loss of part or all of the reservoir. For this case the solutions are similar to those for the embankment.

Another potential threat from seepage through the reservoir bottom or abutments, and to a lesser degree through the embankment and its foundations, is the possibility of a temporary rise in the downstream groundwater levels because of seepage. The time required for seepage from the impoundment to result in a rise in the groundwater depends on the permeability of the materials through which the seepage must flow. If the groundwater rises, basement flooding and similar problems could occur. Potential solutions to these problems include upstream impervious blankets to prevent the seepage from entering pervious zones, seepage cutoff walls, downstream pumping wells to collect the seepage before it reaches areas where it can cause damage, and increased hydraulic capacity of the spillway and outlet works to shorten the duration of impoundment.

Geotechnical considerations for appurtenant structures include stability and settlement/heave. As for the embankment itself, the appurtenant structures should be evaluated for all reasonable loading conditions. For intake structures, retaining walls and other facilities subjected to lateral loads, the evaluation of stability should address both overturning and sliding. For structures such as retaining walls, the earth pressures are affected by drainage behind the structures. The geotechnical engineer must insure that his recommended pressures are consistent with the drainage provisions incorporated into the design.

It is important that appurtenant structures that pass through the embankment, such as pipe outlet works, be designed in such a manner that they do not promote seepage and piping along the structure-soil interface. In recent practice, the use of seepage collars around such structures has been giving way to the specification of more careful compaction near the structure and the use of filter zones near the downstream ends to control and collect any seepage along the structure.

Design studies should include evaluations of whether or not the proposed sources of embankment material are adequate. After the materials have been selected and the availability of sufficient quantities has been established, the placement specifications should be formulated in a manner to insure that the in-place materials will have the properties assumed in the design. The equipment and procedures specified for embankment construction should be suitable for the type of material and should be as consistent as possible with local practice and with locally-available equipment.

X. REFERENCES

American Association of State Highway and Transportation Officials (AASHTO), *Standard specifications for highway bridges.* 12th ed.

American Concrete Institute, *Building code requirements for reinforced concrete.* (ASI 318), ACI.

American Concrete Pipe Association. (1960). *Jacking reinforced concrete pipe lines.* ACPA, Arlington, VA.

American Concrete Pipe Association. (1970). *Concrete pipe design manual.* 7th ed. rev., ACPA, Arlington, VA.

American Concrete Pipe Association. (1988). *Concrete pipe handbook.* 3rd ed., ACPA, Arlington, VA.

American Iron and Steel Institute. (1971). *Handbook of steel drainage and highway construction products.* 1st ed., AISI.

American Iron and Steel Institute. (1980). *Modern sewer design.* 1st ed.

American Railway Engineering Association. (1981–82). *Manual for railway engineering.* AREA, Washington, D.C.

American Society of Civil Engineers. (1969). "Design and Construction of Sanitary and Storm Sewers." ASCE Manual of Practice No. 37, WPCF Manual No. 9, New York, NY.

American Society of Civil Engineers. (1982). "Gravity sanitary sewer design and construction." *Manuals and Reports on Engineering Practice—No. 60*, ASCE, New York, NY.

American Society for Testing and Materials. "Standard specification for resilient connectors between reinforced concrete manhole structures and pipe." *ASTM C 923*,

American Society for Testing and Materials. "Standard specification for reinforced concrete culvert, storm drain, and sewer pipe." *ASTM C 76*, Philadelphia, PA.

American Society for Testing and Materials. "Standard specification for reinforced concrete D-load culvert, storm drain, and sewer pipe." *ASTM C 655*, Philadelphia, PA.

American Society for Testing and Materials. "Standard specification for joints for circular concrete sewer and culvert pipe using rubber gaskets." *ASTM C 443*, Philadelphia, PA.

American Society for Testing and Materials. "Standard specification for precast reinforced concrete box sections for culverts, storm drains, and sewers." *ASTM C 789*, Philadelphia, PA.

Griffith, J.S. and Keeney, C. (1967). "Load bearing characteristics of bedding materials for sewer pipe." *Water Pollution Cntrl Fed.*, 39.

Harell, R.F. and Keeney, C. (1976). "Loads on buried conduit—A ten-year study." *Water Pollution Cntrl Fed.*, 48, August.

Howard, A.K. (1989). "Modulus of soul reaction (E') values for buried flexible pipe." *Journal of the Geotechnical Engineering Division*, 103, (GT1), Proceeding paper 12700.

Howard, A.K. (1989). *Prediction of flexible pipe deflection.* U.S Bureau of Reclamation, Denver, CO.

"Jacked-in-place pipe drainage." *Contractors and Engrs Mnthly*, 45, March 1948.

Jumikis, A.R. (1969). "Stress distribution tables of soil under concentrated loads."

Engineering Research Publication No. 48, Rutgers University, New Brunswick, NJ, 233.

Jumikis, A.R. (1971). "Vertical stress tables for uniformly distributed loads on soil." *Engineering Research Publication No. 52*, Rutgers University, New Brunswick, NJ, 495.

Marston, A. and Anderson, A.O. (1913). "The theory of loads on pipes in ditches and tests of cement and clay drain tile and sewer pipe." *Bulletin No. 31*, Iowa Engineering Experimental Station, Ames, IA.

Marston, A. (1930). "The theory of external loads on closed conduits in the light of the latest experiments." *Bulletin No. 96*, Iowa Engineering Experimental Station.

Moser A.P., Watkins, R.K., and Shupe, O.K. (1977). "Design and performance of PVC pipes subjected to external soil pressure. Utah State University, Logan, UT.

Parmelee, R.A. (1977). "A new design method for buried concrete pipe." *Concrete Pipe and the Soil-Structural System*, ASTM STP 630, 105–118.

Portland Cement Association. (1951). "Vertical pressure on culverts under wheel loads on concrete pavement slabs." *Publ. No. ST-65*, PCA, Skokie, IL.

Portland Cement Association. (1975). "Concrete culverts and conduits." *Publication No. EB061.02W*, PCA, Skokie, IL.

Proctor, R.V., and White, T.L. *Rock tunneling with steel supports*. Commercial Shearing and Stamping Co.

Schlick, W.J. (1932). "Loads on pipe in wide ditches." *Bulletin No. 108*, Iowa Engineering Experimental Station, Ames, IA.

Seaman, D.J. (1979). "Trench backfill compaction controls bumpy streets." *Water and Sewage Works*.

Shrock, B.J. (1978). "Installation of fiberglass pipe." *Journal of the Transportation Division*, ASCE, 104 (TE6), Proceeding Paper 14175,

Sikora, E.J. (1980). "Load factors and non-destructive testing of clay pipe." *Water Pollution Cntrl Fed*.

"Soil resistance to moving pipes and shafts." (1948). *Proc of the Second Intl Conf on Soil Mech and Foundation Engrg 7*, 149.

Spangler, M.G. (1991). "The structural design of flexible pipe culverts. *Bulletin No. 153*, Iowa Engineering Experiment, Ames, IA.

Spangler, M.G. (1956). "Stresses in pressure pipelines and protective casing pipes." *Journal of the Structural Division*, ASCE, 82 (ST5) Proc. Paper 1054.

Spangler, M.G., and Hennessy, R.L. (1946). "A method of computing live loads transmitted to underground conduits." *Proc, 26th Annual Meeting, Highway Research Board*, 179.

Taylor, R.K. (1971). "Final report on induced trench method of culvert installation." *Project IHR-77*, State of Illinois, Department of Public Works and Buildings, Division of Highways, Springfield, IL

Townsend, M. (1966). *Corrugated metal pipe culverts—structural design criteria and recommended installation practices.* Bureau of Public Roads, U.S. Govt. Printing Office, Washington, D.C.

U.S. Army Corps of Engineers. *Report of test tunnel.* Part I, Vols 1 and 2, Garrison Dam and Reservoir.

U.S. Department of the Interior. (1960). *Design of small dams.* 1st ed., Bureau of Reclamation, Water Resources Technical Publication, U.S. Government Printing Office, Washington, DC.

Von Iterson, F.K. Th. (1948). "Earth pressure in mining." *Proc of the Second Intl Conference on Soil Mech and Foundation Engrg*, 3.

Wenzel, T.H. and Parmelee, R.A. (1977). "Computer-aided structural analysis and design of concrete pipe." *Concrete Pipe and the Soil-Structure System*, ASTM STP 630.

Watkins, R.K., and Spangler, M.G. (1958). "Some characteristics of the modulus of passive resistance of soil: A study in similitude." *Highway Research Board Proceedings*, 37, Washington, D.C.

Zicaro, J.P. (1990). "Flexible pipe design." *Proc, American Society of Civil Engineers Intl Conf on Pipeline Design and Installation, American Society of Civil Engineers, New York, March, 1990*, New York, NY.

Chapter 15

CONSTRUCTION CONTRACT DOCUMENTS

I. INTRODUCTION

The purpose of contract documents is to portray clearly by words and drawings the nature and extent of work to be performed, the known or anticipated conditions under which the work is to be executed, contractual requirements, the rights and responsibilities of the various parties to the contract (including the engineer), and the basis for payment (ASCE 1982). Contract documents typically consist of bidding requirements, bid forms, contract forms, conditions of the contract, specifications, addenda, and construction drawings (plans). The contract drawings, conditions of the contract, specifications and other components of the contract documents collectively define the work to be undertaken by the contractor. The drawings, conditions, and specifications are complementary; what is called for by one is to be executed as if called for by all (Abbott 1963).

This chapter provides an overview of the elements of an acceptable package of contract documents for stormwater facility construction, and closes with a checklist, which has been adapted from the Construction Specifications Institute Manual of Practice (Wright Water Engineers 1986). Thoughtful and meticulous assembly of contract documents can be of paramount importance in reducing or eliminating the professional liability and insurance problems that confront every public works engineer and private consultant.

II. CONTRACT DRAWINGS

Contract drawings must depict graphically the project's layout and the work to be done to the owner, to project reviewers, to the bidders, and later to the construction observer and the contractor. Drawings normally consist of plan and profile views and cross-sections, and include both general and detailed drawings. General drawings provide an overview of the facility being constructed and are drawn at a relatively small scale.

Drawings are usually prepared by the design engineer, and together with the written contract conditions and specifications, form the physical dimensional description for the contractor's estimate and for the construction contract itself. Properly formulated design drawings stand on their own and require the contractor to spend relatively little time searching the technical specifications for a description of a particular task.

A. Drawing Preparation

Contract drawings generally are prepared on a medium that facilitates reproduction. For storm sewers or open channels, printed plan and profile blank sheets are available, or they may be specially printed and titled to reduce drafting time. The plan view should be drawn on the top half of the sheet with the profile provided directly below.

Contract drawings should be prepared carefully in a neat, legible fashion. Hastily produced, sloppy drawings can lead to mistakes during construction. Engineers are advised to develop in-house procedures for checking every aspect of the drawings.

B. Contents

The most logical arrangement for a set of contract drawings proceeds from general views to more specific views.

1. Title Sheet

The title sheet should identify the project with the following minimum information:

(a) Project name.
(b) Contract number.
(c) Federal or state agency project number (if applicable).
(d) Owner's name.
(e) Owner's officials, key people, or dignitaries.
(f) Design engineer's name.
(g) Engineer's project number.
(h) Plan set number (for distribution records).
(i) Professional engineer's seal and signature.

2. Title Blocks

Each sheet should have a title block, which includes:

(a) Sheet title and number.
(b) Project name.
(c) Federal or state agency project number (if applicable).

  (d) Owner's name.
  (e) Design engineer's name.
  (f) Engineer's project number.
  (g) Scale (if not provided elsewhere on drawing).
  (h) Date.
  (i) Designer, drafter, and checker identification.
  (j) Revisions block.
  (k) Sign-off space for owner's chief engineer or district superintendent, if
      applicable.

3. Index/Legend

Contract drawings should contain an index listing all the drawings
in the set by title and drawing number in the order of presentation.
This index should normally be located on the drawing following the
title sheet, although for small projects the index can be on the title
sheet.

A legend should be included to provide a description of all symbols
used on the drawings.

4. Location Map

A location map should be provided showing the location of all work
in relation to surrounding features and major routes of access to the
project site. This map should either be on the title sheet or on the
index/legend sheet.

5. Subsoil and Groundwater Information

The sampling locations, logs, and laboratory analyses of soil borings
obtained during the design phase of a project should be included,
together with a Design Summary Report giving all of the assumptions
used by the designer in making the contract documents.

When possible, drawings and specifications should indicate where
special construction is required to accommodate known subsurface conditions. Similar precautions apply to groundwater. With the exceptions
of swamp or valley bottom conditions in which submerged excavation
is a certainty, a note that groundwater may be encountered during
excavation is advisable.

6. Survey Control Data

Necessary survey control information should be shown. Baseline
bearings and distances should be included with references to permanent
physical features. Vertical control points, or bench marks, should be
indicated, and the datum plane used for determining these elevations
defined. A note indicating the dates of the ground survey and aerial
photography should be included. Frequently, there will be inconsistencies among county, city, and U.S. Geological Survey datum planes.
When this is the case, the drawings should indicate which datum plane

is relevant, and should provide instructions for converting to other datum planes if necessary.

7. Closed Conduit, Open Channel, and Detention Pond Stormwater Plans and Profiles

Plan and profile views are normally presented on the same sheet, and stationing of the plan and profile should coincide. A continuous strip map usually indicates the locations of all work in relation to surface topography and existing facilities. The location of underground and overhead utilities that cross or are near the proposed construction route also should be shown, as should any other structure that could affect construction.

Surveyed baseline stationing off the alignment may also be given on the plans, but it should not be substituted for stationing along the center line of the sewer or channel.

Stationing indicated on construction drawings for location of manholes and wye-branches or house connections should be considered approximate only and so noted. Locations of junction structures must be given on construction drawings. Match lines should be used and should be easily identifiable. Special construction requirements, such as sheeting to be left in place, should be shown on the drawings. Where interference with other structures exists, explanatory cross-sections and notes should be included. Such cross-sections, often enlarged in scale, should be identified as to specific location and, if possible, should be placed on the plan/profile drawing near the relevant section.

For storm sewers, the profile is a convenient place to show pipe size, length and slope, the strength or type of pipe, the locations of special structures and appurtenances, and crossings of utilities and other drainage pipes. Drawings for open channels provide such characteristics as bottom and top width, side slopes and depth, on a reach by reach basis, or should refer the reader to cross-sections elsewhere in the drawings.

8. Profile and Section Views

When common pipe sizes and materials are specified, no sewer cross-sections need be shown. For cast-in-place concrete sections, however, complete dimensions and all reinforcing steel should be shown in the drawings (ASCE 1982).

Sectional views of open channels, detention ponds, or related dams or embankments should always be provided, and at multiple locations if necessary. Of particular importance for open channels are the transition sections. There should be at least one cross-section for each outlet or control section.

9. Details

Separate sheets depicting details normally follow the plan/profile sheets.

10. Special Details

Details not covered on standard detail sheets should be given on sheets entitled "Special Details." As an example, for sewer projects the following would be included:

(a) Special structures—full details so that the finished work can be structurally sound and hydraulically correct.
(b) Special castings—sufficient details for the manufacturers to prepare shop drawings. Standard casting items, such as manhole frames, covers, and steps, can be identified by reference to a manufacturer's catalog number in the specifications.
(c) Restoration—complete details for pavement, sidewalk, and curb repairs.

11. Record Drawings

During project construction, the contractor should measure and record the "as-installed" locations of all appurtenances being buried that may have to be located in the future. All changes from the original plans, and locations of unexpected rock, seepage, or other unanticipated natural problems, should also be recorded. Extreme deviations in quantity or character of work can entitle the contractor to scope and payment modifications, so it is very important for the inspector to keep accurate logs and records to document the job fully in the event of a dispute.

Contract drawings should be revised to reflect field information after the project is completed and a notation such as "Record Drawing—based on approximate field observations" should be made on each sheet. Sets of such revised drawings should become a part of the owner's and the engineer's permanent records.

III. SPECIFICATIONS

The sample bidding forms, agreement forms, conditions of the contract, and descriptions of expected construction details and minimum criteria for technical performance or material to be provided, is referred to collectively as the Specifications.

Engineers preparing the specifications should consult the format recommended by the Construction Specifications Institute (CSI undated) as a checklist to assure that all components are provided. Although some flexibility on the part of the engineer is acceptable, failure to adhere to a reasonably consistent format can lead to specifications that are confusing or otherwise inadequate. The components of the specifications should be standard in nature and readily understood by all potential bidders. The major components of the specifications are:

(a) Addenda.
(b) Bidding requirements.

(c) Contract forms.
(d) Conditions of the contract.
(e) Detailed specifications.

These documents should be assembled in this order and should be prefaced with a cover page, title page, and table of contents. It is desirable to have all components of the specifications bound in a single volume, though for large projects, that may not be possible. In such cases a "standard specification" should be developed and bound separately. It should then be incorporated by reference in multiple-volume specifications, each pertaining to a single component of the project.

Documents contained in the specifications set forth the details of the contractual agreement between the contractor and the owner. The documents describe the work to be done (complementing the information provided on the drawings), establish the method of payment, set forth the details for performance of the work including necessary time schedules and requirements for insurance, permits, and licenses, and delineate responsibilities of the various parties involved with the project.

A. Addenda

After the initial set of plans and specifications have been issued to bidders, the necessity to modify contract documents frequently arises. Changes may be made to:

(a) Correct errors and omissions.
(b) Clarify questions raised by bidders.
(c) Issue additions and deletions.
(d) Describe changes in design based on new geotechnical, hydrologic, or other information.
(e) Comply with revisions in the requirements of the owner.

Procedures for issuing addenda are described in the instructions to bidders, and a space for the bidder to acknowledge receipt of addenda is provided on the bid form. Addenda must be issued sufficiently in advance of the bid opening to give bidders time to use the information in bid preparation. An addendum issued after contract award usually results in a change of the contract price and requires the issuance of a contract "change order." The engineer must always obtain the owner's approval before authorizing an addendum.

B. Bidding Requirements

The specifications should contain all bidding requirements, including the invitation to bid, instructions to bidders, and a bid form.

1. Invitation to Bid

The purpose of the invitation to bid (or advertisement) is to inform prospective bidders that a contract is to be awarded and that bids are being solicited by the owner. The advertisement should be included in local, regional, or national newspapers (depending on the size of the project), as well as in periodicals or other journals with wide circulation to potential bidders. It may be mailed to contractors with proven capability to conduct the work, and may also be posted in a public place, such as a post office or municipal building. The advertisement should be published sufficiently in advance of the bid opening to allow time to prepare estimates, obtain prices and sub-bids for specialty work, and to make other arrangements necessary to arrive at the bid amounts. One month is generally considered the minimum time acceptable between issuance of an advertisement and bid opening.

The advertisement should be brief and clearly written. Essential elements include:

(a) Brief description of the work and its location. This should be written to attract the attention of only those interested and qualified to bid.
(b) Name and address of the owner.
(c) Name and address of person authorized to receive bids.
(d) A clear description of any statutory requirements regarding preference to local contractors, labor and materials, and/or set-aside programs.
(e) The place, date, and hour of the opening of bids.
(f) Principal items of the work with approximate quantities involved. This informs the contractor immediately whether his equipment, organization, and experience are suitable for the work.
(g) Bid deposit. The amount of the bid deposit and whether it is to be cash, a certified check, or a bid bond should be stated, as well as the provisions for the return of the bid deposit to unsuccessful bidders.
(h) Information relative to the plans and specifications. It should be stated where the plans and specifications may be obtained or examined. This will usually be in the offices of the owner or the engineer, or sometimes in the offices of the contractor or trade associations. Charges or deposits required for the plans and specifications should be noted, together with provisions for refund when the documents are returned.
(i) Name of the engineer and the owner or their authorized representatives. In the latter case, the authorization should be stated.

2. Instructions to Bidders

The instructions to bidders, also known as information for bidders, is a document that furnishes information on the unique features of the work and detailed instructions on the procedure to be followed in submitting bids. This is desirable to assure that all bidders receive uniform treatment and to provide a common basis for bid preparation. The information given is similar in character to that in the advertisement, but is more explicit and in greater detail.

Instructions to bidders should do the following:

(a) Summarize the major components of the specifications.
(b) Describe technical aspects of the bidding process. For example, who from the bidding company is required to sign the proposal?

(c) State whether the bids are to be on a lump-sum or unit-price basis, and whether they are for the entire project or certain parts only. Note also should be made of any alternate bids requested.

(d) Stipulate requirements for an accompanying bid bond, if any, including the nature and amount of the bond.

(e) Describe the amount and type of performance and labor and material payment bonds required.

(f) Describe procedures to be followed if alternate or substitute materials or processes from those described in the specifications are to be proposed.

(g) Describe any required elements of a statement of competency, including such items as descriptions of work performed in the previous five years, descriptions of equipment available to use on the project, or recent financial statements.

(h) Request or require documentation of contractor's familiarization with the work in question, including not only the physical work to be performed but also the applicable federal, state and municipal laws, regulations and ordinances pertaining to labor, materials, specifications, and contract matters that may affect the proposed work.

(i) Inform the contractor that the owner reserves the right to revise or amend any one of the stated parts of the contract documents prior to the date set for opening the proposals.

(j) Inform the bidder of the time for completion of the work and of the method of payment.

(k) Announce the time and place for proposal submission, the packaging required for the proposal, and any other unique bidding factors.

(l) Outline procedures for bid submission or bidder changes or withdrawals.

(m) Describe procedures to be followed for acceptance of a proposal by the owner, and the owner's reservation of rights to reject any or all proposals, to waive inconsistencies and informalities, and to award the contract on the basis of the owner's determination of the lowest responsible bidder (i.e., the one that best serves the interests of the owner).

(n) Inform bidders of the requirements regarding start-up time, insurance, and other factors. These should be described in the final section of the instruction to bidders.

3. Bid Form

The purpose of providing a bid form is to insure that all bidders submit prices on a uniform basis, so as to facilitate comparison. The bid form is advantageous to both the owner and the bidder, because the form tends to insure accuracy and prevent omission. The bid form should contain the following elements:

(a) Price for which the contractor offers to perform the specified work.

(b) Time of completion.

(c) Bid Deposit.

(d) Agreement by contractor to post required performance and labor and material payment bonds upon award of the contract.

(e) List of addenda to the plans and specifications that were considered when the bid was prepared.
(f) List of subcontractors.
(g) Experience record, financial statement, and plant and equipment questionnaire, when required.
(h) Declaration that no fraud or collusion exists, with particular reference to illegal relationships between the bidder and representatives of the owner, pooling of bids by several bidders, straw-man bids submitted by an employee or other representative of the bidder, and similar illegal acts.
(i) Statement that the site has been examined and that the plans and specifications are understood by the bidder.
(j) Signature and witnesses.

4. Responsibility for Accuracy of Bidding Information

On some types of stormwater-related construction work, the amounts of bids will depend on local conditions at the site, some of which cannot be precisely determined in advance. For example, it may be impossible to obtain a sufficient number of soil borings to adequately assess the existing subsurface conditions. Available geotechnical information may not reveal all conditions and may be misleading or subject to incorrect interpretation. Similarly, unexpected high groundwater table conditions can increase the difficulty of a job.

The construction of hydraulic engineering projects is frequently a risky undertaking, and the three parties involved in the planning, design, and construction of hydraulic works (the owner, the designer, and the contractor), each have certain responsibilities they must assume.

The owner desires a functional project designed and constructed in accordance with generally-accepted industry standards. The owner should be willing to pay for the work, including any overruns due to differing site conditions. He should not expect the designer or the contractor to finance the project when cost overruns were not caused by their errors or omissions.

The designer has a responsibility to provide a cost-effective design and to disclose all of the assumptions used in preparing that design. He should be responsible for any errors or omissions made in the course of the design.

The contractor has the responsibility to make a fair bid on the project and to perform the work in a timely manner. He should be responsible for any costs due to his negligence.

These simple responsibilities and relationships between the three parties can deteriorate to the point where the parties become adversaries. These problems may be alleviated by incorporating a "Design Summary Report" (which gives all project design assumptions) into the contract documents and by including three significant features in the contract documents. These are a Pre-qualification of Bidders, Escrow Bid Documentation, and the Disputes Review Board.

The Pre-qualification of Bidders and their subcontractors is used to demonstrate whether or not the contractor and his staff have previous experience on similar projects. Those who do not show the requisite experience should be eliminated from the bidding process.

The Escrow Bid Documentation is used to determine equitable price adjustments for extra work and changed conditions. This documentation consists of a detailed breakdown of how the Contractor prepared his bid including all assumptions used in preparing his estimate. This documentation is then held in escrow for use in settling any disputes during the course of the work.

The Disputes Review Board is a board comprised of three highly qualified members, none of whom are present or former employees of the concerned parties, one appointed by the owner, one appointed by the contractor, and a third appointed by the owner and the contractor. The third member serves as chairman. The Board meets regularly at the construction site and reviews and makes recommendations regarding disputes between the owner and the contractor.

A sample specification for use in the Pre-qualification of Bidders, Escrow Bid Documentation, and developing a Disputes Review Board is included in Westfall (1987).

The engineer's estimate of quantities is given solely for the purposes of indicating the scope of work and for comparing bids for unit price contracts. The unit prices in the bid are binding on the contractor until the variation in estimated quantities threshold is met. It also has been held that the engineer's estimate is a representation, which, if grossly in error, may provide sufficient grounds for the contractor to obtain a change in unit prices.

Without unusually detailed engineering analysis, it is impossible to quantify construction quantities precisely for large stormwater detention ponds, channel widening projects, or even storm sewer projects. Nonetheless, the goal of the engineer's estimates is to provide a reasonable assessment of the probable magnitude of the project. If quantity requirements in the field exceed the engineer's estimates by about 25% (this percentage is subject to variation), it may be necessary to readjust the contract unit prices. The contract documents should provide for such readjustment.

C. Contract Forms

The basic essentials of a valid contract consist of:

(a) The conditions precedent to the agreement.
(b) A statement of the work to be done.
(c) The time in which it is to be done.
(d) The compensation to be paid for its performance.
(e) Signatures of the contracting parties affirming their agreement as to the conditions imposed by the contract.

These items, together with the plans and specifications, will rarely produce a contract sufficient in scope for complex stormwater projects. It will be necessary for the contract to cover such items as liquidated damages, bonus clauses, escalation clauses, changes or extra work, and other factors. No major construction contract should be entered into until counsel for both the owner and the contractor are satisfied with all contract provisions.

1. Form of Contract

Contracts for construction work follow legal practice and assume many different forms, determined principally by the laws of the local jurisdiction and state in which the contract is executed. The contract must cover all items contained within the specifications.

There are two basic components of the contract. The first contains the basic articles, such as the scope of work, compensation, and completion time, which together are referred to as the "Agreement." The second part of the contract is referred to as the "Conditions" of the contract, which are discussed later.

2. Surety Bonds

Contractors are generally required to furnish surety bonds as a guarantee of faithful performance of the contract and payment of bills for labor and materials. Surety bonds are usually issued by a bonding company, although private individuals may serve as sureties or the contractor may furnish his own surety by depositing acceptable property or collateral with the owner.

There are three parties to the execution of a bond:

(a) The principal (contractor) on whose behalf the bond is written and whose performance is guaranteed.
(b) The obligee (owner and persons furnishing labor and materials on the project) in whose favor the bond is written.
(c) The surety (bonding company) who acts as guarantor for the principal and who is obligated to make good to the obligee any default on the part of the principal.

The three most common types of surety bonds are bid bonds, performance bonds, and payment bonds. The bid bond is submitted with the proposal and basically guarantees that the bidder will enter into the contract if his proposal is accepted. If the contractor has signed the contract and has received authorization to commence work from the owner, and then reneges on his obligation and does not start work, the owner is paid by the bonding company an amount equivalent to the bid bond. The performance bond guarantees that the contractor will perform the contract in accordance with the stipulations of the contract. The performance bond is usually established for the full amount of the contract. The payment bond assures that the contractor will pay all expenses he incurs as the contract is performed, thereby rendering the owner harmless for claims and liens that would arise if the contractor defaulted on his financial obligations. The payment bond should be at least 50% of the value of the contract price.

3. *Special Forms*

Unusual aspects may be associated with the contract and may warrant the use of special contract forms. Such forms are referred to as "special forms" and should be attached to the general contract. Special forms will often be associated with federal, state or local government work.

D. Conditions of the Contract

The conditions of the contract fulfil many important functions, including the following:

(a) Provide definitions of important words, terms, and phrases.
(b) Provide an outline of the contract documents.
(c) Establish the rights and responsibilities of the owner, the contractor, subcontractors, and other salient parties.
(d) Provide instructions on how to implement provisions of the contract.
(e) Discuss such aspects as work supervision, changes in work, claims for extra cost, delays and extensions of time, owner's right to terminate contract, insurance bonds, damages, and other matters.

Although the conditions of the contract often constitute the bulk of the contract documents, they should be administrative in nature and should not include detailed specifications for materials or workmanship. These should be provided in the technical specifications.

Construction contracts generally include both general conditions and supplementary (special) conditions. General conditions cover aspects of job management unrelated to the specific project, but which apply to all construction projects, such as insurance requirements, rights and responsibilities of the contractor and of the owner, or bond requirements. Supplementary conditions relate to the particular project, including such things as special environmental considerations associated with the project, provisions for obtaining water at the site, the role of governmental entities, and other such factors.

E. Detailed Specifications

Written instructions that accompany the project drawings are referred to as "Detailed Specifications." The drawings and detailed specifications provide a complete summary of the technical requirements of the work to be performed. Because most contracts stipulate that, in the cases of conflict between specifications and drawings, the provisions of the specifications will govern, it is of great importance that the engineer develop specifications that are clear, concise, and comprehensive.

There are two categories of detailed specifications: (1) general provisions, which apply to the work as a whole and cover such things as a summary of the work to be performed, measurement of payment, quality control, and related subjects; and (2) technical provisions, which describe technical details for construction processes such as earthwork, boring and jacking, or tunnelling, and for materials such as steel, concrete, metals, or grouted riprap.

The Construction Specifications Institute lists nine items that should be included in the general provisions of the specifications (CSI undated):

(a) Summary of work.
(b) Alternatives.
(c) Measurement of payment.
(d) Project meetings.
(e) Submittals.
(f) Quality control.
(g) Temporary facilities and controls.
(h) Materials and equipment.
(i) Project close-out.

A tenth item, safety, should also be included, with proper references to appropriate federal or state OSHA requirements.

Other items can include electrical service, overhead structures, surveys, lines and grades, access to site, pre-construction conference, time for completion of work, inspection requirements, protection of public and private property, water for construction, erosion control, wildlife mitigation efforts, or dust control.

The technical provisions contain the detailed instructions necessary to obtain the desired quality and service in the finished product. In addition to providing detailed instructions, these portions of the specifications also provide for inspection and testing during construction to assure that the project is constructed in accordance with the contract documents. The technical provisions of the specifications must be developed in close harmony with the drawings to avoid possible conflicts.

Materials are commonly specified by reference to the specifications of the American Society for Testing and Materials (ASTM), American National Standards Institute (ANSI), American Concrete Institute, American Water Works Association, or other similar organizations. Although frequent references are made in various sections of the technical provisions to standard specifications, it is recommended that the complete title, serial number, and date of issuance or revision for each standard specifications be cited in full in the general provisions.

The Construction Specifications Institute provides twenty-one categories of site-work specifications, as well as specifications for concrete, metals, and finishes. These are provided in the checklist at the end of this chapter.

There are three broad categories of technical specifications:

(a) Specifications for materials and workmanship that place responsibility on the contractor for furnishing materials and workmanship that will result in a structure of suitable character.
(b) Specifications for the overall performance of the finished product, when the desired operating characteristics of the facility can be measured by specific tests; these specifications are frequently used for machinery such as pumps and motors.
(c) Specifications used for construction work based on the selection of proprietary products in the open market. No control can be exerted over the manufacturer of proprietary articles, and the specifications may merely identify a desired item known to be satisfactory for the purpose intended. Quality and performance tests or standards may also be prescribed.

The most important principle in specification writing is: within the bounds of reasonableness, only the desired results should be specified, and the contractor should be allowed maximum flexibility to obtain the desired results. Standards of workmanship should be described in specific terms when feasible, but specification of construction methods and safety procedures should always be avoided.

IV. CHECKLIST

The following checklist can be used as a guide for determining the completeness of construction documents for stormwater projects:

A. Bidding Requirements

1. Invitation to Bid

(a) Identification of owner or contracting agency.
(b) Name of project, contract number, or other positive means of identification.
(c) Time and place for receipt and opening of bids.
(d) Brief description of work to be performed.
(e) When and where contract documents may be examined.
(f) When and where contract documents may be obtained, and the deposits and refunds therefor.
(g) Amount and character of any required bid deposit.
(h) Reference to further instructions and legal requirements contained in the related documents.
(i) Statement of owner's right to reject any or all bids.
(j) Contractor's registration requirements.
(k) Bidder's pre-qualifications, if required.
(l) Reference to special federal or state aid financing requirements.

2. Instructions to Bidders

(a) Instructions regarding bid form to include, at a minimum, the format of preparation, signature(s) required, time and place for submittal; instructions on alternatives or options; and data and formal documents to accompany bids.

(b) Bid security requirements and conditions regarding return, retention, and forfeiture.

(c) Requirements for bidders to examine the documents and the site of the work.

(d) Required use of stated quantities in unit price contracts.

(e) Withdrawals or modifications of bid after submittal.

(f) Rejection of bids and disqualification of bidders.

(g) Evaluation of bids.

(h) Award and execution of contract.

(i) Actions to be taken in case of failure of bidder to execute contract.

(j) Instructions pertaining to subcontractors.

(k) Instructions relative to resolution of ambiguities and discrepancies during the bid period.

(l) Contract bonding requirements.

(m) Governing laws and regulations.

3. Bid Form

(a) Identification of contract.

(b) Acknowledgment of receipt of addenda.

(c) Bid prices (lump sum or unit prices).

(d) Construction time or completion date.

(e) Amount of liquidated damages.

(f) Financial statement.

(g) Experience and equipment statements.

(h) Subcontractor listing.

(i) Contractor's statement of ownership.

(j) Contractor's signature and seal.

(k) Non-collusion affidavit.

(l) Consent of surety.

B. Contract Forms

1. Form of Agreement

(a) Identification of principal parties.

(b) Date of execution.

(c) Project description and identification.

(d) Contract amount with reference to the contractor's bid.

(e) Contract time.

(f) Liquidated damage clause, if any.

(g) Progress payment provisions.

(h) List of documents comprising the contract.

(i) Authentication with signatures and seals.

2. *Bonds*

(a) Performance bond.
(b) Labor and material payment bonds.
(c) Maintenance and guarantee bonds (if required).

3. *Special Forms*

C. Conditions of Contract

1. *General Conditions*

2. *Supplementary Conditions*

Perhaps combined with General Requirements of the Specifications to avoid redundancy.

D. Detailed Specifications

1. *General Provisions*

(a) Summary of work.
(b) Alternatives.
(c) Measurement of completed work and payment (many officials include this in the particular work item).
(d) Project meetings.
(e) Submittals.
(f) Quality control.
(g) Temporary facilities and controls (protection).
(h) Material and equipment.
(i) Project closeout.

2. *Technical Provisions—Site Work and Materials (see also ASCE 1969)*

(a) Existing utilities and underground structures.
    (1) Protection.
    (2) Relocation.
(b) Clearing.
    (1) Tree removal.
    (2) Pavement removal.
(c) Earthwork.
    (1) Excavating, backfill, and compacting.
    (2) Limits on trench width.
    (3) Spoil placement.
    (4) Preparation of trench bottom.
    (5) Pipe bedding.
(d) Pipe boring and jacking.

   (e) Tunnelling.
      (1) Excavating.
      (2) Casing installation.
   (f) Sheeting and shoring.
   (g) Rock excavation.
      (1) Definition of rock.
      (2) Excavation.
      (3) Blasting limitations and controls.
   (h) Site drainage.
   (i) Paving and surfacing.
      (1) Streets and roadways.
      (2) Sidewalks.
   (j) Highways and railroad crossings.
   (k) Piping materials and jointing.
   (l) Manholes and appurtenances.
  (m) Pipe laying.
      (1) Control of alignment.
      (2) Control of grade.
   (n) Service connections.
   (o) Connections to existing sewers.
   (p) Connections between different pipe materials.
   (q) Concrete encasement or cradles.
   (r) Sewer paralleling water main.
   (s) Sewer crossing water main.
   (t) Repair of damaged utility services.
   (u) Acceptance tests.
      (1) Infiltration.
      (2) Exfiltration.
      (3) Smoke.
      (4) Air.
   (v) Concrete (CSI Division 3).
      (1) Forms.
      (2) Concrete reinforcement.
      (3) Cast-in-place concrete.
      (4) Concrete curing.
  (w) Metal Fabrications (CSI Division 5).
   (x) Finishes (CSI Division 9).
      (1) Painting.
      (2) Waterproofing.

V. REFERENCES

Abbott, R.W. (1963). *Engineering contracts and specifications.* 4th ed., John Wiley and Sons, New York, NY.

American Society of Civil Engineers. (1969). "Design and construction of sanitary and storm sewers." *Manuals and Reports on Engineering Practice—No. 37, WPCF MOP. No. 9*, ASCE, New York, NY.

American Society of Civil Engineers. (1982). "Gravity sanitary sewer design and construction." *Manuals and Reports on Engineering Practice—No. 60*, ASCE, New York, NY.

Construction Specifications Institute: (a) *Manual of Practice* (2 volumes), (b) *Specification Document Series*, Divisions 1–6.

Westfall, D.E. (1987). "Management of risks during construction." *Seminar on Contracting and Construction of Large and Small Hydraulic Works, International Water Resources Association*.

Wright Water Engineers Inc. (1986). *Standard contract general conditions*. Denver, Colorado.

Chapter 16

CONSTRUCTION METHODS

I. INTRODUCTION

The design and construction of drainage systems are so interdependent that knowledge of one is essential to the competent performance of the other. The intent of this chapter is to introduce the engineer to some common construction techniques and to encourage their consideration in the design and development of the construction documents. Local conditions and proper concern for applicable regulations may dictate variations, and the ingenuity of the owner, engineer, and contractor must be accommodated and encouraged if construction costs are to be minimized and a quality job is to result.

Commencement of the construction phase normally introduces a new party, the contractor, to the project. At this stage of the project the division of responsibility and liability must be understood by all. The role of the engineer will normally change from active direction and performance of the design to that of professional and technical observation during construction.

The engineer's representative on the construction site should not be expected to duplicate the detailed inspection of material and workmanship properly delegated to the manufacturer, supplier, and contractor. In fact, the engineer should not agree to monitor, direct, or in any manner have control over the contractor's work, the means, methods, techniques, sequences, or procedures of construction. The engineer's role should be one of identifying and correcting problems as a result of misunderstanding or misinterpreting the contract documents.

Preconstruction conferences are helpful in deciding whether the contractor's proposed operations are compatible with contract requirements and whether they will result in finished construction acceptable to the owner. These joint meetings of the owner, engineer, and contractor should result in definite construction schedules and administrative procedures to be followed throughout the duration of the construction

Denver, Colorado—Careful attention must be paid to the quality of facility construction. This parking lot was supposed to drain into the adjoining field. Instead, the field drains into the parking lot.

contract. On larger jobs, the geotechnical engineer familiar with the project design should be available to answer questions regarding any special construction techniques. The meeting agenda should include items such as progress schedules, progress payment format and details, method of making submittals for review, and channels of communication. All of these aspects of construction should be settled before construction begins. Where special permits have been issued for construction, specifically in the areas of wetlands or other environmentally sensitive issues, copies of such permits and limiting conditions should be made available to all parties at the meeting.

II. CONSTRUCTION SURVEYS

A. General

Baselines and benchmarks for storm drainage system alignment and grade control should be established along the route of the proposed construction by the engineer, or by the contractor if the work is reviewed by the engineer. All control points should be referenced adequately to permanent objects located outside normal construction limits.

Control points should be identified with both a name and description and should provide both vertical and horizontal control with elevations and coordinates.

B. Right of Way

Acquisition of easement or fee title to rights-of-way to provide adequate working space for construction projects should be completed and rights-of-way cleared before project construction begins. Access, materials storage, and equipment movement space needs should be anticipated.

Project construction in urban areas often requires removal and replacement of fences, landscaping, and even buildings. To avoid construction delays and adverse public reaction, such actions should be pre-arranged with adjacent property owners.

Rights-of-way needed for maintenance may differ from those required for construction. For instance, maintenance vehicle access points may be needed, and should be designed for minimum interference with drainage facility performance.

C. Preliminary Layouts

Prior to the start of any work, work areas, clearing limits, and pavement cuts should be laid out to give proper recognition to, and protection for, adjacent properties. Limits of temporary and permanent easements should be carefully delineated in the field so their relation to permanent improvements can be verified. Access roads, detours, bypasses, and protective fences or barricades also should be laid out and constructed as required in advance of construction. All layout work, if done by the contractor, should be reviewed by the engineer before any demolition or construction begins.

D. Setting Line and Grade

The transfer of line and grade from control points to the construction work is normally the responsibility of the contractor, with spot checks by the engineer as work progresses. The preservation of stakes or other line and grade references provided by the engineer is similarly the responsibility of the contractor. In most cases, there is a charge for re-establishing stakes carelessly destroyed by the contractor, and the charge is stated as part of the contract agreement.

In general, the line and grade for the sewer may be set by one or a combination of the following methods:

(a) Stakes, spikes, or crosses set on the surface at an offset from the sewer centerline.
(b) Stakes set in the trench bottom along the sewer line as the rough grade for the sewer is completed.
(c) Elevations given for the finished trench grade and sewer invert while sewer laying progresses.

(d) Laser beam.
(e) Stakes set on the surface at an offset from the top of slope indicating cut or fill and slopes.

Method (a) generally is used for small diameter sewers. Methods (b) and (c) are used for large sewers or where sloped trench walls result in top-of-trench widths too great for practical use of short offsets or batter boards. Method (d) is independent of the size of sewer. Method (e) is used for open channel conveyance systems that have sloped banks or deep excavations where side slopes are specified prior to construction.

In method (a) stakes, spikes, or crosses are set on the opposite side of the trench from which excavated materials are to be placed and at a uniform offset, insofar as practicable, from the sewer centerline.

The line and grade may be transferred to the bottom of the sewer trench by the use of batter boards, tape and level, or patented bar tape and plumb bob unit.

Batter boards and batter board supports must be suspended firmly across the trench and be adequate to span the excavation without measurable deflection. If the spanning member is to be the batter board, it is set level at an even foot (or other convenient unit of measurement) above the sewer. Preferably, the spanning member is used as a support only and a 1 inch batter board is nailed to it with one edge in a true vertical plane at the centerline of the storm sewer. A nail then is driven in the vertical edge of the batter board at an even foot above sewer grade. A string line is drawn taut across at least three batter boards.

The sewer centerline is then transferred to the trench bottom with a heavy plumb bob held lightly against the string line. Grade is transferred to the sewer invert with a grade rod equipped with a suitable metal foot to extend into the end of the pipe. For steep grades, it is advisable to fasten a bullseye level to the grade rod to assure that the rod is held plumb. For ease in reading, the grade rod may be marked at subgrade, finish grade, and invert grade. The line and grade of the string line should be checked by observation for possible error in cuts or in establishing the batter boards. Periodic inspection should be made during sewer laying to insure that the set line and grade have not been disturbed.

Another method of setting grade is from offset crosses or stakes, or from offset batter boards and double string lines and the use of a grade rod with a target near the top. When the sewer invert is on grade, a sighting between grade rod and two or more consecutive offset bars or the double string line will show correct alignment.

The transfer of surface references to stakes along the trench bottom is in some instances permitted, but the use of batter boards is preferred. If stakes are established along the trench bottom, a string line should be drawn between not less than three points and checked in the manner used for batter boards.

When trench walls are not sheeted but sloped to prevent caving, line-and-grade stakes are set in the trench bottom as the excavation proceeds. This procedure requires a field party to be at the job site almost constantly.

Another method, applicable to large diameter sewers or monolithic sections of sewers on flat grades, requires the line and grade for each pipe length or form sections to be set by means of a transit and level from either on top or inside of the completed conduit.

In the construction of large sewer sections in an open trench, both line and grade may be set at or near the trench bottom. Line points and benchmarks may be established on cross bracing where such bracing is in place and rigidly set. Later, alignment and grade must be determined by checking the setting of the forms.

A method quite widely used is laser beam control. A laser is a device that projects a narrow beam of light down the centerline of the sewer pipe. It is usually set up in the invert of a manhole and then aligned horizontally. The proper slope is established by adjusting a dial on the machine and aiming the laser. A check elevation should be set about 100 feet from the manhole to assure that the proper slope is being maintained by the beam of light. A target set in the pipe centerline is then used to align the end of each pipe section. Care should be exercised in the use of the laser since temperature affects the aiming of the unit.

Where tunnel construction is an extension of a sewer of sufficient size without change of alignment, the initial line and grade for the tunnel work may be established by extending lines and grades through and forward from the completed portion.

When tunnelling begins from an isolated shaft, great care must be taken in transferring line and grade from the surface. If tunnelling from any one shaft extends more than several hundred feet from the shaft, and especially if the alignment is curvilinear, it may be desirable to verify the vertical and horizontal alignment after each two or three advances. Normal deviations in vertical and horizontal alignment are anticipated and can be adjusted for in the final tunnel lining process.

Slope stakes are generally used to provide horizontal and vertical information to equipment operators on large scale open excavations. These are frequently open channel sections but may be large sewer lines or box culverts requiring large scale excavations due to their size or depth. The stakes are located at an offset outside the top of the intersection of the excavation with the existing ground surface. Stakes are marked with centerline station, the offset, the required cut or fill to the excavation bottom and a slope at which the bank is to be constructed. The tops of both banks are slope staked in this fashion at regular intervals along the proposed centerline.

When the excavation is sufficiently large, it also may be possible to stake the sewer centerline and the toe of slope. This will provide the operator with additional control with which to perform the construction. These stakes will be destroyed prior to finish grading, but will provide enough information for rough grading. These stakes are marked with centerline station, their location on the bottom (i.e. centerline, toe of slope, etc.) and the appropriate cut or fill.

III. SAFETY

A. General

Safety on the construction site is a critical concern of all participants in the project. The responsibility for safety on the project site generally lies with the contractor who has direct control of the construction activities. However, all who are on the site have an interest in maintaining a safe working environment. A safety memo or manual setting forth procedures to be followed during an emergency should be present at all construction sites.

Many organizations such as the Federal OSHA and similar State organizations have developed standards specifying minimum provisions for site safety. Of greatest interest on sewer projects are those provisions dealing with excavating, trench supports, and tunnelling and blasting. Hard hats and other personal safety equipment should be required.

B. Excavation

The most critical safety issue for excavations is the stability of open cut slopes. In no case should an unsupported slope be excavated at an angle steeper than the material's angle of repose. If right-of-way or other constraints prohibit the flattening of slopes, shoring and bracing or trench boxes (where allowed) should be provided to allow for the maintenance of safe working conditions.

Precautions should also be taken to prevent unwarranted access to excavations. If necessary, fences, barriers or guardrails should be installed to protect unwary passersby.

C. Tunnelling

In addition to normal safety-related issues associated with movement of materials and equipment, precautions must be taken to insure that adequate ventilation and roof and wall support are provided in tunnelling operations.

Ventilation takes on additional importance when underground operations generate large volumes of dust or during blasting operations. The buildup of flammable, explosive, or toxic gasses such as methane, natural, gas or sulphur dioxide must be prevented. The venting and disposal of tainted air from the tunnel must meet local air and safety standards.

D. Blasting

"Pre-blast" surveys are mandatory wherever the construction adjoins existing residences or other developed properties. Permission should be obtained, if possible, to enter homes for videotaping and/or photographs. The likelihood of claims in this area is great.

The transport, handling, and storage of explosives must be carefully directed and supervised to maintain safe conditions. The handling of blasting caps is also critical and must be performed by personnel certified under local ordinances.

IV. SITE PREPARATION

The amount of site preparation required may be quite variable. In some cases a significant portion of project cost may be expended on site preparation.

Several owners, engineers, and contractors have adopted a practice of assembling extensive photographic or videotape evidence of pre-construction condition of sidewalks, driveways, street surfaces, building facades, etc., to minimize post-construction claims by residents and others for construction-related damages. These evaluations are performed prior to the commencement of any construction activities and might be classified as pre-construction reconnaissance.

Operations that may properly be classified as site preparation are clearing and grubbing; construction of access roads, detours, and by-passes; control of existing drainage; location, protection, or relocation of existing utilities; and pavement cutting. The extent and diversity of these operations make extensive additional discussion thereof impractical here. Note, however, that the success of the contractor in keeping the project on schedule depends to a great degree on the thoroughness of the planning and execution of the site preparation work.

A. Clearing and Grubbing

Clearing and grubbing operations are intended to remove unwanted vegetation and unusable surface materials from the site prior to the commencement of other construction operations. The work will generally consist of the removal of all surface objects (debris), trees, stumps, brush, vines, and other objects not designated to remain. It will often include the demolition of existing structures within the storm sewer alignment to facilitate construction. The contractor should be responsible for the proper disposal of materials from the clearing and grubbing operations unless the contract documents indicate reusing some of the cleared materials.

B. Access

Access to the site should be coordinated between the contractor, the engineer, and the owner to minimize construction problems and impacts on adjacent property owners and public streets. On large projects and on some complex smaller projects, the contractor should be encouraged to analyze the access to various sites within the construction limits to anticipate and resolve problems with material delivery and equipment movement.

C. Utilities

In all excavations, extreme care should be taken to properly locate, support and protect existing utilities, particularly those underground. The owners of the utilities should be contacted before the start of excavation. During design, the utilities should be field-located by staking for horizontal location and excavation where necessary for vertical location. Potential conflicts with existing utilities should be addressed during the design of the project. The contractor should be required to notify all utility owners and to field-verify the location of all utilities prior to the commencement of construction (many utilities will provide an on-site representative to work with and assist the contractor). Any deviations from the lines and grades shown on the drawings should be brought to the immediate attention of the engineer.

Project designs should identify and provide specific construction procedures for utility crossing protection. Often overlooked are "unusual" utilities such as petroleum or gas pipelines, underground electrical distribution, underground cables belonging to private or commercial communications systems or telemetry. Larger conflicting utility installations such as telephone junction boxes, electric utility transformer stations, high pressure gas transmission lines, etc, should be identified and relocated by their owners or under separate contract. If possible, relocations should be completed prior to sewer construction activities to minimize delays or disruption of the sewer construction.

It may even be infeasible to adjust or relocate some existing utilities. Such situations must be identified during project design. There is no excuse for the discovery of such situations during project construction.

Even with these precautions, damage or interruption is a possibility, and drainage system managers should have contingency measures planned and ready for execution.

V. EXCAVATION

A. General

With favorable ground conditions excavation may be accomplished in a single operation; under more adverse conditions it may require several steps. In these circumstances, excavation operations can include stripping, drilling and blasting, and trenching and shoring. In unusual circumstances or complex excavations, hauling and stockpiling of excavated materials may be a necessary part of the excavation process. Excavation work should not be commenced until completion of the site preparation operations.

B. Stripping

Stripping may be advantageous or required as a first step in excavation for a variety of reasons, the most common of which are:

(a) To remove topsoil or other materials to be saved and used for site restoration;
(b) To remove material unsatisfactory for backfill to insure its separation from usable excavated soils;
(c) To remove material having a low bearing value to a depth where there is material capable of supporting heavy construction equipment;
(d) To make it easier to charge drill holes.

In the case of topsoil removal, provisions for the stockpiling of the usable topsoil must be made. This material should be free of large roots or stumps and should be suitable for redistribution subsequent to final grading operations. Stockpiles should be located outside the limits of excavation and as near as possible to the point of final placement.

Unsatisfactory materials should be disposed of in approved on-site locations, as directed by the engineer or owner, or off-site at a location of the contractor's choosing in accordance with applicable local regulations. The off-site location should be reviewed by the engineer.

C. Large Scale Excavation

For large open channels, large diameter pipes, box culverts or extremely deep storm sewer construction, it may be practical and feasible to excavate with earthmoving equipment. This procedure involves operations very similar to those used in grading or embankment construction. Large scrapers may be used to bring the excavation down to a level where more precise excavation is practical. Other alternative methods such as power shovels, draglines, and belt loaders, in conjunction with specialized hauling equipment, may be used in lieu of scrapers. Such equipment may be preferred under certain combinations of excavated materials, haul routes and haul distances.

In some cases, it may be necessary to distinguish between general soils and firmer materials, such as rock, which may require some preliminary breaking up prior to excavation. In this case different procedures must be established for handling the material.

Because of the large volume of materials generally excavated by this method, it is often necessary to carefully evaluate stockpile locations, haul lengths, and haul routes. This will often be the contractor's responsibility and should be addressed during the bidding and preconstruction phases of the work. After award of the contract, the engineer should provide assistance to the contractor in locating stockpiles and in identifying materials suitable for backfill.

D. Small-Scale Excavation

Where large earthmoving equipment is not appropriate and trenching operations are also not suitable, smaller excavation equipment may be useful. This type of application might include the installation of culverts, manholes, unusual channel structures or special construction encountered when using earthmoving equipment.

Self-loading scraper being used for channel excavation.

The type of equipment most often used are backhoes, excavators, front-end loaders,and bulldozers. The preferred equipment is most often dictated by type of facility and nature of material to be excavated. Rubber-tired backhoes are frequently used when the extent of excavation is small. Tracked backhoes or excavators allow excavating without the use of stabilizing legs and can be used to great advantage for trenching. Culvert crossings, manholes and junction structures, and areas with limited access are situations where excavators are most frequently used.

These machines are available with bucket capacities varying from ⅜–3 cubic yards and more. They are convenient for the excavation of trenches with widths exceeding 2' and to depths down to 25'. They are the most satisfactory equipment for excavation in loosened rock.

The excavator may also be used with a cable sling for lowering sewer pipe into the trench. Where safety requirements and the soil do not require sheeting and bracing, this method is very economical. When sheeting and bracing must follow the excavation closely, the use of an excavator for excavation and a crane for placement of sewer pipe is a common practice.

Front-end loaders are commonly used where there is a relatively wide area that requires excavation, but the excavation volume is not large. This might be the case in the installation of box culvert road crossings or short segments of large diameter conduits or open channels where the use of large earthmoving equipment is not justified.

In wide, deep trenches, the front-end loader has sometimes been used as an auxiliary to an excavator or clamshell. In this arrangement the loader excavates the upper part of the trench, leaving the bottom bench or benches for the excavator or clamshell that completes the excavation.

The principal use of loaders in many storm sewer construction projects, however, is in transporting sewer pipe, manhole sections, and other appurtenant structures and granular bedding and backfill material to the trenches.

Bulldozers are generally useful when rippable rock is encountered on the site. The bulldozer can be used to fracture the rock and then move it short distances away from the excavation. In general, it will not be practical to use bulldozers to move material any significant distance. In these cases front-end loaders and trucks will be required to transport the material to a suitable stockpile or disposal site.

E. Dredging

Dredging is an operation commonly performed beneath the water table, through open water or in stream crossings. When it is infeasible to de-water the area, construction by dredging may be required. This generally involves the use of draglines.

The dragline is also useful under some other circumstances. In open country for stream crossings or in a wide right-of-way, it may be feasible to do a large part of the excavation by means of a dragline, allowing the sides of the trenches to acquire their natural slope. In cases of very deep trench excavation, on the order of 30–50', the dragline has been used for the upper part of the excavation, with a backhoe operating at an intermediate level. By rotating the backhoe, the material thus excavated can be relayed to the dragline, which then lifts it to the spoil bank or to trucks at the surface.

F. Rock Excavation

When rock cannot be ripped with bulldozers or other mechanical equipment, drilling and blasting must be used. In addition, some shales and softer rocks, which may be ripped in open excavation, will require blasting before they can be removed in confined areas.

Normally the most economic method will involve pre-shooting; that is, drilling and shooting rock before removal of overburden. In some instances the presence of wet granular materials above the rock ledge will necessitate stripping before drilling, since holes cannot be kept open through the overburden to permit placing of explosive charges.

For narrow trenches in soft rock, a single row of drill holes may be sufficient. One or more additional rows may be required in harder rock, or for wider trenches. To reduce overbreak and improve bottom fragmentation, time delays should be used in blasting for trenches. In tight quarters, trench walls can be pre-split, with the material in between blasted in successive short rounds to an open face to produce minimum vibration.

It must be recognized that there will be a minimum feasible trench width varying with the rock formation, and in the case of small storm sewers, it may be necessary to design the conduit for the extreme loads of the positive projecting condition.

All ground and air pressures that result from blasting should be recorded on a sealed cassette seismograph. Surveys of adjacent structures for the presence of cracks before blasting commences should be considered. Blasting should be done only by persons experienced in such operations.

VI. OPEN-TRENCH CONSTRUCTION

A. Trench Dimensions

As noted in Chapter 14, the trench at and below the top of the storm sewer should be only as wide as necessary for proper installation and backfilling, consistent with safety requirements. The contract must provide for alternate methods or require corrective measures to be employed by the contractor if allowable trench widths are exceeded through overshooting to rock, caving of earth trenches, or over-excavation. The width of trench from a plane 1 foot above the top of the storm sewer to the ground surface is related primarily to its effect on the safety of the workmen who must enter the trench and on adjoining facilities, such as other utilities, surface improvements, and nearby structures.

In undeveloped subdivisions and in open country, economic considerations often justify sloping the sides of the trench for earth stability from a plane 1' above the top of the finished storm sewer to the ground surface. This eliminates placing, maintaining, and removing substantial amounts of temporary sheeting and bracing, although safety regulations may make some type of sheeting or bracing mandatory.

In improved streets, on the other hand, it may be desirable to restrict the trench width to protect existing facilities and reduce the cost of surface restoration. Available working space, traffic conditions, and economics will all influence this decision.

B. Excavation Procedures

The method and equipment used for excavating the trench will depend on the type of material to be removed, the depth, the amount of space available for operation of equipment and storage of excavated material, and prevailing practice in the area. Ordinarily the choice of method and equipment rests with the contractor. However, various types of equipment have practical and real limitations regarding minimum trench widths and depths. The contractor is obligated, therefore, to use only that equipment capable of meeting trench width limitations imposed by pipe strength requirements, or for other reasons set forth in the technical specifications.

Spoil should be placed sufficiently back from the edge of the excavation to prevent caving of the trench wall and to permit safe access along the trench. With sheeted trenches, a minimum distance of 3 feet from the edge of the sheeting to the toe of the spoil bank will normally provide safe and adequate access. Under such conditions the supports must be designed for the added surcharge. In unsupported trenches the minimum distance from the vertical projection of the trench wall to the toe of the spoil bank normally should be not less than one-half the total depth of excavation. In most soils, this distance will be greater in order to provide safe access beyond the sloped trench walls.

Trenching machines are machines generally used for shallow trenches less than 5' deep. For installation of small storm sewers in cohesive soils, the trenching machine can make rapid progress at low cost.

When the protection of other underground structures or soil conditions require close sheeting and the use of vertical-lift equipment, the clamshell bucket is used. In very deep trenches where two-stage excavation is required, the backhoe is sometimes used in combination with the clamshell, with the backhoe advancing the upper part of the excavation and the clamshell following for the lower. Sheeting and bracing of the upper part are installed as required prior to the excavation of the lower part, and the installation of the lower-stage of sheeting.

C. Sheeting and Bracing

Trench sheeting and bracing should be adequate to prevent cave-in of the trench walls or subsidence of areas adjacent to the trench, and to prevent sloughing of the base of the excavation from water seepage. Contracts normally stipulate that the contractor is responsible for the adequacy of any required sheeting and bracing. The strength design of the system of supports should be based on the materials encountered. Sheeting and bracing always must comply with applicable safety requirements.

For wider and deeper trenches a system of wales and cross struts of heavy timber (or steel sections) often is used. Sheeting is installed outside the horizontal wales as required to maintain the stability of the trench walls. Jacks mounted on one end of the cross struts maintain pressure against the wales and sheeting.

In some soil conditions it has been found economical and practical to use steel trench shields that are pulled forward as sewer pipe laying progresses. Care must be exercised in pulling shields forward so as not to drag or otherwise disturb the previously laid pipe sections or to create conditions not assumed in calculating trench loads.

In non-cohesive soils containing considerable groundwater, it may be necessary to use continuous steel sheet piling to prevent excessive soil movement. Such steel piling sometimes extends several feet below the bottom of the trench unless the lower part of the trench is in firm material.

Excavator being used in trenching operation.

In some soils, steel sheet piling can be used with a backhoe operation for the upper part of excavation, but the piling usually needs to be braced before the excavation has reached its full depth. The remaining excavation is performed by vertical-lift equipment such as a clamshell.

Another means of trench sheeting occasionally adopted involves the use of vertical H-beams as "soldier beams" with horizontal wooden lagging. This is sometimes advantageous for trenches under existing overhead viaducts where overhead clearances are low and spread footings lie alongside the trench walls. The vertical beam can be tilted and driven. As excavation progresses downward, the lagging is installed between adjacent pairs of soldier beams. For deep trenches with limited overhead clearances the soldier beams can be delivered to the site in shorter lengths and their ends field-welded as driving progresses.

The removal of sheeting following pipe laying may affect the load on the pipe or adjacent structures (see Chapter 14). This possibility must be considered during the design phase. If removal is to be required or permitted, appropriate directions must be included in the technical specifications for proper removal of sheeting and placement of backfill to thoroughly fill the voids thus created.

VII. TUNNELLING

Tunnelling is considered to be any construction method that results in the placement or construction of an underground conduit without

Trench box used in trench excavation.

continuous disturbance of the ground surface, and includes the various forms of jacking of prefabricated units from shaft or pit locations. Tunnelling methods applicable to storm sewer construction can be classified generally as:

    (a) Auger or boring method.
    (b) Jacking of preformed steel or concrete pipe.
    (c) Mining methods.

A. Auger or Boring Method

In sizes less than 36″ diameter, rigid steel or concrete pipe can be pushed for reasonable distances through the ground and the earth removed by mechanical means under the control of an operator at the shaft or pit location. Several types of earth augers are available, and some contractors specialize in this type of operation. Augers as large as 72″ have been used, but for sizes above 36″ considerable care must be exercised to avoid overbreak. In the case of concrete pipe, it may be necessary to use an auger with a special head having a diameter equal to the outside diameter of the pipe being placed.

The presence of well-cemented soils is a serious deterrent to this method of installation. If such soils are expected, particularly when sewer pipes smaller than 36″ are to be placed, it may be more economical first to install an oversize lining by conventional tunnel or jacking methods. The sewer pipe then can be placed within the liner pipe and the remaining space backfilled with sand, cement grout, or concrete.

B. Jacking

Although the limits will vary with geographic locations and soil conditions, finished interior diameters of 30–108 inch are the generally accepted limits for pipe jacking. Excavation and removal of the excavated material is done by machine or manually, augmented with air spades, special knives, etc. The most commonly used materials for such jacking operations are reinforced concrete or smooth steel pipe. The pipe selected for jacking must be strong enough to withstand the loads exerted by the jacking process.

The usual procedure is to equip the leading edge with a cutter or shoe to protect the sewer pipe. As succeeding lengths of pipe are added between the leading sewer pipe and the jacks and the sewer pipe jacked forward, soil is excavated and removed through the sewer pipe. Material is trimmed with care and excavation does not precede the jacking operations more than necessary. Such a method usually results in minimum disturbance of the natural soils adjacent to the sewer pipe.

When jacking, contractors have sometimes found it desirable to coat the outside of the pipe with a lubricant, such as bentonite, to reduce frictional resistance. In some instances this lubricant has been applied through pressure fittings installed in the wall of the leading pipe. Grout holes sometimes are provided in the walls of the pipes for use in filling outside voids. Protective joint spacers are used to prevent damage to pipe joints. Because soil friction may increase with time, it is desirable to continue jacking operations without interruption until completed.

In all jacking operations it is important that the direction of jacking be carefully established prior to the start of work and checked periodically during the work. Guide rails must be installed in the bottom of the jacking pit or shaft. In the case of a large pipe it is desirable to have such rails carefully set in a concrete slab. The number and capacity of the jacks used depend primarily on the size and length of the pipe to be placed and the type of soil encountered. Backstops must be strong enough and large enough to distribute the maximum loading of the jacks to the soil behind them.

Jacking large-diameter reinforced concrete pipe under a railroad.

In some cases long sewer lines have been installed by jacking from a series of shaft locations spaced along the line of the sewer pipe.

C. Mining Methods

Tunnels with finished interior dimensions of 5' or larger in clay or granular materials ordinarily are built either with the use of tunnel shields or with boring machines, or by open-face mining with or without some breasting. Rock tunnels normally are excavated open-face by conventional mining methods or with boring machines.

1. Tunnel Shields

In clays, silts, sands and gravels, especially in built-up city areas, it will usually be necessary to use tunnel shields for tunnelling operation. Compressed air also may be required to control the entry of water into the tunnel if the phreatic line is above the tunnel invert and the soils lack adequate cohesion.

With a shield it is necessary to install a primary lining of sufficient strength to support the surrounding earth and to provide a progressive backstop for the jacks that advance the shield. The lining may be installed against the earth and the annular space between the lining and the earth filled with pea gravel and grout. Alternatively, the tunnel lining may be expanded against the earth as the shield is advanced. The latter method practically eliminates the need for grouting the annular opening.

2. Boring Machines

Tunnel boring machines, also called digger shields or mechanical moles, have been developed for tunnel excavation in clay and rock. They usually have cutters mounted on a rotating head which is advanced into the heading. A conveyor system moves muck away from the tunnel face. Machines may be braced against the walls of the excavation or against previously placed tunnel lining. Some machines also are equipped with shields. Machines have been used successfully in the construction of tunnels in clay up to 25' in diameter and in rock up to 36 feet in diameter. Machines are most useful in fairly long runs through generally similar material.

3. Open-Face Mining Without Shields

Where the ground allows the use of open-face mining methods it is often more economical to use segmental supports of wood or steel for the sides and top of the tunnel only. The need for compressed air or breast boards in the tunnel heading will depend on the type of soil and amount of moisture or groundwater. The geotechnical report will generally be the best guide to geology and groundwater conditions to be encountered. The particular combination of geology and groundwater will determine the need to use compressed air.

4. Primary or Temporary Lining

Materials used for primary lining are usually steel, wood or a combination of the two. Linings also may be made of segmental precast concrete, stamped steel or cast iron.

Some engineers and contractors prefer to use continuous rings of liner plates having sufficient section modulus to resist the earth pressures without use of special structure ribs or rings. A circular lining formed of such plates becomes a compression ring and has some inherent stability not equalled by horseshoe-shaped supports. Soil conditions and the contractor's preference determine the choice of such a lining. If design of liner plate support is based on the assumption that plates will act as a compression ring, immediate grouting behind liner plates or immediate expansion of the lining is required to insure uniform loading. In any event, voids behind liner plates should be grouted or the lining expanded prior to subsidence of the overburden.

Tunnel lining used in open face mining operation.

5. Tunnel Excavating Equipment

The type of excavating equipment or tools used in tunnelling depends on the kind of material to be excavated and the work space available. Pneumatic spades and special knives are used widely in excavating clay. Drilling and blasting are usually employed in rock tunnels. In the case of shale, roadheaders and undercutting machines like those used in coal mining have been used to advantage.

6. Shafts

Where tunnels are of considerable length, one or more construction shafts may be necessary. On important thoroughfares these shafts are better located in an adjacent side street or vacant lot, with access to the work provided through a short connecting entry tunnel.

Offset shaft locations are especially desirable when soil conditions require the use of compressed air. In such a case only one air lock in the entry tunnel will be required. Shafts generally are located so that tunnelling in both directions is possible. Construction shafts on long tunnels typically are spaced 1,200–2,500 feet apart. Factors tending to affect this spacing are the need for compressed air, the size of tunnel, and the depth below ground.

Shafts should be large enough to permit the installation of an electric hoist. Such equipment should be used only for the handling of material, with separate personnel lifts in deep shafts. Hoisting in shafts by means of a crane may be permitted when the length of the tunnel is short and safety precautions are taken to prevent engine exhaust from entering the shaft and tunnel.

Tunnel drainage is normally discharged from the shaft, and the shaft must be equipped with some form of collection sump and drainage pump.

7. Compressed Air Equipment and Locks

Compressed-air equipment for tunnelling should have sufficient capacity to maintain a pressure that will balance the hydrostatic pressures in the soil at tunnel depth.

The equipment includes compressors, air receivers, piping, control valves, air locks, main and emergency locks, bulkhead walls, gages, etc. Separate locks should be provided for materials and personnel. Generally for long, large diameter tunnels, electrically operated compressors are used with two independent sources of power. Standby compressors in many cases are either diesel or gasoline powered.

8. Ventilating Air

In compressed air tunnels, air must be circulated in sufficient quantity to permit the work to be done without danger or excessive discomfort.

In free air tunnels, the ventilation rate must be adequate to clear the tunnel of gases in a maximum of 15 minutes if explosives are used. Rates also must be adequate to dilute exhaust of permissible diesel equipment to safe limits. A minimum of 200 cubic feet/minute of fresh air per employee underground should be provided. In cold weather it may be necessary to condition ventilating air to prevent excessive fogging at the heading. Air should be monitored constantly for toxic or flammable gases and airborne contaminants. A record of all tests should be maintained.

VIII. DEWATERING

Storm sewers or other storm drainage systems generally lie in or near the lowest point of the drainageway. This provides for the optimum drainage to the facilities, but also can cause problems during construction. Handling flows entering the construction area is frequently one of the biggest problems confronting the contractor during drainage improvement projects. The contractor is generally solely responsible for the dewatering of the work area; however, it is frequently advisable for the engineer to review the contractor's dewatering plan.

Dewatering of excavations and trenches is necessary to provide proper working conditions for the construction or installation of storm sewers. Where possible, the most economical means is the diversion of upstream inflows. This allows for the elimination of flow before it enters the excavation. This can be accomplished by physical diversion of the inflow through cutoffs and diversion channels or pumping around the project site. Where upstream flows cannot be diverted, or where groundwater enters the excavation, it must be removed.

Excavations should be dewatered for concrete placement and sewer pipe laying, and they should be kept continuously dewatered for as long as necessary. Unfortunately, the disposal of large quantities of water from this operation, in the absence of existing storm drains or adjacent water courses, may present problems. The possibility of draining the water through the completed sewer to a permissible point of discharge may be considered when other means of disposal are unavailable. Sufficient precautions must be taken to prevent scour of freshly placed concrete or mortar.

Crushed stone or gravel, possibly in combination with a geotextile, may be used as a sub-drain to facilitate drainage to trench or sump pumps. It is good practice to provide clay dams in the sub-drain to minimize the possibility of excessive groundwater flows undercutting the sewer foundation.

An excessive quantity of water, particularly when it creates an unstable soil condition, may require the use of a well-point system. A system of this type consists of a series of perforated pipes driven or jetted into the water-bearing strata on either side of the sewer trench and connected to a pump by a header pipe. The equipment for a well-point system is expensive and specialized. General contractors often seek the help of special dewatering contractors for such work. Well-point systems must be run continuously to avoid disturbing the excavated trench bottom by uplift pressure.

When excavating in coarse water-bearing material, turbine well pumps may be used to lower the water table. Chemical or cement grouting and freezing of the soil adjacent to the excavation have been used in extremely unstable water-bearing strata.

Water from all types of dewatering systems should be checked periodically to assure that fine-grained material is not being removed from beneath the pipe. This might cause future pipe settlement.

Care must be exercised to insure that property damage, including silt deposits in sewers and on streets, does not result from the disposal of diverted drainage. The water control plan should be developed and implemented in a manner that does not impact the quality or quantity of water in downstream drainageways.

IX. FOUNDATIONS

The foundation of the storm sewer is critical to the structural integrity of the facility. Firm cohesive soils provide adequate sewer pipe foundations when properly prepared. Occasionally the trench bottom may be shaped to fit the sewer pipe barrel and holes dug to receive projecting joint elements. It is often a practice to over-excavate and backfill with granular material, such as crushed stone, crushed slag, or gravel, to provide uniform bedding of the sewer pipe. Such granular bedding is used because it is both practical and economical.

In very soft bottoms it is frequently necessary to first overexcavate to greater depths and stabilize the trench bottom by the addition of gravel or crushed slag or rock compacted to receive the load. The stabilizing material must be graded to prevent movement of the subgrade up into the stabilizing base, and the base into the bedding material. There is increasing use of specialized filter fabrics to prevent this movement. The required stabilization depth should be determined by tests and observations on the job.

Where the trench bottom cannot be stabilized satisfactorily with a crushed rock or gravel bed, and where limited and intermittent areas of unequal settlement are anticipated, a timber cribbing, piling, or reinforced concrete cradle may be necessary.

Where the bottom of the trench is rock, it must be overexcavated to make room for an adequate bedding of granular material, which will uniformly support the conduit. The trench bottom must be cleaned of shattered and decomposed rock or shale prior to placement of bedding.

In some instances, a sewer pipe must be constructed for considerable distances in areas that are subject to subsidence. If the subsidence is shallow, consideration should be given to constructing the sewer on a timber platform or reinforced concrete cradle supported by piling. The sewer's support should be adequate to sustain the weight of the full sewer and backfill. Piling in this case is sometimes driven to grade with a follower prior to making the excavation. This practice avoids subsidence of trench walls resulting from pile driving vibrations. Extreme care must be taken to locate all underground structures.

X. BACKFILLING

Backfilling is an important consideration in construction. The methods and equipment used in placing fill must be selected to provide the appropriate character and compaction of the fill. The method of backfilling varies with the width of the excavation, the character of the materials excavated, the method of excavation, and the degree of compaction required.

A. Degree of Compaction

For trench backfill in improved streets or streets programmed for immediate paving, a high degree of compaction is normally required. In less important streets or in sparsely inhabited subdivisions where flexible macadam roadways are used, a more moderate specification for backfilling may be justified. Along sewers in open country, it may be sufficient to mound the trench and, after natural settlement, return to regrade the area. For general backfill of channel slopes or structural backfill, compaction should be based on structural criteria. The stability of slopes is often dependent on the degree of compaction. Structures often require a specified degree of compaction to satisfy structural design assumptions.

The degree of compaction required is generally expressed as some relation to the laboratory maximum density at optimum moisture content. The two most common tests used to determine maximum density of cohesive soils are ASTM D-698 (standard Proctor density) and ASTM D-1557 (modified Proctor density). For cohesionless soils, ASTM D-4253 and D-4254 are more appropriate tests to determine the maximum density and optimum moisture.

The field density is generally measured by the sand-cone method (ASTM D-1556), the balloon method (ASTM D-2167), or nuclear density meter (ASTM D-2992 and D-3017). For cohesive soils, the degree of compaction required should generally be expressed as the relative compaction. This is defined as the ratio of the field dry density to the laboratory maximum dry density and is expressed as a percentage. For cohesionless free-draining soils that do not exhibit a well-defined moisture/density relationship under impact compaction, vibrating compaction is more appropriate. The degree of compaction required for these soils is expressed as the relative density and is determined by the relationship between the field density, the minimum density, and the maximum density.

$$\frac{D_{max}(D_{field} - D_{min})}{D_{field}(D_{max} - D_{min})} \times 100 \qquad (16\text{-}1)$$

B. Trench Backfilling Sequence

Backfilling should proceed immediately on curing of trench-made joints and after the concrete cradle, arch, or other structures gain sufficient strength to withstand loads without damage. In areas of construction not requiring work beyond pipe placement, backfilling should immediately follow pipe placement.

Backfill generally is specified as consisting of three zones with separate and distinct criteria for each: The first zone (pipe zone) extends from the foundation material to 12" above the top of the sewer pipe or structure; an intermediate zone generally contains the major volume of the fill; and the upper zone consists of pavement subgrade, finish grading materials, topsoil, etc.

The first zone should consist of selected materials placed by hand or by suitable equipment in such a manner as not to disturb the sewer pipe, and compacted to a density consistent with design assumptions. In some instances the material used for granular bedding is brought above the sewer to insure high density backfill with minimum compactive effort. When installing flexible pipes, attention must be given to proper placement and compaction of the haunching material from the base of the pipe to the springline. When high water tables are anticipated, backfill materials without substantial voids are required to prevent soil migration.

Compaction of the intermediate zone is usually controlled by the location of the trench. Under traffic areas or other improved existing surfaces, a high degree of compaction may be required. In undeveloped areas, little compaction may be required. In general, the degree of compaction required will affect the choice of material. The use of excavated material, if suitable, is usually desirable in areas subject to frost heave so that excavated areas will move no more and no less than undisturbed areas.

Depth and compaction of the upper zone are dependent on the type of finish surface to be provided. If the construction area is to be seeded or sodded, the upper 18" may consist of 14" of select material slightly mounded over the trench and lightly rolled, covered by 4" of top soil. If the area is to be paved, the upper zone must be constructed to the proper elevation for receiving base and paving courses under conditions matching design assumptions for the subgrade. If the trench backfill is completed in advance of paving, the top 6" of the upper zone should be scarified and recompacted prior to paving. In such cases, it may be necessary to install a temporary surface to be replaced at a later date with permanent pavement.

Before and during the backfilling of an excavation, precautions should be taken to prevent flotation of pipelines due to the entry of large quantities of water into the trench. The buoyant forces may affect the vertical alignment of the pipe. A check of the hydrostatic pressures causing uplift is advisable for sewers in areas of high groundwater.

There are cases where pipe is laid without excavation (there is no trenching—rather the pipe is laid on a prepared base and fill material is then placed around the pipe). Such pipes frequently are of large diameter, and particular attention must be given to backfilling in the haunch areas and beside the pipe to provide lateral support to prevent compression failures. Construction loads in these cases can far exceed normal service loads.

In all cases, the level of compaction required is dictated by the design assumptions with respect to pipe and structure foundations, the location of the excavation, and the character of backfill materials. A thorough geotechnical analysis should be performed, and the results used to determine the final degree of compaction necessary to achieve the proper installation.

C. Methods of Compaction

Cohesive materials with high clay content are characterized by small particle size and low internal friction. They have small ranges of moisture content over which they may be compacted satisfactorily and are very impervious when compacted. Because of the strong adhesion of the soil particles, high pressures must be exerted to shear the adhesive forces and remold the particles into a dense soil mass. This dictates the use of impact-type equipment for most satisfactory results. In confined areas, pneumatic tampers and engine-driven rammers may give good results. The upper portion of narrow trenches can be consolidated by self-propelled rammers. In wide excavations sheeps-foot rollers may be used; if the degree of compaction required is not high, bulldozers and loaders may be used to compact the fill.

Regardless of equipment used, the soil must be near optimum moisture content and compacted in multiple lifts generally not exceeding 8" in loose depth if satisfactory results are to be obtained. The trench bottom must be free of excessive water before the first lift of backfill is placed.

If the material has a high moisture content at the time of excavation, some preparation of the material probably will be required before spreading in the trench. This may include pulverizing, drying, or blending with dry or granular materials to improve placement and consolidation.

Tamping foot self-propelled compactor compacting backfill.

As noted earlier, cohesionless materials are best compacted using vibratory equipment. Moisture content at the time of compaction is not so critical, and consolidation is effected by reducing the surface friction between particles thus allowing them to rearrange in a more compact mass. In confined areas, vibratory plates give the best results. For wider excavations vibratory rollers are most satisfactory. Again, if the degree of compaction required is not high, and if layers are thin, the vibration imparted by dozer or loader tracks may result in satisfactory consolidation.

In some areas water is used to consolidate granular materials. Unless the fill is saturated and immersion vibrators are used, the degree and uniformity of compaction cannot be controlled closely. With some materials, adequate compaction may be obtained by draining water used to saturate or puddle fill through drains constructed in structure walls. These drains are capped after the backfill has drained.

Sometimes the material removed from the excavation may be entirely unsatisfactory for backfill. In this case, selected materials must be hauled in from other sources. Cohesive materials, noncohesive materials, or a combination of these may be used, but an assessment must be made of the possible change in groundwater movement that the use of outside

Vibratory steel drum compacting subgrade of low flow channel.

materials may cause. For example, the use of cohesive materials to backfill a trench in rock could result in a dam impervious to groundwater travelling in rock faults, seams, and crevices. On the other hand, granular materials placed in a clay trench could result in a very effective sub-drain.

XI. PIPE STORM SEWERS

A. Storm Sewer Pipe Quality

Storm sewer pipe inspection is properly conducted by the manufacturer and by independent testing and inspection laboratories. Moreover, with pipe storm sewers, transportation charges may constitute a substantial portion of material costs, and as a result inspection at the pipe plant is usually desirable. Inspection may consist of visual inspection of workmanship, surface finish, and markings; physical check of length, thickness, diameter, and joint integrity and tolerances; proof of crushing strength (rigid pipe) or pipe stiffness (flexible pipe), design materials tests, and tests of representative specimens. If three-edge bearing tests are not used on precast concrete pipe, core or cylinder tests should be required. Standard cylinder tests are not practical with the mixes used in some manufacturing methods and core tests are generally used. Cores also permit checking tolerances on placement of reinforcing cages.

Storm sewer pipe suppliers should furnish certificates of compliance with specifications that can be easily checked as the storm sewer pipe arrives at the site. Storm sewer pipe also should be checked visually at time of delivery for possible damage in transit, and again as it is laid for damage in storage or handling.

B. Storm Sewer Pipe Handling

Care must be exercised in handling and bedding all precast storm sewer pipe, regardless of cross-sectional shape. All phases of construction should be undertaken to insure that, insofar as practical, pipe is installed as designed. Pipe should be handled during delivery in a manner that eliminates any possibility of high impact or point loading due to dropping or impacting, with care taken always to protect the joints.

C. Storm Sewer Pipe Placement

Storm sewer pipe should be laid on a firm but slightly yielding bedding, true to line and grade, with uniform bearing under the full length of its barrel, without break from structure to structure, and with the socket ends of bell and spigot or tongue and groove storm sewer pipe joint facing upgrade. Storm sewer pipe should be supported free of the bedding during the jointing process to avoid disturbance of the subgrade. A suitable excavation should be made to receive sewer pipe bells and

Installation of large diameter storm sewer.

joint collars where applicable so that the bottom reaction and support are confined only to the pipe barrel. Adjustments to line and grade should be made by scraping away or adding adequately compacted foundation material under the pipe and not by using wedges and blocks or by beating on the pipe.

Extreme care should be taken in jointing to insure that the bell and spigot are clean and free of any foreign materials. Joint materials vary with the type of storm sewer pipe used. All pipe joints should be made properly using the jointing materials and methods specified. All pipe joints should be sufficiently tight to meet infiltration or exfiltration tests.

In large diameter storm sewers with compression-type joints, considerable force will be required to insert the spigot fully into the bell. Come-alongs and winches or the crane itself may be rigged to provide the necessary force. Inserts should be used to prevent the storm sewer pipe from being thrust completely home prior to checking gasket location. After the gasket is checked, the inserts can be removed and the joint completed.

The operation of equipment over small diameter storm sewer pipe, or other actions that would otherwise disturb any conduit after pipe jointing, must not be permitted.

At the close of each day's work, or when storm sewer pipe is not being laid, the end of the pipe should be protected by a close-fitting stopper to keep the pipe clean and to prevent unwanted access into the pipe, with adequate precautions taken to overcome possible uplift. The elevation of the last storm sewer pipe placed should be checked the next morning before work resumes.

If the storm sewer pipe load carrying capacity is increased with either arch or total encasement, contraction joints should be provided at regular intervals in the encasement coincident with the pipe joints to increase flexibility of the encased conduit.

D. Manholes and Inlets

The two primary appurtenances to pipe storm sewer construction are manholes and inlets. These appurtenances are essential to the proper functioning of storm sewer systems. The materials most commonly used for manhole and inlet construction include precast concrete sections and cast-in-place concrete.

Proper construction methods are important in the installation of manholes and inlets and, as with the storm sewer itself, proper backfill compaction is necessary.

XII. OPEN CHANNELS

The excavation for open channels may be accomplished by various types of earth-moving equipment, depending on the size of the facility. Small channels may be excavated to grade by backhoes or excavators whereas large channels may be excavated by large earth-moving equipment.

A. Trapezoidal Channels

Trapezoidal channels are normally lined with grass, concrete, riprap or gabions. A trapezoidal channel that is to be lined with concrete should be excavated to very close tolerances to control the amount of concrete. A grass-lined channel normally will have topsoil placed on it to bring it to grade. Riprap or gabion lined channels need to be overexcavated to allow for the design thickness of the riprap or gabions being used.

Where a concrete lining is used, the bottom is usually a poured and screeded slab. Care must be taken with the placement of concrete to assure that reinforcing steel or wire mesh is properly positioned. Some concrete channels can be constructed with special slip form pavers, or using pre-cast units. In the latter case, the bottom is usually a poured slab with a stub and key-way for the wall to rest on. Any impervious lining should be provided with weep holes for pressure equalization.

B. Rectangular Channels

Rectangular channels are usually constructed where space is limited. The type of rectangular structure will dictate which construction method is to be used. Space limitations, cost, and visual appearance often dictate the materials used for the rectangular channel. Alternative materials to cast-in-place concrete for wall construction are often timber or metal cribbing, soil cement, rock gabions, reinforced earth, precast concrete, or sheet piling. Where cast-in-place concrete is used, the quantity, placement, and cleanliness of the reinforcing steel must be carefully checked.

C. Low-Flow Channels

In conjunction with either rectangular or trapezoidal channels, low-flow channels (or trickle channels) are often constructed to contain the base flows within the bottom of the channel. These channels are typically lower in elevation than the primary channel and, as a result, may be able to provide drainage for the subgrade beneath the structures.

D. Structures

Open channel storm drainage systems often include appurtenances such as drop structures, culvert crossings and siphons. During the excavation for structures, a geotechnical engineer should be available for inspection and recommendations regarding the foundation material and suitability of the excavated material for backfill. If the foundation material is inadequate for the bearing pressure required, over-excavation and replacement with satisfactory materials may be necessary. Backfilling should be done in a manner that will not block weep holes. To prevent plugging, a granular filter, sometimes with a geotextile covering, should be placed behind the weep holes.

1. Drop Structures

Drop structures should be constructed with care to assure that flows will not undermine the structure. They may be constructed of concrete, riprap, gabions, sheet piling, or some combination thereof. A concrete structure must follow good structural construction practices with adequate control of the foundation material and placement of the steel and concrete. Cutoff or crest walls are normally constructed along the length of the structure perpendicular to the flow to prevent flows from undermining the structure. Proper bedding beneath the structure is important to provide an adequate foundation for construction as well as to control subsurface water.

Riprap or gabion placement and grading must be controlled so that the design integrity is not sacrificed. Placement of riprap or gabions below the drop structure is often necessary to provide additional energy dissipation of the high flow velocities. Care should be taken to use adequately sized riprap or gabions to withstand these velocities.

2. Culverts and Siphons

As with drop structures, care needs to be taken with the construction of culverts and siphons to assure structural integrity, including adequate foundation and bedding. Once again, a geotechnical engineer should be available for inspection and recommendations regarding the foundation material and suitability of the excavated material for backfill.

Headwalls and wingwalls are structural components that require careful construction. Warped wingwalls are often provided in conjunction with culverts or siphons in open channels to provide a smoother transition of flow. Care must be taken when backfilling these walls as they vary in section from a vertical wingwall to a sloping concrete lining.

Debris walls are often constructed in conjunction with box culverts, bridges and siphons for the purpose of preventing the accumulation of debris from plugging or damaging the structure. Debris walls normally consist of an extension upstream of the concrete wall or pier dividing the cells of the box culvert, bridge, or siphon. These walls will of course be parallel to the direction of flow. Care must be exercised to provide a hardened surface (often steel) along the projecting face. The structural integrity of these walls is critical to ensure that the wall can withstand vibration and the forces acting upon it by accumulated debris and high flows.

XIII. RIPRAP

Riprap is one of the most common materials used for erosion protection in earth-lined channels. Riprap protection can be classified into two basic types, grouted and non-grouted:

A. Non-Grouted Riprap

Non-grouted riprap or dumped riprap is placed by means of a backhoe, excavator, or loader and is not bonded together by any artificial means, such as concrete, grout, or shotcrete. Therefore, to assure that an area of riprap will retain its structural integrity, the rock must be properly graded. If an adequate amount of smaller rocks is not provided in the riprap, voids may result, which will create an unstable mass that may be subject to failure. Dumped riprap is not recommended on banks with a slope steeper than 2:1.

The type of rock or riprap installed should be carefully reviewed before installation. It is often advisable for the engineer to visit the quarry prior to commencement of construction operations. The rock should be quarried rock that is fractured and has a specific gravity of at least 2.5. If cobbles are used instead of fractured rock, they may lack the capability of interlocking together to make the solid mass required. Angularity enhances the interlocking capabilities and is normally specified.

Riprap with a mean size of less than nine inches may be installed, covered with soil and then seeded to create a more natural appearance. This type of installation does not require bedding beneath the layer of riprap. Riprap with a mean size greater than 9" usually requires bedding (0.75–1.5 inch crushed rock). The bedding allows free passage of water through the rock without allowing it to wash out the fine-grained materials beneath the riprap layer. If site conditions warrant, a layer of filter material may be placed beneath the bedding to further decrease the leaching of fines from the soil. The filter material may be either a geotextile or a properly designed granular filter. If a geotextile is used, the bedding material should be placed with sufficient care to protect the geotextile from ultraviolet light exposure and damage during riprap placement.

When placing riprap, great care should be taken that the riprap is properly toed into the channel bottom. The toe or bottom of the riprap blanket should be below the elevation of ultimate degradation or at least a depth equal to the thickness of the riprap layer below the channel bottom. If the riprap is not toed in, the flow of water in the stream may undermine the riprap, leading to a failure of the entire protected bank.

If a smoother surface is desired, it may be necessary to densify and smooth the riprap layer. This may be accomplished by rolling or plating the riprap. Plating riprap involves slapping the surface with a piece of steel plating (weighing approximately 5,000 pounds) or the back of an excavator bucket. The objective is to produce a reasonably smooth surface along the riprap. This practice will produce a tight, uniform blanket of rock with greater stability because of reduced drag on the individual stones and an increase in the angle of repose produced by the interlocking mass of rock. The smooth surface can also decrease the hydraulic roughness of the channel.

B. Grouted Riprap

Grouted riprap is bonded together as a single mass using concrete or grout. This type of riprap is generally used when the channel banks are steeper than 2:1, when channel velocities are excessive, when turbulence is high, or when available stone is of insufficient size to satisfy hydraulic conditions.

As with dumped riprap, the gradation of the riprap for grouting is important, but for different reasons. The rocks should be large enough (mean size greater than 12 inches) so that the rocks will protrude high enough above the grout to effectively reduce the energy head of the water flowing in the channel. Also, the gradation should be such that the smaller rocks (less than six inches) are removed to allow the grout to fully penetrate the entire layer of the riprap.

Riprap that is to be grouted should be placed on bedding and possibly filter material. The bedding allows a free flow of water through the layer below the riprap. Weep holes should be installed to allow the ground water to be released from the bedding.

Full depth penetration of the grout is required to form a solid mass. To achieve this, several conditions are necessary. A high slump grout, as high as ten inches, should be used so that the grout flows into each void. During delivery of the grout, a low pressure pump with a minimum nozzle diameter nozzle of 2–4″ should be used to direct the grout

Grouting riprap with low-pressure grout pump.

to each void between the rocks at a manageable rate. When a small nozzle is used, smaller aggregate should be specified to prevent clogging. Some additives can also be included in the mix to increase its plasticity. If full depth penetration is still not achieved, then the grout may need to be vibrated.

Before grouting, the rocks should be washed to sluice the fines to the bottom of the riprap which will allow the grout to bond to the rocks. After placing the grout, a curing compound should be applied.

Some construction practices may be used to increase the aesthetic properties of the grouted riprap. Following the application of grout, it may be brushed with a small broom to give a smooth finish. Also, care should be taken during grouting to allow a minimum of grout to splash on top of the exposed rocks. Another method of improving the apppearance of the grouted riprap is to insert a dye into the grout to achieve a desired color.

XIV. EROSION AND SEDIMENT CONTROL

Clearing and stripping of land for storm drainage systems, if not properly conducted, may result in high localized erosion rates with subsequent deposition and damage to off-site properties. An erosion and sediment control plan can reduce the erosion and sediment deposition process to an acceptable level.

Several techniques (best management practices) are available for the control of erosion and sedimentation. Their effectiveness is often dependent upon the characteristics of the soils at the site as well as the topography, drainage, vegetation and other site features. Construction activities should be planned and completed in such a manner so that the exposed area of disturbed land is minimized, and so that the land is disturbed for the shortest possible period. Local regulations often detail acceptable practices for erosion and sediment control.

XV. SPECIAL CONSTRUCTION

A. Railroad Crossings

Storm sewers at times must be constructed under railroad tracks, which may be at street grade or on an existing railroad viaduct. Crossing of tracks at grade or on an embankment is usually accomplished most economically by jacking, boring, tunnelling, or a combination thereof. Usually, to satisfy railroad criteria, a casing pipe is installed and the storm sewer pipe is then placed inside.

When the distance from the base of rail to the top of the storm sewer is insufficient to allow jacking or tunnelling (usually less than one diameter clearance), other construction means, which must be coordinated with the railroad, may be required. It may be necessary to provide a bypass and remove the affected tracks or, for local service lines, to remove the tracks and interrupt service during an open-cut operation. A temporary structure for support of the railroad tracks may also be a solution, with the storm sewer constructed in an open trench below that structure.

Construction of storm sewers under existing railroad viaducts involves a wide variety of methods, depending on the size of the storm sewer, its location in plan and elevation with respect to viaduct footing, type of footings, the nature of the soil, and the requirements of the railroad.

Where the soil is stable and the storm sewer is of sufficient size and is located satisfactorily with respect to viaduct footings, tunnelling may be both safe and economical. When the proposed storm sewer does not meet these criteria, close coordination with the railroad is essential and special methods of sheeting and bracing must be devised. To prevent subsequent movement of soil beneath the footings, all sheeting and bracing should be left in place.

In all cases, early planning with the railroad authorities is essential, since they generally have extensive design, inspection, and permit requirements.

B. Principal Traffic Arteries

Residential and secondary traffic arteries can usually be closed to traffic during the construction of storm sewer crossings. On heavily travelled streets and highways where public safety and convenience are major factors, it may be necessary to use detours, tunnelling, or jacking methods for the crossing.

When required, traffic movements across trenches can be accommodated by temporary decking. Trenches of narrow or medium width can be spanned with prefabricated decks placed on steel or timber mudsills or soldier piles at the edges of the trench. Where the top of the trench is wider than 16–20′ temporary piling for end support, and in some cases center support, may be required.

C. Outfall Structures

Storm sewer outfalls and headwalls may be located above or below surface water levels. When they are partly submerged, it is necessary to provide some form of cofferdam during construction. In shallow water, an earth dike or timber piling may be sufficient to maintain a dry excavation. In deep water, steel sheet piling cofferdams are desirable. Usually a single wall cofferdam with adequate bracing is sufficient, but in excessive depths at the banks of main navigation channels, a double wall may be required. Standard cofferdam design and construction practices should govern.

XVI. CONSTRUCTION RECORDS

It is generally the responsibility of the contractor to record details of construction as accomplished in the field. These data should be transferred to the engineer for incorporation into a final revision of the contract drawings, so that they may be available for future use.

XVII. REFERENCES

American Concrete Pipe Association. (1970). *Concrete pipe design manual.* 7th ed., rev., Arlington, VA

American Concrete Pipe Association. (1988). *Concrete pipe handbook.* 3rd ed., Arlington, VA.

American Iron and Steel Institute. (1971). *Handbook of steel drainage and highway construction products.* 1st ed., AISI,

American Society of Civil Engineers. (1969). "Design and construction of sanitary and storm sewers." *Manuals and Reports on Engineering Practice—No. 37,* ASCE, New York, NY.

American Society of Civil Engineers. (1982). "Gravity sanitary sewer design and construction." *Manuals and Reports on Engineering Practice—No. 60,* New York, NY.

American Water Works Association. (1979). *Installation of concrete pipe. Manual M9,* Denver, CO.

Associated General Contractors. *Manual of Accident Prevention in Construction.* Washington, D.C.

Bickel, J. and Kuesel, T. (1982). *Tunnel engineering handbook.* Van Nostrand Reinhold Co., New York, NY.

Bouchard, H. and Moffit, F. (1965). *Surveying.* International Textbook Company, Scranton, PA.

E.I. du Pont de Nemours and Company, Inc. (1958). *Blasters handbook.* Wilmington, DE.

Linsley, R. and Franzini, J. (1972). *Water resources engineering.* McGraw-Hill Book Co., Heightstown, NJ

National Fire Protection Association. "Code for the manufacture, transportation, storage and use of explosives and blasting agents." *Publ. No. 495,* NFPA.

Oregon Department of Transportation. (undated). *Keyed riprap.* Federal Highway Administration, Region 15, Arlington, VA.

Portland Cement Association. (1968). *Design and construction of concrete sewers.* Chicago, IL

Urban Drainage and Flood Control District. (1984). *Urban storm drainage criteria manual.* Regional Council of Governments, Denver, CO.

U.S. Department of Commerce. (1929). "Mine gases and methods of detecting them." *Miner's Circ. 33*, Bureau of Mines, U.S. Government Printing Office, Washington, D.C.

U.S. Department of Commerce. (1928). "Protection against mine gases." *Miner's Circ. 35*, Bureau of Mines, U.S. Government Printing Office, Washington, D.C.

U.S. Department of Commerce. (1960). "Safety with mobile diesel powered equipment underground. *Publication RI5616*, Bureau of Mines, U.S. Government Printing Office, Washington, D.C.

U.S. Department of Commerce. (1941). "Some essential safety factors of tunnelling." *Bulletin 439*, Bureau of Mines, U.S. Government Printing Office, Washington, D.C.

U.S. Department of Commerce. (1935). "Engineering factors in the ventilation of metal mines. *Bulletin 385*, Bureau of Mines, U.S. Government Printing Office, Washington, D.C.

U.S. Department of the Interior. (1960). *Design of small dams*. Bureau of Reclamation, Water Resources Technical Publication, U.S. Government Printing Office, Washington, DC.

U.S. Department of Interior. (1980). *Earth manual*. Water and Power Resources Service, Washington, D.C.

U.S. Department of Interior. (1982). *Construction safety standards*. Bureau of Reclamation, Denver, CO.

U.S. Department of Transportation. (1976). *Corrugated metal pipe, structural design criteria and recommended installation practice*. Federal Highway Administration, Washington, D.C.

Appendix A
PLANNING AND DESIGN EXAMPLES

This appendix includes two examples, both of which are related to planning. These examples were selected to illustrate principles detailed in the Manual, and to demonstrate how they may be practically applied. The first is an example of a plan to control existing and future impacts of stormwater pollution on the water quality of two drinking water reservoirs in Newport News, Virginia. The second describes the planning, design and construction of a multi-purpose flood control facility in Valparaiso, Indiana.

I. RESERVOIR WATER QUALITY PROTECTION— NEWPORT NEWS, VA

A. Introduction

Newport News, a city in southeastern Virginia, faced a problem of water supply reservoir contamination due to nonpoint source pollution stemming from urban development. This example describes a plan to control the existing and future impacts of stormwater pollution on the water quality in the two terminal water supply reservoirs of the Newport News waterworks system, Lee Hall and Harwood's Mill, whose watersheds are 15.8 mi^2 and 9.5 mi^2, respectively (see Figure A.1).

The plan uses "best management practices" (BMPs). There are two basic approaches possible to manage urban nonpoint pollution with structural BMPs:

    (a) Onsite Control, which involves the construction of individual BMPs (e.g., wet detention basins, infiltration facilities), designed and constructed by individual developers on each development site.
    (b) Regional BMP Master Plan, which involves the strategic location of BMP facilities to simultaneously control nonpoint pollution loadings from multiple development projects. These facilities would be con-

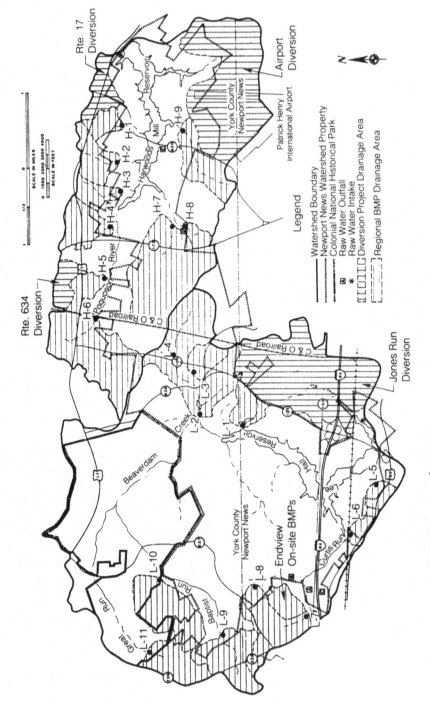

Figure A.1.—Regional BMP master plan.

structed and operated by local government. Capital costs would be recovered by pro-rata contributions by the developers, or by other financing mechanisms.

Virginia has operated local stormwater management programs for peak flow since the mid-1970s, and local governments typically elected to use the onsite control approach because it seemed to require little advance planning and appeared relatively easy to administer. Now, however, many local governments in Virginia are giving greater consideration to the regional approach. The main advantages of this approach include:

(a) Reduction in Capital and O&M Costs: Larger facilities have economies of scale, in terms of costs of construction, land acquisition and engineering design. EPA NURP studies indicate that average annual O&M costs for detention basin BMPs are about 3%–5% of base construction costs, and regional facilities should achieve corresponding economies of scale. Moreover, larger facilities can more easily be designed for ease of maintenance.
(b) Increase in Land Development Opportunities: Less total land is needed for large facilities, and the regional approach typically relies on existing floodplain areas for much of the reservoir surface area.
(c) Increased Opportunities for Recreational Use: Regional facilities are larger, and generally can be designed to provide convenient access.
(d) Opportunity to Manage Nonpoint Pollution Impacts from Existing Development: Some existing development may be located within regional BMP watersheds, and nonpoint control will be achieved.
(e) Improved Peak Flow Control: Because they can be strategically located within a watershed, regional facilities can be sited to ensure more reliable peak flow control than randomly-located onsite controls.

Disadvantages of the regional approach include:

(a) Local government must perform advance studies to locate and develop preliminary designs for such facilities.
(b) Local governments must finance, design, and build the regional facilities before most development occurs. Reimbursement by developers will be delayed until building is complete.
(c) Local governments may incur extraordinary maintenance responsibilities if BMPs are perceived as primarily recreational facilities, which merit commensurate water quality protection.

B. Regional BMP Siting Considerations

Figure A.1 also shows the locations of the 24 regional BMP facilities recommended in the study, designated by the prefix "L" and "H" for the Lee Hall and Harwood's Mill watersheds respectively (for four of the highly commercialized/industrialized watersheds, the recommended plan was to divert this runoff to another receiving water). Factors considered in their location were:

(a) Type of BMP (two types were considered): (1) infiltration controls, which divert stormwater runoff into the soil where pollutant removal occurs through such processes as filtration, adsorption, and oxidation by soil microorganisms; and (2) detention controls, which remove pollutants by sedimentation and, in the case of wet detention basins, through physical, chemical, and biological processes which take place in the basin's permanent pool.

(b) Drainage area restrictions: In this case, large enough to serve more than one development, but not so large as to require a permit under the State's Dam Safety program (which would cause undue delays and greater administrative costs).

(c) Land availability (with highest priority given to sites owned by Newport News Waterworks) and Tributary Land Use: Criteria included appropriate drainage area; sufficient amount of proposed and/or existing urban development; design pool elevations that did not impact on upstream structures, roads or other facilities; adequate access for inspection and maintenance.

(d) Ability to control future land use.

(e) Encroachment upon existing urban areas, roads and utilities.

(f) Comprehensive coverage—In addition to the 20 wet detention basin BMPs selected, major diversion projects were recommended for four areas along watershed boundaries to supplement the detention facility network.

C. General Design Criteria

Because this was a conceptual planning study, only general design criteria were specified. At the time of preliminary design, each facility should receive detailed analyses. Some key design parameters include:

(a) Permanent Pool.
   (1) Storage requirements (volume)—Determined in this study on the basis of land use.
   (2) Depth—Shallow enough to minimize thermal stratification and maintain aerobic bottom conditions, but deep enough to prevent light stimulation of rooted aquatics in the open water zone. Mean depth chosen is about 1–3 meters, and maximum depth to prevent stratification of about 5–6 meters.
   (3) Side Slopes—No steeper than 4:1 (for safety), and planted from 2' below to 1' above the permanent pool control elevation for ecologic balance and safety.

(b) Earth embankment—Suggested criteria taken from U.S. Soil Conservation Service Standards and Specifications (1981). These include minimum top widths of 6' (for embankments <10' high) to 8' (for embankments between 10' and 15' high); combined upstream and downstream side slopes not < 5:1, with neither slope > 2:1; freeboard above the permanent pool of at least 2' (with a spillway to carry the 100-year runoff); and a foundation cutoff to prevent leakage from the reservoir.

(c) Outlet structures—Riser pipe or standard precast unit.

(d) Maintenance easement—Permanent easement at least 15' in width around the perimeter of the basin, measured from the maximum elevation of the storage pool.

D. Recommended Regional BMP Facilities Plan

The general characteristics of each of the recommended BMP facilities are shown in Table A.1. Sites L-1 and L-2 are existing facilities that would be dredged and maintained as BMP facilities. In addition, there are four diversions. With respect to the four watershed diversions noted above, the Jones Run diversion would bypass runoff from a 921-acre area to a point downstream of Lee Hall dam, and would include a dry detention basin upstream of the railroad crossing of Jones Run (sized for the 100-year storm). The Route 17 diversion would bypass runoff from a commercial and residential area along Route 17 to a point below Harwood's Mill dam. The Airport diversion would bypass runoff from a 1,130-acre area programmed primarily from industrial development. The Route 634 diversion would bypass runoff from 104 acres of proposed industrial development into the Chisman Creek watershed.

E. Water Quality Benefits

The impacts of the regional BMP plan were studied using several water quality models. Eutrophication impacts were evaluated with the Rast, Jones, and Lee input/output model (1983) and the Jones and Bachman input/output model (1976). Heavy metals impacts were evaluated with the Dillon and Rigler input/output model (1974). The recommended plan is projected to achieve mean concentrations of chlorophyll-a, nutrients and heavy metals which are lower than those associated either with the onsite BMP approach, or with the existing land use (because regional BMPs will capture runoff from existing, as well as new, development, and the released water will be of better quality than that of the current combined runoff from existing development/undeveloped land).

F. Project Priorities

To assist in the development of an implementation program, the 24 projects were prioritized based on the following criteria.

(a) Existing development—Regional BMPs serving currently developed areas get a higher priority than relatively undeveloped areas.
(b) Future land use—Areas proposed for intensive urban development merit a higher priority.
(c) Location in the watershed—Slug loadings from thunderstorms pose a greater potential threat if the discharge point is relatively close to the raw water intake in the reservoir, and BMPs with discharges near intakes are of higher priority than those discharging to reservoir headwaters.
(d) Permitting—Sites requiring a dam safety permit may merit a higher priority because of lead times.
(e) Land acquisition—Sites on private property merit higher priority than sites on public lands, because of development potential.

TABLE A.1. Regional BMP's: Wet Detention Basins.

Site (1)	Drainage Area (acres) (2)	Storage Volume (acre-ft) (3)	Priority (4)
Harwood's Mill			
H-1	86	22.9	B
H-2	161	33.0	B
H-3	107	17.6	B
H-4	163	29.2	A
H-5	110	19.2	A
H-6	564	86.3	A
H-7	626	72.9	D
H-8	67	19.2	C
H-9	228	32.8	B
Lee Hall			
L-1	117	14.4	C
L-2	806	55.0	B
L-3	114	27.3	D
L-4	318	47.2	D
L-5	45	11.3	B
L-6	69	7.9	B
L-7	230	36.0	A
L-8	75	10.9	B
L-9	348	53.3	C
L-10	358	45.5	C
L-11	226	25–40	C

Assigned priorities for each of the regional BMPs are shown in Table A-1. The Jones Run and Route 17 diversions were assigned priority "A," the Airport diversion was assigned priority "B," and the Route 634 diversion was assigned priority "C."

G. Cost Estimates

Construction costs were estimated using a regression equation derived for the Washington, D.C. region (U.S. Environmental Protection Agency, 1983) that relates construction cost to storage volume, and O&M costs were estimated to be 3%-5% of base construction costs.

H. References

Camp Dresser & McKee. (1986). "Reservoir water quality protection study." *Phase 3 Report*, City of Newport News Department of Public Utilities, Annandale, VA.

Dillon, P.J. and Rigler, F.H. (1974). "A test of simple nutrient budget model predicting the phosphorus content of lake waters." *Journal of the Fisheries Research Board of Canada*. 31.

Jones, J.R., and Bachman, R.W. (1976). "Prediction of phosphorus and chlorophyll levels in lakes." *Journal of the Water Pollution Control Federation*, 48 (9).

Rast, W., Jones, R., and Lee, G.F. (1983). "Predictive capability of U.S. OECD phosphorus loading-eutrophication response models." *Journal of the Water Pollution Control Federation*, 55 (7).

U.S. Soil Conservation Service. (1981). "Maryland standards and specifications for ponds." *SCS Std./Spec. No. 378*, U.S.S.C.S., Washington, D.C.

II. CASE STUDY OF A MULTIPURPOSE FLOOD CONTROL FACILITY

A. Introduction

Successful implementation of a multi-purpose project typically requires a series of complex planning, design, financing, and construction steps involving many public and private sector participants. The complexity of the planning-through-construction process can mean that even technically feasible and economically attractive projects involving storm water detention may never be achieved.

This case study describes the planning, design, financing, and construction of a flood control-recreation facility, together with related public works projects. Although the case study is project-specific, many aspects can be extrapolated to other potential multi-purpose facilities.

B. Recent History of Infrastructure Needs

1. Description of City and Area

The flood control-recreation project and related watershed described in this case study are within and near the City of Valparaiso, Indiana. Valparaiso has a population of 22,000 and is economically related to, but not contiguous with, the large Chicago metropolitan area.

Project area climate is characterized by markedly different seasons with corresponding variations in temperature and in precipitation type, amount, and intensity. Surface water related problems such as flooding and erosion fluctuate during the year, but are generally most severe during spring and summer.

2. Flood Control Needs

Serious flooding occurred in recent years, most notably in June and July, 1981 and again in July, 1983. In each instance, overland flooding and sewer backup caused widespread damage and disruption. The last rainfall was the most severe. A total of 7.3 inches of rain fell in 27 hours and 6- and 13-hour portions of the event had recurrence intervals in excess of 100 years. Flooding problems were especially serious in the Smith Ditch watershed (see Figure A.2).

In October 1983, the city commissioned the preparation of a comprehensive flood control plan for the Smith Ditch watershed and one other problematic watershed. The planning project was completed in June 1984 (Donohue 1984).

The following guidelines were developed to provide the basis for and give direction to the planning process, and were used in the development and evaluation of alternatives:

(a) Configure and size facilities to store or convey runoff from the 100-year recurrence interval, 6-hour rainfall occurring under future land use conditions (the 6-hour duration was selected on the basis of sensitivity analyses).
(b) Resolve all flood problems as close to their point of origin as possible and avoid shifting problems from one location in the watershed to another.
(c) Begin to resolve sanitary and combined sewer backup problems by addressing the extensive and serious surface flooding problem.
(d) Favor gravity inflow and outflow for detention/retention facilities.
(e) Give preference to a few large, publicly owned and maintained detention/retention facilities rather than many small privately owned facilities.
(f) Consider the recreational and aesthetic aspects of potential detention/retention facilities.

Two digital computer models, HEC-1 and HEC-2, developed by the U.S. Army Corps of Engineers' Hydrologic Engineering Center and available in the public domain, were used to analyze the system and

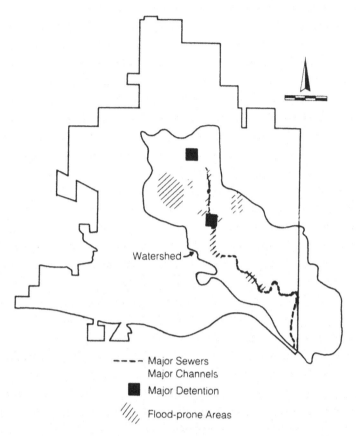

Figure A.2. — Smith Ditch watershed.

explore alternative solutions. HEC-1 was used to determine the volume and timing of run off from the land surface to watershed swales, channels, storm sewers, and detention/retention facilities, and to route that runoff downstream through the conveyance and storage system thereby producing a series of discharges at predetermined locations. HEC-2 was used to calculate flood stages for selected reaches in the open channel system. Figure A.3 shows input to, output from, and inter-relationships between the two computer models.

The relationship between important components of the watershed system and the computer programs used to simulate them are illustrated in Figure A.4. All of the land surface, most of the conveyance system, and all detention/retention facilities were modeled with HEC-1. HEC-2 was applied to those channel reaches requiring detailed hydraulic analyses. The total of 19 simulations that were conducted included existing and future land use, a historic rainfall event, and rainfall events over a range of recurrence intervals.

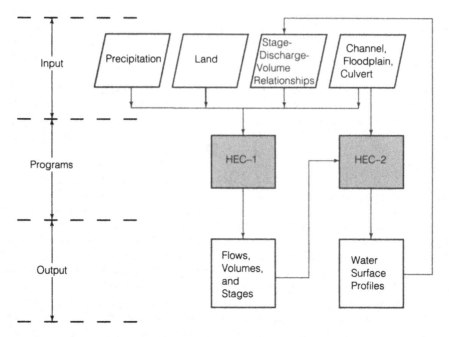

Figure A.3.—Hydrologic-hydraulic model composed of two HEC programs.

The diagnosis of the hydrologic-hydraulic system using simulation indicated that existing detention facilities were undersized, and major channels and conduits through built-up areas had insufficient capacity. Future urbanization would aggravate flooding in scattered locations throughout the watershed.

Various alternative structural flood control facilities were considered in the planning process. For example, the technical and economic feasibility of diverting flood water from the upper portion of the Smith Ditch watershed to the contiguous watershed on the west was screened. This alternative was rejected because of the high cost, as well as the fact that it would be in conflict with one of the planning guidelines established early in the project, which called for resolving flooding problems as near as possible to the point of origin and not moving them from one area to another.

Recommended major improvements, as illustrated in Figure A.5, include the construction of two new detention facilities (one of which is the subject of this paper), modification of two existing detention facilities and extensive channel cleaning and maintenance. The plan also recommended that the city develop and implement a comprehensive program for operation and maintenance of sewers and channels; that owners or renters of residential or commercial property consider purchasing flood insurance; that the city review, codify, and expand its

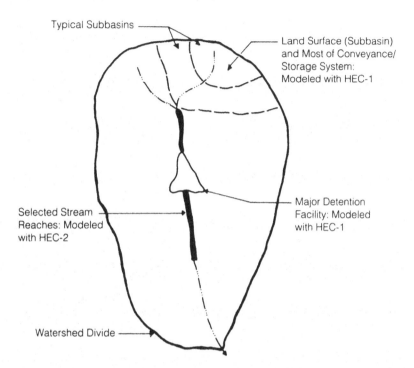

Figure A.4. — *Watershed components and portions of the HEC-1 and HEC-2 model used to simulate them.*

storm water management regulations; and that the city, possibly in cooperation with the surrounding county, develop a current comprehensive stormwater management plan for all watersheds in the Valparaiso area.

Since issuance of the plan, the recommended channel cleaning has been completed, the two existing detention facilities have been enlarged as recommended, and the larger of the two recommended detention facilities has been designed and constructed. Partly as a result of the plan, the city has also taken a lead role in seeking state legislation to enable an Indiana municipality to establish a stormwater management utility. The legislation was adopted and became effective September 1, 1987 (Indiana 1987). Other structural and non-structural recommendations are under consideration by the city.

As shown in Figure A.6, most of that portion of the Smith Ditch watershed subject to flooding has been extensively urbanized. Residential and commercial land uses cover most of the upper two-thirds of the watershed. Undeveloped lands consist mostly of privately held property intended for development.

At the time of the planning project, the fairgrounds site, which is located approximately two-thirds up the Smith Ditch watershed, was

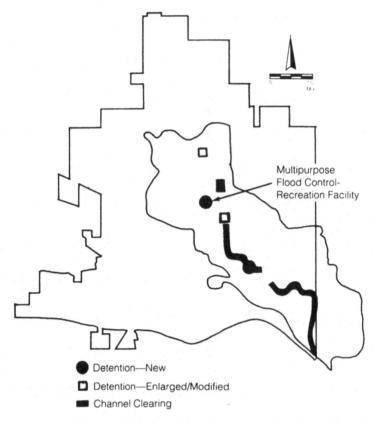

Figure A.5. —*Recommended major improvements in Smith Ditch watershed.*

one of the few undeveloped publicly-owned parcels in the watershed. The site, which was then owned by the Board of County Commissioners and operated by the Porter County Fair Board, was last used as a fair site in the summer of 1984. A new site outside of, but close to, the city was being developed and would be ready for use in the summer of 1985. Therefore, at the time of the flood control planning project, the fairground site was potentially available for other uses, including storm water control.

Figure A.7 illustrates the initial concept for the single purpose, off-channel fairgrounds detention facility. As initially envisioned and, from a flood control perspective, as ultimately constructed, the fairgrounds detention facility consists of four major components. The first component is the diversion structure and sewer to carry flow from Smith Ditch at McCord Road into the fairgrounds. The second component is the 90 acre-foot detention facility initially envisioned as occupying about eleven acres of the 27-acre site. The third component is a system of sewers generally flowing from the west and designed to carry storm water

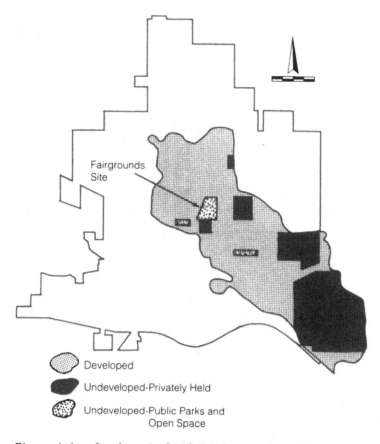

Figure A.6.—*Land use in Smith Ditch watershed, 1984.*

from a currently combined sewer area into the detention facility. The fourth and final component is an outlet control, which would slowly release temporarily stored water back into Smith Ditch and would also control flow from the west. Inflow to and outflow from the fairgrounds detention facility would be by gravity, and pumping or automatic controls would not be required.

With respect to flood control, the fairgrounds detention facility will mitigate surface flooding in the residential area immediately to the east along Smith Ditch, reduce surface and basement flooding in the residential and scattered commercial areas to the west, and relieve flooding along Smith Ditch downstream of the facility.

3. Recreation Needs

Approximately coincident with the growing concern with flooding problems, the potential recreation needs of the city were receiving

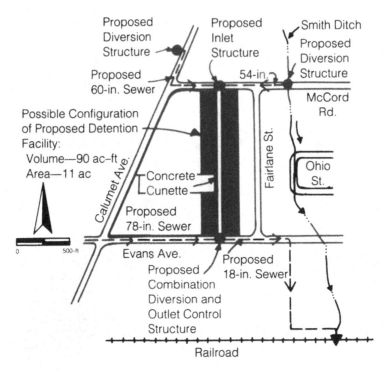

Figure A.7.—Initial concept for off-channel detention.

increased attention. Population growth and expanding interest in year-round outdoor recreation were placing increased pressure on existing facilities and, at the same time, reducing the availability of open space for development of new recreation facilities. Furthermore, eligibility for federal and state recreation funds required an updated recreation master plan. Accordingly, the Park and Recreation Board commissioned the preparation of a park master plan. The park planning effort, which was carried out from February through August 1984 (Earth Plan Associates 1984), was approximately coincident with the flood control planning effort.

Numerous guidelines were established to provide overall direction to the recreation master planning effort. In retrospect, some of these influenced the eventual decision to develop the fairgrounds site as a multi-purpose flood control/recreation facility. Examples of these guidelines were acquiring land for public park purposes in areas of the city currently void of recreation facilities, establishing cooperative agreements with private and public organizations for joint use of properties, and providing recreation programs for all age groups.

Early in the recreation planning program the fairgrounds site was recognized as offering great potential for a large central city park facility.

Because the park planning project was conducted parallel to and with knowledge of the flood control planning project, the park master plan recommended that a major recreation facility be developed at the fairground site and that it be configured to permit joint use for flood control.

The coincident occurrence of the flood control planning program and the recreation planning program was a catalyst for development of a multi-purpose project. However, the success of the project depended upon the willingness of public officials responsible for flood control and for recreation (and their respective consultants) to work together.

The initial concept for a flood control/recreation development of the fairground site is illustrated in Figure A.8. Recreation facilities included four softball diamonds, three soccer fields, a general purpose athletic field using the existing grandstand for spectator viewing, picnic and playground areas, an ice skating rink, a circumferential jogging/walking/biking trail, parking areas and public use and service buildings. The initial concept envisioned excavating the approximate eastern one-third to one-half of the site to a uniform level to provide the necessary 90 acre-feet of flood control storage.

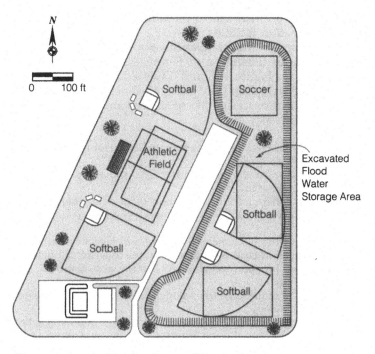

Figure A.8.—Initial flood control/recreation plan.

4. Highway Needs

Beginning in the mid-1940's, city, county, regional, state, and federal officials began discussing ways to route north-south highway traffic around the City of Valparaiso. The heavy volume of through traffic (trucks and automobiles having no origin or destination in the city) was the principal concern.

These efforts culminated in 1979 with the completion of a final design study report (Indiana State Highway Commission 1979). A new ten-mile long north-south section of State Road Highway 49 would be constructed. The four-lane limited-access highway would include a by-pass along the eastern edge of Valparaiso. Because of a relatively large number of grade separations, extensive fill (several million yards) would be required for the Valparaiso segment.

The first construction contract was let in May 1983 for portions of the bypass north of Valparaiso. Construction on the segment immediately east of Valparaiso began in April 1985.

C. Designing the Multi-Purpose Project

The decision by the City of Valparaiso and the City Park and Recreation Department to design a multi-purpose flood control/recreation facility at the fairgrounds site was made in the summer of 1985. The two consulting firms that originally prepared the flood control and recreation plans were retained to provide design services for the flood control and recreation facilities.

A refined, integrated design was developed incorporating flood control requirements as illustrated in Figure A.7, and recreation requirements as shown in Figure A.8 and Table A.2. The project team, consisting of city engineering and city park personnel, flood control engineers, and park and recreation designers considered and resolved many factors during design such as:

(a) Flattening of side slopes on all excavated areas for ease of maintenance and for aesthetic quality.
(b) Grading of all recreation field surfaces to provide positive and rapid surface and subsurface drainage.
(c) Raising the minimum outlet elevation of the detention facility based on a field survey of local storm sewers specially conducted for the design.
(d) Elimination of the old, unsightly fairground bleachers and rearrangement of recreation facilities, particularly the four softball fields, to provide improved function and less costly concession, restroom, and other service facilities.
(e) Saving of most of the large trees along the east side and in the northeast corner of the site.
(f) Using terraces—relatively flat areas at slightly different elevations—to add topographic variety and interest, in plan and in section, to an otherwise flat or severely excavated site and to decrease the frequency of flooding of some sports facilities.

TABLE A.2. Recreation Facilities

1. *Garden Area*—Located at the southeast corner of site, this area will include trees, shrubs, floral displays, bench seating, lawn development, split rail fence, and entrance sign.

2. *Pavilion*—Renovate existing structure as a landmark to the Old Fairgrounds and as a visual focal point for park. Includes 5000 square feet of seasonal space that can be used for events such as family reunions, wedding receptions, square dances, art shows, and teen dances. Includes restroom and kitchen facilities.

3. *Concession/Restroom/Supervisory Building*—Construct a centrally located building that will accommodate the entire park. Because this central location will also be a visual focal point it is designed to complement the renovated Pavilion and will include extensive landscaping.

4. *Maintenance Building*—Dismantle the existing metal building and use materials at another location. A new maintenance facility will be constructed on the northeast corner and will be designed to complement on-site architecture and be landscaped to blend into surroundings.

5. *Softball Fields*—Provide four fields, two lighted, to accommodate primarily men's, women's and co-ed softball with capability to serve youth baseball/softball.

6. *Soccer Fields/Special Events Field*—Provide one class A field that will serve youth, school and adult soccer and football needs as well as serve as a staging area for a variety of special events. One class B soccer field will be used for special events when available.

7. *Walk/Jog Path*—Consists of one mile of surfaced path for use in all seasons.

8. *Picnic Area*—Includes a wooded area with picnic tables and playground equipment.

9. *Landscaping*—Provide extensive landscaping along the site perimeter with trees, shrubs and flowers.

10. *Parking*—Construct three major paved parking lots to accommodate over 250 cars.

11. *Maintenance - Equipment*—Acquire mowers, aerifier, liners, seeder and small truck to serve the facility.

The design process, including consideration of the preceding factors, led to the final design illustrated in Figure A.9. A set of terraces cover the site, and the originally flat site now has a total relief of 13.0 feet with the lowest area occurring at the outlet in the southeast corner.

Least used portions of the facility, such as one of the overflow soccer areas, are on the lowest levels. In contrast, frequently used areas, such as lighted softball fields, are on the highest terraces. Frequency and duration of storm water storage for all areas of the site are presented in Figure A.10 (assuming stormwater loads corresponding to ultimate watershed development). As indicated in Figure A.10, the southeast or lowest corner of the facility will be inundated at least once per year for a duration of about two days. In contrast, the softball diamonds will be inundated for a fraction of a day once every ten or more years depending on their specific elevation.

Very close attention was given to the hydraulic function of the facility. The diversion structure on Smith Ditch will divert all flows above 60

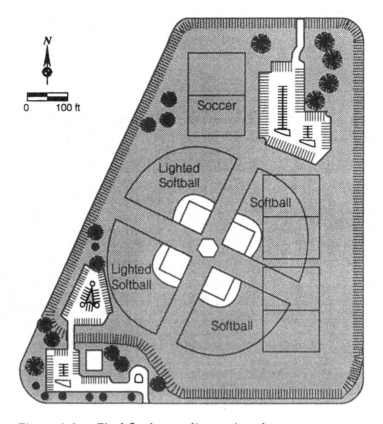

Figure A.9.—Final flood control/recreation plan.

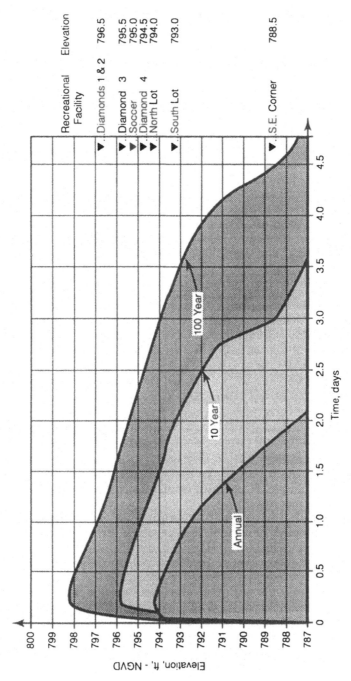

Figure A.10. — *Frequency and duration of storm water storage.*

cfs into the detention facility. Flows of less than six cfs that enter the facility from the north will be carried to the outlet control structure in an 18-inch diameter southerly flowing sewer unobtrusively located along the eastern edge of the facility. Flows below six cfs from the west along Evans Avenue will bypass the facility and go directly to Smith Ditch. Excess flows from the west will be temporarily diverted into storage.

An extensive network of 4–18 inch diameter perforated and corrugated polyethylene pipe under drains and sewers provides subsurface drainage of all recreation areas. All recreation surfaces have a surface slope of at least 1% to encourage rapid drainage after rainfall or flooding. Because of the careful attention given to grading and subsurface drainage, even the lower recreation facilities on this site will probably be available for use more often than other single-purpose facilities scattered around the city.

The final design provides numerous and varied active and passive recreation opportunities. The design also includes on-site provisions for service and maintenance facilities and equipment. The value of the constructed project was about $2.4 million.

D. Financing the Multi-Purpose Project

A major public works project like this is typically financed through sale of bonds. Although a bond issue was used to finance much of the cost of the facility, other innovative means of finance were also used as summarized in Figure A.11 and discussed in the following sections.

1. Land Acquisition

The City and the Park and Recreation Department jointly committed to proceed with the multi-purpose project in summer 1985. The city obtained an appraisal of the 27-acre fairgrounds site and began to negotiate with the County for site acquisition. The city purchased the site in August 1985 for $300,000 to be paid in amounts of $50,000 per year at zero interest. As a part of the negotiations, the city agreed to give the County first right of refusal on a City-owned parking lot close to the County court house.

The city and the Valparaiso Community Schools then negotiated an agreement whereby the school system would give the city 11.3 acres of land valued at $254,000. This land had been held by the Valparaiso Community Schools to meet anticipated future education needs. In return for the land, the city agreed to provide certain recreation facilities at the new fairgrounds facility or elsewhere for use by the school system. The effect of this exchange was to offset most of the cost incurred by the city in purchasing the fairgrounds inasmuch as the City could sell the land acquired from Valparaiso Community Schools as a means of recovering the fairgrounds purchase price or other costs incurred in providing the additional recreation services.

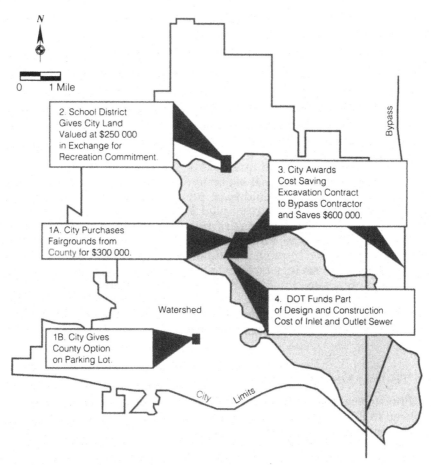

N

0   1 Mile

2. School District
Gives City Land
Valued at $250 000
in Exchange for
Recreation Commitment.

3. City Awards
Cost Saving
Excavation Contract
to Bypass Contractor
and Saves $600 000.

1A. City Purchases
Fairgrounds from
County for $300 000.

Watershed

4. DOT Funds Part
of Design and Construction
Cost of Inlet and Outlet Sewer

1B. City Gives
County Option
on Parking Lot.

Bypass

City   Limits

Figure A.11.—Financing.

2. *Excavation*

As noted earlier, the construction of the State Road 49 bypass in the
vicinity of Valparaiso, which was under way in mid-1985, created a
strong demand for fill material. The fill areas were 1.6 or more miles
away from the fairground site suggesting that a long haul would cause
high excavation costs. However, the city began to investigate the possibility of providing fill for the highway project because projected site
excavation costs were a large part—about one third—of the total cost
of the project.

As a result of exploring the possibility of tying fairground site excavation needs to bypass fill needs, the city advertised the availability

of up to 210,000 cubic yards of material at the fairgrounds site and received six bids. Listed in decreasing order, the per cubic yard bid prices were $5.05, $5.00, $3.85, $3.29, $2.60, and $1.02. The lowest bid was submitted by a contractor involved in the bypass highway project. The excavation contract was awarded to the low bid contractor who eventually removed 180,500 cubic yards from the site for $1.02 per cubic yard ($184,100) and hauled it to the highway project to meet his contractual obligations with the State Department of Highways. Because of the opportunity to begin excavating in August 1985 at considerable cost savings, the design of the excavation portion of the site began immediately and proceeded quickly.

Assuming that the original estimate of $4.20 per cubic yard for excavation, in 1985 costs, would have prevailed in absence of the tie-in to the highway project, the city saved $574,000 by moving quickly and integrating the fairgrounds construction with the bypass construction. In exchange for a time extension, the excavate and haul contractor subsequently agreed to provide additional services at no additional cost to the city. These savings included stripping and stockpiling topsoil at the fairgrounds site, rough grading, replacing topsoil on the steeper side slopes, seeding and mulching the slopes, and demolishing and removing the concrete grandstand. These supplemental services represented an additional savings of at least $20,000 to the city.

3. Tie-In to Street Improvement Project

Prior to the planning of the fairgrounds project and completely unrelated to it, the city, state, and federal government had been designing and preparing financing for improvements to a portion of Calumet Avenue which borders the fairgrounds site to the west. Included were necessary drainage facilities leading to a large sewer to carry stormwater from Calumet Avenue to Smith Ditch.

Construction of the storm sewer to Smith Ditch would have negated some of the flood control benefits being provided by the fairgrounds project and would have resulted in unnecessarily expensive sewer construction from Calumet Avenue to Smith Ditch. Furthermore, the fairgrounds facility was sized to accommodate future development runoff from areas west of the site.

Accordingly, the drainage component of the Calumet Avenue highway improvement project was altered to direct runoff into the fairgrounds facility. This reduced the cost of the street improvement project.

Federal and state transportation funds are being used to pay for some of the design and construction costs associated with the outlet control. City design and construction costs were reduced $90,000 as a result of federal and state participation.

4. Other Savings

The city used its new Project Management Office (PMO) system for this project, thus saving the usual additional fee for a general contractor. City personnel performed full time inspection and had full time control. The PMO approach yielded further cost savings.

E. Constructing the Multi-Purpose Project

Excavation began abruptly in August 1985 as a result of the demand for fill at the nearby highway bypass project. Flood control and recreation facilities were both first used in 1990.

The city staff is designing a separate storm sewer for the now partially combined sewer residential area west of the fairgrounds site. The city is also designing major storm sewers to convey stormwater from the residential area to the west directly into the fairgrounds facility.

F. Building Community Support for the Multi-Purpose Project

Numerous public information events and activities, beginning early in the planning process, were conducted for the flood control and recreation aspects of the project. Early activities included a neighborhood meeting and field reconnaissance of flood prone areas by project engineers during which numerous personal contacts were made with citizens. Other public information events and activities included presentations at City Council meetings, presentations at Park Board meetings and newspaper articles.

Interest in flood control works tends to rise and fall with the flood waters. If the fairgrounds project had been for the single purpose of flood control, community interest may not have been continuously sustained, in spite of the public information efforts, at a high enough level to see the project through to implementation. The recreation features of the project helped to broaden and sustain community interest. Project implementation benefited from the sense of urgency created by both the availability of the fairgrounds site (because of the imminent relocation of the annual county fair) and by the immediate need for major amounts of fill at the bypass highway project.

G. Summary and Conclusions

Multi-purpose public works projects often "look good on paper" because they can be shown to be technically feasible and economically attractive. However, implementation of such projects is difficult because of the planning, design, financing, and construction steps required, and the need to involve many participants.

The case study presented in this paper suggests ways to deal with the complexities inherent in multi-purpose public works projects. Key factors to the success of the fairgrounds flood control-recreation project were sound technical design utilizing watershed computer simulation, a continuous public information effort, imaginative financing and trading, and quickly seizing unexpected, mutually beneficial opportunities.

H. References

Donohue and Associates, Inc. (1984). *Smith Ditch lagoon no. 1 and hotter lagoon investigation*. Valparaiso, IN.

Earth Plan Associates. *Park Master Plan and Management Strategies*. Valparaiso, IN.

Indiana State Highway Commission. (1979). "Final design study report—STH 49." *Project No. ST 165–24*, Division of Planning, Valparaiso, IN.

Indiana, State of. (1987). "Chapter 5: Department of Storm Water Management." *Indiana Code—House Enrolled Act No. 1853*.

INDEX